Automated Driving

Daniel Watzenig • Martin Horn
Editors

Automated Driving

Safer and More Efficient Future Driving

Springer

Editors
Daniel Watzenig
Virtual Vehicle Research Center
and Graz University of Technology
Institute of Electrical Measurement
and Measurement Signal Processing
Graz, Austria

Martin Horn
Institute of Automation and Control
Graz University of Technology
Graz, Austria

ISBN 978-3-319-31893-6 ISBN 978-3-319-31895-0 (eBook)
DOI 10.1007/978-3-319-31895-0

Library of Congress Control Number: 2016949638

Printed on acid-free paper

This Springer imprint is published by Springer Nature
The registered company is Springer International Publishing AG Switzerland

Foreword

Automation of processes and products is a key step into next generation of our economy. Due to the high economic impact of transportation industry and especially of the automotive sector, automated driving and all related ingredients act as strong technology drivers for many other application domains. Therefore, this book with a comprehensive treatment of the relevant elements is very helpful for the advancement of the domain itself, but will also help to spread the technological and procedural essentials towards other applications.

Automation of vehicles is enabled by digitization, which means by the pervasive penetration of interconnected digital technology into our products and systems. This allows us to provide better integration, achieve higher performance, and provide the required intelligence and communication to the environment, especially to the human user.

In aeronautic systems, a high degree of automation already exists and has a certain track record, ranging from many assistance systems up to fully automated aircrafts, including all phases of operation, from taxiing to take-off, cruising, and landing. Sense and avoid is implemented, but based on well-established procedural rules and safety margins. Further on the participants (pilots) are well trained and subject to regular checks. Automotive automation could learn a lot from this profound source of experience, but it has to adapt to its own requirements and operational conditions: The distances are much shorter, the reaction time is very short, and the analysis of the environmental situation is much more complex.

New technologies will gain relevance in the automated vehicles: Software will play a key role, especially in embedded real-time systems. Smart components and subsystems will exchange huge amounts of data, on board of the vehicle, and also with the infrastructure. We will need communication links with high levels of availability and high immunity against undesired penetration from outside. Thus, safety and security will become mutually dependent couples. High performance computing capacities will be needed to process the incoming information and drive the decisions to control the vehicle in a safe way. Despite all this high tech, the human user shall always understand what is going on and have the chance to control the system as ultima ratio decider.

Mastering this chain of new technologies will be a key enabler for families of products in many other domains. This book will provide the reader with considerations and key insight how to achieve the different steps. Therefore, we have to thank the editors and the authors for this comprehensive work!

Heinrich Daembkes

Vice President at Airbus Defence and Space
Former Vice President at Daimler Research Centre for High Frequency Electronics
President of the ARTEMIS Industry Association
Professor at the University of Ulm

Preface

Automated vehicle technology has the potential to be a game changer on the roads, altering the face of motoring as we experience it today. Many benefits are expected, ranging from improved safety, reduced congestion, lower stress for car occupants, social inclusion, lower emissions, and better road utilization due to optimal integration of private and public transport.

Over the last decade, vehicle automation has attracted considerable attention of the public, governments, the scientific community, and the industry around the world, mainly driven by the evolution of driver assistance and active safety. Many cars sold today are already capable of some level of automation while higher automated prototype vehicles are continuously tested on public roads, especially in the United States, Europe, and Japan. Automated driving technology has arrived rapidly on the market, and the future deployment is expected to accelerate over the next years. As a matter of fact, most of the core technologies required for fully autonomous driving (SAE level 5) are available today, many are mature, and some are already on the way being deployed in commercially available vehicles.

During the last three years, public authorities from many countries presented action and innovation plans to facilitate the development and stepwise introduction of automated vehicles. Those plans cover actions for a multitude of technical and non-technical aspects that need to be taken into account. In particular, road safety is expected to substantially improve with vehicle automation since more than 90 % of the crashes involve human errors. However, most often driving involves no crashes. The ultimate safety test for automated vehicles will have to point out how well they can replicate the crash-free performance of human drivers especially at the level of partial and conditional automation within mixed traffic.

In order to move towards a significant market penetration of automated vehicles, different technological configurations have to be considered. Most of the state-of-practice vehicles and prototypes rely on in-vehicle sensor platforms and require little digital infrastructure communication, while a greater connectivity between vehicles and their infrastructure is identified to be beneficial. This entails the development of common communication protocols, encrypted security standards, and investment in new types of infrastructure or upgrading existing ones. Nevertheless, both models require accurate digital representations of their environment at any time, for any weather condition, and in any traffic situation.

Beyond the technical issues, several action items for faster introduction of automated vehicles have to be solved by the governments in order to ensure the full compatibility with the public expectations regarding legal responsibility, safety, and privacy. Authorities have to create the legal framework to remove liability traps, to encourage test regions, to review long-term infrastructure investments, to provide open access, and to set up legal frameworks for inter-car communication.

The challenging technical topics still to be solved and the constant endeavours around the world to drive this exciting technology forward have motivated the creation of this book. In order to get a balanced view on the state of development, the editors have invited authors from different stakeholders including public authorities, car manufacturers, suppliers, and research organizations. Within this book, the state of practice and the state of the art of automated driving building blocks are extensively reviewed and future trends are envisioned. The book encompasses the importance of control engineering, recent advances in environment sensing and perception, in-vehicle architectures, and dependable power computing as well as active and functional safety in automated driving. Furthermore, we have put a strong focus on the validation and testing of automated driving functions. A sampling of relevant industrially driven research projects and industrial initiatives concludes the book.

We strongly believe that this book on automated driving provides an overview of current and emerging technical challenges in that field and gives deep insights into industrial demands. We hope that the reader will be inspired by the different technical articles, selected project summaries, and introductions of renowned national and European initiatives.

Finally, we would like to express our sincere appreciation and gratitude to all authors and co-authors, who made the publication of this book possible. We are grateful to Silvia Schilgerius at Springer for her professionalism and support.

Graz, Austria

Martin Horn
Daniel Watzenig

Contents

Contributors

Dirk Abel Institute of Automatic Control RWTH Aachen University, Aachen, Germany

Josef Affenzeller AVL List GmbH, Graz, Austria

Bassam Alrifaee Institute of Automatic Control RWTH Aachen University, Aachen, Germany

Angelos Amditis Institute of Communications and Computer Systems, Athens, Greece

Aria Etemad Volkswagen AG, Wolfsburg, Germany

Alexandre Armand Renault S.A.S., Technocentre de Guyancourt, Research Department, France

Fredrik Asplund KTH Royal Institute of Technology, Stockholm, Sweden

Phil Barber Jaguar Land Rover, Coventry, UK

Jan Becker Robert Bosch LLC, Palo Alto, CA, USA

Sagar Behere KTH Royal Institute of Technology, Stockholm, Sweden

Carl Bergenhem Qamcom Research and Technology, Gothenburg, Sweden

Gert Blom City of Helmond, Helmond, Netherlands

Jan Micha Borrmann FZI Research Center for Information Technology, Karlsruhe, Germany

Wolfgang Bösch TU Graz, Graz, Austria

Olof Bridal VOLVO Group Trucks Technology, Gothenburg, Sweden

Oliver Bringmann Wilhelm-Schickard-Institute for Computer Science, University of Tübingen, Tübingen, Germany

Lukas Bulwahn Lukas Bulwahn, BMW AG, Munich, Germany

Klaas Burgdorf Volvo Car Group, Gothenburg, Sweden

Erik Coelingh Volvo Car Group, Gothenburg, Sweden

Gerardo Daalderop NXP Semiconductors, Leuven, Belgium

Stefano De Pinto University of Surrey, Guildford, UK

Arnaud de La Fortelle Centre for Robotics, MINES ParisTech—PSL Research University, Paris, France

Christian Ellersdorfer Institute of Vehicle Safety, Graz University of Technology, Graz, Austria

Mehdi Farid BMW AG, Munich, Germany

Udo Gleich Daimler AG, Ulm, Germany

Stephanie Grubmüller VIRTUAL VEHICLE Research Center, Graz, Austria

Thomas Helmer BMW AG, Munich, Germany

Michael Helmle Robert Bosch GmbH, Gerlingen-Schillerhöhe, Germany

Frank-Josef Heßeler Institute of Automatic Control RWTH Aachen University, Aachen, Germany

Kai Homeier Volkswagen Group Research, Wolfsburg, Germany

Matthias Hoppe Institute of Automatic Control RWTH Aachen University, Aachen, Germany

Martin Horn Institute of Automation and Control, Graz University of Technology, Graz, Austria

Richard Huber Kober Engineering Deutschland GmbH, Stuttgart, Germany

Alexander Huesmann BMW AG, Munich, Germany

Javier Ibanez-Guzman Renault, Boulogne-Billancourt, France

Rolf Johansson SP Technical Research Institute of Sweden, Borås, Sweden

Ronald Kates REK Consulting, Otterfing, Germany

Viktor Kaznov Scania CV AB, Södertälje, Sweden

Werner Kober Kober Engineering GmbH, Ilz, Austria

Klaus Kompass BMW Group, Munich, Germany

Jens Krause Volkswagen Group Research, Wolfsburg, Germany

Wolfgang Kriegler Austrian Association for Advanced Propulsion Systems, Vienna, Austria

Thomas Kühbeck BMW AG, Munich, Germany

Andrea Leitner Virtual Vehicle Research Center, Graz, Austria

Ignacio Llatser Technische Universität Dresden, Dresden, Germany

Ali Marjovi École Polytechnique Fédérale de Lausanne, Lausanne, Switzerland

Helmut Martin Virtual Vehicle Research Center, Graz, Austria

Alcherio Martinoli École Polytechnique Fédérale de Lausanne, Lausanne, Switzerland

Markus Maurer Institute of Control Engineering, Technische Universität Braunschweig, Braunschweig, Germany

Holger H. Meinel Daimler AG, Ulm, Germany

Till Menzel Institute of Control Engineering, Technische Universität Braunschweig, Braunschweig, Germany

Andreas Moser Dr. Steffan Datentechnik, Linz, Austria

Elke Muhrer BMW AG, Munich, Germany

Iñaki Navarro École Polytechnique Fédérale de Lausanne, Lausanne, Switzerland

Pavel Nedoma ŠKODA AUTO a.s., Mladá Boleslav, Czech Republic

Jürgen Niehaus SafeTRANS, Oldenburg, Germany

Jonas Nilsson Volvo Car Corporation, Gothenburg, Sweden

Ralf Oberfell Daimler AG, Stuttgart, Germany

Marcus Obst BASELABS GmbH, Chemnitz, Germany

Tilman Ochs BMW AG, Munich, Germany

Anders Ödblom Volvo Car Group, Gothenburg, Sweden

Sebastian Ottlik FZI Research Center for Information Technology, Karlsruhe, Germany

Panagiotis Pantazopoulos Institute of Communications and Computer Systems, Athens, Greece

Michael Paulweber AVL List GmbH, Graz, Germany

Oliver Pink Robert Bosch GmbH, Gerlingen-Schillerhöhe, Germany

Jiri Plihal ÚTIA AV ČR, v.v.i., Praha 8, Czech Republic

Hendrik Post Robert Bosch GmbH, Gerlingen-Schillerhöhe, Germany

Patrick Pype NXP Semiconductors, Leuven, Belgium

Xiangjun Qian Centre for Robotics, MINES ParisTech—PSL Research University, Paris, France

Matthias Reiter Institute of Automatic Control RWTH Aachen University, Aachen, Germany

Per Roos Scania CV AB, Södertälje, Sweden

Wolfgang Rosenstiel Wilhelm-Schickard-Institute for Computer Science, University of Tübingen, Tübingen, Germany

Marcus Rothoff Volvo Car Group, Gothenburg, Sweden

Astrid Rupp Institute of Automation and Control, Graz University of Technology, Graz, Austria

Thomas Schaller BMW Group, Munich, Germany

Markus Schratter Virtual Vehicle Research Center, Graz, Austria

Robin Schubert BASELABS GmbH, Chemnitz, Germany

Fabian Schuldt Institute of Control Engineering, Technische Universität Braunschweig, Braunschweig, Germany

Eva Schulz-Kamm NXP Semiconductors, Leuven, Belgium

Julian Simader Dr. Steffan Datentechnik, Linz, Austria

Andreas Söderberg SP Technical Research Institute of Sweden, Borås, Sweden

Aldo Sorniotti University of Surrey, Guildford, UK

Hermann Steffan Institute of Vehicle Safety, Graz University of Technology, Graz, Austria

Michaela Steinhoff IAV GmbH, Berlin, Germany

Michael Stolz VIRTUAL VEHICLE Research Center, Graz, Austria

Johan Svahn Scania CV AB, Södertälje, Sweden

Martin Törngren KTH Royal Institute of Technology, Stockholm, Sweden

Jörgen Tryggvesson Comentor, Gothenburg, Sweden

Kurt Tschabuschnig Magna Steyr Engineering AG & Co KG, Graz, Austria

Simon Ulbrich Institute of Control Engineering, Technische Universität Braunschweig, Braunschweig, Germany

Stig Ursing Semcon, Gothenburg, Sweden

Milos Vasic École Polytechnique Fédérale de Lausanne, Lausanne, Switzerland

Trent Victor Volvo Car Group, Gothenburg, Sweden

Alexander Viehl FZI Research Center for Information Technology, Karlsruhe, Germany

Maximilian von Grafenstein Alexander von Humboldt Institute für Internet und Gesellschaft, Berlin, Germany

Walther Wachenfeld Institute of Automotive Engineering at TU Darmstadt, Darmstadt, Germany

Eckhard Walters NXP Semiconductors, Leuven, Belgium

Lei Wang BMW AG, Munich, Germany

Fredrik Warg SP Technical Research Institute of Sweden, Borås, Sweden

Daniel Watzenig Virtual Vehicle Research Center and Graz University of Technology, Institute of Electrical Measurement and Measurement Signal Processing, Graz, Austria

Sasha Westermann Hamburg Port Authorities, Hamburg, Germany

Hermann Winner Institute of Automotive Engineering at TU Darmstadt, Darmstadt, Germany

Stefan Winter Austrian Association for Advanced Propulsion Systems, Vienna, Austria

Franz Wotawa Institute for Software Technology, Technische Universität Graz, Graz, Austria

Clément Zinoune Renault, Boulogne-Billancourt, France

Part I
Introduction

Chapter 1
Introduction to Automated Driving

Daniel Watzenig and Martin Horn

1.1 Introduction

Innovation and action plans of public authorities (see, e.g., www.gov.uk) around the world aim to clear the way for a stepwise introduction of automated vehicles. The plans comprise a number of aspects like standardization, testing, safety, or in-vehicle technology. It is expected that automated driving will [1].

- Improve safety by reducing human driving errors
- Significantly contribute to the optimization of traffic flow
- Help to reduce fuel consumption and CO_2 emissions
- Enhance the mobility of elderly people and unconfident drivers

Several forecasts predict a limited availability of automated driving functions in 2020 (partial and conditional automation) and a wide availability by 2040 including high and full automation [2–4]; see Sect. 1.2 Today's advanced driver assistance systems (ADAS) like Automatic Cruise Control (ACC), Lane Departure Warning (LDW), or Pedestrian Detection (PD) will form the backbone of tomorrow's mobility. Vehicles will communicate with each other and with infrastructure [5]. Vehicle-to-Vehicle (V2V) communication allows vehicles to exchange relevant information like local traffic data (e.g., nearby accidents) and about their driving intention. Vehicle-to-Infrastructure (V2I) communication will be used to optimize the road network usage and thereby helps to reduce environmental pollution. The

D. Watzenig (✉)
Virtual Vehicle Research Center and Graz University of Technology, Institute of Electrical Measurement and Measurement Signal Processing, Graz, Austria
e-mail: daniel.watzenig@v2c2.at

M. Horn
Institute of Automation and Control, Graz University of Technology, Graz, Austria
e-mail: martin.horn@tugraz.at

D. Watzenig, M. Horn (eds.), *Automated Driving*,
DOI 10.1007/978-3-319-31895-0_1

role allocation between human drivers and automated driving systems in this scenario is specified by the six levels of driving automation (see www.sae.org)

This chapter is organized as follows: Sect. 1.2 briefly introduces the different levels of automation as defined by SAE J3016 [6]. The three layers specifying the key technologies of automated vehicles are outlined in Sect. 1.3. Section 1.4 summarizes the research challenges, which have to be mastered in order to enable automated driving; Sect. 1.5 concludes the chapter and gives an overview of the topics covered by the book.

1.2 Levels of Automation

Figure 1.1 briefly summarizes the levels of automation according to the definitions of SAE J3016 [6]. They span from *no automation* to *full automation*. The table gives a narrative definition of the respective levels as well as responsibilities (steering, monitoring, fallback scenario) and specifies minimum requirements. The term "system" in this context refers to the respective driver assistance or automated driving system. It should be noted that warning and momentary intervention systems are excluded as they have no impact on the driver's role in performing the driving task. Furthermore, the table shows the correspondence between the levels of the German Federal Highway Research Institute (BASt) and those of the US National Highway Traffic Safety Administration (NHTSA) in its "Preliminary Statement of Policy Concerning Automated Vehicles" in 2013.

Between levels 2 ("Partial Automation") and 3 ("Conditional Automation") is a key distinction as in the latter case the system performs the entire dynamic driving task (execution of steering, acceleration, deceleration, *and* monitoring of environment).

Level	Name	Narrative definition	Execution of steering and acceleration/ deceleration	Monitoring of driving environment	Fallback performance of *dynamic driving task*	System capability (*driving modes*)	BASt level	NHTSA level
***Human driver* monitors the driving environment**								
0	No Automation	the full-time performance by the *human driver* of all aspects of the *dynamic driving task*, even when enhanced by warning or intervention systems	Human driver	Human driver	Human driver	n/a	Driver only	0
1	Driver Assistance	the *driving mode*-specific execution by a driver assistance system of either steering or acceleration/deceleration using information about the driving environment and with the expectation that the *human driver* perform all remaining aspects of the *dynamic driving task*	Human driver and system	Human driver	Human driver	Some driving modes	Assisted	1
2	Partial Automation	the *driving mode*-specific execution by one or more driver assistance systems of both steering and acceleration/deceleration using information about the driving environment and with the expectation that the *human driver* perform all remaining aspects of the *dynamic driving task*	**System**	Human driver	Human driver	Some driving modes	Partially automated	2
***Automated driving system* ("system") monitors the driving environment**								
3	Conditional Automation	the *driving mode*-specific performance by an *automated driving system* of all aspects of the *dynamic driving task* with the expectation that the *human driver* will respond appropriately to a *request to intervene*	System	**System**	Human driver	Some driving modes	Highly automated	3
4	High Automation	the *driving mode*-specific performance by an *automated driving system* of all aspects of the *dynamic driving task*, even if a *human driver* does not respond appropriately to a *request to intervene*	System	System	**System**	Some driving modes	Fully automated	3/4
5	Full Automation	the full-time performance by an *automated driving system* of all aspects of the *dynamic driving task* under all roadway and environmental conditions that can be managed by a *human driver*	System	System	System	**All driving modes**	-	

Fig. 1.1 Summary of levels of driving automation for on-road vehicles [6]

Table 1.1 Current and future systems/functions for vehicle automation [1–4, 7]

Level of automation	Current and future vehicle automation systems and functions	Market introduction
0	Lane change assist (LCA)	Available
0	Lane departure warning (LDW)	Available
0	Front collision warning (FCW)	Available
0	Park distance control (PDC)	Available
1	Adaptive cruise control (ACC)	Available
1	Park assist (PA)	2016
1	Lane keeping assist (LKA)	Available
2	Park assistance	Available
2	Traffic jam assist	2016
3	Traffic jam chauffeur	2017
3	Motorway chauffeur (MWC)	2019
4	Highway pilot	2020+
4	Piloted parking	2020+
5	Robot taxi (fully automated private vehicle)	2030+

In contrast to level 4 ("High Automation"), the driver is expected to be ready for taking over control (within a predefined time period) at all times in level 3.

Table 1.1 gives a brief overview of already introduced driver assistance systems (for both passenger and commercial vehicles) and of systems which are on the way to enter the market. For an in-depth discussion and analysis of the different driver assistance systems and automated driving functions listed below, the following supplementary readings are recommended: [8–12].

1.2.1 Scenarios and Impact of Automation Levels 2–5 According to [7]

Level 2: Partial Automation

- Improved driving comfort (driver is actively engaged)
- Increased safety is expected
- Increase in energy efficiency and traffic throughput (if cooperative)
- Scenario: highways with limited access, traffic jam assistant

Level 3: Conditional Automation

- Improved comfort and convenience
- Impact on safety strongly depends on ability to retake control in emergency conditions
- Increase in energy efficiency and traffic throughput (if cooperative)
- Scenario: traffic jam chauffeur, highway chauffeur

Level 4: High Automation

- Drastically improved driving comfort
- Increased safety improvement due to automatic transition to minimal risk conditions
- Further increase in energy efficiency and traffic throughput due to close-coupled platooning (cooperative)
- Scenario: trucks on dedicated truck lanes, automated valet parking, highway pilot

Level 5: Full Automation

- Ultimate comfort and convenience
- Efficiency gains in energy and road network usage
- Scenario: electronic chauffeur service, driverless urban goods pickup, and delivery service

1.3 Building Blocks for Automated Driving: Key Technologies

The building blocks for automated driving are shown in Fig. 1.2. They constitute three layers covering vehicle control (layer 1), sensing (layer 2), and processing and decision-making (layer 3).

Vehicles capable of (highly) automated driving are controlled agents integrating environment perception and modeling, localization and map generation, path planning, and decision-making; see Fig. 1.3.

The block "Environment Perception and Modeling" provides a real-time model of the surrounding environment. The required data is collected from the environment

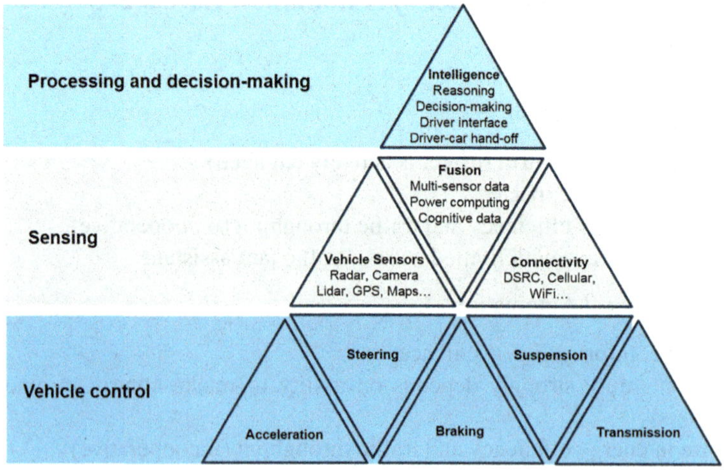

Fig. 1.2 Building blocks for automated driving

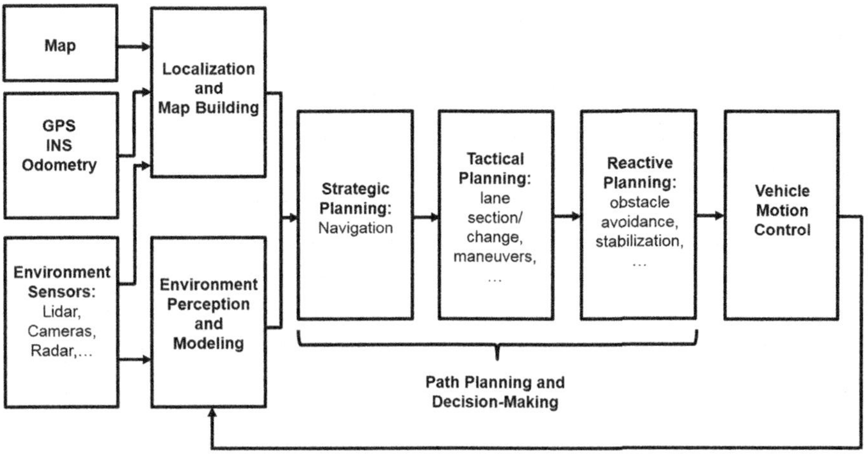

Fig. 1.3 Environment perception and modeling based on multi-sensor fusion [13]

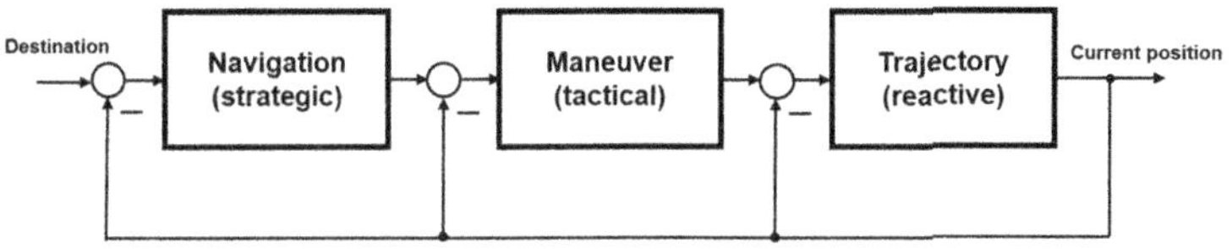

Fig. 1.4 Control structure for path planning and decision-making based on [13]

sensors like cameras, radar, lidar, or ultrasonic sensors. From the gained data original data features are extracted such as lane edges, lane markings, traffic signs, and vehicle types (passenger, commercial…). Semantic objects are recognized using classifiers, and scenarios, driving context, and vehicle positions can be computed.

The blocks "Localization and Map Building" symbolize the vehicle's ability to generate a global map via combination of local maps, global information, and data from environment models. In the context of automated driving, the term "Localization" refers to the estimation of road geometries or of the vehicle position with respect to roads in known or unknown maps.

The block "Path Planning and Decision-Making" guarantees that the vehicle is operated in accordance with the requirements to be met such as safety and legal aspects. The purpose of the three blocks making up "Path Planning and Decision-Making" is to find an optimal (in the sense of safety, speed, distance, energy saving) path in the road space from an initial position to the desired destination while avoiding collisions. Figure 1.4 illustrates the principal structure of these three blocks. It includes a strategic component (global path planning, e.g., navigation), a tactical component (e.g., lane selection), and a reactive component (local path planning, e.g., obstacle avoidance). Decision-making consists of behavioral reasoning and

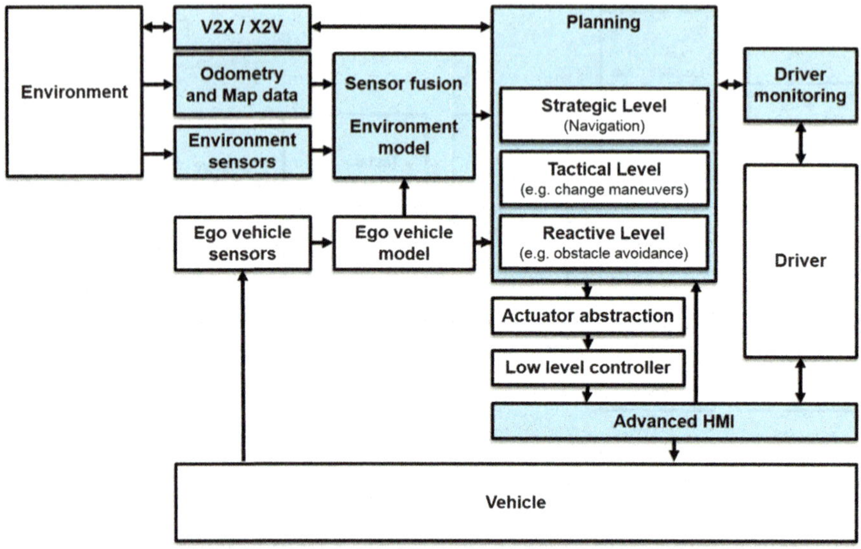

Fig. 1.5 Building blocks of automated driving illustrating the interaction between environment, vehicle, and driver (especially for SAE level 3 and 4)

adaptive mission planning by incorporating new observations and by generating and implementing new rules.

Finally, vehicle automation requires lateral and longitudinal control of motion with respect to the desired objectives and constraints. This task is reflected by the block labeled "Vehicle Motion Control" in Fig. 1.3.

Putting it all together, the overall vehicle architecture to enable automated driving which includes additional control blocks, new sensors, and advanced human–machine interaction (HMI) can be envisioned as shown in Fig. 1.5. The blue boxes indicate new and emerging technology bricks—compared to conventional driving—but also those fields of action where advances are strongly required and expected.

1.4 Enabling Automated Driving: The Research Challenges

Over the last two decades, automated driving has been a challenging research topic. However, despite tremendous improvements in sensor technology, pattern recognition techniques, robust signal processing, control system design, computational power (multi-core and many-core technology), communication bandwidth (Ethernet...), V2X, and other system technology areas, market introduction of a fully automated vehicle that is capable of unsupervised driving in an unstructured environment still remains a long-term goal. Even for structured environments,

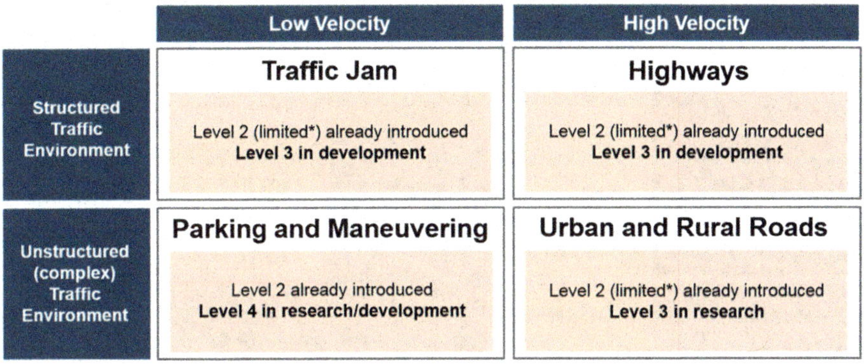

Fig. 1.6 Overview of the maturity of different automated driving functions according to the SAE levels [17]. Currently, SAE level 2 functions are still limited [The current UN Regulation No. 79 (Steering equipment) allows only corrective steering (lateral assistance) above 10 km/h. Due to that, steering capabilities of today's level 2 functions is still limited]

further research is needed to exploit the full potential of road transport automation. In order to be accepted by drivers and other stakeholders, automated vehicles must be reliable and significantly safer than today's driving baseline. A couple of roadmaps for automated road transport has been released by different stakeholders over the last 2 years giving a comprehensive view on the proposed way forward [2–4, 14–16].

As the technology for automated driving becomes more and more advanced, the research focus is shifting towards the emerging issues for their implementation. It is common sense from the industrial point of view that automated driving will be introduced following a stepwise approach. Figure 1.6 illustrates that the maturity of automated driving functions at low speeds in a structured environment is already quite high while high speed maneuvers in an unstructured and complex environment—ultimately fully automated driving—is still facing many technological (and legal) challenges.

Figure 1.7 classifies the automated driving roadmap in midterm (3–6 years) and long-term (7–12 years and beyond) challenges in alignment with the SAE automation levels. The typical approach to enter the market consists of three steps: technological research (long-term relevant), followed by piloting, field operational tests, and large-scale demonstrators (midterm scenarios) and completed by industrialization (market introduction). It is worth mentioning that commercial vehicles follow different roadmaps since other use cases have priority (e.g., platooning).

In the following, we give an overview of fields of action that have to be addressed in order to succeed in terms of technology readiness and public acceptance[1]:

[1]Without raising a claim of completeness.

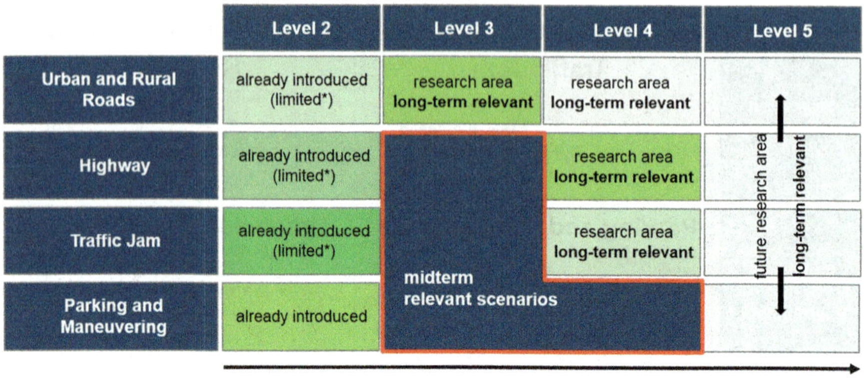

Fig. 1.7 Overview of the classification of midterm (development, industrial research) and long-term (applied and fundamental research) challenges [17]

1.4.1 Demonstrating Safety, Reliability, and Robustness

Automated vehicles by nature rely on sensing, planning, reasoning, and acting (or re-acting). A suite of vehicle sensors based on different sensing modalities (radar, lidar, camera, ultrasonic, GPS . . .) along with external sources (V2X) and detailed digital maps gather raw data of the vehicle environment, driving situation, and ambient conditions. Sophisticated algorithms interpret the data, process it, and convert it to commands for the actuators (steering, braking). Within this complex chain of sensing–processing–controlling–actuating, several types of failures can occur and have to be avoided:

- Sensors fail to recognize and respond to a hazard (false negative), e.g., the vehicle should brake but it does not (in terms of an emergency braking system).
- Sensors detect and respond to a nonexistent event (false positive), e.g., the vehicle brakes without any reason.
- System performance is degraded due to inoperable sensors or the vehicle is completely inoperable.

Consequently, there is a strong need for an independent and reproducible validation of automated vehicles. Without a traceable demonstration of the maturity (technological readiness), reliability, and safety, the societal acceptance will lack. Reliable and safe enough means rare enough, i.e., the failure probability rate should be less than 10^{-n}/h. If somehow an adequate sample size n can be argued—which is currently not the case—then appropriate safety and reliability can be demonstrated by driving (testing) in the order of 10^n h. For $n > 5$, this becomes effectively infeasible since effort and related costs increase tremendously.

As a result and based on current fail-safe and fault-tolerant architectures, automated driving needs further explorations in order to ensure a safe and fail-operational driving (concept of redundancy). For that, we need to:

- Evaluate failure modes and their impact.
- Investigate the enhanced capabilities that can self-detect, self-adapt, self-learn, and predict failures/situations in an evolving traffic scenario.
- Explore what additional requirements might be needed for fail-operational driving in terms of software, hardware, V2X communication, advanced HMI ...
- Determine required safety levels (update the ISO26262 regarding fail-operational systems).
- Develop methodologies for testing to demonstrate safety and reliability (mixed traffic should have midterm priority). Recent publications pointed out that approximately 100 million road driving kilometers are required to prove that automated vehicles are as safe as manually driven ones [18].
- Provide standardized and certified test procedures and test environments for fail-operational vehicles for any ambient condition (traffic and weather).

Furthermore, automated driving has to be considered a self-learning and adapting system based on field operational tests in order to enhance the number of relevant scenarios and to dynamically extend the list of requirements. This feedback loop of real world data will support agile development and agile validation of automated driving functions and increases safety and reliability. A set of basic scenarios and related parameters for each level auf automation should be defined in accordance with certification bodies across countries.

1.4.2 Demonstrating Security and Privacy

Consumers expect privacy and security in their cars. Consequently, the collection, processing, and linking of data have to be in accordance to the laws of privacy. At present, a lot of personalized data is already collected via navigation systems, smartphones, or during vehicle maintenance. Automated vehicles are capable of recording and providing large amounts of data that might assist crash investigations and accident reconstructions. Such data is of high relevance for improving active safety systems and system reliability but also for resolving liability issues. Existing accident data bases such as the GIDAS project (German in-depth accident study) should be updated continuously and extended by automated driving information (SAE level classification of involved vehicles, driver/automated mode...).

Furthermore, cyber security (vulnerability to hacking) has to be considered in order to avoid that the vehicle or driver lose control due to hacking attacks. Car manufacturers and their suppliers are in agreement that V2I and V2V communication, i.e., communication protocols, have to be developed with security embedded along the entire development phase (typical automotive v-cycle). Recent investigations show that nearly all modern vehicles have some sort of wireless connection that could potentially be used by hackers to remotely access their critical systems [19]. In addition to security weaknesses, many car companies are collecting detailed location data from their cars and often use unsecure transmission paths.

1.4.3 Dependable Power Computing

In order to fully deploy the opportunities of next generation ADAS technology, greater on-board computing power based on multi-core and many-core technology is strongly required. Approximately, 1 GB of data will need to be processed each second in the real-time operating system of the vehicle. This cognitive data will need to be analyzed and mined fast enough that decisions can be made and vehicle actions can be performed. The vehicle has to react to changes in the environment, i.e., adapt itself to evolving scenarios, in less than a second based on a range of variables such as vehicle speed, road, and weather conditions. Furthermore, the vehicle will need to account for unpredictable behavior of vulnerable road users and other cars while in the city and to gauge the flow of traffic, e.g., to merge onto a motorway. The current automotive software architecture (AUTOSAR) for electronic control units which has primarily been developed for vehicle control functions is facing its limits when it comes to real-time cognitive data processing. The AUTOSAR development cooperation is well aware of that and has already taken action. The AUTOSAR Adaptive Platform (configurable and adaptive runtime environment layer) is currently under development, mainly driven by BMW, Daimler, Renault, Continental, and Bosch and will be available on midterm scale. Edge computing (networked and distributed computing) will also play an important role in the future. Nevertheless, there is also a strong need for hardware standardization ("plug and play") and service-oriented architectures.

1.4.4 Human Factor (SAE Level 3/4)

It is still an open question how much time is needed for the driver to reengage. Currently, the range of 5–10 s is considered depending on the vehicle speed. As a major requirement, taking back control must happen safely and seamlessly with minimal risk for vehicle occupants and road users. Research has also to be undertaken for the vehicle–driver interface and related communication methods in order to ensure a safe transition between automated and non-automated mode. This includes overriding strategies, i.e., the driver can always override the automated driving mode and regain control. Driver awareness also plays an important role since an increased driver distraction and loss of skills due to reduced driver tasks and experience is expected. Consequently, driver training in particular for SAE level 3 and 4 vehicles will have to be deliberated. In this context, the development of human factor software tools for testing and evaluating driver and vehicle performance for different SAE levels will be essential.

1.4.5 Environment Modeling and Perception

Perceiving the vehicle environment in real-time and in a very accurate and reliable manner as well as adapting to evolving scenarios is seen as one of the key enablers for automated driving. Along with an increased robustness of sensors, several research fields can be identified including

- Reliable image processing and decision-making (object recognition, tracking, classification, interpretation)
- Accurate road representation and positioning (strongly relying on high resolution digital maps)
- Advanced sensor fusion technologies
- Common and standardized architectures
- Quality assessment including functional safety of both (embedded) software and hardware
- Plausibility checking, monitoring, onboard diagnosis, and sensor failure detection

Furthermore, a couple of research questions has to be answered rather quickly since the vehicle intelligence (in particular traceable decision-making required for SAE levels 3–5) heavily depends on environment modeling and virtualization:

- Capturing of artifacts of the vehicle environment (traffic situation) and required confidence level
- Prediction of the situation (traffic) evolvement
- Integrity of such models
- Dealing with uncertain information

1.4.6 Vehicle Control and Actuation

Vehicle automation directly relates to fail-operational control and actuation (redundancy in hardware and software). Recent advances in in-vehicle computation enable to run more sophisticated, model-based, and robust control algorithms in real-time and onboard. The core of an automated vehicle is its intelligence (path planning, path tracking, prediction, reasoning, decision-making, and control) embedded in a flexible and re-configurable architecture. Although control theory and algorithms are well developed, advances are still needed. These include:

- Controllability of the execution of planned maneuvers at any time and always maintaining fail-operational behavior of the vehicle

- Integration and management of uncertainties (e.g., map-based uncertainties)
- Dealing with sensor insufficiency and related misinterpretations (sensor false positive)
- Dealing with sparse data in case of unreliable sensor data
- Dealing with big data in real-time
- Cooperative control and planning (interaction with other road users)
- Interaction with digital infrastructure
- Driver–vehicle interaction

1.4.7 Digital Infrastructure

Currently, almost all car manufacturers worldwide have released research prototypes at different SAE levels. Even though it is well understood and agreed that the embedding in a digital infrastructure is inevitable (V2X and X2V communication), a common development roadmap is still not existing. Vehicle research and infrastructure research follow parallel paths; however, joining forces will also drive innovations forward. From the infrastructure point of view, several topics have to be addressed [16]:

- Traffic management in particular within urban environments (informative vs. influencing systems).
- Digital infrastructure (e.g., 50 Mbit/s by 2018 in Germany, nationwide)
- Data backbone, cloud data, and cloud computing (fog computing, i.e., several clouds with handover depending on driving route)
- Standards for intelligent roads
- Interaction of vehicles and their infrastructure
  - Open-source data cloud for geographic and mobility data, digital radio board (Germany: DAB+ to retrieve detailed and locally precise data in real-time)
  - Swarm intelligence
  - High-precision digital maps (from the infrastructure point of view)
  - Intelligent communication to traffic lights, traffic signs, signaling
- Standardization of IT security
- Privacy (collection, processing, linking of data) according to data privacy laws
- Legal constraints (international and national frameworks, approval of vehicles, technical maintenance and monitoring, driver training)

All of the above-mentioned fields of activity require further investigations to exploit the potential and benefits in order to ensure a safe, reliable, acceptable, and understandable (to the driver and other stakeholders), and secure behavior of the automated vehicle embedded in its intelligent environment at all times. Along with the technological advances, progress legislation, liability, and insurance are obligatory and urgently needed (for details see, e.g., [8, 20–22]).

1.5 Conclusion

This introductory chapter gives a brief overview of the current state-of-the-art of automated driving according to the SAE J3016 levels of automation and based on different recently published roadmaps. We also discuss and list relevant midterm and long-term development steps and innovation fields as well as related research challenges to be faced in order to foster societal acceptance of automated road transport. The following and carefully selected articles of this book will give an in-depth analysis of the currently most relevant technological topics from the perspective of the industry but also from renowned research institutions. The different chapters cover recent advances in environment sensing, sensor fusion, perception, decision-making, and predictive control methods for path planning and tracking. Furthermore, the increasing importance of in-vehicle architectures and embedded power computing is addressed. A major part of the book is dedicated to the demonstration and assessment of reliability and functional safety of automated vehicles along with security requirements. A sampling of ongoing collaborative European research projects and European initiatives finally reflects the strong European movement to automated driving.

References

1. Tech.AD, in *Annual Conference on Automated Driving, Conference Proceedings*, Berlin, Feb 2015
2. Automated Driving Roadmap, European Road Transport Research Advisory Council (ERTRAC) (2015)
3. Roadmap on Smart Systems For Automated Driving, European Technology Platform on Smart Systems Integration (EPoSS) (2015)
4. iMobility Forum, Automation in Road Transport, version 1.0, May 2013
5. NHTSA and US Department of Transportation, US Department of Transportation Announces Decision to Move Forward with Vehicle-to-Vehicle Communication Technology for Light Vehicles (2014)
6. SAE, Taxonomy and Definitions for Terms Related to On-Road Motor Vehicle Automated Driving Systems, J3016, SAE International Standard (2014)
7. S.E. Shladover, Road Vehicle Automation: Reality, Hype And Policy Implications, California PATH Program, University of California, Berkeley, 9 Apr 2015
8. M. Maurer, J.C. Gerdes, B. Lenz, H. Winner, *Autonomes Fahren—Technische, Rechtliche und Gesellschaftliche Aspekte* (Springer Vieweg, Heidelberg, 2015) (German) (Open Access)
9. N. Bizon, L. Dascalescu, N.M. Tabatabaei, *Autonomous Vehicles* (Nova Science Publishers, New York, 2014)
10. H. Cheng, *Autonomous Intelligent Vehicles—Theory, Algorithms, and Implementation* (Springer, Heidelberg, 2011)
11. U. Oezguener, T. Acarman, K. Redmill, *Autonomous Ground Vehicles* (Artech House, Boston, 2011)
12. H. Winner, S. Hakuli, G. Wolf, *Handbuch Fahrerassistenzsysteme*, 2nd edn (Vieweg-Teubner, Wiesbaden, 2012) (German)
13. J. Becker, Driver Assistance and Automated Driving, Lecture, ME302, Stanford University, USA

14. 2020 Roadmap, European New Car Assessment Programme (EURO NCAP) Jun 2014
15. J. Carlson, ADAS Evolving: New Developments in Hardware and Software on the Road to Autonomous Driving, HIS Automotive, Spring Media Briefing, Detroit, 19 Mar 2015
16. German Federal Government, Strategie Automatisiertes und Vernetztes Fahren, Strategic Paper, Jun 2015
17. International Organization of Motor Vehicle Manufacturers, ITS/AD-04-14, Berlin, Jun 2015
18. H. Winner, W. Wachenfeld, Absicherung Automatischen Fahrens, 6. FAS Tagung, Munich, Nov 2013
19. A. Greenberg, Markey Car Security Report, Feb 2015
20. R.K. Jurgen, *Autonomous Vehicles for Safer Driving* (SAE International, Warrendale, 2013)
21. G. Meyer, S. Beiker, *Road Vehicle Automation*. Lecture Notes in Mobility (Springer, Heidelberg, 2014)
22. G. Meyer, S. Beiker, *Road Vehicle Automation 2*. Lecture Notes in Mobility (Springer, Heidelberg, 2015)

Chapter 2
Privacy and Security in Autonomous Vehicles

Patrick Pype, Gerardo Daalderop, Eva Schulz-Kamm, Eckhard Walters, and Maximilian von Grafenstein

2.1 Introduction and Definition of Terminology

Since the early 1920s, people have envisioned cars 1 day driving them, free of human intervention. We have now reached a point where the technology exists to make this a reality. In less than 10 years, we can expect to see autonomous vehicles on our roads, fundamentally changing what it means to drive. While the technical challenges of getting a car to navigate on its own have been largely addressed, that is not the only criteria required for success.

No autonomous driving program will ever make it out of trial phase unless issues surrounding privacy, safety, and security have been addressed. In this chapter, we use the term "safety" for the correct system functioning of the car and the protection of the people in the car, mainly to ensure avoidance of car and/or traffic accidents. The term "privacy" is about the protection of a person and his/her behavior, meaning that he/she is able to control the risks for his or her rights to privacy, freedom, or equality caused by the processing of data related to him or her. The term "security" defines a situation where the integrity, confidentiality, and availability of data are guaranteed. Security particularly serves here, referring to the sum of measures taken, to ensure safety and privacy in its meaning described before. Therefore, a holistic development view on security, safety, and privacy is required, as they are all interacting, and cannot be seen completely independent.

P. Pype (✉) • G. Daalderop • E. Schulz-Kamm • E. Walters
NXP Semiconductors, Leuven, Belgium
e-mail: patrick.pype@nxp.com

M. von Grafenstein
Alexander von Humboldt Institute für Internet und Gesellschaft, Berlin, Germany

D. Watzenig, M. Horn (eds.), *Automated Driving*,
DOI 10.1007/978-3-319-31895-0_2

Consumers, governments, and businesses need to be assured that autonomous vehicles are safe from hacks, viruses, and other malicious elements that could cause widespread damage. Just a serious incident with one vehicle may be enough to ensure that autonomous vehicles never make it to our roads, so technologists must be sure that what they have created is safe and secure.

Privacy is also front of mind for consumers. Concern about what data is being collected on individuals and how it is being used, is widespread, particularly relating to new technology. The autonomous cars of tomorrow will have to take these serious concerns into account and use data in a way that is useful, but not damaging to individuals. Guardedness will be even more a strong differentiator for OEMs to get a competitive advantage in the sharing economies of the future.

2.2 Principles of Autonomous Driving

2.2.1 Technological Principles

From a technological perspective, driverless cars will be enabled by connectivity, raising the necessity to achieve the next level of performance and reliability in security, safety, privacy protection, and traffic efficiency.

This will be achieved through the design and development of next-generation automotive components, subsystems, and connected platform architectures. All of these must be based on the following critical-trust principles:

1. Secure maintenance of safety, comfort, productivity, and mobility services in a privacy conserving way.
2. Health management (monitoring and upgrading).
3. Open platform principles and architectures utilizing trusted components and trusted intra and extra vehicular networks, including seamless cooperation of different communication paradigms and high data rate sensor networking, e.g., using Automotive Ethernet.
4. Trusted cloud services for diagnostics, prognostics, and the monitoring and upgrading of security, functionality, and reliability.
5. All highlighting the need for security, privacy frameworks, and implementations on the basis of security and privacy by design, as well as "need to know" principles.

This is illustrated in Fig. 2.1, where security is seen to be a system-property spanning from components to cloud solutions. It is to be noted that for some of the components physical security is essential (so protection to side-channel attacks and physical reverse engineering), since trustworthiness of the services and information provided by complete infrastructure solutions can be at stake. This is for example the case with the V2X communication subsystems which are based on public key security infrastructures.

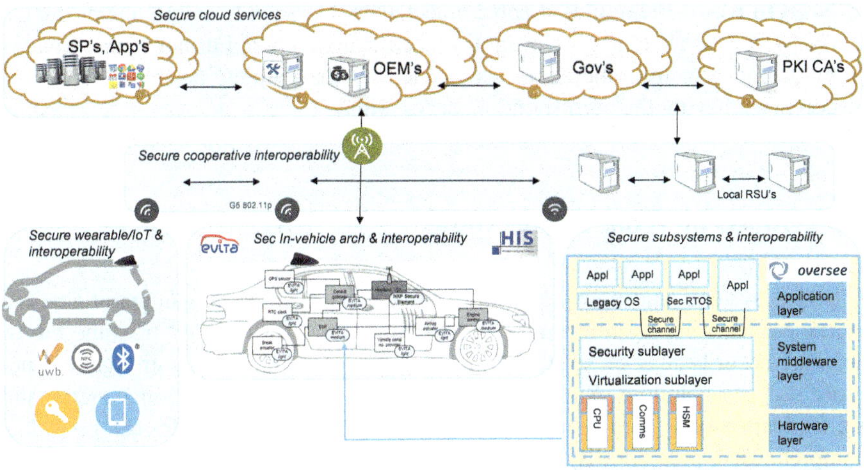

Fig. 2.1 The vehicle and its components, integrated in the Internet-of-Things, including security requirements

2.2.2 *Data Principles*

Legal uncertainty with regard to security and privacy currently disadvantages Europe's ability to take the lead for intelligent, connected, or autonomous driving solutions. Therefore, Europe should take a proactive role in defining and implementing a strategy how to create impetus for a secure and privacy respecting Internet of Things (IoT), supporting this with leading state-of-the-art technology solutions.

The aim is to facilitate automotive IoT components and subsystems to seamlessly work together across the various connectivity standards and compiling big data on their operations, including:

- Vehicle-to-Anything (V2X)/802.11p Wireless
- Radar
- Radio and Satellite Broadcast
- Automotive Telematics Onboard Unit Platforms

This also applies the concept of data governance, referring to the needs of private companies for immaterial property rights and business secrets as well as of citizens for privacy and safety, providing a best rule of access to data. In doing so, we must focus on three data principles:

- **Security by Design**—All devices connected to the IoT shall be protected via ensuring the devices and communication are intrinsically secure, achieved through a.o. incorporating secure coding and encrypted communications.
- **Privacy by Design**—Privacy is safeguarded not only by normative expectations such as by law but already on the infrastructural technological level.

- **Need to Know Principle**—Each connected entity can only communicate data to the IoT that is absolutely relevant for its core application. Further to the privacy-by-design principle, the access is granted by the entities, be it a person or a company, having the corresponding access rights.

2.3 Status of What Exists Today

Different systems and networks within the car have different vulnerabilities and attack points. They therefore require different levels of security. In some cases, software security may be sufficient, but other cases require much stronger tamper proof security solutions. Similarly, solutions like intrusion detection systems and authentication of messages and secure firmware over-the-air updates can provide a comprehensive protection system. Physical security at component level is needed in order to protect against advanced side-channel attacks and physical reverse engineering.

As more vehicles leverage V2X to eliminate accidents, ease congestion, and reduce emissions, they are creating new challenges for the industry. With more cars communicating wirelessly, vehicles are now opening up gateways for car systems to be accessed and manipulated.

Hackers are able to use these communications channels to gain direct control of cars and as a result wreak potential havoc on the roadways and even create mass accidents. For example, causing multiple cars to brake suddenly can cause a mass accident. It is therefore critical that vehicles are able to detect malicious data such as viruses or intrusions and authenticate incoming messages.

In the past year, several leading automotive manufacturers have fallen victim to connected car hacks. In summer 2015, two security experts hacked into a Jeep Cherokee via its internet-connected entertainment system. As a result, they were able to cut the vehicle's transmission, bringing it to a grinding halt on a busy American highway—all from the comfort of their sofa. A month later, researchers at the Usenix security conference demonstrated how they could access critical functions of a Corvette through a wirelessly connected device commonly used for tracking by insurers and trucking fleets.

Despite these risks, vulnerability to hacking is not sufficiently covered in the main automotive industry standards. ISO26262, for example, only determines Automotive Safety Integrity levels for passenger cars and light utility vehicles based on the reduction of systematic failures caused by human error and random failures caused by factors such as aging or thermal wear-out.

Leading automotive manufacturers such as Ford, GM, Nissan, Mazda, Honda, VW, Audi, Daimler, Hyundai Motor, and Kia are now cooperating to drive common security approaches and standards. For example, the CAMP consortium (Crash Avoidance Metrics Partnership consortium) in America has been set up to build a secure system in cooperation with the U.S. Department of Transportation. CAMP has already spent several years researching and testing a variety of security solutions. Similar initiatives are the Car 2 Car Communication Consortium (C2CC) and the C-ITS Advisory Group in Europe and the ITS Connect Promotion Consortium in Japan. Regarding privacy requirements, there are similar initiatives. For example, members of the German Union of Automobile Industry (VDA) provided for data protection principles for connected cars. These privacy-by-design principles shall ensure safe and transparent processing of personal data in order to meet legal responsibility and the customers' trust.

Though connected vehicles are still a long way from mass adoption, the recent hacking incidents are an important lesson for the automotive industry. They highlight that there needs to be a solid foundation of security, privacy, and trust to fully take advantage of innovations in connected and self-driving vehicles. A large part of this relies on the entire ecosystem coming together, manufacturers, technology suppliers, regulators, etc., to make sure that the physical and digital safety of drivers is prioritized, so vehicles will be robust enough to stand up to hacking attempts and any further abuse of data.

2.4 Future Expectations for Autonomous Driving

The transformation towards autonomous driving is already well underway, thanks to several features of automated driving (adaptive cruise control, lane-warning, etc.). Over the next few years, the car as we know will transform from a simple mode of transport to a personalized mobile information hub—fully connected to the outside world.

Innovations are already helping to create a more enjoyable, customized driving experience for consumers as well as making driving safer and more enjoyable. But this is just the beginning. Let's take a glimpse of what the future of driving looks like and how privacy and security will work in the age of the autonomous car.

One of the biggest changes brought on by autonomous driving will be the shift from car ownership to a share/rental model. Already car sharing schemes like DriveNow, Car2Go, or ZipCar are changing urban driving habits to a more pay-by-use model. This trend will accelerate as autonomous driving becomes more commonplace. Cars will be able to be booked in advance from a provider or available on demand via smartphone (or wearable) apps. The autonomous car will drive directly to the customer's house and wait outside. One can think about "uber taxis" without driver. It will be important to ensure that only the approved driver has access to the autonomous vehicle. Solutions in place today, like RFID tags for example, will be able to securely identify users via their smartphone.

With the advent of mass car sharing comes enhanced risk of data being mistakenly shared between users. The autonomous car sharing schemes of tomorrow will use data collected about users to enhance their experience. Data will be stored on:

- Recent or favorite destinations
- Personal details on the user, including payment details
- Preferences relating to seating, temperature, music, etc.

This data also awakens the interests of many further parties such as insurance companies or providers of commercial products and services. Insurance companies are able, for example, to adapt the insurance policies to the personal driving behavior of drivers. Providers of commercial products and services will increasingly seek to advertise their offers based on the location and personal preferences of the drivers.

Securing and protecting this personal data will be vital to maintaining consumer trust, as proven by several high profile incidents of data theft/loss. UK Internet provider TalkTalk, for example, suffered huge reputational and financial damage following a cyber attack in which customers personal details were stolen.

Autonomous vehicles will travel on the roads using a complex system of radar and V2X technologies which gather information about its surroundings, providing the onboard computers with an image of obstacles or dangers the driverless system must avoid.

The cars will know the route and automatically stream onto the motorway to "platoon"—hooking themselves to another group of cars heading in the same direction. Safe speed and distance is maintained via the onboard V2X technology. As well as communicating with vehicles around it, the device also speaks to other connected infrastructure.

These "communications" are enabled by V2X resulting in ad hoc data exchange networks between the vehicle and environment—in other words, independent, self-organizing networks of mobile users.

As with any other wireless network, communication is exposed to security risks that must be guarded against in order to prevent access from hackers and other potential threats. To do this, firstly the quality and integrity of data has to be ensured. Intelligent vehicles must be able to detect whether data has been altered and falsified for any reason when collected or transmitted. Wrong or defective data can block the applications on which they are based or render them ineffective—in the worst case becoming a genuine safety risk.

For instance, if inaccurate data misleads a vehicle into incorrectly recognizing the speed of the vehicle driving ahead of it, there could be fatal consequences. So mechanisms need to be integrated that can detect bad data, remove it from the communication circuit, or destroy it entirely. Automotive solutions will encrypt, authenticate, and secure data at a chip level. Using a set of security keys, the car can determine if the data really originates from a specific, trustworthy vehicle.

The same principle applies to other devices and services, which the vehicle may interact with. For example, there is no reason why the autonomous cars should not be able to interact with a café en route, making orders and sending payments in advance. Maintaining a secure connection will be essential to safeguard the reputations of both the vendor and the car manufacturer. Service providers as well as consumers will not buy into a system unless they are confident it is secure and privacy preserving.

2.5 Building Social Trust

We have already mentioned the importance of consumer trust, but it is worth looking at in greater depth as it is vital to the future of autonomous vehicles. The world is becoming an increasingly networked place. From billboards to bus stops, lamppost to cars and roads, infrastructure is becoming smarter and more connected. Many people feel scared or at least insecure about the consequences. Is the technology reliable? What happens with my data? Companies and legislators have to face up to these concerns.

According to the latest surveys, half of all Internet users do not regard their data as secure on the Internet. For businesses, customer trust and revocation of trust are important elements of business models.

No matter how well the technology works, without consumer trust it will never achieve widespread adoption. This is especially true in case of autonomous cars. The technical challenges of driving will eventually be solved but the much harder task of convincing drivers to let go of the wheel will take longer. Without trust, there is no or very slow market uptake.

Figure 2.2 shows the concerns of new-car buyers regarding data privacy and hacking.

To build trust, customers need to be actively kept in the loop. They have to be able to understand the potential security vulnerabilities in the car and also understand how they are protected. This approach also benefits companies because strong security is a real cost factor. People will not be willing to pay for good security if they don't understand how it is benefiting them. But if the awareness is raised, they will appreciate activities ensuring strong security even at higher costs.

Another important element to build trust is ensuring people and companies have complete sovereignty over their risks resulting from the processing of personal data. Connected cars and the entire Internet of Things will only be trusted if the data involved is solely used for the value added purposes offered to the consumer in a most transparent way. Furthermore, applying data minimization, i.e., by only collecting data that is really needed for the application, and nothing else, may increase people's trust.

This requires absolute certainty in knowing which other devices have access to data and transparency about what they are using it for. They also need to be reassured that data is only accessed when necessary. For example, health records may be kept in autonomous cars in case of an accident but only emergency services should have access to these, not any other non-medical party.

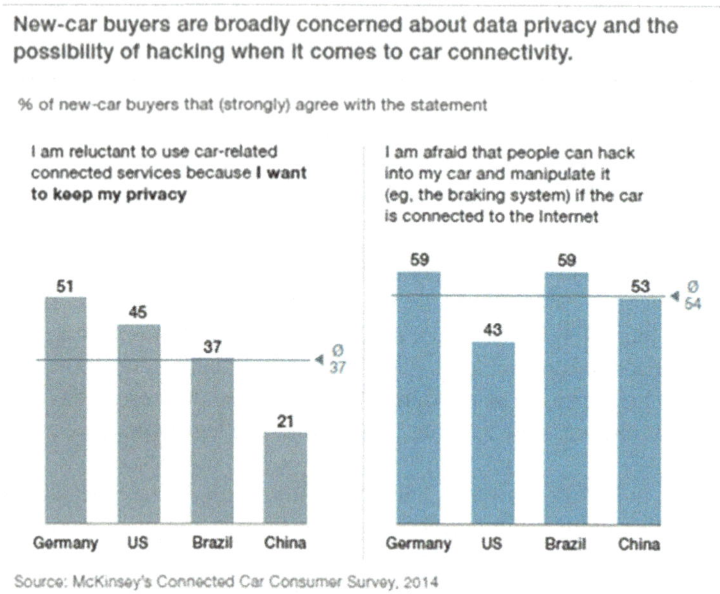

Fig. 2.2 Concerns of new-car buyers regarding data privacy and hacking

In a sharing economy consumers eventually acting as "prosumers," who are offering mobility services themselves, hold even more influence and power. So companies will very quickly learn that they can only be successful with a new deal approach based on cooperation. Ultimately consumers will decide more freely and volatile which trusted company they give their money to. In the end, only ethical and secure companies will survive.

When it comes to improving security, cross-industry and international collaboration is essential, all involved players need to be engaged for best-in-class solutions. By adhering to the aforementioned autonomous driving principles many of these issues can be overcome and trust can be built and maintained.

2.6 Impact on Industry

Maintaining privacy and security in autonomous vehicles is a complex issue that requires the support and cooperation of several industries in the complete value chain from semiconductor companies to TIER1's and OEMs. When considering the implications of onboard connectivity, the need for security, and the possibility of cars driving themselves, it becomes clear that a cooperation is needed comprising the industry in the value chain, universities and research institutes. NXP Semiconductors and the Alexander von Humboldt Institute have started initiatives in this field and are engaging with other partners to work towards complete solutions. In addition, other interest groups need to be actively involved, as, e.g., governments, insurance companies, and consumer groups (as, e.g., ADAC, ANWB, Touring, etc.).

Currently, a gap exists between governments and technology companies. This stems from a fundamental difference in how they traditionally operate. Governments are used to sharing information with the general public and being collaborative and open when it comes to new projects that have the potential to affect the lives of city residents. For democratic governments transparency is vital and they are very used to being held to account.

On the other hand, technology companies have a much more guarded approach to working. They come from a background where secrecy is essential to protecting risky R&D investments, valuable know-how, and sensitive intellectual property. They only unveil their solutions—even only partially—when they are ready to bring them to the market, as opposed to government projects which are generally transparent from the start and invite for active participation and feedback.

These systemic and cultural differences can cause friction between the two parties, and this slows down the development of a potentially joint innovation project like autonomous vehicles, which is reliant on the two working together. Ultimately without the government collaboration, autonomous vehicles can't be approved for use on the road so a new way of working needs to be established that suits both parties. They must enter into a deliberative process where the outcomes are joint strategies and memorandums, which include both, the technological and social aspects of autonomous driving.

Self-driving vehicles will also have big implications for the insurance industry. Today, car insurance is set up to protect drivers financially after an involvement in an accident, as a result of their own error or someone else's. The advent of autonomous driving changes this model completely. Firstly, road traffic accidents should be almost eliminated completely so the risk of drivers and passengers will virtually nearly disappear. As the human element comes out of the equation, question marks also remain around liability.

Insurance companies will likely have to shift liability from consumers to manufacturers, which has already been seen considering Volvo saying that it will accept full responsibility for driverless car crashes. They will also have to look more into pay-as-you-drive models for insuring cars. Ultimately insurance companies will need to change their business model and in order to do this, they will need to work with the government and manufacturers to develop best practices. Without sufficient insurance measures in place, autonomous driving will falter.

There needs to be a change of thinking and a new coordinated approach between governments, technology companies, and insurers. If all three parties collaborate to create joint strategies, which encompass technological and social aspects, autonomous vehicles can be on the market in just few years.

2.7 Next Steps

Security, privacy, and issues around consumer trust are the biggest hurdles to overcome autonomous driving. Ultimately, consumers need to trust this technology with not only their lives but also the lives of their families.

In order for this trust to be built, the entire ecosystem must come together, manufacturers, technology suppliers, governments, and insurers to take the following steps.

1. **Multi-stakeholder dialogue**
   Establish a coherent multi-stakeholder dialogue including representatives of end-user interests (e.g., data protection authorities and/or consumer protection organizations) in order to develop solutions that respect societal needs and expectations regarding access to data.

   These groups should be used in order to gain broad acceptance for privacy and security by design measures, which fully addresses the concerns of the individuals with regard to privacy and security of data and assets, as well as the organizations.
2. **Establish global principles and standards**
   The second step is to build on the dialogue to establish a condensed set of implementable principles and standards of security and privacy by design.

   They need to be applied to all areas including design, organizational structures, and international legislative requirements. A global consensus will also need to be reached on reliable certification processes and trust providing privacy policies.
3. **Develop modular system solutions**
   Establish architectures and prototypes of system-within-systems solutions that address critical security and privacy by design with the following features:

   (a) User-selectable opt-in configuration modes
   (b) Based upon interoperable, modular principles (software/hardware) with standard interfaces
   (c) Automotive certified with built-in privacy-preserving features that restrict access to data
   (d) Allowing separation of systems for cost-effective multi-vendor component sourcing

   Separation of systems also protects against hacks that target different, less critical areas of the car in order to reach more essential components. E.g., targeting the digital radio to affect the highly safety critical braking system.
4. **Next Gen secure components**
   Develop next generation components with inherent privacy protection and secure authentication comparable to current banking standards.

These secure components will combine high performance with the latest security principles to maintain protection over its lifecycle, mitigating impact of software attacks, fault attacks side channel attacks, and physical attacks including physical reverse engineering. The components and software will be flexible and open, to an extent, to ensure that they can be easily updated and upgraded.

5. **Impactful applications**
   Important to the entire process will be to demonstrate and communicate the capabilities of the system solutions and the security and privacy by design solution strategies through a number of societal and business use cases.

   Through this process, seals of "quality and trust" can be developed for certified secure solutions.

2.8 Conclusions

Based on recent car hacking incidents, all stakeholders in the car industry have become aware that security and privacy protection become more and more important in cars, which are integrated in the Internet-of-Things and/or which are autonomously driving. This will require the inclusion of security elements and privacy protect features from components to electronic systems and the complete car architecture. Furthermore, a close cooperation between all stakeholders involved in the value chain must be strengthened, including the interaction with the citizens, through a multi-stakeholder dialogue. Finally, it is also of crucial importance for all parties to work together towards a social trust model in the evolution from communicating cars to fully autonomous cars.

Chapter 3
Automated Driving from the View of Technical Standards

Stephanie Grubmüller, Jiri Plihal, and Pavel Nedoma

3.1 Introduction

Nowadays, assisted driving systems such as Adaptive Cruise Control (ACC) are state-of-the-art. If activated by the driver, these systems support the driver in specific driving situations and control either the longitudinal or lateral vehicle movement. In case of an unstable behavior of the system, a high availability of the driver is needed to take over control. Automated driving systems allow the driver to be absent from the active driving task for a certain duration of time. These systems control the vehicle's longitudinal and lateral movement simultaneously. The degree of automation of a driving function ranges from partial automation and permanent supervising by the driver to full automation and no supervising activity by the driver [1].

The first partially automated functions are expected to be introduced into day-to-day traffic by 2016 [2]. These functions require data from outside the vehicle, which leads to the term of connected driving. Improving traffic information for each traffic participant, for automated vehicles, real-time Vehicle-to-Vehicle (V2V) communication and Vehicle-to-Infrastructure (V2I) communication are mandatory [3]. The further development of automated and connected driving raise new requirements to guarantee a safe system in respect to the reconsideration of the

S. Grubmüller (✉)
VIRTUAL VEHICLE Research Center, Inffeldgasse 21a/II, 8010 Graz, Austria
e-mail: Stephanie.Grubmueller@v2c2.at

J. Plihal
ÚTIA AV ČR, v.v.i., Pod Vodárenskou věží 4, 182 08 Praha 8, Czech Republic
e-mail: j.plihal@volny.cz

P. Nedoma
ŠKODA AUTO a.s., V. Klementa 869, 293 60 Mladá Boleslav, Czech Republic
e-mail: Pavel.Nedoma@skoda-auto.cz

D. Watzenig, M. Horn (eds.), *Automated Driving*,
DOI 10.1007/978-3-319-31895-0_3

technical realization of a vehicle's subsystem as well as consideration of appropriate data sources and communication. This leads to the adoption or establishment of vehicle-related or communication-related technical standards. Infrastructure and vehicles equipped with high-quality data communication components combined with the positive cooperation between the automotive industry, researchers, and the government will result in competent backgrounds for the intended innovation, development, and use of self-driving cars.

This chapter gives an overview of different Standard Developing Organizations (SDOs) related to the automotive domain and automated driving. The levels of automation of different standards are introduced. Finally, the future steps of standardization and requirements on the vehicle systems and environment are summarized.

3.2 Standard Developing Organizations

SDOs are defined by [4] as: "SDOs include professional societies, industry, and trade associations and membership organizations who develop standards within their area of expertise. They may develop standards with their own members or in cooperation with other SDOs and interested parties." There are mainly the following SDOs related to automated driving systems:

- International Organization for Standardization[1] (ISO) consists of a network of national standards bodies which represents ISO in their countries. The central secretariat is located in Geneva, Switzerland. Regarding automated driving, two Technical Committees (TC) are of interest:
  - ISO/TC204 (WG1, WG 3, WG9, WG14, WG16, WG18)
  - ISO/TC22 (SC03, SC32, SC33, SC39)
- European Telecommunications Standard Institute[2] (ETSI) is recognized as a European Standards Organization by the European Union and produces globally-applicable Information and Communications Technologies (ICT)
- Society of Automotive Engineers (SAE) International[3] is an American United States (US)-based professional association and standards organization focused on various transport industries such as automotive, aerospace, and commercial vehicles.
- European Committee for Standardization[4] (CEN) brings together the national standardization bodies of 33 European countries.

[1]http://www.iso.org

[2]http://www.etsi.org

[3]http://www.sae.org

[4]https://www.cen.eu/

– Institute of Electrical and Electronics Engineers[5] (IEEE) is the world's largest technical professional association located in the USA.

The key international legislation relating to road safety is based on the two treaties: First, "The Vienna Convention on Road Traffic," and second "The Vienna Convention on Road Signs and Signals." Both were signed by 73 parties in Vienna in 1968.

3.3 Standard Developing Organizations

Depending on the standards, different scales of levels of automation of on-road vehicles exist. The automatic functions will be increased step-by-step and will need to be tested in practice.

In Fig. 3.1 all levels of automation according to the SAE J3016 [5] standard are given. The scale of the SAE level ranges from level 0, which indicates the full-time performance by the human driver, to level 5, which defines the full-time performance by an automated driving system. At lower levels, like levels 0–2, the driver is the main actor and is responsible for making decisions and monitoring

SAE level	Name	Narrative Definition	Execution of Steering and Acceleration/ Deceleration	*Monitoring* of Driving Environment	Fallback Performance of *Dynamic Driving Task*	System Capability *(Driving Modes)*
Human driver **monitors the driving environment**						
0	No Automation	the full-time performance by the *human driver* of all aspects of the *dynamic driving task*, even when enhanced by warning or intervention systems	Human driver	Human driver	Human driver	n/a
1	Driver Assistance	the *driving mode*-specific execution by a driver assistance system of either steering or acceleration/deceleration using information about the driving environment and with the expectation that the *human driver* perform all remaining aspects of the *dynamic driving task*	Human driver and system	Human driver	Human driver	Some driving modes
2	Partial Automation	the *driving mode*-specific execution by one or more driver assistance systems of both steering and acceleration/deceleration using information about the driving environment and with the expectation that the *human driver* perform all remaining aspects of the *dynamic driving task*	**System**	Human driver	Human driver	Some driving modes
Automated driving system **("system") monitors the driving environment**						
3	Conditional Automation	the *driving mode*-specific performance by an *automated driving system* of all aspects of the dynamic driving task with the expectation that the *human driver* will respond appropriately to a *request to intervene*	System	**System**	Human driver	Some driving modes
4	High Automation	the *driving mode*-specific performance by an automated driving system of all aspects of the *dynamic driving task*, even if a *human driver* does not respond appropriately to a *request to intervene*	System	System	**System**	Some driving modes
5	Full Automation	the full-time performance by an *automated driving system* of all aspects of the *dynamic driving task* under all roadway and environmental conditions that can be managed by a *human driver*	System	System	System	**All driving modes**

Fig. 3.1 Levels of automation according to SAE J3016 [5], Copyright © 2014 SAE International

[5] https://www.ieee.org

Level	0	1	2	3	4	5
SAE	No Automation	Driver Assistance	Partial Automation	Conditional Automation	High Automation	Full Automation
NHTSA	No Automation	Function-specific Automation	Combined Function Automation	Limited Self-Driving Automation	Full Self-Driving Automation	
BASt	Driver only	Driver Assistance	Partial Automation	Highl Automation	Full Automation	

Fig. 3.2 Levels of automation defined by the different standards according to SAE J306 [5], BASt working group "Legal Consequences of an Increase of Vehicle Automation" [7], and NHTSA [8]

the system all the time while driving. At level 2 the driver is responsible for handing over control of the vehicle guidance in case of a fault. For higher levels of automation, such as levels 3–5, the system will monitor the driving environment and with increasing levels the number of functionalities and decision-making processes adopted by the system grows.

For levels 0–2 the driver's attention lies fully on the driving task. If a fault occurs, the human driver will have to react within a second. At level 3 the reaction time of the human driver increases up to several seconds. Finally, for automation levels 4 and 5 the vehicle reacts independently during the driving task and the reaction time of the human driver is extended to several minutes [6]. With respect to international standards of automated driving, J306 is the major standard.

Figure 3.2 shows the correspondence between the SAE levels and the levels of automation developed by the Germany Federal Highway Research Institute[6] (BASt) and those described by the US National Highway Traffic Safety Administration[7] (NHTSA) [5]. The NHTSA standard integrates the high and full automation of the SAE standard in one level, whereas the BASt standard defines five levels and the last level is equivalent to Level 4 of the SAE standard.

3.4 Standard Developing Organizations

The introduction of a set of driving functions and increasing automation level requires a set of specifications and requirements depending on whether the vehicle system and/or the communication between the vehicle and other vehicles or the infrastructure are affected. These specifications are not standardized at the moment. As a consequence, an adaption of all related technical standards is needed. Two

[6]http://www.bast.de

[7]http://www.nhtsa.gov/

categories of technical standards may be distinguished: (1) vehicle-related, (2) communication-related.

3.4.1 Vehicle-Related Standards

In [1], an expectation of the introduction of highly automated functions on Germany's market is given. The Traffic Jam Pilot is expected to be deployed to the market first, followed by the Highway Pilot. In traffic congestion driving situations, the Traffic Jam Pilot will provide automated vehicle guidance. The Highway Pilot will be the extension of the Traffic Jam Pilot for higher velocities. Both will be restricted to highway-like environments at first. The authors emphasize that a more complex environment, as in urban areas, demands higher requirements on perception, situation recognition, and decision making than automated driving in a well-defined environment, as, e.g., a freeway. Consequently, whether it is an urban or highway scenario, the adaption or expansion of the vehicle system's architecture by new requirements compromises the vehicle safety. In case an unforeseen event occurs or the system boundaries are reached, the system will be responsible for switching into a safe state and handling the situation. This scenario will raise additional software requirements.

Besides, in case of a hardware fault, the system must guarantee safe operation despite the reduction in functionality. Hence, new requirements on the hardware components of the vehicle system will also appear [9]. Different use cases have to be covered by the automated functionality; for example, special weather conditions such as snowfall, as mentioned in [10], can cause a limited sight of the sensors, which leads to a decision-making based on the surroundings as to how the system needs to behave.

The functional safety standard ISO 26262 [11], published in 2011, offers a general approach for the development of E/E systems. The aforementioned use cases regarding automated driving are not currently covered by the functional safety standard ISO26262. The next version of the standard will be published in 2018. At the moment, the inclusion of the automated driving topic in the standard is under discussion. Concerning the safety of automated functionalities in on-road vehicles, new standards for specification and architecture are needed as shown by the use cases in this section.

3.4.2 Communication-Related Standards

Over the years, a range of Intelligent Transport Systems (ITS) will support the driver and finally the system in performing the driving task. ITS covers Advanced Driver Assistant Systems (ADAS), in-vehicle information systems, and roadside

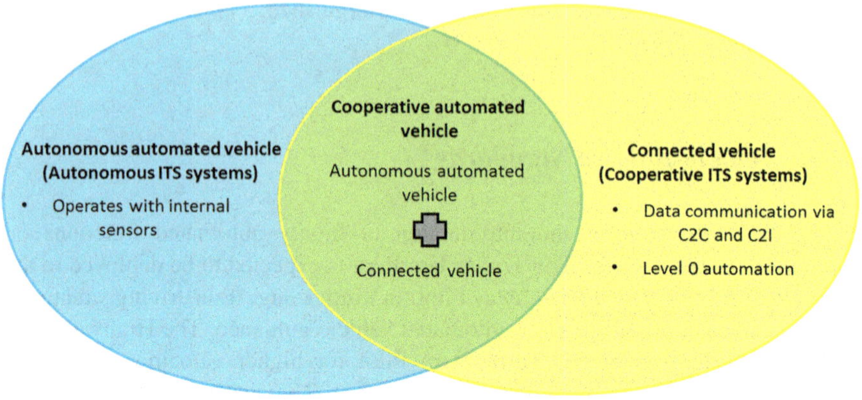

Fig. 3.3 Illustration of relation of connected, autonomous automated vehicles and cooperative automated vehicles according to [2]

telematics [12]. The basis for an ITS are three main technologies: (1) information, (2) communication, and (3) satellite technology.

Autonomous automated vehicles operate isolated from other vehicles and use only satellite data, internal sensors, and radars for guidance of the vehicle, as shown in Fig. 3.3. Without any wireless V2C or V2I, also called V2X, communication technology, the functionalities of the autonomous automated vehicle are limited. This kind of vehicle needs vehicle-based standards. On the other hand, connected vehicles provide such data communication but do not offer any automated functionality. In case a decision regarding the automated guidance of the vehicle is made on the vehicle's sensor data and data provided from outside, the system is called connected or cooperative automated vehicle.

In [3] is stated that V2V data communication improves the traffic information of each traffic participant considerably. As an example, preceding vehicles are able to inform following vehicles about traffic jams in real time. If an automated function is activated, the vehicle can adapt its velocity precisely. Furthermore, due to the V2I data exchange, the automated function is able to align the guidance of the vehicle based on data provided by surrounding information, such as velocity limitations or lighting signals. A cooperative system has the potential to increase the traffic efficiency, enhance the traffic safety, and decrease emissions.

Starting from the year 2014, Fig. 3.4 shows an expectation of the gradual development from connected to cooperative vehicles, from lower levels of automation, and finally to fully cooperative automated vehicles in urban areas by 2030. The ongoing deployment of ADAS, like automated parking or adaptive cruise control, will improve the degree of automation.

On highways, the behavior and diversity of the traffic participants as well as the road course provide good conditions for automated driving, as mentioned in Chap. 15. Since highway scenarios are easier to predict, full automation for that use case is predicted to be available sooner. The driving situation in urban areas is unpredictable

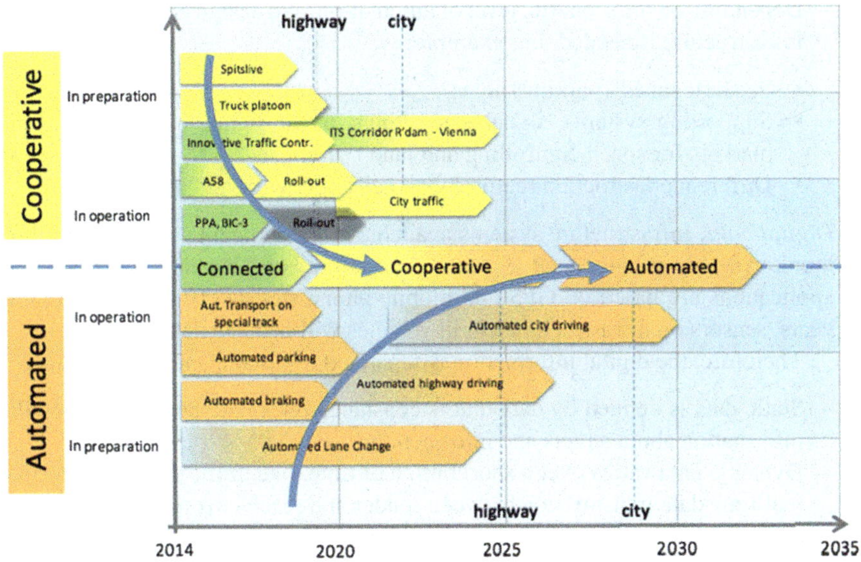

Fig. 3.4 Future development of automated and cooperative systems [2]

due to different traffic participants, intersections, etc. [13, 14]. Deploying automated driving in urban areas, the system will need to be extended, for example with an advanced perception system, traffic rules interpreter, and a decision system [15].

As a result, [3] defined fields of action to deploy cooperative automated driving in Germany. Three of them are important concerning standardization: infrastructure, IT security and data privacy, as well as legislation.

Infrastructure Cooperative driving requires high-speed data transmission. Therefore, sensors on, e.g., buildings which return traffic information need to be updated. Germany's government demands a comprehensive broadband basic supply of at least 50 Mbit/s by 2018. Furthermore, new standards for the digitalization of the federal road network need to be developed.

In [2], the author deals with deploying automated driving on roads in the Netherlands and the new demands on the infrastructure. He divides the requirements into two categories: (1) physical elements and (2) digital data.

1. *Physical elements* of the existing road infrastructure which have to be modified or extended:

   – Roads and road markings: Each type of road, e.g., highway and urban road, needs to be clearly defined and marked.
   – Signs and traffic management systems: All signs and traffic lights are required to be on-line due to the usage of determination of possible acceleration, deceleration, and maximum speed of the automated vehicle.

- Depending on the scenario, level of automation, and design choices, additional infrastructure is needed. For example:
  - Separate lanes, (parking) areas
  - Supporting systems, signage, markings, traffic lights, barriers, magnetic markers for speed controlling and lane keeping, etc.
  - Different construction requirements (different wear and tear)

2. *Digital data*: Infrastructure systems available at the moment are used for traffic monitoring and management, e.g., cameras and radars. Currently, the main V2I applications are based on GPS and mobile phone signals [16]. In the next few years, sensors on infrastructure and in vehicles will become data sources.
   Therefore, the digital information is separated into static and dynamic data:

   - Static data is defined by data that is constant over a long period of time, like information about streets and intersections.
   - Dynamic data varies over a short period of time, like traffic data. An effective real-time data transmission between sender and receiver is mandatory.

IT Security Increasing automation and connection of vehicles is a reason for the increase in the amount of data. Unprotected access to this data may influence the vehicle safety in a bad way. As an example, fake messages produced during a cyber attack could cause severe damage, which is why data encoding is important. Automobile manufacturers and tier 1 suppliers must ensure secure data encryption and communication. Validation by external organizations and a certain certification of the systems have to be considered. Hence, initiatives for cyber security for cars already exist [17]:

- ETSI TS 102 941:2012—ITS; Security; Trust and Privacy Management
- SAE J3061: Cybersecurity Guidebook for Cyber-Physical Automotive Systems
- Information Technology-Promotion Agency (IPA), Japan: Approaches for Vehicle Information Security

These standards do not, however, consider automated driving. Further, how IT security could affect functional safety is not considered, either. It is necessary to extend the ISO 26262 as a result of the increase of digital data in automated and connected vehicles. Moreover, an improvement and concretization of IT security and encryption standards need to be coordinated permanently.

Privacy Regarding data processing, generation, and linking techniques for data, it will be required that anonymization be employed more and more. The policy of data privacy must always be followed.

Legislation Concerning international legislation, the Vienna Convention implies that only a human being is able to act as a driver. That article must be extended to treat the system driver as the equivalent of a human driver. Also, the increase in the maximum velocity for automated driving up to 130 km/h is mandatory. Additionally, in national legislation, traffic laws need to be adapted to enable automated driving.

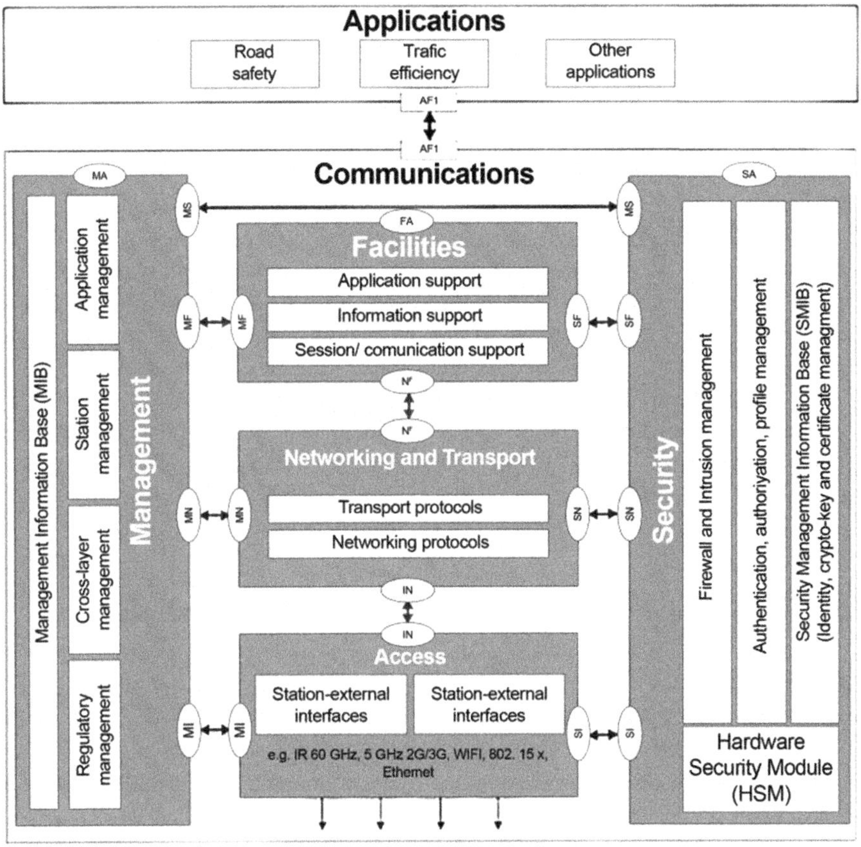

Fig. 3.5 ITS station reference architecture [18]

From the government point of view, the four fields of action give a rough overview of future work in respect to data transmission and legislation. From the technical point of view, communication systems need to be determined.

The ITS system focuses on the vehicular communication. A reference architecture of an ITS station is supported by ISO 21217:2014 [18] and shown Fig. 3.5.

The author in [19] compares different ITS standardization referring to the vehicular communication. Standards have been developed in parallel in the USA and Europe. They are commonly named C-ITS [20] in Europe and Dedicated Short Range Communication (DSRC) [21] in the USA, both relying on the WiFi standard IEEE 802.11. There are significant similarities between both, whereas an ITS communication in Japan operates at a different frequency (700 MHz). In the area of C-ITS, in Europe, ETSI and CEN with their TCs ITS and 278, in strong cooperation with ISO TC 204, are the major SDOs. In the USA, the main SDOs are IEEE and SAE with their working groups IEEE 802.11 Wireless LAN and SAE 1609 DSRC. DSRC combines IEEE 802.11 and 1609 standards, which is also known as Wireless Access for Vehicular Environment (WAVE). The core standards for DSRC are SAE J2736 and IEEE 1609 and 802.11. The core standards for C-ITS are listed as ETSI TS 103 175, 102 687, 102 724, 102 941, 103 097, 102 539, EN 302 663, and 302 636 and for the CEN/ISO TS 19 321 and 19091. The release 1 of standards for vehicular communication was published in 2014.

Regarding automated driving, both C-ITS and DSRC have to be extended. Concerning V2X communication, release 1 supports up to the level of partial automation. High aggregate information is transmitted. For higher levels of automation, sensor data with focus on latency, data rate, and reliability have to be considered, as well. Furthermore, data is sent out to all participants in close environments. Regarding cooperative maneuvering, numbers of vehicles may align the next maneuvers (e.g., for lane merging). Also, future systems, such as the 5G cellular system, need to be taken into account in future standards.

Concerning road management, the ISO 14825 [22] specifies data models for geographic databases for ITS applications. The next version GDF5.1 will support automated driving systems.

3.5 Road Map of Automation and Future Standardization

The ISO/TC204 WG14 is responsible for ITS—vehicle/roadway warning and control systems. The latest outcome was presented in Hangzhou in April 2015 [23]. The acceptance of particular steps regarding automated driving in the different regions is shown in [23]. A forecast of deploying automated driving and related ITS in the USA, Europe, and Japan is shown.

It is expected that Europe and Japan's development in that field will be driven more by automobile manufacturers and tier 1 suppliers than in the USA, where a regulatory process has been initiated [20].

Regarding the EU, serious deployment of V2I communication is demanded as a basis for the Highway Pilot. That system of automation level 3 is expected to be on the market by 2020 and fully automated driving functions, such as the Auto Pilot, by 2025. In contrast, the USA gives only a forecast of level 2 systems, which are expected by General Motors by the beginning of 2017. The whole V2X communication will be available in the middle of 2019, as also expected in Europe. Japan expects the deployment of level 3 automation on roads with intersections by the middle of 2019.

The ISO TC204/WG14 prepared an expectation of the introduction of automated driving functions and the related V2X communication on the international market. According to that plan, first level 3 functions will be deployed by 2017, the Highway Pilot by 2019. Therefore, these expectations are mandatory for future standardization of ISO/TC204 and other SDOs. It represents when standards are needed, namely before relevant systems are seriously deployed.

References

1. O. Pink et al., Automated driving on public roads: Experiences in real traffic, in it – Information Technology 2015, **57**(4), 223–230, (De Gruyter Oldenbourg, 2015). doi:10.1515/itit-2015-0010
2. T. Alkim, Infrastructure Requirements for Automated Driving Systems. Presentation presented at the ISO/TC204, Hangzhou, 23 Apr 2015
3. Bundesministerium für Verkehr und digitale Infrastruktur, Strategie automatisiertes und vernetztes Fahren: Leitanbieter bleiben, Leitmarkt werden, Regelbetrieb einleiten (2015), https://www.bmvi.de/SharedDocs/DE/Publikationen/StB/broschuere-strategie-automatisiertes-vernetztes-fahren.pdf?__blob=publicationFile. Accessed 9 Nov 2015
4. ANSI American National Standards Institute, www.StandardsPortal.org a resource for global trade (2015), http://www.standardsportal.org/usa_en/resources/glossary.aspx. Accessed 24 Aug 2015
5. SAE International On-Road Automated Vehicle Standards Committee, Standard J3016: Taxonomy and Definitions for Terms Related to On-Road Motor Vehicle Automated Driving Systems (2014)
6. J. Dokic et al., European Roadmap Smart Systems for Automated Driving. EPoSS: European Technology Platform on Smart Systems Integration (2015), http://www.smart-systems-integration.org/public/documents/publications/EPoSSRoadmap_Smart Systems for Automated Driving_V2_April 2015.pdf . Accessed 9 Nov 2015
7. T.M. Gasser et al., Legal Consequences of an Increase in Vehicle Automation (2012), http://bast.opus.hbz-nrw.de/volltexte/2013/723/pdf/Legal_consequences_of_an_increase_in_vehicle_automation.pdf. Accessed 20 Nov 2015
8. National Highway Traffic Safety Administration, National Highway Traffic Safety Administration (2013), http://www.nhtsa.gov/staticfiles/rulemaking/pdf/Automated_Vehicles_Policy.pdf. Accessed 20 Nov 2015
9. M. Helmle et al., Transient System Safety and Architectural Requirement for Partly Highly Automated Driving Functions, in *Electric & Electronic Systems in Hybrid and Electric Vehicles and Electrical Energy Management* (2014)
10. J. Becker, M. Helmle, Architecture and system safety requirements for automated driving, in *Road Vehicle Automation 2*, ed. by G. Meyer, S. Beiker. Lecture Notes in Mobility, vol. 2 (Springer, Cham, 2015), pp. 37–48
11. International Organization for Standardization, Road Vehicles—Functional Safety. ISO 26262, 2011 (2011)
12. T. Vaa, Modelling Driver Behaviour on Basis of Emotions and Feelings: Intelligent Transport Systems And Behavioural Adaptations, in *Modelling Driver Behaviour in Automotive Environments*, ed. by P.C. Cacciabue (Springer, London, 2007), pp. 208–232
13. M. Campbell et al., Autonomous driving in urban environments: Approaches, lessons and challenges. Philos. Trans. R. Soc. Math. Phys. Eng. Sci. **1928**, 4649–4672 (2010). doi:10.1098/rsta.2010.0110
14. J. Levinson et al., Towards Fully Autonomous Driving: Systems and Algorithms, in *Intelligent Vehicles Symposium (IV)*, KIT Germany, Baden-Baden, 5–9 Jun 2011
15. M. Czubenko, Z. Kowalczuk, A. Ordys, Autonomous Driver Based on an Intelligent System of Decision-Making, in *Cognitive Computation* [Online]. Accessed 26 Aug 2015
16. M. Van Schijndel-deNooij et al., Definition of Necessary Vehicle and Infrastructure Systems for Automated Driving, Study Report (2011), http://vra-net.eu/wp-content/uploads/2014/12/SMART_2010-0064-study-report-final_V1-2.pdf. Accessed 27 Aug 2015
17. E. Schoitsch, Approaches Towards Safe and Secure Mixed-Criticality Systems, in *Functional Safety Community*, Graz, 15 June 2015
18. International Organization for Standardization, Intelligent Transport Systems—Communications Access for Land Mobiles (CALM)—Architecture. ISO 21217, 2014 (2014)

19. A. Festag, Standards for vehicular communication—From IEEE 802.11p to 5G. Elektrotechnik Informationstechnik **132**(7), 409–416 (2015). doi:10.1007/s00502-015-0343-0
20. A. Festag, Cooperative intelligent transport systems (C-ITS) standards in Europe. IEEE Commun. Mag. **12**(52), 166–172 (2014)
21. J.B. Kenney, Dedicated short-range communications (DSRC) standards in the United States. Proc. IEEE **99**(7), 1162–1182 (2011)
22. International Organization for Standardization, Intelligent Transport Systems—Geographic Data Files (GDF)—GDF5.0. ISO 14825, 2011 (2011)
23. M. Misumi, *ISO TC204/WG14 Near Term Standardization Items*. Presentation presented at the ISO/TC204 WG14 Meeting, Hangzhou, Apr 2015
24. United Nations, 19. Convention on Road Traffic, in United Treaty Collection. United Nations Conference on Road Traffic—Chapter XI. Transport and Communications B. Road Traffic, Vienna (1968), https://treaties.un.org/doc/Treaties/1977/05/19770524%2000-13%20AM/Ch_XI_B_19.pdf. Accessed 9 Nov 2015
25. United Nations, 20. Convention on road signs and traffic, in United Treaty Collection. United Nations Conference on Road Traffic—Chapter XI. Transport and Communications B. Road Traffic, Vienna (1968), https://treaties.un.org/doc/Treaties/1978/06/19780606%2000-35%20AM/Ch_XI_B_20.pdf. Accessed 9 Nov 2015

Part II
The Importance of Control for Automated Driving

Chapter 4
Survey on Control Schemes for Automated Driving on Highways

Astrid Rupp and Michael Stolz

4.1 Introduction

Over the past two decades, autonomous driving has attracted the attention of several research groups. The main goal of the control design for automated or autonomous vehicles is an increased safety by means of reduced accidents. Moreover, traffic utilization and energy efficiency are considered as additional control objectives. An early publication on control issues on automated highway systems has been published in [34]. Much research has been conducted and many vehicles have been equipped with automated driver assistance functions since that time. Adaptive cruise control (ACC), which controls the longitudinal vehicle speed with respect to a target vehicle or a desired velocity, is widely used in common cars. Lane keeping assist (LKA) and lane departure warning are available lateral assistance functions. The assistance is achieved by active steering and braking. In contrast, the warning functions only alert the driver by giving haptic or acoustic signals. A major topic of recent years is a lane change assist (LCA) function. This function must compute and track a safe and comfortable trajectory for the lane change maneuver. The combination of the mentioned functions yields advanced assistance functions for highway driving, e.g., the combination of ACC and LKA allows the vehicle to drive highly automated in its own lane. Adding a function capable of lane changes results in the so-called Motorway Chauffeur or Highway Chauffeur. Another highly automated assistance function responsible for efficient and safe driving is platooning. A platoon consists of several vehicles driving with a closer

A. Rupp (✉)
Institute of Automation and Control, Graz University of Technology, Graz, Austria
e-mail: astrid.rupp@tugraz.at

M. Stolz
VIRTUAL VEHICLE Research Center, Graz, Austria
e-mail: michael.stolz@v2c2.at

D. Watzenig, M. Horn (eds.), *Automated Driving*,
DOI 10.1007/978-3-319-31895-0_4

headway for traffic congestion relief. The first vehicle is the leader and may be controlled by a human driver, the following vehicles track this leader autonomously.

The control task of a highly automated vehicle is to drive safely and efficiently from one point to another while avoiding obstacles and infeasible maneuvers.

This task is usually divided into three sublevels [91]: first, the mission planning level computes the shortest or best path from point A to point B. The second level constitutes the behavioral planning, which is the decision unit of the vehicle. This level is responsible for processing environment data and computing the possible lanes and goal points for the local motion planner. Motion planning is the third and lowest level. It computes the best trajectory based on the information from the behavioral planner. The tracking controller takes care of tracking this trajectory and is the focus of this work.

This survey is organized as follows: the remainder of Sect. 4.1 states the mathematical problem of tracking a trajectory on a highway with a highly automated vehicle. Sections 4.2–4.6 treat the controller design using different control techniques, namely fuzzy control Sect. 4.2, linear state feedback control Sect. 4.3, sliding mode control Sect. 4.4, model predictive control Sect. 4.5, and other concepts Sect. 4.6. In Sect. 4.7, tracking controllers implemented in autonomous vehicle prototypes are briefly discussed. A comparison is performed in Sect. 4.8, analyzing the performance of some of the presented controllers and finally, an outlook is given.

4.1.1 Problem Statement

To reduce complexity of the control tasks during highway driving, the dynamics of a highly automated vehicle is usually divided into longitudinal and lateral motion. The longitudinal behavior is often modeled by a simple first order system, while the lateral behavior is of a more complex nature. There are two common models for the lateral dynamics: the kinematic bicycle model and the dynamic bicycle model. Both are summarized briefly in this section, for details, refer to [74] or [83].

The simple kinematic bicycle model assumes left and right wheels being collapsed into single wheels at the center of the front and rear axle. The wheels are assumed to have no lateral slip and only the front wheel is steerable. These assumptions are fulfilled for low driving speeds and the model can be used in case of zero velocity. Figure 4.1 shows the kinematic bicycle model.

This single track model can be written as

$$\begin{bmatrix} \dot{x} \\ \dot{y} \\ \dot{\theta} \end{bmatrix} = \begin{bmatrix} \cos(\theta) \\ \sin(\theta) \\ L^{-1} \tan(\delta) \end{bmatrix} v \,, \tag{4.1}$$

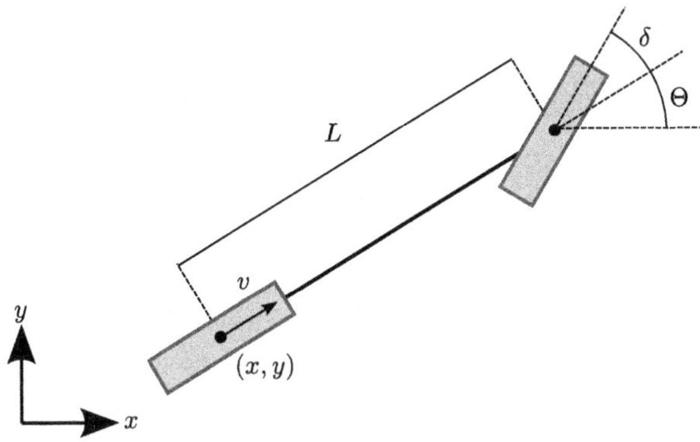

Fig. 4.1 Kinematic bicycle model with the heading θ, the steering angle δ, and the global coordinates of the rear wheels x, y. The wheels are connected by a shaft of length L. The velocity vector at the center of the rear wheels is v

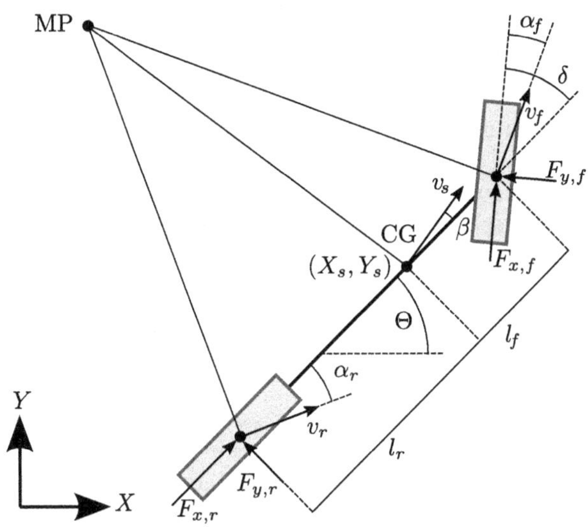

Fig. 4.2 Dynamic bicycle model, which considers forces acting at the wheels: $F_{x,f}$ and $F_{y,f}$ are the longitudinal and lateral forces of the front wheel, $F_{x,r}$ and $F_{y,r}$ describe the forces acting on the rear wheel. The slip angles of the center of gravity (CG) and the wheels are denoted by β, α_r, and α_f

where the input variables v and δ are the longitudinal velocity of the rear wheel and the steering angle. Symbols x, y are the global coordinates and θ is the heading of the car with respect to the global x axis.

The dynamic bicycle model includes wheel slipping effects due to lateral forces on the vehicle (Fig. 4.2).

The nonlinear dynamic bicycle model is given by the relations

$$\dot{v}_y = \frac{-c_f\alpha_f \cos(\delta) - c_r\alpha_r}{m} - v_x r\,, \quad \dot{r} = \frac{-l_f c_f \alpha_f \cos(\delta) + l_r c_r \alpha_r}{I_z} \tag{4.2}$$

at the center of gravity, where v_x and v_y are the velocities in a vehicle *local* coordinate system in longitudinal and lateral direction. Symbol r denotes the yaw rate. The coefficients c_i correspond to the cornering stiffness and α_i to the slip angle of the wheels, $i \in \{f, r\}$. m denotes the vehicle mass and I_z its yaw inertia. l_f and l_r are the distances from the wheels to the center of gravity. Applying small angle assumptions, a linearized state-space description can be formulated as

$$\begin{bmatrix} \dot{y} \\ \ddot{y} \\ \dot{\theta} \\ \ddot{\theta} \end{bmatrix} = \begin{bmatrix} 0 & 1 & 0 & 0 \\ 0 & -\frac{c_f+c_r}{mv_x} & 0 & -v_x - \frac{c_f l_f - c_r l_r}{mv_x} \\ 0 & 0 & 0 & 1 \\ 0 & -\frac{c_f l_f - c_r l_r}{I_z v_x} & 0 & -\frac{c_f l_f^2 - c_r l_r^2}{I_z v_x} \end{bmatrix} \begin{bmatrix} y \\ \dot{y} \\ \theta \\ \dot{\theta} \end{bmatrix} + \begin{bmatrix} 0 \\ \frac{c_f}{m} \\ 0 \\ \frac{c_f l_f}{I_z} \end{bmatrix} \delta\,, \tag{4.3}$$

which is of the linear form $\dot{\mathbf{z}} = \mathbf{A}\mathbf{z} + \mathbf{b}u$ for constant velocity v_x. For many tracking tasks, a reformulation in path and/or error coordinates is advantageous. For more detail, see [74, 83].

An important prerequisite for these tracking tasks is a trajectory generation. The following subsection tackles one way to generate such reference trajectories.

4.1.2 Trajectory Generation

For automated driving, trajectory generation is essential for maneuvering safely and efficiently on a highway or in urban scenarios. In contrast to the basic assistance functions such as ACC and LKA, which can be executed without reference trajectory (they have other references, i.e., target vehicles or target lanes), LCA requires a smooth path in order to guarantee a safe lane change.

A survey on trajectory generation can be found in [28]. One can distinguish between two main areas where trajectory generation arises:

1. Robotics: in the navigation task of robots in known or unknown environments, grids are usually used to decompose the sensor data with information of obstacle-free and obstacle-occupied space into cells. Neighboring cells are connected by weighted edges and a graph is constructed with the initial condition as starting node. A graph-based algorithm is then applied to find the path with the lowest cost from the starting point to an end point. This path consists of several segments that are tracked one after another, which yields non-smooth behavior. A smoothing algorithm thus needs to be applied for the use in autonomous driving, in order to avoid discontinuities in trajectories.

The involved kinematics and dynamics of the system are ignored by most algorithms, which may result in infeasible solutions for motorway driving. Another drawback of these approaches is the memory storage required for building the graph and the necessity of post-smoothing. Moreover, the lanes of the highway are typically available in the form of "drive corridors" from sensor fusion, which makes the search for a possible path unnecessary.

2. Control theory: optimal theory is used to find trajectories of a dynamical system that minimize a given cost function given an initial condition and an end condition. The dynamics of the system is taken into account explicitly, the generated solutions are feasible with respect to the model. Obstacles can be considered by formulating constraints or by rejecting trajectories that result in collisions.

The control theory domain offers several possibilities to deal with the trajectory generation task. Polynomial approaches or model predictive control (MPC) are employed in many publications, e.g., [46, 96] and [18]. Since the polynomial approach is simple and very popular, its main points are described in brief below.

4.1.2.1 Polynomial Approach

One common straightforward approach for lane changing trajectories is the calculation of fifth order polynomials, since they can establish smooth transitions between the vehicle's actual and desired final positions. In most cases, the lateral behavior is defined by the polynomial and the longitudinal trajectories are chosen based on simpler methods.

Assuming constant longitudinal velocity for simplicity, the lateral trajectory generation process consists of three main steps:

1. End point definition:

   In order to avoid obstacles and to allow efficient driving, a set of end positions on the road is defined as goals of the local trajectories, as described, for instance, in [91]. The lane centers of the possible lanes are typically included. Additionally, other end points may be considered for safety purposes, e.g., if a truck overlaps the lane markings slightly, the vehicle shall keep a safe distance and prefer driving more on the far side of the lane (but still in the lane). This non-centered driving when adjacent to large traffic participants also resembles human driving strategy.
2. Generation of trajectories:

   Given initial and end conditions for position, velocity, and acceleration, one can choose a fifth order polynomial in order to minimize a cost function consisting of the lateral jerk, which is the third time derivative of the position. The initial conditions are provided by sensor data or observers, while the end conditions are chosen as desired. As described, e.g., in [41], the coefficients of the polynomials can be computed based on this problem formulation and the trajectories from the starting point to each end point are generated. Changing the cost function results in a different computation of the coefficients of the polynomial. Since each generated trajectory is evaluated by another cost function, several aspects such as safety and efficiency can be incorporated in the last step.

3. Evaluation of trajectories:

   The best choice in respect to a specific cost function is selected from the set of generated trajectories. The cost function typically includes obstacle avoidance, choice of lane, and lateral position in the desired lane, among others. Obstacle avoidance is certainly the crucial task in this step. If a trajectory causes a collision, it is rejected by setting the cost to infinity. A prediction of the ego-vehicle and all obstacles is thus needed. If no collision is detected, the cost can be computed based on the distance, given, e.g., by an ellipsoid around the obstacle as in [41]. For the other cost components, the trajectory is evaluated at certain time stamps with respect to certain aspects, e.g.,

   - desired lane (defined by the driver, default is the rightmost lane in Europe)
   - lateral position in current lane
   - jerk and/or acceleration
   - desired longitudinal velocity
   - stationarity

   Finally, all cost components are summed up and the trajectory with the least cost is chosen. However, this may result in high-frequency switching if the computation is executed at every time step. A filter can be added in order to ensure smooth and efficient behavior.

In many trajectory generation approaches, the longitudinal behavior is not constant but defined by the so-called velocity profile. This profile allows combinations of constant velocities, acceleration or deceleration, or other fixed functions of time to be taken into consideration. The evaluation process is then executed based on this velocity, see, e.g., [56]. Other approaches consider the generation of longitudinal and lateral behavior simultaneously, e.g., [100].

The output of the trajectory planning algorithm is the references $x(t), y(t), \theta(t)$, $\kappa(t)$, where $t \in [t_0, t_f]$ and κ denotes the path's curvature. Tracking of these reference trajectories is a control objective and is described in the next section.

4.1.3 *Control Concepts*

Controlling a vehicle's motion is a crucial task for advanced driving assistance functions. In the context of a highly automated vehicle many control tasks need to be considered: e.g., lateral stability control and driving at the limits for collision avoidance. The focus in this survey, however, lies on comfortable driving on highways only. In order to concentrate on control approaches it is additionally assumed that a smooth and feasible reference trajectory is available and that the vehicle is able to track this in respect to dynamic constraints such as maximum steering angle and maximum steering rate.

In the subsequent comparison of the different approaches that will be summarized below, important requirements from the implementation perspective are introduced. We pay special attention in the discussion that follows to the following controller properties:

- real-time capability: the control law must be executed on an embedded control unit within a defined and guaranteed calculation time.
- parametrization: tuning of parameters should be straightforward.
- structure (specific vs. general): the controller should work on different vehicles.
- robustness: since parameter uncertainties exist such as unknown load or road surfaces, and external disturbances such as side wind force and inclination of road, robust performance must be ensured.
- nonlinearities/dependence on vehicle speed: the controller must work from 0 (in the event of traffic jams) to at least 130 km/h.

These requirements typically result in a two degree of freedom controller consisting of a feed-forward term based on the reference trajectory, and a feedback controller responsible for disturbance rejection. Not all approaches will satisfy the requirements stated above. Nevertheless, some approaches can be used for simulation purposes only.

In some control concepts, the so-called look-ahead distance (LAD) is introduced in order to improve controller performance. Not the actual position with respect to the reference, but the position a certain distance ahead is considered for the control task. The choice of this distance is not a trivial matter: by choosing a greater value, the vehicle tends to cut corners, which is not desirable. However, for greater values of LAD robustness and smoother behavior can be achieved.

There are many different control schemes that are applied to the tracking task. In this survey, we want to highlight the benefits and drawback of these schemes. Before listing previous work, a short introduction to the different controllers is given.

4.1.3.1 PID Control

The PID controller is a simple control law which takes into account the error variable (P as "proportional"), the integral (I as "integral"), and the derivative of the error variable (D as "derivative") see, e.g., [2]. Various adjustment rules exist for the parameters of such controllers, e.g., [69]. The control law in time-domain is given by

$$u(t) = k_p e(t) + k_i \int_{t_0}^{t} e(\tau) d\tau + k_d \frac{d}{dt} e(t) \,. \tag{4.4}$$

The parameters k_p, k_i, k_d can either be fixed or computed by a scheme called "gain scheduling," which updates the parameters based on the velocity or other scheduling variables. An ideal PID controller cannot be implemented in reality, therefore the derivative part is replaced by a DT1-transfer function in practice. The main advantage of this controller is its generic applicability, i.e., it is not necessary to know the mathematical model of the plant. However, the integral part may be troublesome, and the D-part of the controller may be sensitive to noise in measurement. In many cases, this simple controller can be outperformed by other control approaches.

4.1.3.2 Fuzzy Control

This heuristic scheme has been introduced in [35] and is similar to a PID controller since it employs the error, its integral, and derivative. Fuzzy control is usually applied in systems where no mathematical model is known or where models are difficult to obtain. It is thus possible to use the controller for nonlinear dynamics and multiple-input multiple-output systems. The input variables are transformed into linguistic variables by using membership functions. The output of the controller is chosen based on fuzzy rules which take an "if-then"-like form. Such a rule can be designed as

if *"lateral position error"* is left **then** *"steering"* is right.

In this case, *"lateral position error"* is the linguistic input variable and *"steering"* the output variable. Tuning of the membership functions for these variables is done manually during driving tests. The controller acts in a manner similar to human behavior, because of the human-like rules. However, the tuning is not straightforward, and stability analysis is hardly possible without mathematical models. Moreover, depending on the number of variables, the rules can become unmanageable.

4.1.3.3 Neural Networks

Artificial neural networks have first been investigated for vehicle dynamic control in [73]. They are typically represented by a system of interconnected neurons, where each connection is assigned a certain weight that is tuned based on training data or on-line. This procedure results in an adaptive net that is capable of learning. The network can be taught to imitate driver reaction if the driver makes an allowance for a specific training phase. A controller is designed based on the model to emerge from this. The main drawbacks of this approach are the need for training data and the fact that no explanation for failure can be given.

4.1.3.4 Linear Quadratic Regulator

The linear quadratic regulator (LQR) uses a linear plant model and optimal control theory to obtain an optimal state feedback controller as described, for instance, in [50]. The control input u is computed by the relation $u(t) = -\mathbf{k}^T\mathbf{x}(t)$, where $\mathbf{x} \in \mathbb{R}^n$ is the states of the system and $\mathbf{k} \in \mathbb{R}^n$ is computed in such a manner that the cost function

$$J[u] = \int_0^\infty (\mathbf{x}^T\mathbf{Q}\mathbf{x} + \mathbf{u}^T\mathbf{R}\mathbf{u})dt \tag{4.5}$$

is minimized with respect to an infinite time horizon. In contrast to the already mentioned control structures, this approach needs the information of a plant model in advance and actual signals of all states during operation. Since this is not given by default, a state observer needs to be implemented.

4.1.3.5 Feedback Linearization

Feedback linearization is a common technique rendering the closed loop system linear with the help of nonlinear compensation, see [33, 49]. Flatness-based approaches are closely related to this linearization, for further details, refer to [53].

4.1.3.6 Sliding Mode Control

The sliding mode control (SMC) approach relies on a variable structure controller and is robust with respect to a specific class of modeling uncertainties and external disturbances [20]. It consists of two main design parts: in a first step, a desired dynamics is defined by the so-called sliding variable, e.g., $s = \dot{e} + \lambda e$, which ensures for $s = 0$ that the error e converges to 0 in finite time, with the weighting factor λ. In the second step, a controller is defined so that this desired dynamics is obtained, for example, $\dot{s} = -k\text{sign}(s)$ with the control parameter k. This switching law may generate chattering in the control input, which can be prevented by using higher-order SMC. A well-known second-order sliding mode controller is the "super-twisting algorithm," [52]. A downside of this approach is that it is derived in continuous-time, but its discrete-time behavior strongly depends on the sampling frequency [62].

4.1.3.7 Model Predictive Control

At each time step an internal model is used to predict the system behavior over a predefined horizon, where an optimal control input sequence is generated that minimizes a certain cost function. This approach allows the consideration of different types of constraints for states and inputs, which is a major advantage of MPC. The main disadvantage is the high computational complexity, which is the reason for its scarce utilization in real-time applications. Additionally, there is need for an explicit model for prediction, and external inputs and disturbances need to be known fairly accurately in advance for the entire prediction horizon [6]. A simplified MPC formulation similar to the one derived in [14] can be represented by

$$\begin{aligned} \min_{u_{0|t},\dots,u_{N_c-1|t}} \quad & \sum_{k=0}^{N_p-1} \mathbf{x}_{k|t}^T \mathbf{Q} \mathbf{x}_{k|t} + \sum_{k=0}^{N_c-1} \mathbf{u}_{k|t}^T \mathbf{R} \mathbf{u}_{k|t} \\ \text{subject to} \quad & \mathbf{x}_{k+1|t} = \mathbf{A}\mathbf{x}_{k|t} + \mathbf{B}\mathbf{u}_{k|t}, \qquad k = 0, 1, \dots, N_p - 1, \\ & \mathbf{x}_{k+1|t}^{\min} \le \mathbf{x}_{k+1|t} \le \mathbf{x}_{k+1|t}^{\max}, \qquad k = 0, 1, \dots, N_p - 1, \\ & \mathbf{u}^{\min} \le \mathbf{u}_{k|t} \le \mathbf{u}^{\max}, \qquad k = 0, 1, \dots, N_c - 1, \\ & \mathbf{u}_k = \mathbf{0}, \qquad N_c \le k \le N_p , \end{aligned} \tag{4.6}$$

where N_c is the control horizon, N_p the prediction horizon, $N_c \le N_p$, and $\mathbf{x}_{k+1|t}^{\min}$, $\mathbf{x}_{k+1|t}^{\max}$, $\mathbf{u}^{\min}$, $\mathbf{u}^{\max}$ the constraints for states and input, respectively. The subscript $k+1|t$ denotes the value of the variable $k+1$ steps ahead of the current time t.

4.1.3.8 H_∞ Control

This robust approach attempts to control a plant affected by modeling uncertainties and parameter variations [82, 102]. An optimization problem needs to be solved once, minimizing the so-called H_∞-norm of a particular transfer function T of the control system. The transfer function T is defined by the control objective (noise rejection, tracking, etc.), the plant, additional uncertainty models, weighting transfer functions, and the feedback control matrix, which is the optimization parameter. For a stable single-input-single-output system, the H_∞-norm is the largest value of the frequency response magnitude. For a stable multi-input-multi-output system it is the largest singular value $\bar{\sigma}$ across all frequencies ω

$$\|T\|_\infty := \sup_{\omega \in \mathbb{R}} \bar{\sigma}[T(j\omega)] \,. \tag{4.7}$$

The advantages are inherent robust stability and robust performance. Depending on the approach (choice of weighting/shaping transfer functions), this design may result in dynamic controllers of high order, which are complex to realize in practice.

4.1.4 Partitioning the Problem

The control design of automated vehicles can be partitioned into the following tasks:

(a) Longitudinal control:

A common approach to control the longitudinal behavior is to divide the controller level into an inner loop for throttle and brake control, and an outer loop for velocity or acceleration tracking. This approach is exploited in many of the listed publications, for an introduction, see [74].

There are typically two fields of research: the first one tackles the ACC functionality, where a desired velocity is to be maintained or a target is to be followed at a safe distance. In [43], safety issues in real world contexts and vehicles equipped with ACC in 2003 are discussed. Since the control task of the ACC is more or less completed, we refer to a review on earlier works on ACC in [8].

The second field deals with platooning, where a leader vehicle shall be followed by the other vehicles. A survey on platooning can be found in [6]. With the assumption of communication between the vehicles in the platoon, the so-called cooperative ACC (CACC) arises, as treated in [81].

(b) Lateral control:

For comfortable automated driving, the central lateral control tasks are lane keeping and lane changing. From a control point of view, the main difference of these two functions is that the LKA tries to keep the vehicle in the middle of the lane, i.e., a stabilization task, while LCA tries to follow a reference trajectory computed by another entity (trajectory planning), which corresponds to a tracking task. In publications from the late 1990s, e.g., [38, 68], magnetic strips are considered to allow localization of the vehicle. In the California PATH program, the lateral sensing system consisted of magnetic markers embedded along the road center line and two sets of magnetometers installed under the front and rear bumpers of the vehicle, see, for example, [40]. At the present time the algorithms are either vision-based [87] or GPS-based [45, 64]. However, the control task for lateral control stays the same: minimize the lateral displacement and the angular error with respect to the reference trajectory (either lane center or polynomial).

(c) Combined control:

Assistance functions for both longitudinal and lateral motion are in general more complex than the control strategies discussed so far. The separate design of the steering angle happens mostly under the assumption that the velocity is constant. This of course is not always the case and may lead to problems. For example, at higher speeds, a smaller steering angle is needed for the same lateral displacement at a certain LAD. However, many publications handle the steering and the velocity inputs separately, using the current velocity of the vehicle in the lateral control design.

The control techniques described in the next sections are partitioned according to problems (a)–(c).

4.2 Fuzzy Control

(a) In [63], a fuzzy controller for throttle and brake control is proposed for a combination of ACC and Stop&Go. Car-following behavior based on fuzzy control is also discussed in [101] and in [90], where the latter proved stability based on the so-called linear matrix inequalities (LMI) conditions.

The longitudinal controller is implemented by a neuro-fuzzy controller in [13]. The output layer is updated on-line via a gradient algorithm. According to the authors, an attractive feature of this method is that it does not require training data or the vehicle longitudinal dynamic model. Simulations show that in comparison to a PID controller, the neuro-fuzzy controller exhibits a smoother performance and hence provides a more comfortable feeling for passengers onboard. In thesis [11], more details on this approach can be found.

The authors of [72] introduced a fuzzy control strategy for CACC in order to achieve human-like driving behavior in a cooperative environment and produce satisfactory results.

(b) In [12], lateral control is conducted via fuzzy PD control. A generic algorithm method is used to optimize the parameters of the fuzzy controller.

A comparison of a fuzzy controller and a control law derived by using the nonlinear kinematics model with a Lyapunov function has been published in [58]. The fuzzy controller demonstrates effective performance.

A simple lane change fuzzy controller with four rules is presented in [64], where two different control strategies are designed: one for straight road driving, the other one for lane change maneuvers. Simulations indicate that the switching between the controllers is smooth and the overtaking maneuver can be executed similar to human performance.

In [71], a cascade architecture for lateral control has been presented. The low-level is governed by a discrete PID controller. The high-level is governed through fuzzy logic, which computes desired steering and angular speed for the PID controller. This control architecture gives good results for different vehicle speeds and curves. The high-level output signals are smoothed out by the lower level PID controller, which avoids undesirable oscillations of the steering wheel that could be particularly dangerous at higher speeds.

A genetic algorithm has been applied to adjust a fuzzy controller automatically in [70], which overcomes the tuning problem of the fuzzy control scheme.

In [93], two controllers based on fuzzy logic are designed for the lateral offset and angle error, respectively. The speed of the vehicle is taken into account and the two outputs are finally weighted and summed up to obtain the steering wheel angle for the actuator.

4.3 Linear State Feedback Control

(b) The lateral dynamics is controlled by using LQR design for the state feedback controller in [74]. A feed-forward term is computed considering the road curvature. A similar LQR approach incorporating constraints on the steering angle has been discussed in [48].

In [40], an output-feedback controller has been derived by the LQR method when only position feedback is available, preserving certain robustness properties such as phase margin and gain margin.

In [87], lane keeping performance on a curved road is improved by adding the integral of the lateral offset error to the state feedback controller. The authors implement a multi-rate Kalman Filter for state estimation in order to allow the LQR controller to operate at the fast update rate of the microprocessor. This multi-rate control scheme can reduce the inter-sample ripples in the yaw rate.

Bilinear matrix inequality (BMI) optimization is presented in [7] to compute a state feedback controller, which is capable of lane keeping as well as obstacle avoidance assistance.

4.4 Sliding Mode Control

(a) In [75], a sliding mode-based longitudinal control design for platoons is presented using a time-variant LAD. Similarly, a sliding mode controller for platooning with wireless communication is demonstrated in [74]. The experiments indicate that the spacing performance and ride quality are superior to human driving skills.

Cruise and longitudinal tracking control of vehicles are considered in [24] and [25]. The first publication exploited suitably designed observers for SMC. The second presented an additional collision avoidance functionality.

SMC for hybrid electric vehicles has been applied for cruise control in [27]. The authors claim that this approach yields good performance, especially when compared to a PID controller.

The authors of [98] focus on a sliding mode controller for platooning due to its robustness to parameter uncertainties and external disturbances.

(b) In [1], the first sliding mode controller for steering a city bus has been published. The author introduced the so-called Ackermann-model for vehicle steering, which is used quite extensively. The sliding mode approach is extended in [38] to a chatter-free performance for automated highway systems. In thesis [39], loop shaping, H_∞ control, and SMC for lateral motion on a tractor semi-trailer are compared. It is shown that the performance of the controllers strongly depends on the choice of the LAD. The robustness of a sliding mode controller has also been mentioned by Hatipoglu et al. [31], where lane changing and lane keeping maneuvers are implemented as different modes. Sliding mode controllers for trajectory tracking or path tracking can also be found in [84–86]. In these papers, the lateral displacement and the heading error of the vehicle are combined in the sliding manifold for the steering input. This combination demands that both variables converge and have been implemented using the kinematic bicycle model as well as four-wheel steering vehicles. In [32], active front steering and dynamic stability control are coordinated. For the steering functionality, a sliding surface for yaw rate tracking is employed. The dynamic stability control is developed independently for emergency maneuvers. A rule-based integration scheme using a fuzzy membership function chooses between the two subsystems. The authors state that this integrated approach leads to improved vehicle stability.

In [88], lateral dynamics are controlled by a super-twisting controller, which is robust to time-varying speed, curvature, and parameter uncertainties, while chattering is avoided. A terminal sliding mode method based on a look-ahead scheme has been presented in [99]. An adaptive algorithm is used to tune the controller parameters.

A sliding mode controller is compared in [16] to a driver-model-based controller, which consists of a PI controller with LAD. Two sliding surfaces are designed: one for the lateral displacement and one for the heading error. Since only one control input is available, these two sliding surfaces are then combined to one sliding variable. The SMC achieves higher accuracy in path tracking, but needs a larger steering input. With a saturation on the steering angle, both controllers yield similar performance. It is also demonstrated that a SMC using preview control does not outperform the conventional SMC.

(c) In [30], a sliding mode controller is implemented by the super-twisting algorithm for the tracking problem of a car-like system called "Robucar." The derived control laws are of discontinuous type and may lead to discontinuous velocities in practice. This difficulty is overcome by taking into account the actuator dynamics. The approach works well for the robot, but needs to be adapted for autonomous vehicles.

In thesis [21], an integrated high-level control for longitudinal and lateral dynamics has been introduced. A higher-dimensional sliding surface is chosen to compute the reference for the low-level controller. In the higher level, the dynamics of the longitudinal and lateral control are coupled, whereas the low-level control treats the dynamics separately. A simple PI controller is used in the feedback loop, a feed-forward block is in charge of compensating for the estimated disturbances.

Discrete-time and continuous-time sliding mode controllers have been investigated in [19] on a four-wheel-robot. The work points out that the discrete-time controller has a lower longitudinal error, while the continuous-time controller has a lower lateral error.

The authors of [97] tested a sliding mode based control on the autonomous vehicle "Kuafu-II," where the driving control system integrates both the longitudinal and lateral controllers. The longitudinal controller is designed as a main speed controller and a space controller for obstacle avoidance. The parameters of the speed controller are computed by an adaptive law, while the space controller is implemented as PI controller with gain scheduling. For low velocities, the lateral controller will use Stanley's law that is discussed in Sect. 4.7 and published in [89]. In higher speed ranges, the controller switches to a sliding mode controller in order to improve tracking accuracy.

4.5 Model Predictive Control

The main advantage of the MPC approach is the possibility for explicitly handling constraints. Moreover, the control design for nonlinear models or combined dynamics is easier than with multiple-input multiple-output control schemes. Since MPC is able to consider highly dynamical systems, many lateral controllers are also combined with longitudinal control. The controllers from the lateral control (b) are thus moved to the combined problem (c). The main drawback has long been the computational complexity, which is why a computation time is stated in many publications. Efficient solvers exist, however, that make use of MPC in real-time applications possible. A little more computation time is not critical, especially in comfortable scenario. Due to the constraint handling, many publications on MPC also treat collision avoidance maneuvers. In these maneuvers, driving at the limits occurs and constraints become active, which induces longer computation times. However, the collision avoidance maneuvers are practically "built-in," which is a very nice property of this control scheme.

(a) Early works using MPC for ACC control tasks have been presented in [5] and [17]. In [55], an MPC strategy has been used for ACC. A low-level controller compensates for vehicle dynamics and tracks the desired trajectory. The high-level MPC minimizes a quadratic cost function that consists of minimal tracking error, low fuel consumption, and accordance with driver dynamic car-following characteristics. Driver longitudinal ride comfort, driver permissible tracking range, and rear-end safety are formulated as linear constraints which are softened to avoid computational infeasibility.

    In [79], a nonlinear MPC approach is investigated using only one control loop (instead of an inner and outer control loop). The overall model is augmented in order to combine distance and speed tracking control.

(c) In [9], a nonlinear model predictive approach (NLMPC) is described, where its performance is checked for a double-lane change maneuver on snow. The performance depends on the horizons of the control scheme and the vehicle speed. The computation time has been investigated using the NPSOL software package and is given for inactive constraints by 0.1 to 0.3 s, for active constraints by 0.1 to 1.6 s. An extension of this work that examines side wind rejection has been published in [47], with an average NPSOL computational time 0.13 s, and in the worst case 0.38 s.

In [23], another extension has been proposed for combined braking and steering, where a linear time-varying (LTV) MPC is recast into a quadratic program. The performance is enhanced when compared to steering only.

NLMPC has also been applied in [3]. Additionally, a cruise speed-profile generator is responsible for adapting the cruise speed at lateral maneuvers. The proposed control law allows the consideration of variable longitudinal speed and sliding during lateral maneuvers. The same author presented a nonlinear longitudinal control strategy considering powertrain dynamics in [4]. A control architecture is introduced which combines the steering and the longitudinal controllers so as to ensure the simultaneous control of longitudinal and lateral motions.

In [51], the authors aim to reduce the computational complexity of the MPC approach. An approximate explicit MPC scheme is demonstrated, where suboptimal controls are used with its computation time being drastically reduced and its accuracy being significantly improved. The authors generate grid points to build nodal state parameter vectors in the state space, for which optimal solutions are to be computed off-line. The nodal state parameter vectors construct polytopes, for which equivalent suboptimal feedback control gains are computed off-line.

Automated lane change maneuvers have also been treated in [67], where the longitudinal and the lateral control is handled by two loosely coupled low complexity quadratic programs.

In [57], an LTV-MPC using sparse clothoid-based path description is implemented. Only a few waypoints are computed to represent the road, which makes the cost function minimization more efficient by allowing larger prediction distances.

4.6 Other Concepts

(b) The so-called H_2 control design for lane keeping has been presented in [80]. LQG and H_∞ have been investigated by Eom et al. [22], where it has been shown that H_∞ is more robust and produces lower control input.

Low energy consumption is accomplished by H_2 control in [42]. Additionally, a H_∞ controller for disturbance rejection is employed. A switching control for the two subsystems was designed for further improving system performance.

In [66], an adaptive output-feedback controller called "self-tuning regulator" is demonstrated. Adaption laws with the so-called parameter reduction are stated. All vehicle parameters are considered unknown and therefore this approach can be realized for a wide variety of vehicles. It is also robust to variations on curvature and lateral wind. The main disadvantage of this approach is the high computational complexity, as has been pointed out in [83].

The self-tuning regulator has been compared to H_∞ control, fuzzy control, and a P controller in [15]. Three different situations are simulated in order to investigate the robustness of the approaches: first, the road friction coefficient is varied. The P controller produces the largest error while the self-tuning regulator has the best response. Second, the longitudinal speed is changed. Similar performances as in the first case have been observed. Third, lateral wind is introduced as disturbance. The P control is less affected by the wind force than the other controllers. The authors state that the self-tuning regulator is one with the best performance, at least from a simulation point of view. However, it is also the most complicated to be implemented.

The authors of [59] propose an active front steering control based on the yaw rate tracking error. They use two PID control loops, where one loop tracks the yaw rate reference signal that is generated by the other. Seven control gains and the LAD must be chosen, where the design parameters are tuned through numerical optimization.

In [44], a steering control algorithm that does not depend on vehicle parameters has been presented. The steering angle is computed in such a manner that a desired yaw rate can be tracked and adapts the corresponding gain in order to compensate the yaw rate error. This control law can be applied for any vehicle without any information on vehicle parameters. However, real-time experiments for automated driving have not yet been tested by the authors.

(c) In one of the earlier publications on autonomous driving [68], an H_∞ based lane keeping controller has been investigated that works for both curved and straight highway sections without knowledge of the radius of the road curvature. Another early approach [10] applies a gain scheduled P controller for active safety systems.

The authors of [45] propose a steering controller based on finite preview optimal control, which is basically an LQR controller plus feed-forward term. The speed controller is designed as a simple proportional control and the desired speed is set so that the lateral acceleration stays below a predefined value. This controller in combination with the steering controller significantly reduces lateral acceleration when compared to constant speed.

In [37], a longitudinal control via PID controller and a lateral control via H_∞ loop shaping have been published. The authors employed a feed-forward term for curvature dependence and also drew attention to the high computational burden this involved. The same research group published an adaptive backstepping approach in [36], using the yaw rate as virtual input. The controller is robust with respect to parameter variation in the matrix $\mathbf{A}$ of the dynamic bicycle model (4.3). However, variations in the input vector $\mathbf{b}$ have not been investigated.

Backstepping is also mentioned in [29], where a simple trajectory planning algorithm for the lane change maneuver is presented and the tracking controller produces satisfactory performance. Coupled longitudinal and lateral backstepping control with a control robustification is discussed in [65].

Another backstepping-based approach has been studied in [95], where the authors consider forward and backward driving. A feedback law with orientation control has been investigated mainly for backward driving, since forward driving is improved when using a feedback law without this orientation control. The same author also describes exact input-output-linearization in his thesis [94], where the nonlinear transformation yields two linear decoupled systems that are stabilized by simple state feedback controllers. The exact linearization shows good performance at higher speeds, while the backstepping-based approaches have smooth behavior also in the starting and stopping processes, since the singularities at standstill are avoided by velocity-independent feedback gains.

A modular approach has been presented in [77]. A dynamic feed-forward term is computed using a model of the system with a state feedback controller and a pre-filter. A simple P controller for lateral displacement and angular error is combined with a disturbance observer that provides steady-state accuracy. Gain scheduling for the controller parameters depending on the velocity is proposed. A lookup table is obtained using a parameter space approach by the so-called gamma and beta stability, as also described in [92].

Flatness-based control for nonlinear vehicle dynamics has been proposed in [26]. A nonlinear bicycle model is used and a flat output is derived. Given the desired trajectory of the flat output, the tracking controller computes the control input for the nonlinear state feedback control block. This block then generates the steering and velocity for the vehicle. The performance has been shown for a lane change maneuver.

In [60] and [61], lateral and longitudinal controllers are combined to perform some coupled maneuvers, such as stop-and-go control with obstacle avoidance and lane change maneuvers. A flatness-based controller computes steering and velocity for the nonlinear bicycle model. Since derivatives of reference and measured vehicle signals are needed in these computations, an algebraic nonlinear estimation is introduced for the numerical differentiation of the signals. The results indicate good performance even under sudden and sharp maneuvers.

The authors of [76] compare a simple kinematic controller consisting of a feed-forward term plus P controller with a flatness-based control approach. This approach combines a flatness-based feed-forward control with an LQR feedback control. Disturbances can be rejected in both cases, but the simple approach has higher overshoot and the flatness-based approach is faster. The main drawback of the flatness-based strategy is that it needs additional information from the vehicle, specifically the side-slip angle and the yaw rate.

4.7 Autonomous Vehicles

This section briefly lists the control structures demonstrated on autonomous vehicle prototypes. The control schemes are primarily standard approaches, since the focus has been on environment perception or trajectory planning.

The team of Stanford [89] implemented the velocity and steering controllers separately for its vehicle "Stanley" at the DARPA Grand Challenge. The velocity control is implemented as a PI controller, while the steering angle is computed by a nonlinear feedback function of the cross-track error, which is known as the "Stanley method." The performance of this controller degrades seriously at higher speeds, as also mentioned in [83].

The controller of "Junior" consists of an MPC strategy with a PID block [54]. The feed-forward term has to be tuned specifically for the vehicle. Longitudinal control for "Leonie" has been presented in [78]. The work introduces the so-called Grip Value, which is an indicator for potential changes of road conditions and is applied as an additional input to the controller.

In [103], research on the autonomous vehicle "Bertha" has been published. A Luenberger observer is designed to estimate the lateral displacement and its derivative with respect to time. The lateral control consists of a feed-forward part for the desired yaw rate that is based on the road curvature at a look-ahead point, and a feedback component realized by a PI controller. Then, an inverse single-track model is used to compute the desired steering angle. An additional observer is responsible for the offset compensation of the steering angle in order to obtain steady-state accuracy.

4.8 Comparison

A comparison of control approaches has been conducted in [83], where different driving courses have been simulated with a maximum velocity of 20 m/s. The Stanley method [89] has been tested and the authors observed that it is not robust to a rapid lane change maneuver for collision avoidance. However, it is well suited for standard driving maneuvers. Optimal control is proposed using LQR, LQR with feed-forward term, and preview control. LQR alone fails the rapid lane change maneuver entirely at higher speed since information about the path is not included. The optimal control with feed-forward term considers curvature and speed, and therefore improves performance. Optimal preview control combines the optimal control with road information, which gives the best performance for highways at (almost) constant speed. The survey demonstrates that depending on the maneuver and the vehicle speed, the controllers show different performance.

In addition to the comparison in the literature a simple example of a highway overtaking maneuver is simulated, using different controllers for lane keeping.

4.8.1 Simulation Example

The simulated overtaking maneuver is executed on a curved road (road radius $R = 1930\,\text{m}$). The longitudinal velocity of the ego-vehicle is $v_x = 120\,\text{km/h}$

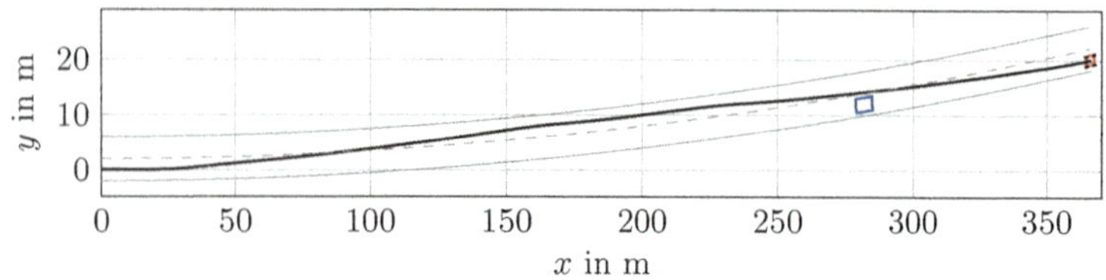

Fig. 4.3 Overtaking maneuver on a curved road at high speeds ($v_x = 120\,\text{km/h}$). The driven trajectory of the ego-vehicle is depicted by the *thick black line*

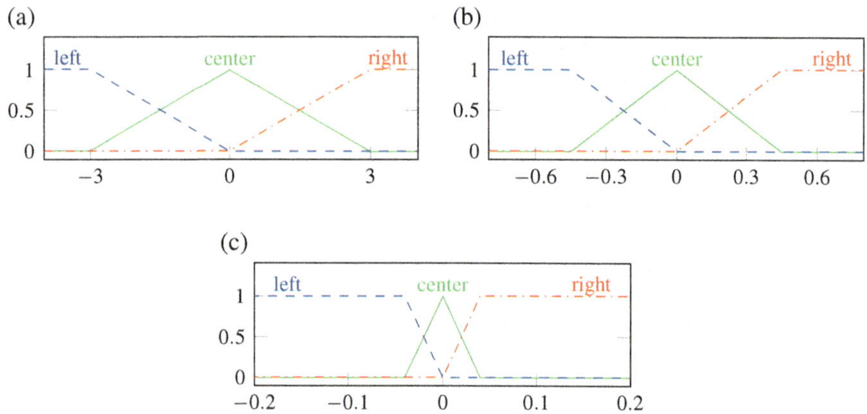

Fig. 4.4 Membership functions of the input and output variables of the fuzzy controller. (**a**) Input lateral error, (**b**) input angular error, (**c**) output steering

while the speed of the other vehicle is 75 km/h. The maneuver is illustrated in Fig. 4.3, marking the ego-vehicle in red and its overtaking trajectory in black. For control and simulation a dynamical bicycle model has been used with the following parameters $m = 1564\,\text{kg}$, $c_r = c_f = 140000\,\text{Nm/rad}$, $l_f = 1.268\,\text{m}$, $l_r = 1.620\,\text{m}$, $I_z = 2230\,\text{kgm}^2$.

Tuning of the controllers has been done in simulation runs with the nominal plant leading to approximately the same maximum steering angle $|\delta| < \delta_{\text{max}} = 0.03\,\text{rad}$.

The parameters of the different controllers have been chosen as follows:

The gain of the Stanley method is $k = 5$, the LQR matrices are $\mathbf{Q} = \text{diag}(1, \mathbf{0})$, $R = 1$. The super-twisting algorithm of the sliding mode controller has the gains $k_1 = 0.2$, $k_2 = 0.0005$. The sliding variables are defined by $s_i = \dot{e}_i + \lambda_i e_i$, $i \in \{y, \theta\}$ with $\lambda_y = 5$ and $\lambda_\theta = 1$. The weight for the combined sliding variable $s = s_y + \lambda s_\theta$ is $\lambda = 0.1$. The membership functions of the fuzzy controller are shown in Fig. 4.4. The four rules were defined according to [64]. Note that the membership functions have been found by trial and error and also that the tuning was time-consuming.

For the MPC controller, the matrix $\mathbf{Q}$ and input weight R are the same as in the LQR cost function. The input constraint has been set to $\delta_{\text{max}} = 0.03\,\text{rad}$, and the control horizon to $N_c = 30$ with a sampling rate $T_c = 0.01\,\text{s}$.

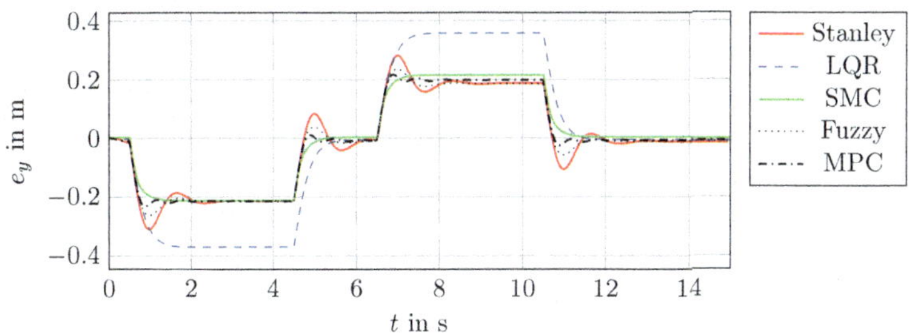

Fig. 4.5 Lateral errors of the different control schemes during an overtaking maneuver on a curved road

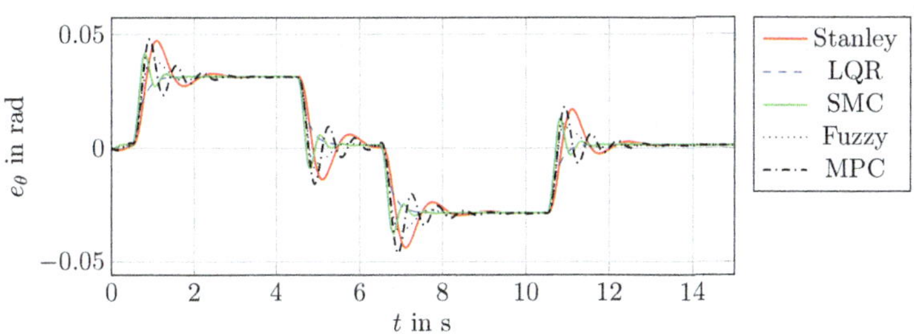

Fig. 4.6 Angular errors of the different control schemes during an overtaking maneuver on a curved road

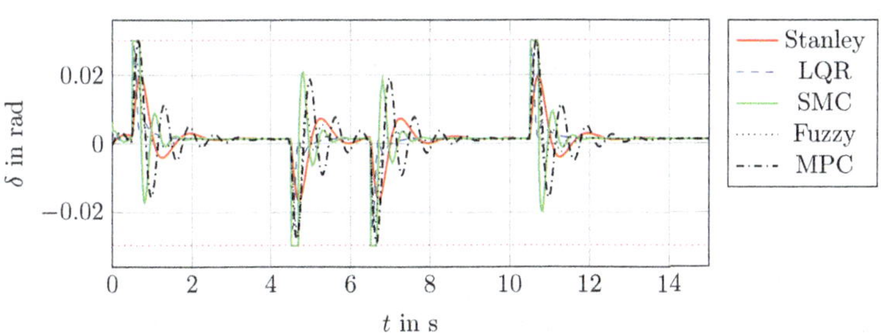

Fig. 4.7 Steering angle of the different control schemes during an overtaking maneuver on a curved road

The lateral displacements of the different controllers are depicted in Fig. 4.5. SMC exhibits the smoothest lateral errors while tracking the dynamic parts of the trajectory, and the LQR approach has the largest error. However, the angular error of the LQR approach has no overshoot as illustrated in Fig. 4.6.

The computed steering angle is shown in Fig. 4.7. As expected, SMC applies higher input values, which has also been observed in [16]. The Stanley method exhibits the smoothest performance.

To sum up, the Stanley method is distinguished by its simplicity and smooth performance during standard driving scenarios. The LQR exhibits low computational complexity, robustness, and simple parameter tuning. SMC is a very robust controller, but its discrete-time performance has to be checked. MPC has a high computational complexity, but good performance and has the ability to handle constraints and include collision avoidance, which make this approach very appealing. For this reason applications based on MPC-algorithms to solve automated driving assistance tasks efficiently are still subject to intensive research.

4.9 Outlook

The latest research trends on control in automated driving can be partitioned into two main research areas, namely

- MPC and efficient solvers, e.g., [14, 57], due to its capability of dynamics and constraint consideration,
- CACC and coordinated maneuvers utilizing information from other traffic partners, e.g., [81, 92], due to the ongoing research on inter-vehicle communication.

When using communication for information exchange between traffic partners, networked control strategies may be exploited to improve the overall traffic performance, see, e.g., [6]. Moreover, automated driving in urban areas poses new challenges: intersections and other traffic participants such as bicyclists and tractors need to be considered and pedestrian safety must be ensured. These topics have been investigated recently and will be of interest in the future. The authors look forward to improvements and new challenges in these fields.

References

1. J. Ackermann, J. Guldner, W. Sienel, R. Steinhauser, V. Utkin, Linear and nonlinear controller design for robust automatic steering. IEEE Trans. Control Syst. Technol. **3**(1), 132–143 (1995)
2. K.J. Astrom, *PID Controllers: Theory, Design and Tuning* (Instrument Society of America, Research Triangle Park, 1995)
3. R. Attia, R. Orjuela, M. Basset, Coupled longitudinal and lateral control strategy improving lateral stability for autonomous vehicle, in *2012 American Control Conference (ACC)* (2012), pp. 6509–6514
4. R. Attia, R. Orjuela, M. Basset, Combined longitudinal and lateral control for automated vehicle guidance. Veh. Syst. Dyn. **52**(2), 261–279 (2014)
5. V.L. Bageshwar, W.L. Garrard, R. Rajamani, Model predictive control of transitional maneuvers for adaptive cruise control vehicles. IEEE Trans. Veh. Technol. **53**(5), 1573–1585 (2004)
6. L. Baskar, B. De Schutter, J. Hellendoorn, Z. Papp, Traffic control and intelligent vehicle highway systems: a survey. IET Intell. Transp. Syst. **5**(1), 38–52 (2011)
7. A. Benine-Neto, S. Scalzi, S. Mammar, M. Netto, Dynamic controller for lane keeping and obstacle avoidance assistance system, in *2010 13th International IEEE Conference on Intelligent Transportation Systems (ITSC)* (2010), pp. 1363–1368

8. S. Björnander, L. Grunske, Adaptive cruise controllers - a literature review. Technical Rep. C4-01 TR M50, Faculty of Information and Communications Technologies, Swinburne University of Technology, Australia, 2008
9. F. Borrelli, P. Falcone, T. Keviczky, J. Asgari, MPC-based approach to active steering for autonomous vehicle systems. Int. J. Veh. Auton. Syst. **3**(2), 265–291 (2005)
10. A. Broggi, M. Bertozzi, A. Fascioli, C.G.L. Bianco, A. Piazzi, The ARGO autonomous vehicle's vision and control systems. Int. J. Intell. Control Syst. **3**(4), 409–441 (1999)
11. L. Cai, Novel algorithms for longitudinal and lateral control for application in autonomous vehicles. PhD thesis, The Hong Kong Polytechnic University, 2003
12. L. Cai, A.B. Rad, W. Chan, K. Cai, A robust fuzzy PD controller for automatic steering control of autonomous vehicles, in *The 12th IEEE International Conference on Fuzzy Systems (FUZZ)*, vol. 1 (IEEE, 2003), pp. 549–554
13. L. Cai, A.B. Rad, W. Chan, An intelligent longitudinal controller for application in semiautonomous vehicles. IEEE Trans. Ind. Electron. **57**(4), 1487–1497 (2010)
14. A. Carvalho, S. Lefévre, G. Schildbach, J. Kong, F. Borrelli, Automated driving: the role of forecasts and uncertainty - a control perspective. Eur. J. Control **24**, 14–32 (2015); SI: {ECC15European} Control Conference
15. S. Chaib, M.S. Netto, S. Mammar, H_∞, adaptive PID and fuzzy control: a comparison of controllers for vehicle lane keeping, in *2004 IEEE Intelligent Vehicles Symposium* (2004), pp. 139–144
16. J.M. Choi, S. Lui, J.K. Hedrick, Human driver model and sliding mode control - road tracking capability of the vehicle model, in *Proceedings of the 2015 European Control Conference (ECC)* (2015), pp. 2137–2142
17. D. Corona, M. Lazar, B. De Schutter, M. Heemels, A hybrid MPC approach to the design of a smart adaptive cruise controller, in *2006 IEEE Computer Aided Control System Design, 2006 IEEE International Symposium on Intelligent Control* (2006), pp. 231–236
18. A. Dotlinger, F. Larcher, R.M. Kennel, Robust receding horizon based trajectory planning, in *2014 IEEE/ASME International Conference on Advanced Intelligent Mechatronics (AIM)* (IEEE, 2014), pp. 845–851
19. B. Dumitrascu, A. Filipescu, V. Minzu, A. Voda, E. Minca, Discrete-time sliding-mode control of four driving-steering wheels autonomous vehicle, in *Proceedings of the 30th Chinese Control Conference* (2011), pp. 3620–3625
20. C. Edwards, S. Spurgeon, *Sliding Mode Control: Theory and Applications* (CRC Press, Boca Raton, 1998)
21. T. Eigel, Integrierte Längs- und Querführung von Personenkraftwagen mittels Sliding-Mode-Regelung. Ph.D. thesis, Technische Universität Carolo-Wilhelmina zu Braunschweig, 2010
22. S. Eom, E. Kim, T. Shin, M. Lee, F. Harashima, The robust controller design for lateral control of vehicles, in *2003 IEEE/ASME International Conference on Advanced Intelligent Mechatronics (AIM)*, vol. 1 (IEEE, 2003), pp. 570–573
23. P. Falcone, F. Borrelli, J. Asgari, H.E. Tseng, D. Hrovat, A model predictive control approach for combined braking and steering in autonomous vehicles, in *Mediterranean Conference on Control & Automation (MED)* (IEEE, 2007), pp. 1–6
24. A. Ferrara, P. Pisu, Minimum sensor second-order sliding mode longitudinal control of passenger vehicles. IEEE Trans. Intell. Transp. Syst. **5**(1), 20–32 (2004)
25. A. Ferrara, C. Vecchio, Second order sliding mode control of vehicles with distributed collision avoidance capabilities. Mechatronics **19**(4), 471–477 (2009)
26. S. Fuchshumer, K. Schlacher, T. Rittenschober, Nonlinear vehicle dynamics control - a flatness based approach, in *44th IEEE Conference on Decision and Control, 2005 and 2005 European Control Conference. CDC-ECC'05* (IEEE, 2005), pp. 6492–6497
27. B. Ganji, A.Z. Kouzani, S.Y. Khoo, M. Shams-Zahraei, Adaptive cruise control of a HEV using sliding mode control. Expert Syst. Appl. **41**(2), 607–615 (2014)
28. C. Goerzen, Z. Kong, B. Mettler, A survey of motion planning algorithms from the perspective of autonomous UAV guidance. J. Intell. Robot. Syst. **57**(1–4), 65–100 (2010)

29. L. Guo, P. Ge, M. Yue, Y. Zhao, Lane changing trajectory planning and tracking controller design for intelligent vehicle running on curved road. Math. Probl. Eng. **2014**, 1–9 (2014)
30. F. Hamerlain, K. Achour, T. Floquet, W. Perruquetti, Trajectory tracking of a car-like robot using second order sliding mode control, in *European Control Conference (ECC)* (IEEE, 2007), pp. 4932–4936
31. C. Hatipoglu, U. Ozguner, K.A. Redmill, Automated lane change controller design. IEEE Trans. Intell. Transp. Syst. **4**(1), 13–22 (2003)
32. J. He, D.A. Crolla, M.C. Levesley, W.J. Manning, Coordination of active steering, driveline, and braking for integrated vehicle dynamics control. Proc. Inst. Mech. Eng. D J. Automob. Eng. **220**(10), 1401–1420 (2006)
33. J.K. Hedrick, A. Girard, Controllability and observability of nonlinear systems, in *Control of Nonlinear Dynamic Systems: Theory and Applications* (University of California, Berkeley, 2005)
34. J.K. Hedrick, M. Tomizuka, P. Varaiya, Control issues in automated highway systems. IEEE Control Syst. **14**(6), 21–32 (1994)
35. T. Hessburg, M. Tomizuka, Fuzzy logic control for lateral vehicle guidance. IEEE Control. Syst. **14**(4), 55–63 (1994)
36. S. Hima, S. Glaser, A. Chaibet, B. Vanholme, Controller design for trajectory tracking of autonomous passenger vehicles, in *2011 14th International IEEE Conference on Intelligent Transportation Systems (ITSC)* (IEEE, 2011), pp. 1459–1464
37. S. Hima, B. Lusseti, B. Vanholme, S. Glaser, S. Mammar, Trajectory tracking for highly automated passenger vehicles, in *International Federation of Automatic Control (IFAC) World Congress* (2011), pp. 12958–12963
38. P. Hingwe, M. Tomizuka, Experimental evaluation of a chatter free sliding mode control for lateral control in AHS, in *Proceedings of the 1997 American Control Conference (ACC)*, vol. 5 (1997), pp. 3365–3369
39. P. Hingwe, J. Wang, M. Tai, M. Tomizuka, *Lateral Control of Heavy Duty Vehicles for Automated Highway System: Experimental Study on a Tractor Semi-Trailer* (California Partners for Advanced Transit and Highways (PATH), Berkeley, 2000)
40. T. Hsiao, M. Tomizuka, Design of position feedback controllers for vehicle lateral motion, in *American Control Conference (ACC)* (IEEE, 2006), pp. 6
41. R. Hult, Tabar, R.S., Path planning for highly automated vehicles. Master's thesis, Chalmers University of Technology, Gothenburg, 2013
42. C. Hwang, S. Han, L. Chang, Trajectory tracking of car-like mobile robots using mixed h_2/h_∞ decentralized variable structure control, in *IEEE International Conference on Mechatronics* (IEEE, 2005), pp. 520–525
43. H.M. Jagtman, E. Wiersma, Driving with adaptive cruise control in the real world, in *Improving Safety by Linking Research with Safety Policy and Management, Proceedings of the 16th ICTCT Workshop* (2003)
44. C. Jung, H. Kim, Y. Son, K. Lee, K. Yi, Parameter adaptive steering control for autonomous driving, in *2014 IEEE 17th International Conference on Intelligent Transportation Systems (ITSC)* (2014), pp. 1462–1467
45. J. Kang, R. Hindiyeh, S. Moon, J.C. Gerdes, K. Yi, Design and testing of a controller for autonomous vehicle path tracking using GPS/INS sensors, in *Proceedings of the 17th IFAC World Congress* (2008), pp. 6–11
46. A. Kelly, B. Nagy, Reactive nonholonomic trajectory generation via parametric optimal control. Int. J. Robot. Res. **22**(7–8), 583–601 (2003)
47. T. Keviczky, P. Falcone, F. Borrelli, J. Asgari, D. Hrovat, Predictive control approach to autonomous vehicle steering, in *American Control Conference (ACC)* (IEEE, 2006), pp. 4670–4675
48. D. Kim, J. Kang, K. Yi, Control strategy for high-speed autonomous driving in structured road, in *2011 14th International IEEE Conference on Intelligent Transportation Systems (ITSC)* (IEEE, 2011), pp. 186–191

49. J. Kosecka, R. Blasi, C.J. Taylor, J. Malik, A comparative study of vision-based lateral control strategies for autonomous highway driving. Int. J. Robot. Res. **18**(5), 442–453 (1999)
50. E. Lavretsky, K.A. Wise, Optimal control and the linear quadratic regulator, in *Robust and Adaptive Control*. Advanced Textbooks in Control and Signal Processing (Springer, London, 2013), pp. 27–50
51. S. Lee, C. Chung, Multilevel approximate model predictive control and its application to autonomous vehicle active steering, in *2013 IEEE 52nd Annual Conference on Decision and Control (CDC)* (2013), pp. 5746–5751
52. A. Levant, Sliding order and sliding accuracy in sliding mode control. Int. J. Control. **58**(6), 1247–1263 (1993)
53. J. Levine, *Analysis and Control of Nonlinear Systems: A Flatness-based Approach* (Springer Science and Business Media, Berlin, 2009)
54. J. Levinson, J. Askeland, J. Becker, J. Dolson, D. Held, S. Kammel, J.Z. Kolter, D. Langer, O. Pink, V. Pratt, M. Sokolsky, G. Stanek, D. Stavens, A. Teichman, M. Werling, S. Thrun, Towards fully autonomous driving: systems and algorithms, in *2011 IEEE Intelligent Vehicles Symposium (IV)* (2011), pp. 163–168
55. S. Li, K. Li, R. Rajamani, J. Wang, Model predictive multi-objective vehicular adaptive cruise control. IEEE Trans. Control Syst. Technol. **19**(3), 556–566 (2011)
56. X. Li, Z. Sun, Q. Zhu, D. Liu, A unified approach to local trajectory planning and control for autonomous driving along a reference path, in *2014 IEEE International Conference on Mechatronics and Automation (ICMA)* (2014), pp. 1716–1721
57. P.F. Lima, M. Trincavelli, J. Mårtensson, B. Wahlberg, Clothoid-based model predictive control for autonomous driving, in *Proceedings of the 2015 European Control Conference (ECC)* (2015), pp. 2988–2995
58. E. Maalouf, M. Saad, H. Saliah, A higher level path tracking controller for a four-wheel differentially steered mobile robot. Robot. Auton. Syst. **54**(1), 23–33 (2006)
59. R. Marino, S. Scalzi, M. Netto, Nested PID steering control for lane keeping in autonomous vehicles. Control Eng. Pract. **19**(12), 1459–1467 (2011)
60. L. Menhour, B. D'Andréa-Novel, C. Boussard, M. Fliess, H. Mounier, Algebraic nonlinear estimation and flatness-based lateral/longitudinal control for automotive vehicles, in *2011 14th International IEEE Conference on Intelligent Transportation Systems (ITSC)* (IEEE, 2011), pp. 463–468
61. L. Menhour, B. d'Andréa Novel, M. Fliess, H. Mounier, Coupled nonlinear vehicle control: flatness-based setting with algebraic estimation techniques. Control Eng. Pract. **22**, 135–146 (2014)
62. G. Monsees, Discrete-time sliding mode control. TU Delft, Delft University of Technology, 2002
63. J. Naranjo, J. Reviejo, C. González, R. Garca Rosa, T. de Pedro, A throttle and brake fuzzy controller: towards the automatic car, in *EUROCAST*, ed. by R. Moreno-Daz, F. Pichler. Lecture Notes in Computer Science, vol. 2809 (Springer, New York, 2003), pp. 291–301
64. J. Naranjo, C. Gonzalez, R. Garca, T. de Pedro, Lane-change fuzzy control in autonomous vehicles for the overtaking maneuver. IEEE Trans. Intell. Transp. Syst. **9**(3), 438–450 (2008)
65. L. Nehaoua, L. Nouveliere, Backstepping based approach for the combined longitudinal-lateral vehicle control, in *2012 IEEE Intelligent Vehicles Symposium (IV)* (2012), pp. 395–400
66. M.S. Netto, S. Chaib, S. Mammar, Lateral adaptive control for vehicle lane keeping, in *Proceedings of the 2004 American Control Conference (ACC)*, vol. 3 (2004), pp. 2693–2698
67. J. Nilsson, M. Brunnström, E. Coelingh, J. Fredriksson, Longitudinal and lateral control for automated lane change maneuvers, in *Proceedings of the 2015 American Control Conference (ACC)* (2015), pp. 1399–1404
68. R.T. O'Brien, P.A. Iglesias, T.J. Urban, Vehicle lateral control for automated highway systems. IEEE Trans. Control Syst. Technol. **4**(3), 266–273 (1996)
69. A. O'Dwyer, *Handbook Of PI And PID Controller Tuning Rules*, 3rd edn. (Imperial College Press, London, 2009)

70. E. Onieva, J.E. Naranjo, V. Milanés, J. Alonso, R. García, J. Pérez, Automatic lateral control for unmanned vehicles via genetic algorithms. Appl. Soft Comput. **11**(1), 1303–1309 (2011)
71. J. Pérez, V. Milanés, E. Onieva, Cascade architecture for lateral control in autonomous vehicles. IEEE Trans. Intell. Transp. Syst. **12**(1), 73–82 (2011)
72. J. Pérez, V. Milanés, J. Godoy, J. Villagra, E. Onieva, Cooperative controllers for highways based on human experience. Expert Syst. Appl. **40**(4), 1024–1033 (2013)
73. D.A. Pomerleau, Neural networks for intelligent vehicles, in *Proceedings of IEEE Conference on Intelligent Vehicles* (1993), pp. 19–24
74. R. Rajamani, *Vehicle Dynamics and Control.* Mechanical Engineering Series (Springer Science, New York, 2006)
75. R. Rajamani, H. Tan, B.K. Law, W. Zhang, Demonstration of integrated longitudinal and lateral control for the operation of automated vehicles in platoons. IEEE Trans. Control Syst. Technol. **8**(4), 695–708 (2000)
76. T. Raste, S. Lüke, A. Eckert, Automated driving, technical approach with motion control architecture. at-Automatisierungstechnik **63**(3), 191–201 (2015)
77. Ch. Rathgeber, F. Winkler, D. Odenthal, S. Muller, Lateral trajectory tracking control for autonomous vehicles, in *2014 European Control Conference (ECC)* (IEEE, 2014), pp. 1024–1029
78. A. Reschka, J. Bohmer, F. Saust, B. Lichte, M. Maurer, Safe, dynamic and comfortable longitudinal control for an autonomous vehicle, in *2012 IEEE Intelligent Vehicles Symposium (IV)* (IEEE, 2012), pp. 346–351
79. P. Shakouri, A. Ordys, M.R. Askari, Adaptive cruise control with stop and go function using the state-dependent nonlinear model predictive control approach. {ISA} Trans. **51**(5), 622–631 (2012)
80. M. Shimakage, S. Satoh, K. Uenuma, H. Mouri, Design of lane-keeping control with steering torque input. JSAE Rev. **23**(3), 317–323 (2002)
81. S.E. Shladover, Recent international activity in cooperative vehicle-highway automation systems. Technical Report FHWA-HRT-12-033, Federal Highway Administration, 2012
82. S. Skogestad, I. Postlethwaite, *Multivariable Feedback Control: Analysis and Design*, vol. 2 (Wiley, New York, 2007)
83. J.M. Snider, Automatic steering methods for autonomous automobile path tracking. Technical Report CMU-RITR-09-08, Robotics Institute, Pittsburgh, 2009
84. R. Solea, U. Nunes, Trajectory planning with velocity planner for fully-automated passenger vehicles, in *2006 IEEE Intelligent Transportation Systems Conference (ITSC)* (2006), pp. 474–480
85. R. Solea, U. Nunes, Trajectory planning and sliding-mode control based trajectory-tracking for cybercars. Integr. Comput. Aided Eng. **14**(1), 33–47 (2007)
86. R. Solea, A. Filipescu, V. Mînzu, S. Filipescu, Sliding-mode trajectory-tracking control for a four-wheel-steering vehicle, in *2010 8th IEEE International Conference on Control and Automation (ICCA)* (IEEE, 2010), pp. 382–387
87. Y.S. Son, W. Kim, S.-H. Lee, C.C. Chung, Robust multi-rate control scheme with predictive virtual lanes for lane-keeping system of autonomous highway driving. IEEE Trans. Veh. Technol. **64**(8), 3378–3391 (2015)
88. G. Tagne, R. Talj, A. Charara, Higher-order sliding mode control for lateral dynamics of autonomous vehicles, with experimental validation, in *2013 IEEE Intelligent Vehicles Symposium (IV)* (IEEE, 2013), pp. 678–683
89. S. Thrun, M. Montemerlo, H. Dahlkamp, D. Stavens, A. Aron, J. Diebel, P. Fong, J. Gale, M. Halpenny, G. Hoffmann, K. Lau, C. Oakley, M. Palatucci, V. Pratt, P. Stang, S. Strohband, C. Dupont, L. Jendrossek, C. Koelen, C. Markey, C. Rummel, J. van Niekerk, E. Jensen, P. Alessandrini, G. Bradski, B. Davies, S. Ettinger, A. Kaehler, A. Nefian, P. Mahoney, Stanley: the robot that won the DARPA grand challenge: research articles. J. Intell. Robot. Syst. **23**(9), 661–692 (2006)
90. P.F. Toulotte, S. Delprat, T.M. Guerra, Longitudinal and lateral control for automatic vehicle following, in *Vehicle Power and Propulsion Conference* (2006), pp. 1–6

91. C. Urmson, J. Anhalt, H. Bae, J.A. Bagnell, C.R. Baker, R.E. Bittner, T. Brown, M. N. Clark, M. Darms, D. Demitrish, J.M. Dolan, D. Duggins, D. Ferguson, T. Galatali, C.M. Geyer, M. Gittleman, S. Harbaugh, M. Hebert, T. Howard, S. Kolski, M. Likhachev, B. Litkouhi, A. Kelly, M. McNaughton, N. Miller, J. Nickolaou, K. Peterson, B. Pilnick, R. Rajkumar, P. Rybski, V. Sadekar, B. Salesky, Y. Seo, S. Singh, J.M. Snider, J.C. Struble, A. Stentz, M. Taylor, W. Whittaker, Z. Wolkowicki, W. Zhang, and J. Ziglar, Autonomous driving in urban environments: boss and the urban challenge. J. Field Robot. **25**(8), 425–466 (2008); Special Issue on the 2007 DARPA Urban Challenge, Part I
92. M. Walter, N. Nitzsche, D. Odenthal, S. Muller, Lateral vehicle guidance control for autonomous and cooperative driving, in *2014 European Control Conference (ECC)* (IEEE, 2014), pp. 2667–2672
93. X. Wang, M. Fu, Y. Yang, H. Ma, Lateral control of autonomous vehicles based on fuzzy logic, in *2013 25th Chinese Control and Decision Conference (CCDC)* (2013), pp. 237–242
94. M. Werling, Ein neues Konzept für die Trajektoriengenerierung und -stabilisierung in zeitkritischen Verkehrsszenarien. Ph.D. thesis, Karlsruher Institut für Technologie, Germany, 2011
95. M. Werling, L. Gröll, G. Bretthauer, Invariant trajectory tracking with a full-size autonomous road vehicle. IEEE Trans. Robot. **26**(4), 758–765 (2010)
96. W. Xu, J. Dolan, A real-time motion planner with trajectory optimization for autonomous vehicles, in *Proceedings of the International Conference on Robotics and Automation* (2012), pp. 2061–2067
97. L. Xu, Y. Wang, H. Sun, J. Xin, N. Zheng, Design and implementation of driving control system for autonomous vehicle, in *2014 IEEE 17th International Conference on Intelligent Transportation Systems (ITSC)* (2014), pp. 22–28
98. Y. Ying, T. Mei, Y. Song, Y. Liu, A sliding mode control approach to longitudinal control of vehicles in a platoon, in *2014 IEEE International Conference on Mechatronics and Automation (ICMA)* (2014), pp. 1509–1514
99. J. Zhang, D. Ren, Lateral control of vehicle for lane keeping in intelligent transportation systems, in *International Conference on Intelligent Human-Machine Systems and Cybernetics (IHMSC)*, vol. 1 (IEEE, 2009), pp. 446–450
100. S. Zhang, W. Deng, Q. Zhao, H. Sun, B. Litkouhi, Dynamic trajectory planning for vehicle autonomous driving, in *2013 IEEE International Conference on Systems, Man, and Cybernetics (SMC)* (IEEE, 2013), pp. 4161–4166
101. P. Zheng, M. McDonald, Application of fuzzy systems in the car-following behaviour analysis, in *Fuzzy Systems and Knowledge Discovery*, ed. by L. Wang, Y. Jin. Lecture Notes in Computer Science, vol. 3613 (Springer, Berlin Heidelberg, 2005), pp. 782–791
102. K. Zhou, J.C. Doyle, K. Glover, *Robust and Optimal Control* (Feher/Prentice Hall Digital/Prentice Hall, Upper Saddle River, 1996)
103. J. Ziegler, P. Bender, M. Schreiber, H. Lategahn, T. Strauss, C. Stiller, T. Dang, U. Franke, N. Appenrodt, C.G. Keller, E. Kaus, R.G. Herrtwich, C. Rabe, D. Pfeiffer, F. Lindner, F. Stein, F. Erbs, M. Enzweiler, C. Knöppel, J. Hipp, M. Haueis, M. Trepte, C. Brenk, A. Tamke, M. Ghanaat, M. Braun, A. Joos, H. Fritz, H. Mock, M. Hein, E. Zeeb, Making Bertha drive - an autonomous journey on a historic route. IEEE Intell. Transp. Syst. Mag. **6**(2), 8–20 (2014)

Chapter 5
Path Tracking for Automated Driving: A Tutorial on Control System Formulations and Ongoing Research

Aldo Sorniotti, Phil Barber, and Stefano De Pinto

Nomenclature

The superscript "*" is used to indicate the complex conjugate transpose.

a:	front semi-wheelbase
$\overline{a}$:	longitudinal distance between the center of gravity and the front end of the vehicle
a_x, $a_{x,\max}$:	longitudinal acceleration, maximum longitudinal acceleration
a_y:	lateral acceleration
A, B, C, D, E:	generic state-space formulation matrices
A_r, B_r, C_r:	state-space matrices for path profile modeling
A_{v}, B_{v}, C_{v}, D_{v}:	state-space matrices for vehicle modeling
A', B':	state-space matrices for modeling the tracking dynamics at the centers of percussion
b:	rear semi-wheelbase
$\overline{b}$:	longitudinal distance between the center of gravity and the rear end of the vehicle
b_δ:	multiplicative factor of steering angle in the yaw acceleration error formulation used for backstepping control design
B_1, B_2, B_3:	matrices of the state-space single-track model formulation

A. Sorniotti (✉) • S. De Pinto
University of Surrey, Guildford, UK
e-mail: a.sorniotti@surrey.ac.uk

P. Barber
Jaguar Land Rover, Coventry, UK

D. Watzenig, M. Horn (eds.), *Automated Driving*,
DOI 10.1007/978-3-319-31895-0_5

$\mathcal{B}$: box used in the formulation of the tube-based model predictive controller

$\bar{c}$: half of vehicle width

$c_{\mathrm{COP,f}}$, $c_{\mathrm{COP,r}}$: coefficients used in the definition of the sliding variables for the front and rear centers of percussion

c: constant gain for sliding mode controller

C_{f}, C_{r},: front and rear cornering stiffnesses

$C_{\mathrm{f},\mu_{\mathrm{c}0}}$, $C_{\mathrm{r},\mu_{\mathrm{c}0}}$: front and rear cornering stiffnesses for the nominal tire-road friction coefficient $\mu_{\mathrm{c}0} = 1$

C_{β}, C_{ψ}, $C_{\Delta\psi}$, C_{κ}: coefficients used for backstepping controller design

C_1, C_2: controller formulations 1 and 2

d: distance from the summit of the bend

$d_{\mathrm{min}_{k,t}}$: minimum distance between the vehicle and the obstacle points calculated at time t and associated with the time k within the tracking horizon

$d_{k,t,j}$: distance between the vehicle and the obstacle point j calculated at time t and associated with the time k within the tracking horizon

d_1, d_2: denominators of controller formulations C_1 and C_2

e, e_k: error, discretised error

D_{r}: damping ratio

$D\,(s_L)$: denominator of the transfer function

f_{a}: function expressing the system dynamics

$f^{dt}_{s_{k,t},\mu_{\mathrm{c}k,t}}$: system model function for the coordinate $s_{k,t}$ and the tire road friction coefficient $\mu_{\mathrm{c}_{k,t}}$ calculated at time t and associated with the time k within the tracking horizon

$F_{\mathrm{b,l}}$, $F_{\mathrm{b,r}}$: braking forces on the left-hand and right-hand sides of the vehicle

$F_{y\mathrm{,f}}$, $F_{y\mathrm{,r}}$: lateral forces at the front and rear axles

$F^{\mathrm{FB}}_{y\mathrm{,f}}$, $F^{\mathrm{FB}}_{y\mathrm{,r}}$: feedback contribution to the reference lateral forces on the front and rear axles

$F^{\mathrm{FFW}}_{y\mathrm{,f}}$, $F^{\mathrm{FFW}}_{y\mathrm{,r}}$: feedforward contribution to the lateral forces on the front and rear axles

$F^{\mathrm{TOT}}_{y\mathrm{,f}}$: sum of the feedforward and feedback contributions to the lateral forces on the front axle

g: gravity

$g_{x\mathrm{,PP}}$, $g_{y\mathrm{,PP}}$, $g_{x\mathrm{,S}}$, $g_{y\mathrm{,S}}$: longitudinal and lateral coordinates of the goal points according to the pure pursuit and Stanley path tracking methods

$g(\xi)$: nonlinear term within the model formulation for the robust tube-based controller design

G_c:	compensator of actuator dynamics
$G_H, \widetilde{M}_H, \widetilde{N}_H$:	matrices used for the coprime factorization of the nominal plant within the H_∞ controller design
$G_{H\Delta}, \Delta_{M_H}, \Delta_{N_H}$:	matrices used for the coprime factorization of the perturbed plant within the H_∞ controller design
G_{l_d}:	virtual sensor look-ahead filter
h:	function for obtaining the outputs starting from the inputs
h_p:	generic parameter considered in the parameter space approach
h':	matrix for modeling the disturbances
$\mathfrak{H}, \wp$:	polytopes used in the definitions of Pontryagin difference and Minkowski sum
$H_\mathrm{C}, H_\mathrm{P}$:	control horizon, prediction horizon
$\mathcal{H}$:	set of parameters in the parameters space approach
i:	imaginary unit
I_z:	yaw moment of inertia
J:	cost function for optimal control
$J_{\mathrm{obs}_{k,t}}$:	cost function at time t and associated with the time k to the predicted distance between the vehicle and the obstacle
J_1, J_2:	tracking performance criteria
k:	discretization step number or step time
k_c,k_int:	controller gain and integrator to keep the steady-state tracking error small
k_ch:	control gain of the chained-form controller used for calculating k_{1_CC}, k_{2_CC}, and k_{3_CC}
k_d:	multiplicative factor of the state vector in the discontinuous control law
k_D:	derivative gain
k_DD:	gain used in the PIDD^2 controller
k_I:	integral gain
k_LK:	control gain in the feedback force contribution on the rear axle, $F_{y,\mathrm{r}}^{\mathrm{FB}}$
k_P:	proportional gain
k_PP:	tuning gain of the pure pursuit algorithm
k_S:	tuning gain of the Stanley method
k_{w1},k_{w2}, k_{ξ_e}:	gains used in the optimal preview steering control law
k_U:	understeer gradient
$k_{\dot\psi}$:	multiplicative factor of $\dot\psi$
$k_{\Delta_{yCG}}$:	multiplicative factor of Δ_{yCG}
$k_{\Delta_{yl_d}}$:	multiplicative factor of Δ_{yl_d}
$k_{\Delta\psi}, k_{\Delta\dot\psi}$:	multiplicative factor of the heading error, multiplicative factor of the yaw rate error

$k_{\Delta\psi,w}, k_{\Delta\dot{\psi},w}, k_{\Delta y,w}$: multiplicative factors of the scalar errors $\Delta\psi_w, \Delta\dot{\psi}_w, \Delta y_w$ in the linear quadratic regulator with preview

$k_{\Delta\psi,w}, k_{\Delta\dot{\psi},w}, k_{\Delta y,w}$: gains of the preview controller, to be multiplied by the weighted values of the heading angle error, yaw rate error and lateral displacement error

$k_{1_{\mathrm{CC}}}, k_{2_{\mathrm{CC}}}, k_{3_{\mathrm{CC}}}$: gains of the chained controller

$k_{1_{\mathrm{LC}}}, k_{2_{\mathrm{LC}}}, k_{3_{\mathrm{LC}}}, k_{4_{\mathrm{LC}}}$: gains of the limit cornering controller

$k_{1_{\mathrm{LQ}}}, k_{2_{\mathrm{LQ}}}, k_{3_{\mathrm{LQ}}}, k_{4_{\mathrm{LQ}}}$: gains of the linear quadratic controller

K: rate of decay of $\Delta y_{l_{\mathrm{d}}}$

K_{d}: gain multiplying $\Delta y_{l_{\mathrm{d}}}$

K_{LC}: linear quadric matrix gain of the limit cornering controller

K_{LQ}: matrix gain of the linear quadratic regulator

$K_{\mathrm{LQ_P}}$: matrix gain of the linear quadratic regulator with preview

K_{OBS}: collision weight

l: wheelbase

l_{d}: look-ahead distance

L: vehicle wheelbase

$L_{\mathrm{Lipschitz}}$: Lipschitz constant

L_{Lyapunov}: Lyapunov function

m: vehicle mass

M: constant big enough to disregard obstacles that do not lie within the vehicle line of sight

$M_{\mathrm{COP,f}}, M_{\mathrm{COP,r}}$: gains to provide robustness against the variation of cornering stiffness

M_u: tuning parameter of the sliding mode controller

M_z: yaw moment

n: counter

n_1, n_2: numerators of controller formulations C_1 and C_2

N: matrix of the Riccati equation for linear quadratic regulator design

p: characteristic polynomial

$p_{x_{k,t,j}}, p_{y_{k,t,j}}$: coordinates of the j-th point of the obstacle in the body frame, calculated at time t and associated with the time k within the tracking horizon

$p_{X_{t,j}}, p_{Y_{t,j}}$: coordinates of the j-th point of the obstacle in the inertial frame at time t

P_{L}: Lyapunov matrix

$P_{t,j}$: j-th point of the obstacle in the inertial frame at time t

P_1, P_2: transfer functions calculated from a linear single-track model of the system, used in the feedforward contribution $\delta_{\mathrm{FFW},1}$

$\widehat{q}$: observer output

$q_{\Delta a_y}, q_{\Delta y_{l_d}}, q_{\Delta\psi}, q_i$: parameters of the filters of the frequency-shaped linear quadratic controller

q_1, q_2, q_3, q_4: extreme operating points for the Γ-stability controller

Q, R: weighting matrices of the cost function formulation

$\mathrm{Reach}_f(\mathcal{S}, \mathcal{W})$: one-step robust reachable set from a given set of states ($\mathcal{S}$)

r: input of the model reference system obtained by scaling the input of the desired trajectory generator

R_{tr}: trajectory radius

$s, \dot{s}$: trajectory coordinate and its time derivative

s_{L}: Laplace operator

S: matrix of the Riccati equation for linear quadratic regulator design

t: time

t_{l_d}: time corresponding to the look-ahead distance (at the current vehicle speed)

t_{r}: time delay due to the driver's reaction

t_{d}: time delay

t_1, t_2: initial and final time values

T: matrix describing the dynamics of the system considering the center of percussion

$T_{\mathrm{b,lf}}, T_{\mathrm{b,rf}}, T_{\mathrm{b,lr}}, T_{\mathrm{b,rr}}$: braking torques of the left front, right front, left rear, and right rear wheels

T_{i}: time constant of the first order filter in the derivative term of the PID controller' and delete the existing line

T_{s}: sampling time

u: input vector in the state-space formulation

$\overline{u}$: vector of the front and rear steering angles

$u_k, \overline{u}_k, \widehat{u}_k(e_k)$: control law, nominal controller, and state feedback control action used in the robust tube-based model predictive controller

u_1, u_2: control outputs of the chained controller

$\mathcal{U}, \overline{\mathcal{U}}$: polyhedra used in the tube-based model predictive control constraints

$v, v_x, v_y, v_{\max}$: vehicle speed, longitudinal and lateral components of vehicle speed, maximum speed

$v_{k,t}$: speed of the vehicle at time k predicted at time t

$\dot{v}_{y,\mathrm{path,CG}}$: time derivative of the lateral velocity of the path

v_0: speed at which the four-wheel-steering controller changes the sign of the steering angle of the rear axle (transition from opposite signs to the same signs of the front and rear steering angles)
V_1, V_2: vehicle dynamics transfer functions
w: system disturbance
$\tilde{w}$: element of $\widetilde{\mathcal{W}}$
$W_d, W_{\dot{\psi}}$: weighting function adopted in the backstepping steering control law
$\mathcal{W}$: polyhedron used in the robust tube-based model predictive control constraints
$\widetilde{\mathcal{W}}$: Minkowski sum of the two polytopes $\mathcal{W}$ and $\mathcal{B}$
x, h: elements of the polytopes used in the definition of Pontryagin difference and Minkowski sum
$x_{\mathrm{COP_f}}, x_{\mathrm{COP_r}}$: coordinates of the front and rear centers of percussion
x_{P}: longitudinal position of a generic point P in the vehicle reference system
$\dot{x}_{\mathrm{ref}}$: reference longitudinal speed
$x_{\mathrm{v}}, y_{\mathrm{v}}$: vehicle positions in the tracking coordinates
X, Y: coordinates in the inertial reference system
$X_{\mathrm{r}}, Y_{\mathrm{r}}, \dot{X}_{\mathrm{r}}, \dot{Y}_{\mathrm{r}}$: positions and velocities of the rear wheel according to the inertial reference system
y, y_k: output and discrete output of the state-space formulation
y_{ri}: disturbance in the form of white noise in the linear quadratic regulator with preview
$\ddot{y}_{l_{\mathrm{d}}}$: lateral acceleration at the look-ahead distance l_{d}
$\ddot{y}_{\mathrm{ref}}$: reference lateral acceleration
Y_{ref}: reference lateral position in the inertial frame
z: complex number used in the z-transform representation
z_1, z_2, z_3, z_4: augmented states for modelling the filter dynamics
$z_{\mathrm{MPC}}, z_{\mathrm{MPCref}}$: system outputs and references in the model predictive controller
$\mathcal{Z}, \mathcal{Z}_{\infty}$: subset of Ξ, minimal robust positively invariant set
α: half of the angular extension of the circular arc defined by the pure pursuit algorithm
α_{r}: slip angle of the rear axle
$\alpha_{\mathrm{ref,f}}$: reference slip angle of the front axle
$\alpha_{\mathrm{f}}^{\mathrm{FFW}}, \alpha_{\mathrm{r}}^{\mathrm{FFW}}$: reference values of the feedforward contributions to the front and rear slip angles
α_1, α_2: functions of the states, reference path and steering wheel input in the chained controller
$\alpha_{1,\mathrm{ST}}, \alpha_{2,\mathrm{ST}}$: tuning constants of the super-twisting controller

β, β_{SS}: vehicle sideslip angle, steady-state vehicle sideslip angle

$\beta_{x,f}$, $\beta_{y,f}$, β_r: normalized longitudinal force on the front axle, normalized lateral force on the front axle, normalized longitudinal force on the rear axle

γ: sensitivity parameter for LQR design

Γ, $\partial\Gamma$: desired region for locating the poles of the closed-loop system, with the corresponding boundary

$\delta, \dot{\delta}$: steering angle and its time derivative. In the case of absence of any subscript, this notation refers to the front axle. In the case of a four-wheel-steering vehicle, the subscript '*f*' is used to indicate the front steering angle and the subscript '*r*' is used to indicate the rear steering angle. The additional subscript '*ss*' is used to indicate steady-state conditions

δ_{eq}: equivalent steering angle in the super-twisting sliding mode formulation

$\delta_{FB,LC,1}$, $\delta_{FB,LC,2}$, $\delta_{FB,LC,3}$: feedback steering angle contributions according to different formulations of the path tracking controllers for limit cornering

$\delta_{FFW,LC}$: feedforward steering angle contribution according to the path tracking controller for limit cornering

$\delta_{FFW,1}$, $\delta_{FFW,2}$, $\delta_{FFW,3}$: feedforward contributions to the reference steering angle according to different formulations

δ_{min}, δ_{max}: minimum steering angle, maximum steering angle

δ_{ST}, $\delta_{ST,1}$, $\delta_{ST,2}$: steering angle contribution of the super-twisting controller, consisting of the contributions $\delta_{ST,1}$ and $\delta_{ST,2}$

$\dot{\delta}_{y_f}$: steering rate contribution depending on the lateral deviation error at the front end of the vehicle

$\dot{\delta}_{\dot{\psi}}$: steering rate contribution depending on vehicle yaw rate

Δa_y: difference between the reference and the actual lateral acceleration

Δu : control input variation

ΔU_t: optimization vector at time *t*

Δy, $\Delta \dot{y}$, $\Delta \ddot{y}$: lateral position error and its first and second time derivatives. They can be calculated at the front axle (hence the subscript "f"), at the rear axle (hence the subscript "r"), at the vehicle center of gravity (hence the subscript "CG"), or at any other point along the longitudinal axis of the vehicle reference

	system (e.g., at the center of percussion or at the look-ahead distance)
$\Delta y_{CG,SS}$:	steady-state value of Δy_{CG}
ΔY_{rms}, ΔY_{max}:	root mean square error on vehicle lateral position, maximum error on the vehicle lateral position
$\Delta\delta_{min}$, $\Delta\delta_{max}$:	minimum and maximum variation of the steering angle
$\Delta\psi$, $\Delta\dot{\psi}$, $\Delta\ddot{\psi}$:	yaw angle (i.e., heading) error and its first and second time derivatives. They can be calculated with respect to the reference path at the front axle (hence the subscript "f"), at the rear axle (hence the subscript "r"), at the vehicle center of gravity (hence the subscript "CG"), or at any other point on the longitudinal axis of the vehicle reference system (e.g., at the centers of percussion)
$\Delta\psi_{CG,SS}$:	steady-state value of the yaw angle error
$\Delta\psi_{rms}$, $\Delta\psi_{max}$:	root mean square error on the yaw angle, maximum error on the yaw angle
$\Delta\psi_w$, $\Delta\dot{\psi}_w$, Δy_w:	scalar values of the weighted average of the heading errors and lateral position error along the preview distance
ε:	small number
$\mathcal{E}$:	subset of Ξ containing $\{0\}$
ϵ, ϵ_{max}:	stability margin and maximum stability margin within the H_∞ controller design
ξ, ξ_{ref}:	state vector, reference state vector
$\xi_{augm.}$:	augmented state vector for the frequency-shaped linear quadratic controller
ξ_k, $\overline{\xi}_k$:	discrete state vector and predicted state vector
ξ_{v_k}:	discrete state vector used for describing the vehicle system in the tracking coordinates
Ξ, $\overline{\Xi}$:	polyhedra used in the tube-based model predictive control constraints
η_r:	corrective coefficient of the rear axle cornering stiffness
κ, $\widehat{\kappa}$:	path curvature and its estimated value. They can be calculated at the front axle (hence the subscript "f"), at the rear axle (hence the subscript "r"), or at any other point along the longitudinal axis of the vehicle reference system (e.g., the front and rear centers of percussion)
κ':	derivative of the path curvature with respect to the trajectory coordinate
λ:	sliding mode lateral position gain

$\lambda_{\Delta a_y}$, $\lambda_{\Delta y_{ld}}$, $\lambda_{\Delta\psi}$: parameters of the filters of the frequency-shaped linear quadratic controller

μ_c: tire-road friction coefficient

μ_f, μ_r: parameters for avoiding chattering in the sliding mode controller

σ, σ_f, σ_r: sliding variable, front sliding variable and rear sliding variable

σ_L: real part of the hyperbola in s_L-plane

σ_{L_0}: parameter of the hyperbola in s_L-plane

τ: integration variable

ϕ_r: road camber angle

ψ, $\dot{\psi}$, $\ddot{\psi}$: vehicle yaw angle and its time derivatives

ψ_{path}, $\dot{\psi}_{path}$: yaw angle corresponding to the reference path and its time derivative. They can be calculated on a point of the reference trajectory corresponding to the vehicle center of gravity (hence $\psi_{path,CG}$ and $\dot{\psi}_{path,CG}$), the front axle (hence $\psi_{path,f}$ and $\dot{\psi}_{path,f}$), the rear axle (hence $\psi_{path,r}$ and $\dot{\psi}_{path,r}$), or any other location on the longitudinal axis of the vehicle reference system (e.g., at the front bumper)

ψ_{ref}, $\dot{\psi}_{ref}$, $\ddot{\psi}_{ref}$: reference yaw angle and its first and second time derivatives

ω_C: pulsation

ω_f, ω_r: angular speeds of the front and rear wheels

ω_{lf} ω_{rf} ω_{lr} ω_{rr}: angular speeds of the left front, right front, left rear and right rear wheels

ω_L: imaginary part of the hyperbola in s_L-plane

ω_{L_0}: parameter of the hyperbola in s_L-plane

$\ominus$: Pontryagin difference of two polytopes

$\bigoplus$: Minkowski sum of two polytopes

5.1 Introduction

In automated driving system architectures (see the classification according to [1]), three layers can be typically defined [2]:

(i) The perception layer, aimed at detecting the conditions of the environment surrounding the vehicle, e.g., by identifying the appropriate lane and the presence of obstacles on the track;
(ii) The reference generation layer, providing the reference signals, e.g., in the form of the reference trajectory to be followed by the vehicle, based on the inputs from the perception layer;

(iii) The control layer, defining the commands required for ensuring the tracking performance of the reference trajectory. These commands are usually expressed in terms of reference steering angles (usually on the front axle only) and traction/braking torques.

This chapter focuses on the control layer and, in particular, the steering control for autonomous driving, also defined as path tracking control. The foundations of path tracking control for autonomous driving date back to well-known theoretical and experimental studies on robotic systems and driver modeling, detailed in several papers and textbooks (e.g., see the driver model descriptions in [3–9]). Moreover, automated driving experiments with different controllers have been conducted since the 1950s and 1960s, by using inductive cables or magnetic markers embedded in roadways to indicate the reference path [10, 11].

This contribution presents a survey of the main control techniques and formulations adopted to ensure that the automatically driven vehicles follow the reference trajectory, including analysis of extreme maneuvering conditions. The discussion will be based on a selection of different control structures, at increasing levels of complexity and performance. The focus will be on whether complex steering controllers are actually beneficial to autonomous driving. This is an important point, considering that Stanley and Sandstorm, the vehicles that obtained the first two places at the DARPA Grand Challenge (2004–2005), used very simple steering control laws based on kinematic vehicle models. In contrast to this, Boss, the autonomous vehicle winning the DARPA Urban Challenge (2007), was characterized by a far more advanced model predictive control strategy [12–15].

The main formulas for the different steering control structures will be concisely provided as a tutorial on the control system implementations, so that the reader can actually appreciate the characteristics of each formulation, and ultimately refer to the original papers in the case of specific interest. Also, the main simulation and experimental results obtained through the implementation of each control structure will be reported and critically analyzed.

The chapter is organized as follows:

- Section 5.2 presents path tracking methods based on simple geometric relationships, and a chained controller relying on a vehicle kinematic model, i.e., developed under the approximation of zero slip angles on the front and rear tires.
- The first part of Sect. 5.3 deals with conventional feedback controllers designed with a simplified dynamic model of the vehicle system, i.e., the well-known linear single-track vehicle model. The second half of Sect. 5.3 discusses relatively simple optimal control formulations, e.g., linear quadratic regulators, without and with feedforward contributions, and including the concept of preview in their most advanced declination. The layout of Sects. 5.2 and 5.3 mostly follows the guideline of a very relevant previous survey work [16], dating back to 2009, which critically assessed path tracking control methods through vehicle simulations with the software package CarSim.
- Section 5.4 discusses a couple of sliding mode formulations, one of them based on the important concept of center of percussion, and briefly mentions

other examples of path tracking controllers, e.g., based on H_∞ control and backstepping control.
- Section 5.5 presents in detail the latest developments in the subject area, through a selection of examples of advanced controllers (i.e., path tracking controllers for autonomous racing and model predictive controllers) from recently published papers, including critical analysis of their specific benefits.
- Section 5.6 provides concluding remarks and ideas for future research on the subject.

5.2 Methods Based on Geometric and Kinematic Relationships

5.2.1 *Pure Pursuit Method*

The most basic path tracking method is represented by the pure pursuit formula, derived by geometrically calculating the curvature of a circular arc (describing an angle 2α in a top view of the single-track model of the system, see Fig. 5.1) that connects the rear axle location to the goal point on the reference trajectory and by applying the well-known Ackerman steering formula, $1/(R_{\mathrm{tr}}\delta) = 1/L$. The goal point has coordinates $(g_{x,\mathrm{PP}}, g_{y,\mathrm{PP}})$ and is located at a look-ahead distance l_d on the reference trajectory, measured from the rear axle. This brings a reference steering angle $\delta(t)$ equal to:

$$\delta(t) = \tan^{-1}\left(\frac{2L\sin\alpha}{l_\mathrm{d}}\right) = \tan^{-1}\left(\frac{2L\sin\alpha}{k_{\mathrm{PP}}v_x(t)}\right) \tag{5.1}$$

It can be shown that the resulting curvature is $\kappa = 1/R_{\mathrm{tr}} = \left(2\Delta y_{l_\mathrm{d}}\right)/l_\mathrm{d}^2$, i.e., this controller acts like a proportional controller, with gain $2/l_\mathrm{d}^2$, on the error Δy_{l_d}, defined as the lateral distance between the x-axis of the vehicle reference system and the goal point $(g_{x,\mathrm{PP}}, g_{y,\mathrm{PP}})$ in Fig. 5.1. As shown in the right term of Eq. (5.1), the look-ahead distance l_d is often scaled as a function of vehicle speed, $v_x(t)$, i.e., $l_\mathrm{d}(t) = k_{\mathrm{PP}}v_x(t)$. In general, low values of l_d result in high precision tracking and low stability. As Eq. (5.1) is based on vehicle kinematics, it can generate significant tracking errors, caused by the absence of consideration of the vehicle sideslip.

5.2.2 *Stanley Method*

Another geometry-based path tracking method is the Stanley method, usually more suitable for medium-high speed driving conditions than the pure pursuit method and adopted by the Stanford University's entry (called Stanley) to the DARPA Grand

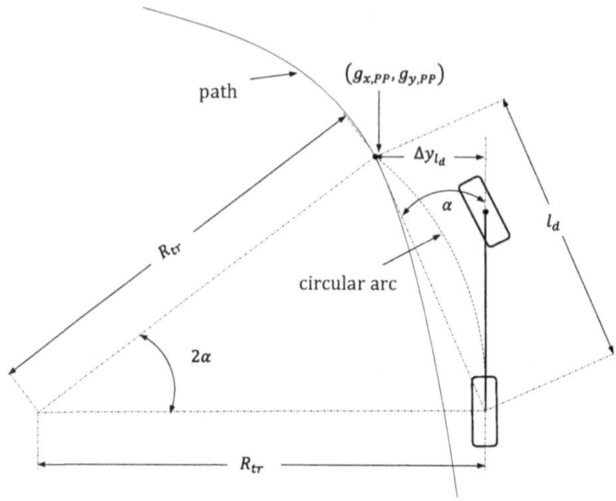

Fig. 5.1 Geometric construction for the derivation of the pure pursuit method (adapted from [16])

Challenge. According to this method (see Fig. 5.2), the steering angle consists of: (i) a component equal to the heading error (i.e., the yaw angle error), $\Delta\psi_{\mathrm{f}} = \psi - \psi_{\mathrm{path,f}}$, where $\psi_{\mathrm{path,f}}$ is measured at the goal point $(g_{x,\mathrm{S}}, g_{y,\mathrm{S}})$ on the reference path; and (ii) a term based on the lateral distance error at the front axle, Δy_{f} (or any other point in the front part of the vehicle), ensuring that the intended trajectory intersects the target path at an approximated distance $v_x(t)/k_{\mathrm{S}}$ from the front axle, with k_{S} being the tuning parameter of the controller:

$$\delta(t) = \Delta\psi_{\mathrm{f}}(t) + \tan^{-1}\left(\frac{k_{\mathrm{S}}\Delta y_{\mathrm{f}}(t)}{v_x(t)}\right) \tag{5.2}$$

Many other variants of geometric path tracking methods can be found in the literature. For example, Wit [18] proposes a vector pursuit path tracking method (for specific details refer directly to [18]).

5.2.3 *Chained Controller Based on Vehicle Kinematics*

Similarly to the previous controllers, the chained controller formulation in [17] (see also [19, 20]) is based on the single-track model of vehicle kinematics, i.e., considering zero slip angles for the front and rear tires. In particular, the hypotheses are that (see Fig. 5.2): (i) the rear wheels move along a direction with angle ψ (i.e., the yaw angle) with respect to the X-axis of the inertial reference system; (ii) the front wheels move along a direction with an angle $\psi - \delta$ (where the signs are according to the conventions in [17]) with respect to the same X-axis; and (iii) a

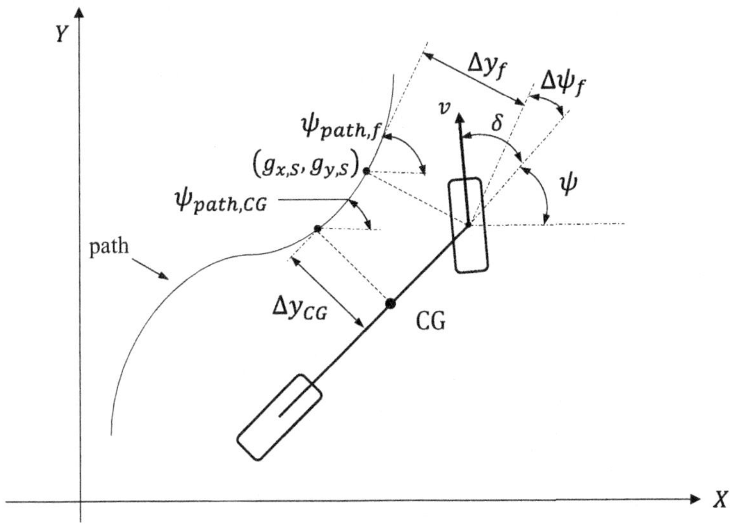

Fig. 5.2 Geometric construction for the derivation of the Stanley method (adapted from [16, 17])

simple kinematic relationship between yaw rate and steering wheel angle is valid, i.e., $\dot{\psi} = v \tan(\delta)/L$. Under (i)–(iii), the kinematic model of the vehicle in matrix form is:

$$\begin{bmatrix} \dot{X}_r \\ \dot{Y}_r \\ \dot{\psi} \\ \dot{\delta} \end{bmatrix} = \begin{bmatrix} \cos(\psi) \\ \sin(\psi) \\ \frac{\tan(\delta)}{L} \\ 0 \end{bmatrix} v + \begin{bmatrix} 0 \\ 0 \\ 0 \\ 1 \end{bmatrix} \dot{\delta} \tag{5.3}$$

By expressing the kinematic model in path coordinates, with s being the trajectory coordinate and κ its curvature, the model formulation becomes:

$$\begin{bmatrix} \dot{s} \\ \Delta\dot{y}_r \\ \Delta\dot{\psi}_r \\ \dot{\delta} \end{bmatrix} = \begin{bmatrix} \frac{\cos(\Delta\psi_r)}{1-\Delta\dot{y}_r\kappa(s)} \\ \sin(\Delta\psi_r) \\ \frac{\tan(\delta)}{L} - \frac{\kappa(s)\cos(\Delta\psi_r)}{1-\Delta\dot{y}_r\kappa(s)} \\ 0 \end{bmatrix} v + \begin{bmatrix} 0 \\ 0 \\ 0 \\ 1 \end{bmatrix} \dot{\delta} \tag{5.4}$$

with $\Delta\psi_r = \psi - \psi_{\text{path,r}}$. Through an appropriate transformation of the system coordinates (the general formulation of the coordinate transformation and the theory are reported in [17] and [19, 20]), it is possible to express this system into a typical two-input chained form with the following structure:

$$\begin{cases} \dot{\xi}_1 = u_1 \\ \dot{\xi}_2 = \xi_3 u_1 \\ \dot{\xi}_3 = \xi_4 u_1 \\ \dot{\xi}_4 = u_2 \end{cases} \tag{5.5}$$

where the change of coordinates is:

$$\begin{cases} \xi_1 = s \\ \xi_2 = \Delta y_{\mathrm{r}} \\ \xi_3 = (1 - \Delta y_{\mathrm{r}} \kappa(s)) \tan(\Delta \psi_{\mathrm{r}}) \\ \xi_4 = -\kappa'(s) \Delta y_{\mathrm{r}} \tan(\Delta \psi_{\mathrm{r}}) + \\ -\kappa(s)\left(1 - \Delta y_{\mathrm{r}} \kappa(s)\right) \frac{1+\sin^2(\Delta \psi_{\mathrm{r}})}{\cos^2(\Delta \psi_{\mathrm{r}})} + \\ + \frac{\left(1-\Delta y_{\mathrm{r}} \kappa(s)\right)^2 \tan(\delta)}{L \cos^3(\Delta \psi_{\mathrm{r}})} \end{cases} \tag{5.6}$$

and the input transformation is:

$$\begin{cases} v = \frac{1-\Delta y_{\mathrm{r}} \kappa(s)}{\cos(\Delta \psi_{\mathrm{r}})} u_1 \\ \dot{\delta} = \alpha_2 \left(u_2 - \alpha_1 u_1\right) \end{cases} \tag{5.7}$$

with $\alpha_1 = \frac{\partial \xi_4}{\partial s} + \frac{\partial \xi_4}{\partial \Delta y_{\mathrm{r}}} (1 - \Delta y_{\mathrm{r}} \kappa(s)) \tan(\Delta \psi) + \frac{\partial \xi_4}{\partial \Delta \psi} \left(\frac{\tan(\delta)\left(1-\Delta y_{\mathrm{r}} \kappa(s)\right)}{L \cos(\Delta \psi)} - \kappa(s) \right)$ and $\alpha_2 = \frac{L \cos^3(\Delta \psi) \cos^2(\delta)}{\left(1-\Delta y_{\mathrm{r}} \kappa(s)\right)^2}$. The controller design in chained systems is carried out in two phases. According to [16], "the first phase assumes that one control input is given, while the additional input is used to stabilize the remaining sub-vector of the system states. The second phase simply consists of specifying the first control input so as to guarantee convergence while maintaining stability." In practical terms, if vehicle speed v is imposed, from Eq. (5.7) it is possible to calculate u_1, and then the steering input is a function of both u_1 and u_2, where u_2 depends on u_1. In the proposed two-input chained structure for path tracking, the two controllers are designed so that for any piecewise-continuous, bounded, and strictly positive (or negative) u_1, u_2 is expressed as:

$$u_2(\xi_2, \xi_3, \xi_4, t) = -k_{1\mathrm{CC}} |u_1(t)| \xi_2 - k_{2\mathrm{CC}} u_1(t) \xi_3 - k_{3\mathrm{CC}} |u_1(t)| \xi_4 \tag{5.8}$$

De Luca et al. [17] suggest imposing $k_{1\mathrm{CC}} = k_{\mathrm{ch}}^3$, $k_{2\mathrm{CC}} = 3k_{\mathrm{ch}}^2$, $k_{3\mathrm{CC}} = 3k_{\mathrm{ch}}$, so that the control system tuning consists of selecting a single gain. The main (and possibly only) benefit of this convolute control formulation is that its straightforward extension allows the automated control of articulated vehicles, including vehicle systems with multiple trailers.

5.3 Methods Based on Conventional Feedback Controllers and Simplified Vehicle Dynamics Models

5.3.1 Simple Feedback Formulations

A significant body of literature adopts simple feedback control structures, such as proportional integral derivative (PID) controllers, to solve the steering control problem for autonomous vehicles. These feedback controllers are usually designed starting from simplified models of the system dynamics, accounting for the fact that the actual vehicle behavior is different from the one predicted by a geometric model, because of (i) the slip angles on the front and rear tires, generally different among the two axles, with the subsequent vehicle understeer gradient in steady-state conditions [21] and (ii) vehicle inertia in transient conditions, providing second order yaw dynamics, with variable equivalent stiffness and damping characteristics as functions of vehicle speed [22, 23].

On the one hand, the relatively weak link between the basic feedback formulations and the respective linearized dynamic models used for control system design, and also the significant level of approximation of these models, do not automatically guarantee improved performance for all driving conditions with respect to the algorithms based on the system geometry alone, presented in Sect. 5.2. On the other hand, despite the existence of multiple state-of-the-art contributions focused on advanced control techniques, e.g., based on model predictive control, relatively simple feedback controllers for path tracking can provide good performance for a variety of operating conditions, if carefully tuned. For example, the ARGO autonomous vehicle prototype developed by the team of Prof. Broggi of the University of Parma was characterized by a proportional (P) controller for steering control [24, 25]. Even in a recent paper [26] authored by the investigators of the European Union FP7 V-Charge project, the path tracking controller is based on a simple proportional controller, with an advanced algorithm for the reference path generation, using an optimization considering the current vehicle position with respect to equivalent electric field lines. The following paragraphs provide an overview of some relatively simple control structures designed through linear single-track vehicle models.

Already in the 1960s in Japan [10, 27], experiments were conducted on a path tracking controller based on a Proportional Derivative (PD) structure on the lateral displacement error between the reference path and the actual path of the vehicle, and a P controller on the yaw angle error. The two contributions were summed together to provide δ. The yaw angle error contribution allowed an improvement in tracking performance with respect to the path deviation feedback contribution alone. This is further confirmed in a more recent paper by Nissan [28], in which a PD controller exclusively based on lateral position error is designed through pole placement, with clear issues in the results, caused by the effects of yaw rate and yaw angle, not addressed by the position error controller.

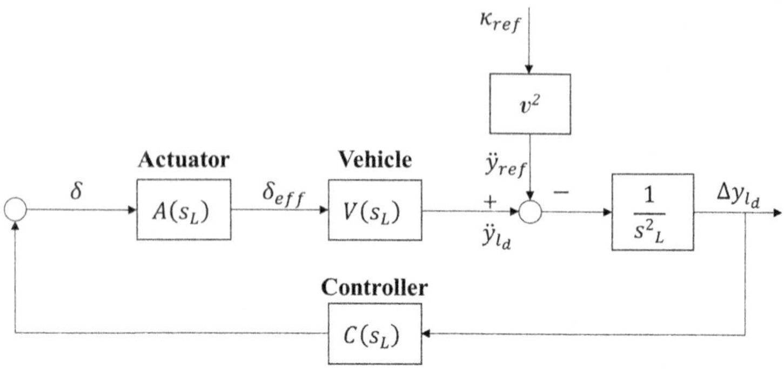

Fig. 5.3 Output feedback path tracking control system based on the lateral displacement error measured at the front bumper (adapted from [29])

More systematically, [29] critically examines the limitations of pure output feedback on the lateral vehicle displacement error measured at the front bumper, $\Delta y_{l_d} = \frac{1}{s_L^2}\left(\ddot{y}_{l_d} - \ddot{y}_{ref}\right)$ (Fig. 5.3). This was the control practice (dictated by necessity) in the initial look-down automated driving system implementations based on devices installed on the road and not based on vision systems implemented on the vehicle, intrinsically allowing look-ahead path tracking control.

The analysis of [29] is based on the lateral acceleration and yaw rate frequency response characteristics for a steering wheel input, obtained from the equations of the linear single-track vehicle model, with the lateral acceleration being considered at a distance l_d in front of the vehicle. In [29], l_d is used to indicate both the installation position of the lateral displacement error sensor and a virtual look-ahead distance.

The transfer functions of vehicle response (i.e., $V_1(s_L) = \dot{\psi}(s_L)/\delta(s_L)$ and $V_2(s_L) = \ddot{y}_{l_d}(s_L)/\delta(s_L)$) are characterized by the same second order denominator, i.e., $D(s_L) = I_z m v^2 s_L^2 + v\left\{I_z(C_f + C_r) + m\left(C_f a^2 + C_r b^2\right)\right\} s_L + m v^2(C_r b - C_f a) + C_f C_r l^2$, with the corresponding damping coefficient being a decreasing function of vehicle speed. In [29], the effect of the tire-road friction coefficient is modeled by imposing $C_f = C_{f,\mu_{c0}}\mu_c$ and $C_r = C_{r,\mu_{c0}}\mu_c$, which, according to the authors of this chapter, can be the object of discussions. The lateral acceleration transfer function has a second order numerator, while the yaw rate transfer function has a first order numerator. In formulas:

$$V_1(s_L) = \frac{\dot{\psi}(s_L)}{\delta(s_L)} = \frac{C_f a m v^2 s_L + C_f C_r L v}{D(s_L)} \quad (5.9)$$

$$V_2(s_L) = \frac{\ddot{y}_{l_d}(s_L)}{\delta(s_L)} = \frac{C_f v^2 (m a l_d + I_z) s_L^2 + C_f C_r L v (l_d + b) s_L + C_f C_r L v^2}{D(s_L)} \quad (5.10)$$

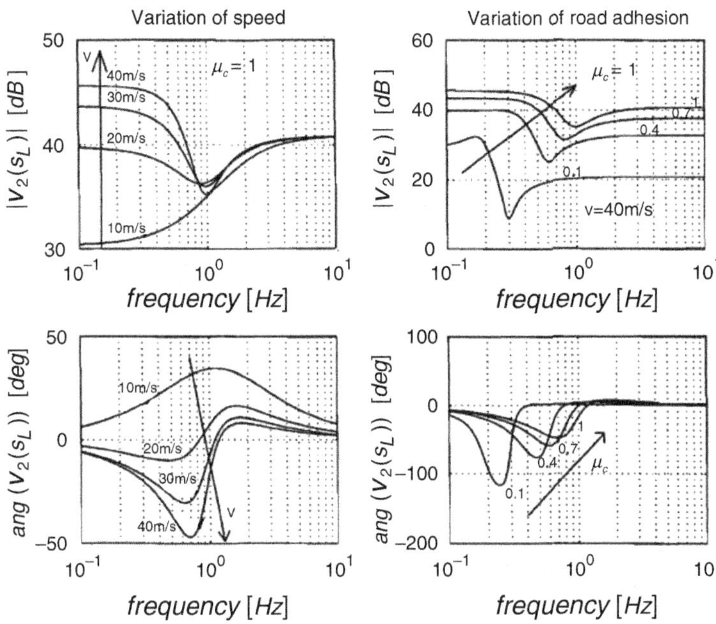

Fig. 5.4 Bode plots of $V_2(s_L)$ for a lateral displacement sensor location on the front bumper (i.e., $l_d = 1.96$ m) (from [29])

Equations (5.9) and (5.10) are the essential plant transfer functions to be considered for linear path tracking control design in the frequency domain. The inclusion of the transfer function of the specific steering actuator is also recommended by the authors of this chapter for any path tracking control design activity, consistently with the practice of many sources in the literature, some of them indicating a typical actuation bandwidth of 5–10 Hz.

Figure 5.4 shows the Bode plots of $V_2(s_L)$ for different vehicle speeds and tire-road friction conditions. In particular: (i) the natural mode of vehicle dynamics is negligible for low speed but significant for high speed; (ii) the frequency of the natural mode is almost independent of speed, but decreases as a function of the friction coefficient; (iii) the steady-state gain depends on both v and μ_c (the latter is true especially at high speed); (iv) the high-frequency gain depends on μ_c but not on v; and (v) the contribution of yaw motion to lateral acceleration diminishes by the inverse of vehicle speed, in favor of the sideslip contribution.

According to [29], realistic control system specifications should be in terms of maximum lateral displacement error at any vehicle speed and robust behavior for $0.5 \leq \mu_c \leq 1$. The variations of μ_c can happen very quickly; therefore, the same controller must be capable of providing robustness for the indicated range of friction conditions. For very low friction values, i.e., for $\mu_c \leq 0.5$, it is expected that the longitudinal controller imposes much lower values of vehicle speed than in normal friction conditions. The important conclusion of the analysis in [29] is that for relatively high vehicle speeds look-down controllers are not sufficient to meet the expected performance requirements.

As a consequence, [29] proposes:

(i) A feedforward steering contribution based on the reference path curvature;
(ii) Feedback control of vehicle absolute motion, i.e., control of yaw rate and lateral acceleration;
(iii) Modification of the system zeros in Eq. (5.10) by increasing the look-ahead distance l_d, which can be done even in look-down systems by having two position error sensors, installed at the front and rear bumpers.

With respect to (iii), Fig. 5.5 shows the increase of the damping ratio of the numerator of the transfer function in Eq. (5.10) as a function of l_d, for different velocities and $\mu_c = 0.5$. The zero pair in the numerator of Eq. (5.10) determines "the undershoot in Fig. 5.4 between 1 Hz and 2 Hz and the distribution of phase lag/lead in this frequency range. Conversely, prescribing a fixed maximum phase lag ... yields look-ahead requirements for different speeds," which is outlined in Fig. 5.6, i.e., the look-ahead distance should be an increasing function of vehicle speed. Section 5.5 will show that the most recent path tracking controllers, specifically implemented for high lateral acceleration conditions, actually include the combination of (i)–(iii).

Hsu and Tomizuka [30] extend the general analysis of [29] to include the specificities of vision-based control systems. In particular, the main conclusions are that (i) Look-ahead distance enhances stability but increases the error and decreases the closed-loop bandwidth; (ii) Higher vehicle speed increases the cross-over frequency and reduces stability; (iii) The time delays caused by the vision system decrease

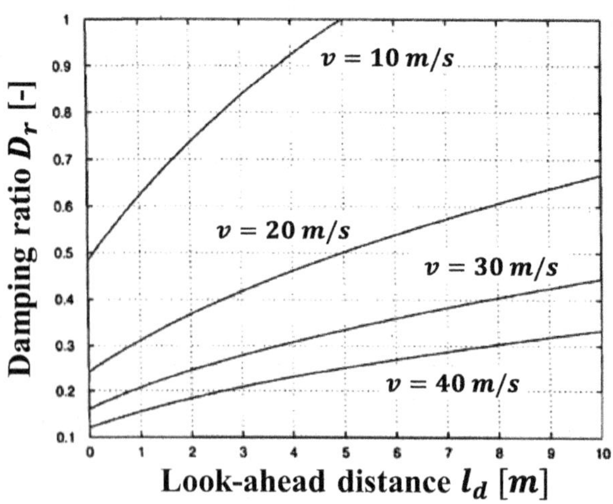

Fig. 5.5 Damping ratio of the numerator of the transfer function in Eq. (5.10) as a function of l_d, for different velocities and $\mu_c = 0.5$ (from [29])

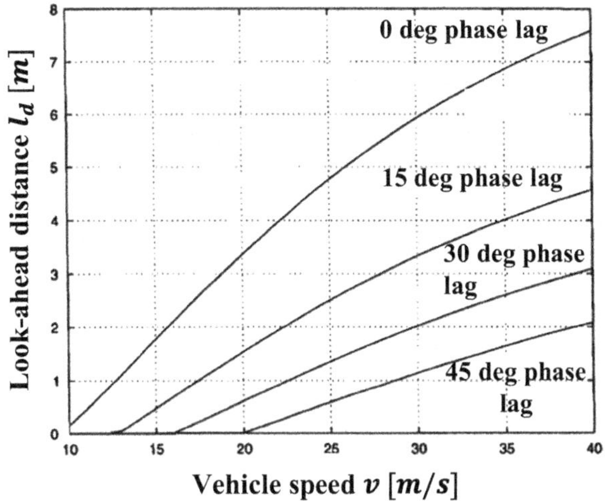

Fig. 5.6 Look-ahead distance (l_d) requirements for prescribed maximum phase lag for different velocities and $\mu_\mathrm{c} = 0.5$ (from [29])

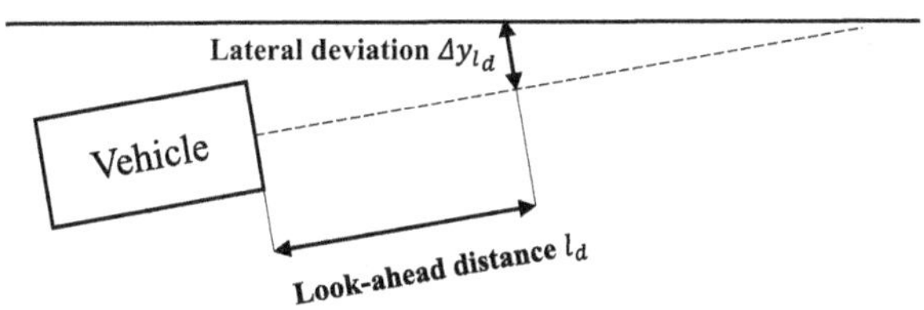

Fig. 5.7 Definition of the look-ahead distance and path deviation (adapted from [31])

stability, through a reduction of the phase margin and equivalent effect of the right-hand-plane zeros appearing in the loop transfer function. Hsu and Tomizuka [30] propose a path tracking controller, consisting of a feedforward contribution of the form $\delta_{\mathrm{FFW},1} = -P_1^{-1}(s_\mathrm{L})\, P_2(s_\mathrm{L})\, e^{l_\mathrm{d} s_\mathrm{L}} \kappa(s_\mathrm{L})$ (with $P_1(s_\mathrm{L}) = \Delta y_\mathrm{CG}(s_\mathrm{L})/\delta(s_\mathrm{L})$ and $P_2(s_\mathrm{L}) = \Delta y_\mathrm{CG}(s_\mathrm{L})/\kappa(s_\mathrm{L})$ being calculated from a linear single-track model of the system), and a PID feedback contribution on the lateral displacement error, receiving Δy_CG as input, with $k_\mathrm{P} = -0.01$, $k_\mathrm{I} = 0$, $k_\mathrm{D} = -0.0074$ and $T_\mathrm{i} = 0.0001$.

Consistently with the conclusions of [29, 30], Tachibana et al. [31] propose a PD controller on the lateral position error at the look-ahead distance l_d, i.e., the position error at a single point in front of the vehicle is monitored, assuming that the heading angle of the vehicle does not change with distance (Fig. 5.7). The control law in discretized form is:

$$\delta_k = k_\mathrm{P}(l_\mathrm{d}, v)\, \Delta y_{l_\mathrm{d},k} + k_\mathrm{D}\left(\Delta y_{l_\mathrm{d},k} - \Delta y_{l_\mathrm{d},k-1}\right) \tag{5.11}$$

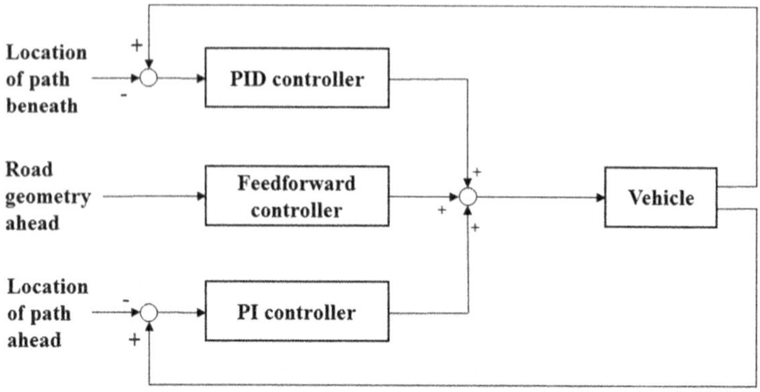

Fig. 5.8 The path tracking control structure proposed in [32] (adapted from the original paper)

Vehicle experiments showed that the control gains had to be varied as functions of vehicle speed and that at 50 km/h optimal look-ahead distances are in the range of 20 m, in the case of a curved trajectory, and 25 m, for a straight reference trajectory.

In [32], based on the Japanese Ministry of Construction Automated Highways System (AHS) project (1995–1996), the path tracking controller (Fig. 5.8) consists of a PID controller on Δy_{CG}, whose output is summed to that of a PI controller on $\Delta y_{l_{\mathrm{d}}}$, and to a feedforward contribution of the form:

$$\delta_{\mathrm{FFW},2} = \left(1 + k_{\mathrm{U}} v^2\right) \frac{L}{R_{\mathrm{tr}}} \tag{5.12}$$

The feedforward contribution accounts for the steady-state vehicle understeer and brings a reduction of the maximum value of Δy_{CG} from about 110 cm (with the feedback contribution only) to about 40 cm (with the feedback and feedforward contributions) during the case study tests. The comment of the authors of [32] is that the feedforward contribution allows achieving a good compromise between stability and tracking characteristics, without large gains of the feedback part. Interestingly, very recent papers by Prof. Gerdes of Stanford University reach the same conclusion, and the respective controllers significantly rely on nonlinear feedforward contributions (see Sect. 5.5). A similar combination of feedforward and feedback contributions was adopted for the anticollision system developed during the PRORETA research project, within a collaboration between Darmstadt University of Technology and Continental [33].

Marino et al. [34] propose a PID control architecture with two nested control blocks. The outer one calculates the reference yaw rate starting from the lateral position error, while the inner loop calculates the reference steering angle in order to track the reference yaw rate. According to the authors of [34], this architecture allows "to design standard PID controls in a multi-variable context." Singular values analysis is used to assess the robustness of the controller with respect to the variation of the main vehicle parameters. The simulation results show improved performance with respect to a model predictive driver model. The experiments on a Peugeot 307 prototype vehicle confirmed the adequate performance of the controller in normal driving conditions (i.e., for relatively low values of lateral acceleration).

Simple feedback formulations with constant gains can be designed to provide the required level of robustness for normal operating conditions. For example, [35] is an important study, focused on the design of a linear controller for an automated bus, with the main specifications being: (i) $|\delta|\leq 40$ deg; (ii) $\left|\dot{\delta}\right| \leq 23$ deg/s; (iii) lateral displacement error not exceeding 0.15 m in transient conditions and 0.02 m in steady state; (iv) lateral acceleration not exceeding 2 m/s^2 for passenger comfort with an ultimate limit of 4 m/s^2 for preventing vehicle rollover; and (v) a natural frequency of the lateral motion not exceeding 1.2 Hz.

The control system is designed through a linear single-track vehicle model, under the hypothesis of a lateral displacement sensor located on the front end of the vehicle, measuring the distance from the reference road path. The steering controller has the following formulation, where the benefit of the yaw rate-related contribution $k_{\dot{\psi}}\dot{\psi}$ is discussed in [35] through root-locus analysis:

$$\dot{\delta} = \dot{\delta}_{y_{\mathrm{f}}} + \dot{\delta}_{\dot{\psi}} = \dot{\delta}_{y_{\mathrm{f}}} - k_{\dot{\psi}}\dot{\psi} \tag{5.13}$$

The fixed controller design must be robust with respect to the variation of vehicle mass (very significant for a bus) and speed. In particular, the controller must provide stability for the four design points indicated in Fig. 5.9. The contribution $\dot{\delta}_{y_{\mathrm{f}}}$ is designed from the frequency domain analysis. In the Laplace domain, $\dot{\delta}_{y_{\mathrm{f}}}(s_{\mathrm{L}})$ is defined as

$$\dot{\delta}_{y_{\mathrm{f}}}(s_{\mathrm{L}}) = \omega_C^3 \frac{k_{\mathrm{DD}}s_{\mathrm{L}}^2 + k_{\mathrm{D}}s_{\mathrm{L}} + k_{\mathrm{P}} + k_{\mathrm{I}}/s_{\mathrm{L}}}{\left(s_{\mathrm{L}}^2 + 2D_{\mathrm{r}}\omega_C s_{\mathrm{L}} + \omega_C^2\right)(s_{\mathrm{L}} + \omega_C)} \Delta y_{\mathrm{f}}(s_{\mathrm{L}}) \tag{5.14}$$

which, according to the authors of [35–37], is a PIDD^2 controller. This control structure was already recommended for high-speed path tracking by the authors of [29].

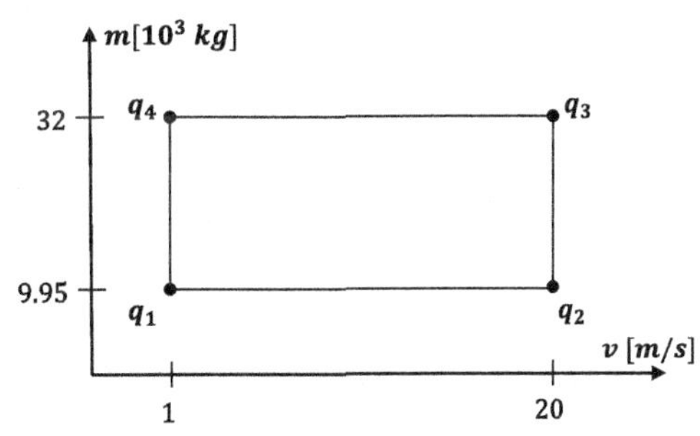

Fig. 5.9 Vehicle operating domain (adapted from [35])

A parameter space approach is used for the design of the compensator in Eq. (5.14). This method allows "to determine the set of parameters $\mathcal{H}$, for which the characteristic polynomial $p(s_L, h_p)$, $h_p \in \mathcal{H}$, is stable. The plant is robustly stable if the operating domain is entirely contained in the set of stable parameters." For the specific problem, Hurwitz stability is not considered sufficient. A hyperbola in the s_L-plane is used to provide the required performance characteristics, i.e., the eigenvalues of the closed-loop system should lie in the region Γ on the left of the boundary $\partial\Gamma$ defined as:

$$\partial\Gamma = \left\{ s_L = \sigma_L + i\omega_L \Big| \left(\frac{\sigma_L}{\sigma_{L0}}\right)^2 - \left(\frac{\omega_L}{\omega_{L0}}\right)^2 = 1, \sigma_L \leq -\sigma_{L0} \right\} \tag{5.15}$$

Values of σ_{L_0} equal to 0.12 and 0.35 are selected, respectively, for low and high velocities, with $\frac{\omega_{L0}}{\sigma_{L0}} = 5$. The Γ-stability boundaries for each of the four extreme operating points (q_1,q_2,q_3,q_4) in Fig. 5.9 are calculated, resulting in four Γ-stable regions. The intersection of the stable regions is the set of controllers stabilizing the four considered plants (e.g., see Fig. 5.10).

The initial selection of the controller parameters [k_{DD} k_D k_P k_I] was further refined in a simulation-based optimization procedure minimizing tracking performance criteria such as:

$$J_1 = \int_{t_1}^{t_2} \Delta y_f^2 dt, J_2 = \max_t |\Delta y_f| \tag{5.16}$$

A similar control design methodology, using the concept of Γ-stability, is presented in [37], this time based on the control of the errors of the front and tail lateral positions, instead of the front position and yaw rate. The option of a feedforward steering angle considering dynamic curvature preview was included in

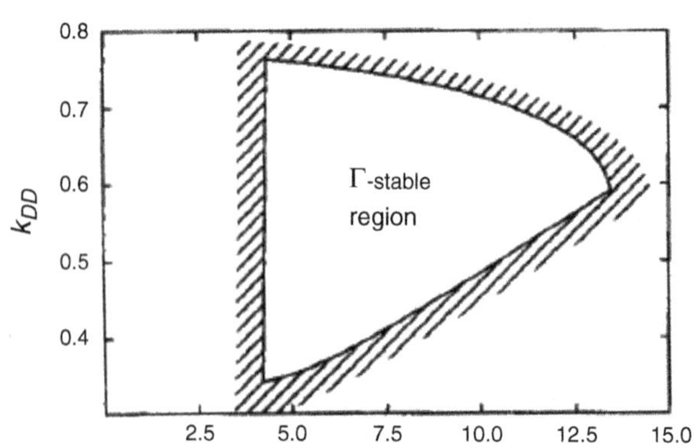

Fig. 5.10 Set of Γ-stabilizing controllers for $\omega_C = 100$, $D_r = 0.5$, $k_I = 3$ and $k_P = 10$ (from [35])

the controller, which was validated with experimental tests in collaboration with the Californian PATH (Partners for Advanced Transportation Technology) center.

Another very relevant study [38], including experiments, assessed the performance of an automated Fiat Brava 1600 ELX. The control system design was based on the commonly used single-track vehicle model (the detailed formulation is provided in the following lines), together with the transfer function of the steering actuator.

Similarly to [35], the objective was to design fixed controllers (e.g., without any form of gain scheduling), capable of robustly stabilizing the plant for: (i) v ranging between 60 km/h and 130 km/h; (ii) m ranging between 1226 kg and 1626 kg; (iii) I_z ranging between 1900 kgm^2 and 2520 kgm^2; and (iv) C_f and C_r ranging between 51 kN/rad and 69 kN/rad, and between 81.6 kN/rad and 110.4 kN/rad, respectively. The performance specifications included $|\Delta y_{CG}| \leq 0.2$ m, $|v_y(t)| \leq 1.5$ m/s, $|\Delta a_y| \leq 3.3$ m/s^2, and consideration of the steering actuator saturation.

Based on previous experimental analyses of human drivers [39] showing that the steering wheel action is applied "on the basis of the distance between the lane and the longitudinal axis of the car at a look-ahead point, " the controller in [38] uses feedback output on $\Delta y_{l_d} = \Delta y_{CG} + \Delta y_{path,l_d}$, with l_d=11.5 m. The controller is designed through classical loop-shaping techniques. In particular, two controllers, $C_1(z) = n_1(z)/d_1(z)$ and $C_2(z) = n_2(z)/d_2(z)$, were experimentally assessed, with numerical values of the controller parameters, in descending powers of z, given by:

$$\begin{aligned} n_1 &= [-7.844, 30.82, -47.37, 35.51, -13.24, 2.388, -0.2273] \\ d_1 &= [1, -4.92, 10.06, -10.96, 6.703, -2.181, 0.2949] \\ n_2 &= [-7.837, 29.03, -44.6, 33.43, -12.46, 2.2, -0.2138] \\ d_2 &= [1, -4.937, 10.13, -11.07, 6.794, -2.218, 0.3008] \end{aligned} \tag{5.17}$$

The values of the gains are reported here for a quick implementation of the controller and reproduction of the results. $C_1(z)$ was designed with the purpose of providing good tracking performance, while $C_2(z)$ was specifically aimed at ride comfort. $C_1(z)$ and $C_2(z)$ were assessed along a curve with 1000 m radius followed by a straight section, at v=100 km/h. The tracking performance results are in Figs. 5.11 and 5.12, which show that $C_2(z)$ provokes higher amplitude and lower frequency oscillations of the lateral offset between the path and the vehicle center of gravity. Overall, the test drivers provided a better assessment for $C_2(z)$, which demonstrates the extreme importance of the human factor in the evaluation of path tracking algorithms.

Tan et al. [40] includes a wide set of experimental results obtained during public demonstrations at various sites and in very different operating conditions, ranging from docking to high speeds and relatively high lateral accelerations. The control system design is based on a simple but reliable controller, with the following form:

$$\delta = -k_c(v) G_c(s_L) \left(k_{int}(s_L) \Delta y + l_d(v) G_{l_d}(s_L) \Delta\psi\right) \tag{5.18}$$

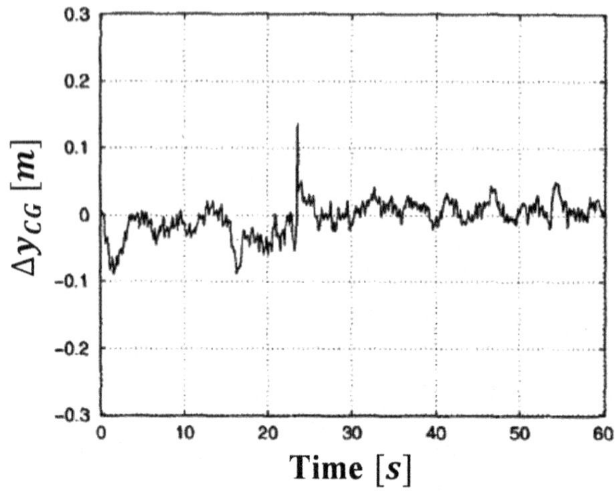

Fig. 5.11 Example of $\Delta y_{CG}(t)$ with the controller $C_1(z)$ (from [38])

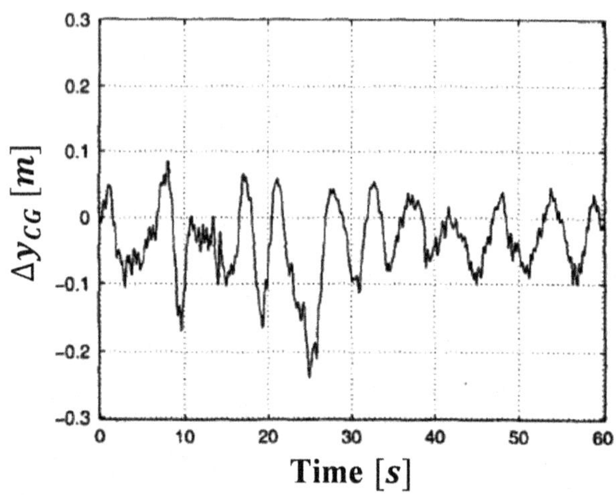

Fig. 5.12 Example of $\Delta y_{CG}(t)$ with the controller $C_2(z)$ (from [38])

where the measured inputs are Δy and $\Delta \psi$. The tuning of the controller is carried out through a linear vehicle model, including roll dynamics. This peculiar (and interesting) choice is justified by the fact that a good match between the model and the experimental results is achievable, in the case of a comfort-oriented tuning of the vehicle suspension system, only through the inclusion of the roll dynamics and actuator dynamics in the model for control system design.

The formal design specifications are: (i) the maximization of the gain margin (with the requirement of being > 2) and phase margin (with a requirement of being > 50 deg) on the open-loop transfer function through the optimal selection of the gain $k_c(v)$ and the preview distance $l_d(v)$ and (ii) to guarantee that the lateral displacement deviation of the closed-loop system shall not exceed a given threshold (reported in Fig. 5.13) for a 1 m/s^2 step input of the reference road acceleration.

In the control structure, $G_c(s_L)$ mainly compensates for the actuator dynamics and consists of a low-frequency integrator and high-frequency roll-off, "to reduce the effects of the steady-state tracking bias and the unwanted excitation of the high-frequency unmodeled actuator dynamics. $G_{l_d}(s_L)$ is made of a high-frequency roll-off portion and a mid-frequency lead-lag filter to limit the look-ahead amplification and to provide extra look-ahead between 0.5 and 2 Hz." In formulas:

$$G_c(s_L) = \frac{25\pi\,(s_L + 0.5\pi)}{(s_L + 0.02\pi)\,(s_L + 25\pi)} \tag{5.19}$$

$$G_{l_d}(s_L) = \frac{20\pi\,(s_L + 0.4\pi)}{(s_L + 0.8\pi)\,(s_L + 10\pi)} \tag{5.20}$$

$k_{int}(s_L)$ is an integrator to keep the steady-state tracking error small. The main controller parameters and performance indicators based on the linear model of the system are reported in Fig. 5.13. Interestingly, $k_c(v)$ and $l_d(v)$ do not significantly change between 15 m/s and 30 m/s, while they are significantly different especially at low speed.

The comprehensive set of experimental results, collected on eight vehicle demonstrators including different number of passengers (from 0 to 4), shows: (i) 0.1 m maximum tracking error for highway driving up to 100 mph with standard deviation of less than 5 cm; (ii) 0.2 m maximum error in transient conditions for 0.5 g maneuvering; (iii) sharp curve tracking with less than 3 cm error; (iv) low speed precision docking with less than 1 cm error; and (v) smooth steering actions, i.e., the passengers feel no observable oscillations.

Based on these experimental proofs and also on the other references mentioned so far, the conclusion of this section is that in the absence of uncertainty and

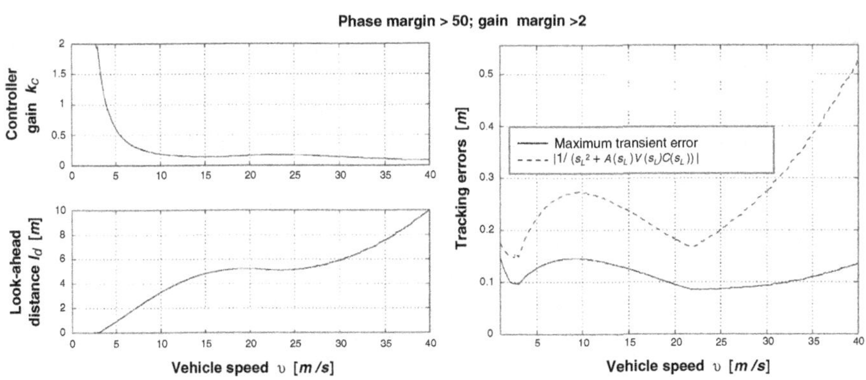

Fig. 5.13 Optimal controller parameters and performance indicators (from [40])

significant disturbances in the road scenario, simple, conventional, reliable, and easily tunable control structures with appropriate gain scheduling are sufficient for providing good performance in all operating conditions, and are actually recommended by the authors of this chapter for vehicle implementation.

5.3.2 Linear Quadratic Regulators

5.3.2.1 Basic Linear Quadratic Formulation

This subsection focuses on the basic mathematical formulation of linear quadratic regulator (LQR) control structures for path tracking. LQRs for path tracking are based on the state-space formulation of the system dynamics. The main building block is represented by the well-known single-track model of the vehicle [4, 21, 22], already used in the previous subsection, with a linear model of the tires, parameterized by their cornering stiffness, and adopting the lateral slip velocity of the vehicle center of gravity, v_y, and the yaw rate, $\dot{\psi}$, as system states. This model is suitable for describing vehicle dynamics at moderate lateral acceleration levels with respect to the available tire-road friction conditions, and at small longitudinal accelerations/decelerations (its equations are actually derived for the condition of constant speed). In formulas:

$$\begin{bmatrix} \dot{v}_y \\ \ddot{\psi} \end{bmatrix} = \begin{bmatrix} -2\frac{(C_f+C_r)}{mv_x} & 2\frac{bC_r-aC_f}{mv_x} - v_x \\ 2\frac{bC_r-aC_f}{I_z v_x} & -2\frac{\left(a^2C_f+b^2C_r\right)}{I_z v_x} \end{bmatrix} \begin{bmatrix} v_y \\ \dot{\psi} \end{bmatrix} + \begin{bmatrix} \frac{2C_f}{m} \\ \frac{2aC_f}{I_z} \end{bmatrix} \delta \tag{5.21}$$

The single-track vehicle model has to be expressed in the states relevant to the implementation of the path tracking controller. The following control variables are usually defined (minor variations are present in the different papers and reports): (i) the lateral position error, Δy_{CG}, i.e., the length of the segment perpendicular to the symmetry plane of the vehicle and connecting the center of gravity with the corresponding point on the reference path (see Fig. 5.2, according to [16]). In some of the papers, the distance is measured along a segment perpendicular to the reference path, rather than perpendicular to the vehicle symmetry plane (i.e., a reference system aligned with the road is adopted) and (ii) the heading angle error, $\Delta\psi_{CG} = \psi - \psi_{path,CG}$ (indicated in Fig. 5.2). The controller formulation would not significantly change if the errors were measured from any other point (different from the center of gravity) located on the longitudinal axis of the vehicle reference system or the road reference system (e.g., in front of the vehicle, in order to generate a basic preview effect).

By considering the approximated system kinematics, it is $\Delta\ddot{y}_{CG} = \left(\dot{v}_y + v_x\dot{\psi}\right) - \dot{v}_{y,path,CG}(s) = \dot{v}_y + v_x\left(\dot{\psi} - \dot{\psi}_{path,CG}(s)\right) = \dot{v}_y + v_x\Delta\dot{\psi}_{CG}$ and $\Delta\dot{y}_{CG} = v_y + v_x\Delta\psi_{CG}$. Equation (5.21) can be transformed into the path coordinates, thus obtaining the state-space formulation directly applicable to the path tracking LQR control system design:

$$\begin{bmatrix} \Delta\dot{y}_{\text{CG}} \\ \Delta\ddot{y}_{\text{CG}} \\ \Delta\dot{\psi}_{\text{CG}} \\ \Delta\ddot{\psi}_{\text{CG}} \end{bmatrix} = \begin{bmatrix} 0 & 1 & 0 & 0 \\ 0 & -\frac{2(C_{\text{f}}+C_{\text{r}})}{mv_x} & 2\frac{C_{\text{f}}+C_{\text{r}}}{mv_x} & 2\frac{bC_{\text{r}}-aC_{\text{f}}}{mv_x} \\ 0 & 0 & 0 & 1 \\ 0 & 2\frac{bC_{\text{r}}-aC_{\text{f}}}{I_z v_x} & 2\frac{aC_{\text{f}}-bC_{\text{r}}}{I_z} & -2\frac{(a^2C_{\text{f}}+b^2C_{\text{r}})}{I_z v_x} \end{bmatrix} \begin{bmatrix} \Delta y_{\text{CG}} \\ \Delta\dot{y}_{\text{CG}} \\ \Delta\psi_{\text{CG}} \\ \Delta\dot{\psi}_{\text{CG}} \end{bmatrix} + \begin{bmatrix} 0 \\ \frac{2C_{\text{f}}}{m} \\ 0 \\ \frac{2aC_{\text{f}}}{I_z} \end{bmatrix} \delta$$
$$+ \begin{bmatrix} 0 \\ 2\frac{bC_{\text{r}}-aC_{\text{f}}}{mv_x} - v_x \\ 0 \\ -2\frac{(a^2C_{\text{f}}+b^2C_{\text{r}})}{I_z v_x} \end{bmatrix} \dot{\psi}_{\text{path,CG}}(s) + \begin{bmatrix} 0 \\ 0 \\ 0 \\ -1 \end{bmatrix} \ddot{\psi}_{\text{path,CG}}(s) \tag{5.22}$$

which corresponds to the canonical form:

$$\dot{\xi} = A\xi + B_1\delta + B_2\dot{\psi}_{\text{path,CG}}(s) + B_3\ddot{\psi}_{\text{path,CG}}(s) \tag{5.23}$$

where $\xi = \left[\Delta y_{\text{CG}}\ \Delta\dot{y}_{\text{CG}}\ \Delta\psi_{CG}\Delta\dot{\psi}_{CG}\right]^{\text{T}}$, and the steering angle δ is the control output. The term in Eqs. (5.22) and (5.23) including the reference yaw acceleration, $\ddot{\psi}_{\text{path,CG}}(s)$, is usually neglected in the literature. The term $B_2\dot{\psi}_{\text{path,CG}}$ represents a disturbance within the control system design.

The system is controllable as the controllability matrix $[B_1\ AB_1\ A^2B_1\ A^3B_1]$ has full rank. The LQR formulation for path tracking is normally based on state feedback regulation, i.e., on the control of the lateral position and velocity errors at the center of gravity, and the control of the heading angle and heading rate errors, with all references set to 0. This means that $\delta = -K_{\text{LQ}}\xi = -k_{1_{\text{LQ}}}\Delta y_{\text{CG}} - k_{2_{\text{LQ}}}\Delta\dot{y}_{\text{CG}} - k_{3_{\text{LQ}}}\Delta\psi_{\text{CG}} - k_{4_{\text{LQ}}}\Delta\dot{\psi}_{\text{CG}}$. The LQR controller minimizes the following quadratic cost function J:

$$J = \int_0^\infty \left(\xi^T Q\xi + u^T Ru\right) dt \tag{5.24}$$

where the diagonal 4×4 weighting matrix Q is selected to define the relative importance of the tracking performance for the different states of the system, while the weighting factor R defines the relative importance of control effort and tracking performance. The gain K_{LQ} can be designed through the well-known algebraic Riccati equation [41]. In formulas:

$$K_{LQ} = R^{-1}B_1^T S \tag{5.25}$$

$$A^T S - SA - \left(SB_1^T + N\right) R^{-1}B_1^T S + Q = 0 \tag{5.26}$$

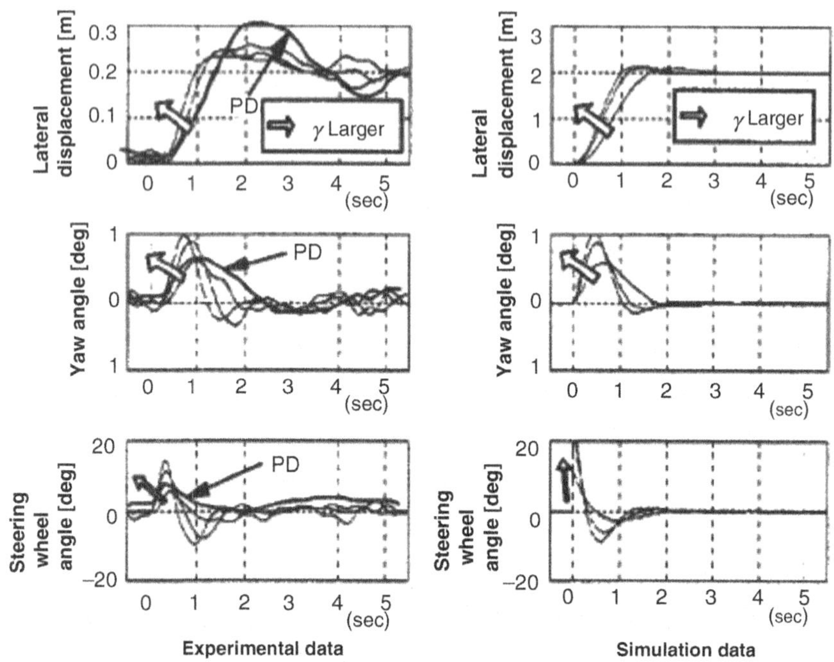

Fig. 5.14 Comparison of PD and LQR controllers for different values of the parameter γ of the LQR (from [28])

thus bringing the following closed-loop formulation of the controlled system dynamics:

$$\dot{\xi} = \left(A - B_1 K_{\mathrm{LQ}}\right) \xi + B_2 \dot{\psi}_{\mathrm{path,CG}}(s) \tag{5.27}$$

The continuous form of the LQR path tracking controller was presented here. The continuous system in Eq. (5.27) can be easily subject to discretization for the design of a discrete LQR controller.

An LQR implementation was experimentally assessed by Nissan in [28] and compared with the performance provided by the PD controller on the lateral position error mentioned in Sect. 5.3.1. The tests (Fig. 5.14) were conducted at 80 km/h and consisted of a step change of 20 cm in the lateral coordinate of the target path. The study included a sensitivity analysis as a function of $\gamma = Q(1,1)/R$, i.e., the ratio between the lateral deviation weighting and the control effort weighting in Eq. (5.24).

When commenting their experimental and simulation results, the authors of [28] state that "with PD control, large overshoot occurred when response was improved and the system became more susceptible to noise. This made it impossible to set the control constants at larger values." In any case, the same authors observe that the model for LQR design exhibits "large modeling errors on the curved segments of the test road, making it unable to track the path accurately on curves." A possible solution to this limitation is to adopt a feedforward contribution based on road curvature, which will be discussed in the next subsection. The alternative proposed in [28] is a Kalman filter based on the vehicle motion equations, combined with a curvature approximation model. The output of the Kalman filter is the curvature estimation, $\widehat{\kappa}$, which is used as a state variable in the augmented LQR scheme.

5.3.2.2 Linear Quadratic Regulator with Feedforward Contribution

The feedback LQR formulation of the controller can be enhanced through a feedforward contribution, aimed at canceling the steady-state lateral position error of the center of gravity, which is a problem especially in the case of curved paths. Hence, the control output and closed-loop system dynamics assume the following shape:

$$\delta = -K_{\mathrm{LQ}}\xi + \delta_{\mathrm{FFW},3} \tag{5.28}$$

$$\dot{\xi} = \left(A - B_1 K_{\mathrm{LQ}}\right)\xi + B_1\delta_{\mathrm{FFW},3} + B_2\dot{\psi}_{\mathrm{path,CG}}(s) \tag{5.29}$$

By manipulating Eq. (5.29), it is possible to obtain the analytical expression of the steady-state errors of the controlled system. By imposing $\Delta y_{\mathrm{CG,SS}} = 0$ and under the hypothesis of a constant radius trajectory, it is:

$$\delta_{\mathrm{FFW},3} = \frac{L}{R_{\mathrm{tr}}} + \left(\frac{mb}{2C_{\mathrm{f}}} - \frac{ma}{2C_{\mathrm{r}}}\right)\frac{a_y}{L} + k_{3_{\mathrm{LQ}}}\Delta\psi_{\mathrm{CG,SS}} = \frac{L}{R_{\mathrm{tr}}} + k_{\mathrm{U}}a_y + k_{3_{\mathrm{LQ}}}\Delta\psi_{\mathrm{CG,SS}} \tag{5.30}$$

where the resulting steady-state value of the heading error becomes:

$$\Delta\psi_{\mathrm{CG,SS}} = -\frac{b}{R_{\mathrm{tr}}} + \frac{a}{2C_{\mathrm{r}}L}\frac{mv_x^2}{R_{\mathrm{tr}}} \tag{5.31}$$

The important conclusion is that $\Delta\psi_{\mathrm{CG,SS}}$ is not controllable through the feedforward contribution of the steering angle, if this is aimed at achieving $\Delta y_{\mathrm{CG,SS}} = 0$.

From a practical viewpoint, as the yaw dynamics of the vehicle are strongly dependent on vehicle speed, i.e., yaw damping is a decreasing function of vehicle speed, a careful scheduling of the feedback controller gains is also required as a function of v_x, in order to provide consistent tracking performance at different vehicle speeds, as already discussed in Sect. 5.3.1.

5.3.2.3 Linear Quadratic Regulator with Preview

The LQR formulations in the previous sections can be significantly enhanced through a preview scheme, i.e., by augmenting the system to include the future profile of the reference path. To this purpose, the reference path is discretized, and a vector with its lateral coordinates is progressively updated at each time step. The vehicle system model in the tracking coordinates (Eq. 5.22) can be discretized as:

$$\begin{cases} \xi_{v_{k+1}} = A_{v_k}\xi_{v_k} + B_{v_k}\delta_k \\ y_{v_k} = C_{v_k}x_{v_k} + D_{v_k}\delta_k \end{cases} \tag{5.32}$$

The future reference path profile, represented by the vector $y_{r_{k+1}}$, is modeled as:

$$y_{r_{k+1}} = A_{r_k}y_{r_k} + B_{r_k}y_{ri_k}$$
$$A_{r_k} = \begin{bmatrix} 0\,1 & 0\,0 & \cdots 0 \\ 0\,0 & 1\,0 & \cdots 0 \\ \vdots\,\vdots & \vdots\,\ddots & \vdots\;\vdots \\ \vdots\,\vdots & \vdots\;\vdots & \ddots\,0 \\ 0\,0 & 0\cdots & \cdots 1 \\ 0\,0 & 0\cdots & \cdots 0 \end{bmatrix}, \quad B_{r_k} = \begin{bmatrix} 0 \\ 0 \\ 0 \\ \vdots \\ 1 \end{bmatrix} \tag{5.33}$$

The reference lateral position of the vehicle is located into a register that is progressively updated at each time step. y_{ri} is considered as a disturbance in the form of a white noise.

The discretized model of the vehicle system, including the reference position coordinates, can be expressed as the combination of the models in Eqs. (5.32) and (5.33):

$$\begin{cases} \xi_{k+1} = A\xi_k + B\delta_k + Ey_{ri_k} \\ y_k = C\xi_k + D\delta_k \end{cases} \tag{5.34}$$

with augmented state vector $\xi_k = [\xi_{v_k}\; y_{r_k}]^{\mathrm{T}}$ (see [16] for the details regarding the formulation of the different matrices). The state feedback control law becomes $u_k = -K_{\mathrm{LQ_P}}\xi_k$, where the states related to the future values of the lateral coordinates of the road have an effect on the control action. The control gain matrix $K_{\mathrm{LQ_P}}$ can be calculated by using the well-known LQR discrete formulation and the respective Riccati equation.

It is interesting to observe that the path tracking control formulation deriving from Eq. (5.34), presented in [16], is very similar to the driver model with preview in [5] (originally not conceived for automated driving), based on the reference steering angle $\delta = k_{\Delta\psi}\Delta\psi + k_{\Delta y_{\mathrm{CG}}}\Delta y_{\mathrm{CG}} + \sum_{k=1}^{n}\Delta y_k$, where the index k refers to points located in front of the vehicle. Actually, the formulation in Eq. (5.34) is identical to the motorbike driver model in [7]. An extension of the linear quadratic formulation with preview could be based on the driver model in [8], which is an enhancement of the driver model in [5], with consideration of the future heading angle errors and their time derivatives, to give $\delta = k_{\Delta\psi,w}\Delta\psi_w + k_{\Delta\dot{\psi},w}\Delta\dot{\psi}_w + k_{\Delta y,w}\Delta y_w$. In this case, the errors $\Delta\psi_w$, $\Delta\dot{\psi}_w$, and Δy_w are scalar variables, calculated as the linear combination of the errors at different points *k*.

5.3.2.4 Frequency-Shaped Linear Quadratic Control

A relevant contribution to the science of LQR path tracking control was provided by Peng and Tomizuka in the PATH framework in the 1990s [42–44]. Their frequency-shaped linear quadratic controller is based on the linear equations of the system, according to the formulation in Eqs. (5.21) and (5.22). The output is represented by the lateral deviation measured by a sensor located in the front part of the vehicle. The main advantages of the frequency-shaped LQR formulation with respect to conventional LQR control are: (i) the robustness on measurement noise at high frequencies and (ii) the possibility of including ride quality explicitly in the performance index (which is very important, as pointed out in some of the references including actual experimental tests and not only computer simulations).

The adopted performance indicator is:

$$J = \frac{1}{2\pi}\int_{-\infty}^{\infty}\left[\Delta a_y^*(j\omega_{\mathrm{L}})\frac{q_{\Delta a_y}^2}{1+\lambda_{\Delta a_y}^2\omega_{\mathrm{L}}^2}\Delta a_y(j\omega_{\mathrm{L}}) + \Delta y_{CG}^*(j\omega_{\mathrm{L}})\frac{q_{\Delta y_{l_{\mathrm{d}}}}^2}{1+\lambda_{\Delta y_{l_{\mathrm{d}}}}^2\omega_{\mathrm{L}}^2}\right.$$
$$\times\Delta y_{CG}(j\omega_{\mathrm{L}}) + \Delta\dot{\psi}^*(j\omega_{\mathrm{L}})\frac{q_{\Delta\psi}^2}{1+\lambda_{\Delta\psi}^2\omega_{\mathrm{L}}^2}\Delta\dot{\psi}(j\omega_{\mathrm{L}})$$
$$\left.+\Delta y_{l_{\mathrm{d}}}^*(j\omega_{\mathrm{L}})\frac{q_i^2}{(j\omega_{\mathrm{L}})^2}\Delta y_{l_{\mathrm{d}}}(j\omega_{\mathrm{L}}) + \delta^*(j\omega_{\mathrm{L}})R\delta(j\omega_{\mathrm{L}})\right]d\omega_{\mathrm{L}} \tag{5.35}$$

where Δa_y is the difference between the reference and the actual lateral acceleration. The coefficients $q_{\Delta a_y}$ and $\lambda_{\Delta a_y}$ are chosen to provide the expected ride quality, while the coefficients of the other three terms are selected to provide responsiveness to the road curvature and robustness with respect to the measurement noise. The disturbance term in the vehicle model, $B_2\dot{\psi}_{\mathrm{path}}$ (i.e., the effect of the curvature, see Eq. (5.27)), is previewed with a preview time $t_{l_{\mathrm{d}}}$.

The problem is solved as a conventional linear quadratic controller after augmenting the system state variables with the states $z_1, \ldots, z_4$ corresponding to the four filters in Eq. (5.35), so that the augmented state vector is $\xi_{\mathrm{augm.}}^T = \left[\Delta y_{\mathrm{CG}}\ \Delta\dot{y}_{\mathrm{CG}}\ \Delta\psi\Delta\dot{\psi}\ z_1\ z_2\ z_3\ z_4\right]$. The minimization of *J* requires the knowledge of the disturbance, i.e., $\dot{\psi}_{\mathrm{path}}$, from the current time to infinity. As this is not practically possible, an exponential decay of the curvature-related disturbance $w(t) = 1/\kappa(t)$ beyond the preview region is assumed. The resulting optimal preview steering control law is expressed by:

$$\delta = -k_{\xi_e}\xi_{\text{augm.}} + \int_0^{t_{l_d}} k_{w1}(l_d)\, w(t+\tau)\, d\tau + k_{w2} w(t + t_{l_d}) \tag{5.36}$$

where the first term is the state feedback controller, and the second and third terms are the preview control terms.

5.4 Other Control Structures for Path Tracking and Remarks

The controllers discussed in Sects. 5.2 and 5.3 are only an arbitrary selection of the very wide literature on the subject. This section presents an overview of the variety of less conventional control structures used for path tracking control.

5.4.1 Sliding Mode Controllers

Many papers (e.g., [35] and [45, 46]) present sliding mode controllers for path tracking. Utkin in [35] proposes a relatively simple first order sliding mode control structure (Fig. 5.15), resulting in a yaw control law of the form:

$$\dot{\delta} = -M_u \operatorname{sgn}(\sigma) \tag{5.37}$$

where the sliding variable, σ, is given by $\sigma = c\Delta\dot{\psi} + \Delta\ddot{\psi}$, with $\Delta\dot{\psi} = \dot{\psi} - \dot{\psi}_{\text{ref}}$. As a consequence, the sliding variable is a function of $\dot{\psi}$, $\ddot{\psi}$, $\dot{\psi}_{\text{ref}}$ and $\ddot{\psi}_{\text{ref}}$. The reference yaw rate is based on the lateral position error, having a time derivative $\Delta\dot{y}_{l_d} = v(\beta + \Delta\psi) + l_d\dot{\psi}$. According to traditional feedback linearization, $\dot{\psi}_{\text{ref}}$

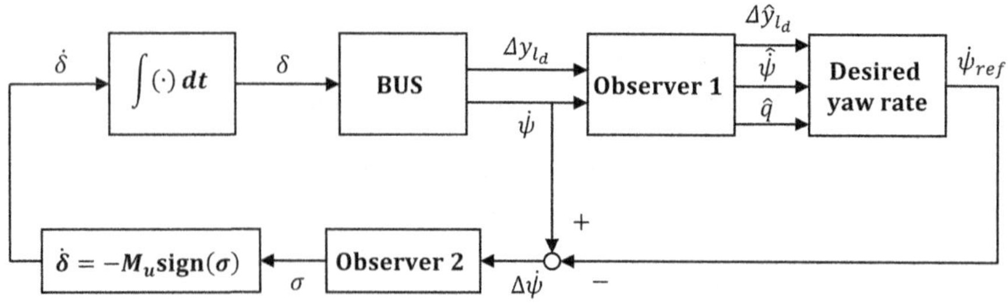

Fig. 5.15 Nonlinear controller structure with observer (adapted from [35])

can be expressed as:

$$\dot{\psi}_{\mathrm{ref}} = -\frac{1}{l_{\mathrm{d}}}\left[v\left(\beta + \Delta\psi\right) + K\Delta y_{l_{\mathrm{d}}}\right] \tag{5.38}$$

with K determining the rate of decay of $\Delta y_{l_{\mathrm{d}}}$. In order to estimate the term $\widehat{q} = v\left(\beta + \Delta\psi\right)$, a dynamic observer (Observer 1 in Fig. 5.15) is implemented. As $\Delta\ddot{\psi}$ is not directly measured but is necessary for the computation of the sliding variable, the time derivative of $\Delta\dot{\psi}$ is computed through a robust observer (Observer 2 in Fig. 5.15), as a simple differentiator would imply a significant risk of chattering. The comparison of the performance of this sliding mode formulation with the continuous controller discussed in Eqs. (5.13)–(5.16), carried out in [35], does not allow clear conclusions.

Particularly relevant, also considering the recent experimental developments at Stanford University to be discussed in Sect. 5.5, is the sliding mode path tracking controller developed in [46], for a four-wheel-steering vehicle, according to the strong tradition of Japanese vehicle engineering in four-wheel-steering systems. The controller is based on the important concept of centers of percussion of the front and rear axles. The centers of percussion (COPs) are located on the symmetry plane of the vehicle. Their longitudinal position with respect to the center of gravity is defined by (see also Fig. 5.16):

$$|x_{\mathrm{COP,f}}| = \frac{I_z}{m\,b},\quad |x_{\mathrm{COP,r}}| = \frac{I_z}{m\,a} \tag{5.39}$$

The four-wheel-steering vehicle layout of [46] allows the independent control of two variables, i.e., in the specific case the lateral position errors at the front and rear centers of percussion, respectively, $\Delta y_{\mathrm{COP,f}}$ and $\Delta y_{\mathrm{COP,r}}$.

The relationship between the vehicle state variables and the time derivatives of the lateral position errors at the centers of percussion is:

$$\begin{cases} \Delta\dot{y}_{\mathrm{COP,f}} = -\left[v\left(\beta + \Delta\psi_{\mathrm{COP,f}}\right) + |x_{\mathrm{COP,f}}|\,\dot{\psi}\right] \\ \Delta\dot{y}_{\mathrm{COP,r}} = -\left[v\left(\beta + \Delta\psi_{\mathrm{COP,r}}\right) - |x_{\mathrm{COP,r}}|\,\dot{\psi}\right] \end{cases} \tag{5.40}$$

This means that the state vector $\xi = \left[\beta\ \dot{\psi}\right]^{\mathrm{T}}$ of a linear single-track vehicle model similar to that reported in Eq. (5.21) can be expressed as a function of $\Delta\dot{y}_{\mathrm{COP}} = [\Delta\dot{y}_{\mathrm{COP,f}}\Delta\dot{y}_{\mathrm{COP,r}}]^{\mathrm{T}}$ and $\Delta\psi_{\mathrm{COP}} = [\Delta\psi_{\mathrm{COP,f}}\Delta\psi_{\mathrm{COP,r}}]^{\mathrm{T}}$:

$$\xi = T\Delta\dot{y}_{\mathrm{COP}} + vT\Delta\psi_{\mathrm{COP}} \tag{5.41}$$

The substitution of Eq. (5.41) into the single-track vehicle model equations leads to the following tracking position dynamics at the centers of percussion:

$$\Delta\ddot{y}_{\mathrm{COP}} = A'\Delta\dot{y}_{\mathrm{COP}} + B'\overline{u} + vA'\Delta\psi_{\mathrm{COP}} + v\dot{\psi}_{\mathrm{path,COP}} + h'w \tag{5.42}$$

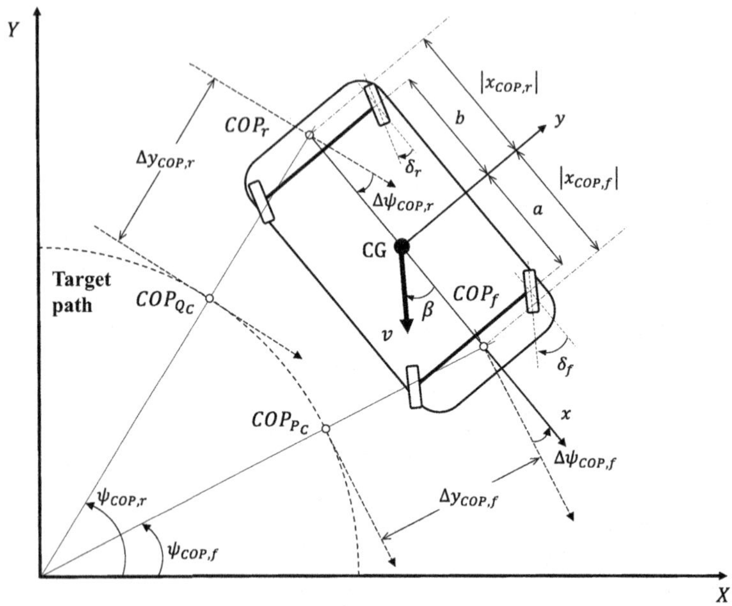

Fig. 5.16 Vehicle model for four-wheel-steering path tracking based on the centers of percussion (adapted from [46]). The sign convention is the one in [46]

where $\dot{\psi}_{\text{path,COP}} = \left[\dot{\psi}_{\text{path,COP,f}}\ \dot{\psi}_{\text{path,COP,r}}\right]^{\text{T}}$ is the vector formulation of the time derivatives of the target path heading angles at the front and rear centers of percussion.

With a proper selection of the feedback controller configuration and by imposing that the two control points are the centers of percussion of the vehicle, Eq. (5.42) simplifies into the form:

$$\Delta\ddot{y}_{\text{COP}} = B'\overline{u} + v\dot{\psi}_{\text{path,COP}} + h'w \tag{5.43}$$

where:

$$B' = \begin{bmatrix} -\frac{2C_{\text{f}}L}{mb} & 0 \\ 0 & -\frac{2C_{\text{r}}L}{ma} \end{bmatrix} \tag{5.44}$$

The diagonal matrix B' very importantly indicates that the position tracking problems at the front and rear centers of percussion are decoupled. Hence, each center of percussion path deviation can be independently controlled by the front and rear steering angles. As a consequence, the control laws for the front and rear axles can be separately designed, as single-input single-output controllers. This means that the lateral displacement dynamics of the front center of percussion are independent from the lateral force of the rear tires, while the lateral displacement dynamics of the rear center of percussion are independent from the lateral force of the front tires. A different selection of the longitudinal position of the control points would imply the design of a multivariable controller.

In the specific case of [46], the resulting sliding mode control laws have the following shape:

$$\begin{cases} \delta_\mathrm{f} = \beta + \frac{a}{v}\dot{\psi} + \frac{c_{\mathrm{COP,f}}}{\kappa_{\mathrm{COP,f}}}\Delta\dot{y}_{\mathrm{COP,f}} + M_{\mathrm{COP,f}}\mathrm{sgn}\left(\sigma_\mathrm{f}\right) \\ \delta_\mathrm{r} = \beta - \frac{b}{v}\dot{\psi} + \frac{c_{\mathrm{COP,r}}}{\kappa_{\mathrm{COP,r}}}\Delta\dot{y}_{\mathrm{COP,r}} + M_{\mathrm{COP,r}}\mathrm{sgn}\left(\sigma_\mathrm{r}\right) \end{cases} \tag{5.45}$$

with the sliding variables σ_f and σ_r defined as $\sigma_i = c_{\mathrm{COP},i}\Delta y_{\mathrm{COP},i} + \Delta\dot{y}_{\mathrm{COP},i}$, with $i = \mathrm{f,r}$. The formulation in Eq. (5.45) requires an estimation of sideslip angle, while $\dot{\psi}$ and $\Delta\dot{y}_{\mathrm{COP},i}$ can be easily measured or estimated. The sliding mode formulation of the specific paper is designed, through the appropriate definition of the gains, $M_{\mathrm{COP,f}}$ and $M_{\mathrm{COP,r}}$, to provide robustness against the variations of cornering stiffness and target path radius, and cross-wind disturbances. In order to prevent chattering, the following approximation of the discontinuous part of the control law is used:

$$\mathrm{sgn}\left(\sigma_i\right) \cong \sin\left(\tan^{-1}\left(\frac{\sigma_i}{\mu_i}\right)\right), \mu_i > 0 \;\; (i = \mathrm{f, r}) \tag{5.46}$$

The resulting steady-state values of the front and rear steering angles, $\delta_{\mathrm{f,ss}}$ and $\delta_{\mathrm{r,SS}}$, are given by:

$$\begin{cases} \delta_{\mathrm{f,SS}} = \pm\frac{1}{R_{\mathrm{tr}}}\left(a - \frac{a-b}{2mab}I_z + \frac{mbv^2}{2C_\mathrm{f}L}\right) \\ \delta_{\mathrm{r,SS}} = \pm\frac{1}{R_{\mathrm{tr}}}\left(-b - \frac{a-b}{2mab}I_z + \frac{mav^2}{2C_\mathrm{r}L}\right) \end{cases} \tag{5.47}$$

In practice, the sign of $\delta_{\mathrm{r,SS}}$ changes at the vehicle speed v_0 (about 55 km/h for the case study vehicle of [46]), i.e., at low speeds the rear wheels steer in counterphase with respect to the front wheels, while the opposite happens for $v > v_0$.

$$v_0 = \sqrt{\frac{2C_\mathrm{r}l}{ma}\left(b + \frac{a-b}{2mab}I_z\right)} \tag{5.48}$$

The four-wheel-steering controller based on the COP concept was validated through CarSim simulations of a cross-wind disturbance situation. Figure 5.17 reports a sample of the simulation results, showing that the maximum deviations at the front and rear control points of the four-wheel-steering vehicle are not influenced by vehicle velocity, while the rear path deviation of the two-wheel-steering vehicle used as a term of comparison increases at exponential rate.

Overall, the main benefit of sliding mode control is the low-complexity of the resulting control law (see Eqs. (5.37) and (5.45)), however: (1) "it needs knowledge about the bounds of the disturbances and uncertainties in advance" [47]; (2) "it is not robust outside the sliding surface" [47]; and (3) it can present chattering.

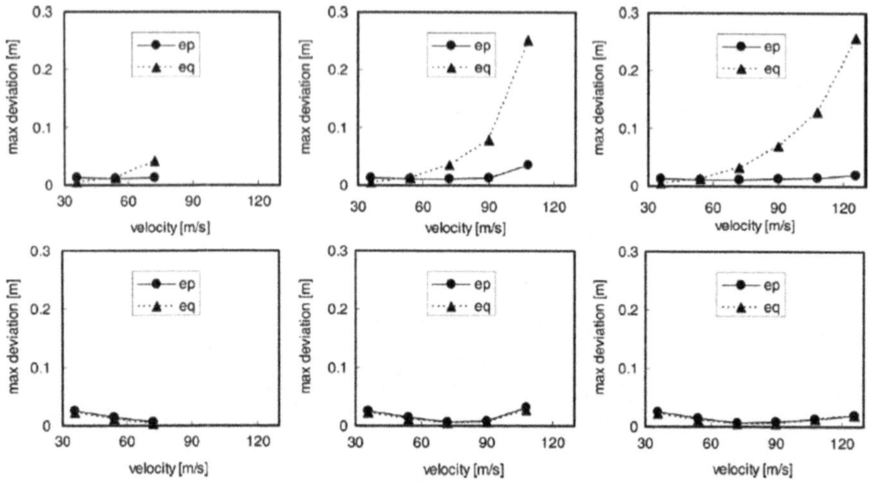

Fig. 5.17 Example of validation results of the controller in [46], in terms of maximum path deviations at the front (indicated as "ep" in the figure) and rear (indicated as "eq" in the figure) centers of percussion. From left to right: $\mu_c = 0.3, 0.5, 0.85$. Upper: two-wheel-steering vehicle, lower: four-wheel-steering vehicle

In order to alleviate (3), Tagne et al. [48] adopted and experimentally validated a super-twisting algorithm of the form $\delta = \delta_{ST} + \delta_{eq}$, where δ_{ST} is the super-twisting contribution:

$$\delta_{ST} = \delta_{ST,1} + \delta_{ST,2}, \begin{cases} \delta_{ST,1} = -\alpha_{1,ST}|\sigma|^{0.5}\mathrm{sign}(\sigma) \\ \dot{\delta}_{ST,2} = -\alpha_{2,ST}\mathrm{sign}(\sigma) \end{cases} \tag{5.49}$$

with $\sigma = \Delta\dot{y}_{CG} + \lambda\Delta y_{CG}$. The equivalent term corresponding to $\dot{\sigma} = 0$ is calculated starting from the equations of the single-track model of the system and is given by:

$$\delta_{eq} = \frac{C_f + C_r}{C_f}\beta + \frac{aC_f - bC_r}{C_f v_x}\dot{\psi} + \frac{mv_x^2}{C_f}\kappa - \frac{m\lambda}{C_f}\Delta\dot{y}_{CG} \tag{5.50}$$

[49, 50] are other useful references in the area of sliding mode control applied to automated driving. The recent paper [47] presents and experimentally validates, including comparison with the super-twisting algorithm of Eqs. (5.49) and (5.50), a path tracking controller based on the theory of immersion and invariance, where the target dynamics for the system are selected during the control design phase, similarly to sliding mode control. The main benefits with respect to sliding mode control are that: (i) "the manifold does not necessarily have to be reached"; (ii) it allows more flexibility in the selection of the target dynamics; and (iii) "it avoids the use of a discontinuous term in the control law."

In the area of control structures with discontinuous control action, [51] experimentally demonstrates a nonlinear controller (in practice a sliding mode controller including saturation functions, even if it is not explicitly called in this way in the original paper), based on the lateral offset at a look-ahead point (estimated by a Kalman filter, together with its time derivative) and the yaw angle error. [52] presents a discontinuous control law of the form $\delta = -(|k_\mathrm{d}\xi| + |r|)\,\mathrm{sign}\left(b^T P_\mathrm{L}(\xi - \xi_\mathrm{ref})\right)$, generated through model-reference control and a Lyapunov approach, with simulation results showing significant chattering, unacceptable for a real vehicle implementation.

5.4.2 Other Control Structures

Many other path tracking controllers were implemented in the literature, covering most of the control structure options.

For example, O'Brien et al. [53] discuss an H_∞ controller based on the theory of loop-shaping [54, 55], with the purpose of providing robustness with respect to the variation of the plant parameters. Given a plant with a co-prime factorization $G_H = \widetilde{M}_H^{-1}\widetilde{N}_H$ [54, 55], the perturbed plant is expressed as $G_{H\Delta} = \left(\widetilde{M}_H + \Delta_{M_H}\right)^{-1}\left(\widetilde{N}_H + \Delta_{N_H}\right)$, where $\|\Delta_{M_H}\ \Delta_{N_H}\|_\infty < \epsilon$ formulates the fact that the actual plant can be different from the nominal plant. The purpose of the H_∞ optimization is to find a controller stabilizing the system and maximizing the value of ϵ. The value of ϵ_max represents a measure of the stability margin (i.e., the so-called robust stability margin) for the nominal system to perturbations in the co-prime factorization of the plant. The paper assesses the performance of the H_∞ controller through simulations with a basic 3-degree-of-freedom vehicle model, including robustness analysis with respect to varying speeds, icy roads, and wind gusts. As the controller is based on two inputs (i.e., the lateral displacement error and yaw angle error) and produces one output, a singular value decomposition analysis [55] would allow discussing its functional controllability. A more recent second example of H_∞ controller implementation and experimental demonstration on a prototype vehicle is included in [56].

Shin and Joo [57] present a backstepping controller design (see [58] for the theory of backstepping), based on a linear single-track vehicle model, in this case with the following states: (i) sideslip angle; (ii) yaw rate; (iii) the difference between the actual and reference yaw angles, $\Delta\psi = \psi - \psi_\mathrm{ref}$; and (iv) the vehicle offset from the center of the lane, Δy_{l_d}. In particular, in the model for control system design, the equation of $\Delta\dot{y}_{l_\mathrm{d}}$ is $\Delta\dot{y}_{l_\mathrm{d}} = v(\beta + \Delta\psi) + l_\mathrm{d}\dot{\psi}$. Hence, the reference yaw rate is defined as

$$\dot{\psi}_\mathrm{ref} = -\frac{v(\beta + \Delta\psi) + K_\mathrm{d}\Delta y_{l_\mathrm{d}}}{l_\mathrm{d}} \tag{5.51}$$

By defining $\Delta\dot{\psi} = \dot{\psi} - \dot{\psi}_\mathrm{ref}$ and substituting into the equation of $\Delta\dot{y}_{l_\mathrm{d}}$, the lateral displacement error dynamics are described by $\Delta\dot{y}_{l_\mathrm{d}} = -K_\mathrm{d}\Delta y_{l_\mathrm{d}} + l_\mathrm{d}\Delta\dot{\psi}$. After differentiation of $\Delta\dot{\psi}$ with respect to time and re-substitution into the single-track model equations, the yaw rate error dynamics are given by Eq. (5.52), based on the expressions of the coefficients C_β, $C_{\dot{\psi}}$, $C_{\Delta\psi}$, C_κ, and b_δ.

$$\begin{aligned}\Delta\ddot{\psi} &= C_\beta\beta + C_{\dot{\psi}}\dot{\psi} + C_{\Delta\psi}\Delta\psi - C_\kappa\kappa + b_\delta\delta \\ C_\beta &= \frac{-\frac{2C_f+2C_r}{m}+vk_d+l_d\frac{2C_rb-2C_fa}{I_z}}{l_d} \\ C_{\dot{\psi}} &= \frac{v\left(-1+\frac{2C_rb-2C_fa}{mv^2}\right)+v+K_dl_d-l_d\frac{2C_rb^2-2C_fa^2}{I_zv}}{l_d} \\ C_\psi &= \frac{k_dv}{l_d} \qquad C_\kappa = \frac{v^2}{l_d} \qquad b_\delta = \frac{2C_fa}{I_z}+\frac{2C_f}{l_dm}\end{aligned} \tag{5.52}$$

The steering control law is then chosen as:

$$\delta = \frac{1}{b_\delta}\left[-\left(C_\beta\beta + C_{\dot{\psi}}\dot{\psi} + C_{\Delta\psi}\Delta\psi - C_\kappa\kappa\right) - \frac{W_dl_d}{W_{\dot{\psi}}}\Delta y_{l_d} - k_{\dot{\psi}}\dot{\psi}\right] \tag{5.53}$$

which brings the following yaw rate error dynamics: $\Delta\ddot{\psi} = -k_{\dot{\psi}}\dot{\psi} - \frac{W_dl_d}{W_{\dot{\psi}}}\Delta y_{l_d}$. The term $k_{\dot{\psi}}\dot{\psi}$ in Eq. (5.53) is used to decouple the lateral displacement and yaw rate error dynamics. The stability of the controller can be demonstrated through the Lyapunov function $L_{\text{Lyapunov}} = \frac{1}{2}\left(W_d\Delta y_{l_d}^2 + W_{\dot{\psi}}\Delta\dot{\psi}^2\right)$.

The controller was assessed in relatively low lateral acceleration conditions, through experimental tests with a vehicle prototype (a Hyundai sedan) on a proving ground consisting of a straight section with a length of about 1.2 km and a curved section with a radius of 260 m. Interestingly, the experimental analysis included the comparison of the lateral offset of the vehicle trajectory with respect to the reference one for the cases of human driving and autonomous driving. In general, the proposed controller "has the properties of deviating outwards in the lane during curve entry and inwards during curve exit, similar to the human driver. This can be reduced by adjusting the look-ahead distance to vary proportionally to the vehicle speed and by tuning the feedforward gain related to the curvature" [57].

Many papers (e.g., [59–62]) present fuzzy logic-based controllers for vehicle path tracking control. Given the general caution with respect to fuzzy control still present in industry (e.g., because of the lack of formal proof of stability), this chapter will not report the detailed descriptions of the available fuzzy implementations. The only note is that Naranjo et al. [62] include experimental results (on a Citroen Berlingo) with a fuzzy controller based on a real-time kinematic differential global positioning system, used as the main sensor for vehicle positioning.

5.4.3 Remarks

Unluckily, apart from [16], assessing geometric controllers and different LQR formulations on a CarSim vehicle model (see Table 5.1 summarizing the main

Table 5.1 Comparison of a selection of the path tracking controller formulations discussed so far (adapted from [16])

Tracking method	Robustness to disturbances	Path requirements	Cutting Corners	Overshooting	Steady state error	Suitable applications
Pure pursuit	Good	None within reason	Significant as speed increases	Moderate as speed increases	Significant as speed increases	Slow driving and/or on discontinuous paths
Stanley	Fair	Continuous curvature	No	Moderate as speed increases	Significant as speed increases	Smooth highway driving and/or parking maneuvers
Kinematic (chained controller)	Poor	Continuous through second derivative of curvature	No	Moderate as speed increases, significant during rapidly changing curvature	Significant as speed increases	Smooth parking maneuvers
LQR with feedforward	Poor	Continuous curvature	No	Significant during rapidly/changing curvature	Minimal until much higher speeds	Smooth high speed urban driving at speed
LQR with preview	Fair	None within reason	Moderate in rapidly changing curvature and/or speed	Moderate in rapidly changing curvature and/or speed	Minimal until much higher speeds	Highway driving at relatively constant speed

conclusions of that study), there is limited available literature objectively comparing the performance of the different path tracking controllers discussed so far in this chapter.

[63] is an exception, presenting a simulation-based comparison of four controllers, i.e., a self-tuning regulator (for the details, see [64]), an H_∞ controller, a fuzzy logic controller, and a P controller (of the form $\delta = k_{\Delta\psi}\Delta\psi + k_{\Delta y_{l_d}}\Delta y_{l_d}$). The assessment includes consideration of the effects of curvature, wind, and variations of vehicle speed and tire-road friction coefficients, along the simulated test track circuit at Satory – Versailles, France. The model used for the comparison is a simple linear single-track vehicle model, with a limited level of realism. The comparison shows that the self-tuning controller provides the best performance, followed by the H_∞ controller and the fuzzy logic controller, which are approximately at the same level (even if the authors of [63] mention that fuzzy control is generally less reliable than a conventional controller), and finally by the proportional controller. However, this important analysis would require further development and level of detail.

5.5 Recent Advances in Path Tracking Control

The conclusion of the comparative study of different path tracking controllers in [16] (see Table 5.1), dating back to 2009, was that the expected evolution of the science of path tracking control would have been in the directions of (a) controllers combining different structures and formulations depending on the operating condition of the vehicle, in order to provide consistently reliable automated driving and (b) model predictive controllers, for autonomous driving even in extreme conditions, for example, at high lateral accelerations.

Based on the literature discussed so far, it is evident that there are already extensive experimental demonstrations of gain scheduled controllers capable of simultaneously providing the required vehicle tracking response for a wide range of speeds and lateral accelerations, and very precise maneuverability in docking conditions. In particular, [40] explicitly mentions that with the specific experimentally validated path tracking controller, based on linear control theory and implemented with realistic vehicle actuators, there is no need for multiple control structures, in order to achieve consistently reliable and comfortable path tracking behavior. On the other hand, the very recent paper [65] still includes kinematic model-based controllers in the analysis and considers them useful in low-speed conditions. Nevertheless, given results such as those in [40], the authors of this chapter do not consider the development of heterogeneous control structures to be a priority or major obstacle for the development of the automated driving agenda.

Two main trends can be observed in the recent research in the subject area of path tracking control:

(i) Development of path tracking controllers characterized by the capability of controlling the vehicle at its cornering limit, for example, even at lateral accelerations of 9.5 m/s^2 (e.g., for automated car racing).
(ii) Progressive increase of the level of sophistication of the implemented control structures, with particular focus on model predictive control, now extensively implemented in simulation and preliminarily demonstrated at the experimental level (which confirms the conclusion of [16]).

The following subsections describe examples of recently published path tracking controllers, with concise critical analyses and discussions.

5.5.1 *Advanced Feedforward and Feedback Controllers for Limit Cornering*

Important recent contributions in the area of path tracking control, mainly developed at Stanford University, are aimed at achieving high path tracking performance at the cornering limit of the vehicle, e.g., at lateral acceleration levels up to 9.5 m/s^2 in high friction conditions (which represents the cornering limit for a typical passenger car), or for extreme combined cornering and braking/traction [66–70]. Particular focus is on the development of feedforward steering formulations, allowing a relaxation of the specifications of the feedback part of the controller and a reduction of the issues related to the effect of measurement disturbances.

For example, Kritayakirana and Gerdes [67] present and experimentally validate a feedforward/feedback steering controller for a two-wheel-steering vehicle, based on the path tracking control of the front center of percussion (COP), already defined in Eq. (5.39). By considering $\Delta\dot{\psi}_{\mathrm{CG}} = \dot{\psi} - \dot{\psi}_{\mathrm{path},CG} = \dot{\psi} - \kappa\dot{s}$ and the yaw moment balance equation of a single-track vehicle model, the time derivative of $\Delta\dot{\psi}_{\mathrm{CG}}$ (i.e., the yaw acceleration error) is:

$$\Delta\ddot{\psi}_{\mathrm{CG}} = \ddot{\psi} - \kappa\ddot{s} - \dot{\kappa}\dot{s} = \frac{aF_{\mathrm{y,f}} - bF_{\mathrm{y,r}}}{I_z} - \kappa\ddot{s} - \dot{\kappa}\dot{s} \tag{5.54}$$

The lateral position error at a generic point P along the x-axis of the vehicle reference system is given by: $\Delta y_{\mathrm{P}} = \Delta y_{\mathrm{CG}} + x_{\mathrm{P}} \sin \Delta\psi_{\mathrm{CG}}$ and its acceleration $\Delta\ddot{y}_{\mathrm{P}}$ is approximated with:

$$\Delta\ddot{y}_{\mathrm{P}} = \frac{F_{\mathrm{y,f}} + F_{\mathrm{y,r}}}{m} - v_x\kappa\dot{s} + x_{\mathrm{P}}\frac{aF_{\mathrm{y,f}} - bF_{\mathrm{y,r}}}{I_z} - x_{\mathrm{P}}\left(\kappa\ddot{s} + \dot{\kappa}\dot{s}\right) \tag{5.55}$$

The issue is that the steering actuator can command the front tire force, but it does not have any direct control over the rear tire force, which, thus, represents a disturbance in Eq. (5.55). However, by imposing $x_{\mathrm{P}} = x_{\mathrm{COP,f}}$ and hence that $\frac{F_{\mathrm{y,r}}}{m} +$

$x_{\mathrm{COP,f}}\frac{-bF_{\mathrm{y,r}}}{I_z} = 0$, Eq. (5.55) can be simplified into:

$$\begin{aligned}\Delta\ddot{y}_{\mathrm{COP,f}} &= \frac{F_{\mathrm{y,f}}+F_{\mathrm{y,r}}}{m} - v_x\kappa\dot{s} + x_{\mathrm{COP,f}}\frac{aF_{\mathrm{y,f}}-bF_{\mathrm{y,r}}}{I_z} - x_{\mathrm{COP,f}}\left(\kappa\ddot{s}+\dot{\kappa}\dot{s}\right)\\ &= \frac{L}{b}\frac{F_{\mathrm{y,f}}}{m} - v_x\kappa\dot{s} - x_{\mathrm{COP,f}}\left(\kappa\ddot{s}+\dot{\kappa}\dot{s}\right)\end{aligned} \tag{5.56}$$

The rear tire force is thus eliminated from the equation of the lateral acceleration error at the control point, which now depends only on $F_{\mathrm{y,f}}$, directly controllable through the steering input. The conclusion, similar to the outcome of the analysis in [46] referred to a four-wheel-steering vehicle, is that at the front COP the effect of the rear tire forces on the lateral position error dynamics can be neglected.

The feedforward contribution of the steering controller in [67] has the purpose of eliminating the dynamics of the lateral acceleration error, i.e., it can be obtained by imposing $\Delta\ddot{y}_{\mathrm{COP,f}} = 0$ in Eq. (5.56). Hence, the feedforward lateral force on the front axle, $F_{\mathrm{y,f}}^{\mathrm{FFW}}$, is given by:

$$F_{\mathrm{y,f}}^{\mathrm{FFW}} = \frac{mb}{L}\left(v_x\kappa\dot{s} + x_{\mathrm{COP,f}}\left(\kappa\ddot{s}+\dot{\kappa}\dot{s}\right)\right) \tag{5.57}$$

Eq. (5.57) allows ideal tracking performance, independently from the rear lateral tire force contribution, provided that the terms related to the reference path can be accurately estimated. By substituting Eq. (5.57) into the yaw moment error equation and introducing the feedback part of the control force on the front axle, $F_{\mathrm{y,f}}^{\mathrm{FB}}$, such that the total control force is $F_{\mathrm{y,f}}^{\mathrm{TOT}} = F_{\mathrm{y,f}}^{\mathrm{FFW}} + F_{\mathrm{y,f}}^{\mathrm{FB}}$, the system dynamics become:

$$\begin{cases}\Delta\ddot{\psi}_{\mathrm{CG}} = \frac{a}{I_z}F_{\mathrm{y,f}}^{\mathrm{FB}} + \frac{v_x}{L}\kappa\dot{s} - \frac{b}{I_z}F_{\mathrm{y,r}} - \frac{b}{L}\left(\kappa\ddot{s}+\dot{\kappa}\dot{s}\right)\\ \Delta\ddot{y}_{\mathrm{COP,f}} = \frac{L}{b}\frac{F_{\mathrm{y,f}}^{\mathrm{FB}}}{m}\end{cases} \tag{5.58}$$

Through the terms $\frac{v_x}{L}\kappa\dot{s} - \frac{b}{L}\left(\kappa\ddot{s}+\dot{\kappa}\dot{s}\right)$, Eq. (5.58) shows that the disturbance caused by the curvature cannot be eliminated from the yaw tracking equation, unless an independent actuator controlling the rear tire force is adopted (as already proposed in [46]).

The objective of the feedback part of the controller in [67, 68] is to provide path tracking and yaw stability even when the rear tires are saturated, while the scenario in which the front tires are saturated is not considered. Note that passenger cars are usually characterized by an understeering behavior for vehicle safety, i.e., the absolute value of the front slip angles is normally larger than the absolute value of the rear slip angles, and the front lateral forces saturate first.

During the design of the feedback controller, the nonlinear behavior of the rear tires is considered through the model proposed in [69], according to which $F_{\mathrm{y,r}} = -2\eta_{\mathrm{r}}C_{\mathrm{r}}\alpha_{\mathrm{r}}$, with η_{r} $(0 \le \eta_{\mathrm{r}} \le 1)$ being a monotonically decreasing function of the absolute value of the rear slip angle. By substituting the control variables into the expression of $\alpha_r \approx \frac{v_y - b\dot{\psi}}{v_x}$, it is $\alpha_r \approx \frac{\Delta\dot{y}_{COP,f} - (b + x_{COP,f})\Delta\dot{\psi}_{CG} - v_x \Delta\psi_{CG} - b\kappa\dot{s}}{v_x}$. By including this formulation into the expression for $F_{y,r}$ and then into the equations of the single-track vehicle model in path coordinates (see also Eq. (5.22)), the state-space formulation of the system is obtained, including the effect of the feedforward controller.

The feedback contribution of the front lateral force is then expressed as a full-state feedback controller:

$$F_{y,f}^{FB} = -K_{LC}\xi = -k_{1_{LC}}\Delta y_{COP,f} - k_{2_{LC}}\Delta\dot{y}_{COP,f} - k_{3_{LC}}\,\Delta\psi_{CG} - k_{4_{LC}}\,\Delta\dot{\psi}_{CG} \quad (5.59)$$

By substituting Eq. (5.59) into the state-space formulation, the closed-loop system equations can be used for control system design:

$$\begin{bmatrix} \Delta\dot{y}_{CG} \\ \Delta\ddot{y}_{CG} \\ \Delta\dot{\psi}_{CG} \\ \Delta\ddot{\psi}_{CG} \end{bmatrix} = \begin{bmatrix} 0 & 1 & 0 & 0 \\ -\frac{k_{1_{LC}}L}{bm} & -\frac{k_{2_{LC}}L}{bm} & -\frac{k_{3_{LC}}L}{bm} & -\frac{k_{4_{LC}}L}{bm} \\ 0 & 0 & 0 & 1 \\ -\frac{k_{1_{LC}}a}{I_z} & -\frac{k_{2_{LC}}a}{I_z} + \frac{2b\eta_r C_r}{I_z v_x} & -\frac{k_{3_{LC}}a}{I_z} - \frac{2b\eta_r C_r}{I_z} & -\frac{k_{4_{LC}}a}{I_z} - \frac{2(b + x_{COP})b\eta_r C_r}{I_z v_x} \end{bmatrix} \times \begin{bmatrix} \Delta y_{CG} \\ \Delta\dot{y}_{CG} \\ \Delta\psi_{CG} \\ \Delta\dot{\psi}_{CG} \end{bmatrix} + \begin{bmatrix} 0 \\ 0 \\ 0 \\ \frac{v_x\kappa\dot{s}}{L} - \frac{b}{L}(\kappa\ddot{s} + \dot{\kappa}\dot{s}) - \frac{2b^2\eta_r C_r \kappa\dot{s}}{I_z v_x} \end{bmatrix} \quad (5.60)$$

In the specific controller of [67], it is $k_{2_{LC}} = 0$. In particular, $F_{y,f}^{FB}$ of Eq. (5.59) is manipulated to become $F_{y,f}^{FB} = -k_{LK}\left(\Delta y_{COP,f} + (l_d - a)\,\Delta\psi_{CG}\right) - k_{\Delta\dot{\psi}}\,\Delta\dot{\psi}_{CG}$, where clearly the control point of the feedback part of the controller is not located at the front center of percussion any more. During the control system implementation, the following values are adopted: $k_{LK} = 4000$ N/m, $l_d = 20$ m and $k_{\Delta\dot{\psi}} = 9500$ Ns/rad. The stability of the control system at the cornering limit is demonstrated through Lyapunov method applied to Eq. (5.60), without considering the disturbance from the curvature (which does not affect stability). Detailed tuning criteria of the feedback control gains are reported in [71].

Starting from the previous formulations of the reference front lateral force, $F_{y,f}^{TOT}$, the Fiala tire model for pure cornering conditions is used to obtain the reference slip angle on the front axle, $\alpha_{ref,f}$. Then, based on the measured vehicle yaw rate and estimated sideslip angle, the reference steering angle δ for the front axle is calculated, based on the kinematic relationship $\delta \approx \beta + \frac{a\dot{\psi}}{v_x} - \alpha_{ref,f}$. In practice, these steps can introduce significant errors in the process, in the absence of very accurate state estimation.

A more recent paper of the same research group [70] presents a further development of the controller discussed so far, as the experimental tests show a suboptimal tracking performance of the controller based on Eqs. (5.57) and (5.59) above 7 m/s^2 of lateral acceleration. The authors suggest considering a simplification of the feedforward contribution in Eq. (5.57), by imposing $x_{\mathrm{COP_f}}(\kappa\ddot{s} + \dot{\kappa}\dot{s}) = 0$ and $\dot{s} = v_x$, in order to reduce vehicle response oscillations and increase damping. As a consequence, the feedforward values of the lateral force at the front and rear axles simply become:

$$\begin{cases} F_{y,\mathrm{f}}^{\mathrm{FFW}} = \frac{mb}{L} v_x^2 \kappa \\ F_{y,\mathrm{r}}^{\mathrm{FFW}} = \frac{ma}{L} v_x^2 \kappa \end{cases} \tag{5.61}$$

which are the steady-state values of lateral force in cornering at the reference lateral acceleration $v_x^2\kappa$. From $F_{y,\mathrm{f}}^{\mathrm{FFW}}$ and $F_{y,\mathrm{r}}^{\mathrm{FFW}}$, it is possible to calculate $\alpha_{\mathrm{f}}^{\mathrm{FFW}}$ and $\alpha_{\mathrm{r}}^{\mathrm{FFW}}$, which are the corresponding slip angles. The new expression of the feedforward steering angle is:

$$\delta_{\mathrm{FFW,LC}} = L\kappa - \alpha_{\mathrm{f}}^{\mathrm{FFW}} + \alpha_{\mathrm{r}}^{\mathrm{FFW}} \tag{5.62}$$

which is the well-known equation of vehicle response in steady-state cornering conditions. The main challenge is to actually implement the controller and state estimators computing the correct values of $\alpha_{\mathrm{f}}^{\mathrm{FFW}}$ and $\alpha_{\mathrm{r}}^{\mathrm{FFW}}$.

In [70], the feedback control law of Eq. (5.59) is simplified into the proportional controller:

$$\delta_{\mathrm{FB,LC,1}} = -k_{\mathrm{P}}\left(\Delta y_{\mathrm{CG}} + l_{\mathrm{d}}\Delta\psi_{\mathrm{CG}}\right) \tag{5.63}$$

The overall control law is, thus, given by $\delta = \delta_{\mathrm{FFW,LC}} + \delta_{\mathrm{FB,LC,1}}$. The steady-state response of the system as a function of v is reported in Fig. 5.18, at lateral accelerations of 3 m/s^2 and 7 m/s^2, where for the first case the calculation is based on a linear vehicle model and for the latter it is based on a nonlinear model.

Since the controller is aimed at eliminating a weighted sum of Δy_{CG} and $\Delta\psi_{\mathrm{CG}}$, the feedback part of the controller actually tries to reduce $\Delta y_{l_{\mathrm{d}}}$, while the steady-state values of Δy_{CG} and $\Delta\psi_{\mathrm{CG}}$ are nonzero. From a physical viewpoint, this is the situation corresponding to Fig. 5.19a. An important observation from Fig. 5.18 is that the steady-state value of path deviation is close to zero for vehicle speeds of 17 m/s and 20 m/s, depending on the considered vehicle model. This means that at these speeds the velocity vector at the center of gravity is tangent to the path and $\Delta\psi_{\mathrm{CG}} \approx -\beta$ (see Fig. 5.19b).

A new form of look-ahead feedback control law is suggested, with the aim of constantly keeping the vehicle in the operating condition of Fig. 5.19b:

$$\delta_{\mathrm{FB,LC,2}} = -k_{\mathrm{P}}\left(\Delta y_{\mathrm{CG}} + l_{\mathrm{d}}\left(\Delta\psi_{\mathrm{CG}} + \beta\right)\right) \tag{5.64}$$

With this control law, the steady-state error Δy_{CG} is significantly reduced (i.e., it is ideally zero when including the feedforward contribution according to Eq. (5.62)); however, the system is now characterized by reduced stability margins, with respect to the feedback control law of Eq. (5.63). This is evident through

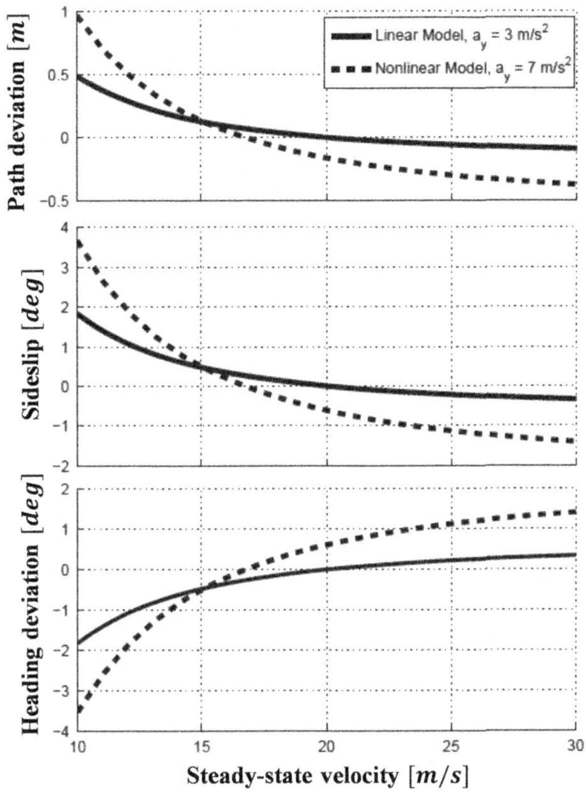

Fig. 5.18 Steady-state path tracking error Δy_{CG}, sideslip angle β, and heading deviation $\Delta\psi_{\mathrm{CG}}$ as a function of vehicle speed (from [70])

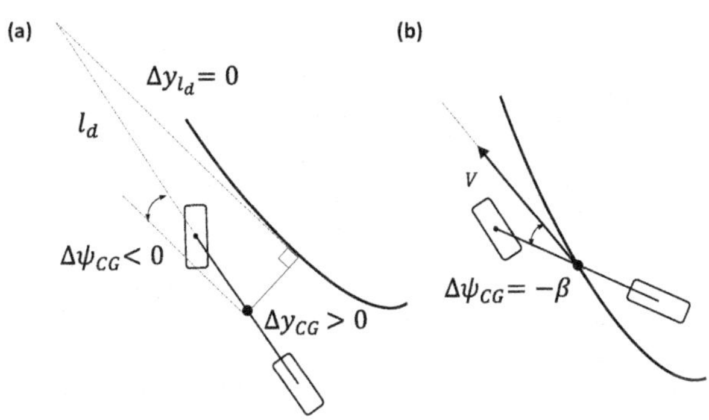

Fig. 5.19 (**a**) Steady-state cornering when the vehicle has lateral position error at the center of gravity but no look-ahead error; (**b**) Steady-state cornering with zero lateral deviation at the center of gravity, which requires the velocity vector to be tangent to the path, i.e., $\Delta\psi_{CG} = -\beta$ (from [70])

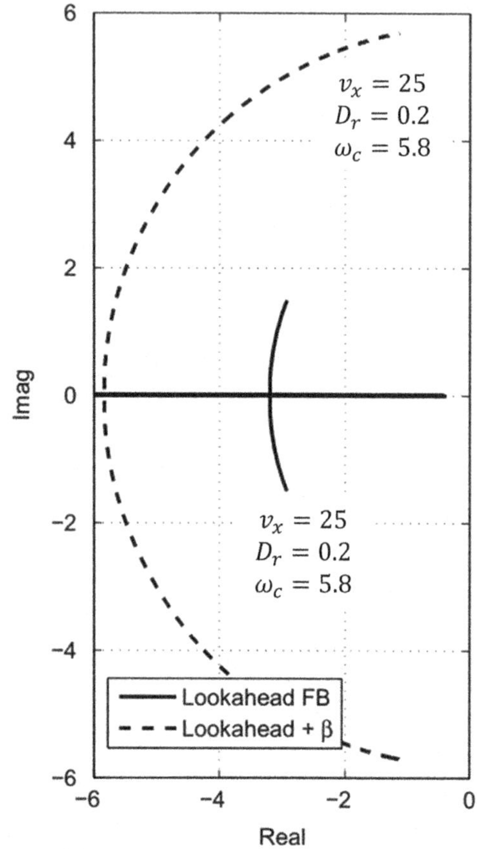

Fig. 5.20 Root-locus analysis for the controllers corresponding to Eqs. (5.62) and (5.63) ("Lookahead FB" in the figure) and Eqs. (5.62) and (5.64) ("Lookahead + β" in the figure), for vehicle speeds varying from 5 m/s to 25 m/s (from [70])

the root-locus analysis of Fig. 5.20, comparing the systems governed by the same feedforward control law (Eq. 5.62), and the feedback laws corresponding to Eqs. (5.63) and (5.64). For example, at 25 m/s the closed-loop steering response with Eq. (5.63) is well damped, with a damping ratio of 0.9, while in the same conditions the damping ratio is only 0.2 with the feedback control of Eq. (5.64).

In order to prevent the stability issues of path tracking enforced through feedback, Kapanja and Gerdes [70] finally propose to eliminate the sideslip-related feedback contribution from Eq. (5.64). Nevertheless, the zero displacement error condition (corresponding to $\Delta\psi_{CG} = -\beta$) is incorporated into the feedback contribution by using β_{SS} (i.e., the expected steady-state value of sideslip angle) instead of the estimated β. Hence, Eq. (5.64) becomes:

$$\begin{aligned}\delta_{FB,LC,3} &= -k_P\left(\Delta y_{CG} + l_d\left(\Delta\psi_{CG} + \beta_{SS}\right)\right)\\ &= -k_P\left(\Delta y_{CG} + l_d\left(\Delta\psi_{CG} + \alpha_r^{FFW} + b\kappa\right)\right)\end{aligned} \quad (5.65)$$

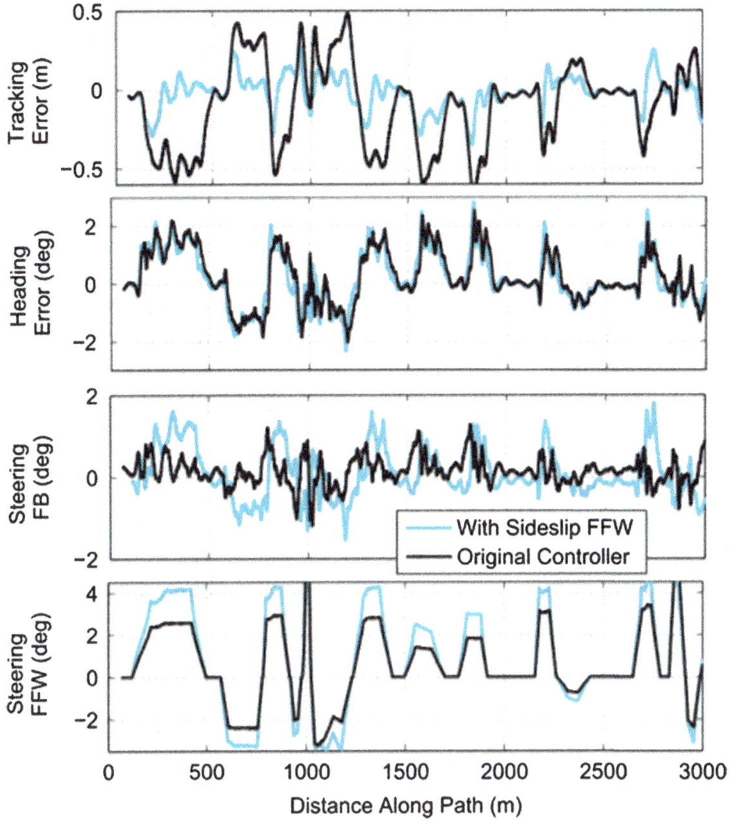

Fig. 5.21 Performance comparison between: (i) the controller corresponding to Eqs. (5.62) and (5.65), "With Sideslip FFW" in the legend; and (ii) the controller corresponding to Eqs. (5.62) and (5.63), "Original Controller" in the legend (from [70])

Actually, the sideslip contribution of Eq. (5.65), i.e., $-k_{\mathrm{P}} l_{\mathrm{d}} \left(\alpha_{\mathrm{r}}^{\mathrm{FFW}} + b\kappa\right)$, is now a feedforward term.

Vehicle experiments with an autonomous Audi TTS were executed at the Thunderhill Raceway Park, in order to compare the performance of (i) the controller corresponding to Eqs. (5.62) and (5.65) and (ii) the controller corresponding to Eqs. (5.62) and (5.63). The experimental results, reported in Figs. 5.21 and 5.22, show that controller (i) implies significantly improved path tracking. In any case, the whole approach needs further developments, as the feedforward contribution is sensitive to system uncertainty (e.g., tire-road friction conditions), which is much more important in the case of a vehicle operating on a real road rather than on a race track.

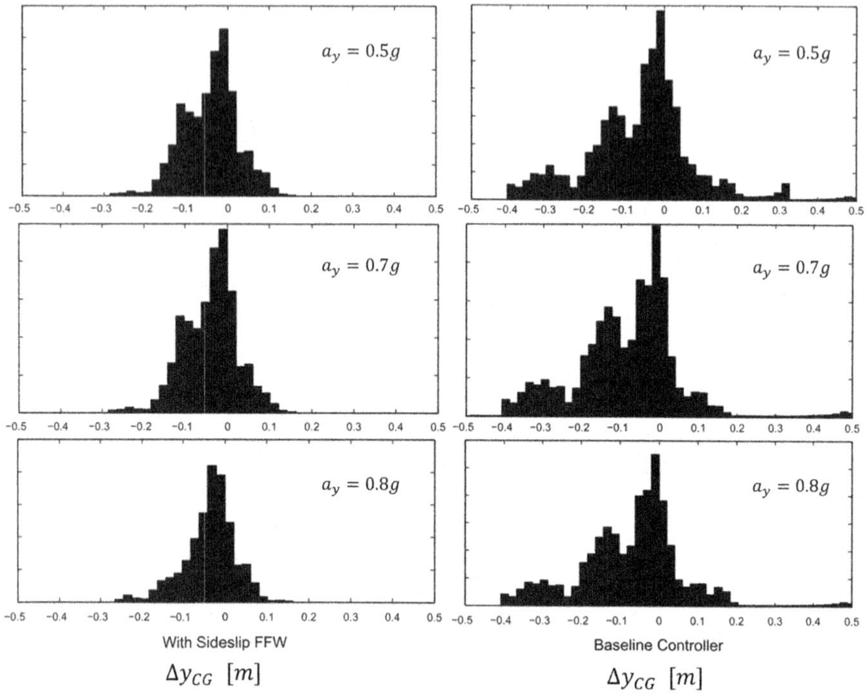

Fig. 5.22 Performance comparison, in terms of probability distribution of Δy_{CG}, between: (i) the controller corresponding to Eqs. (5.62) and (5.65), "With Sideslip FFW" in the legend; and (ii) the controller corresponding to Eqs. (5.62) and (5.63), "Baseline Controller" in the legend (from [70])

5.5.2 *Model Predictive Control*

With respect to the path tracking formulations discussed in the previous sections, model predictive control [72–74] brings the following benefits:

- Inclusion of constraints on inputs and states;
- Systematic approach to the control problem, with the possibility of considering multiple actuators and models at different levels of complexity within the same control design framework;
- Enhanced tracking performance at medium-high lateral accelerations and during emergency conditions, depending on the complexity level of the selected model for control system design.

Extensive literature provides simulation and experimental results of model predictive control applications for path tracking. For example, one of the first attempts is presented in [75], with a path tracking model predictive controller based on a single-track vehicle model. This adopts a nonlinear tire model with constant cornering stiffness and lateral force saturation at the value corresponding to the estimated tire-road friction coefficient.

Falcone et al. [76] discuss and compare three model predictive control formulations. The first one (here called Controller A) is based on a nonlinear single-track vehicle model. This considers constant vertical load on the front and rear axles and uses Pacejka magic formula [77], under the hypothesis of zero longitudinal slip ratio (i.e., pure cornering conditions). The model is expressed in the form:

$$\xi_{k+1} = f_{s(k),\mu(k)}^{dt}\left(\xi_k, \Delta u_k\right) \tag{5.66a}$$

$$u_k = u_{k-1} + \Delta u_k \tag{5.66b}$$

The system output is $z_{\mathrm{MPC}_k} = [\psi_k \; y_k]^{\mathrm{T}}$. The cost function to be minimized is:

$$J\left(\xi_t, \Delta U_t\right) = \sum_{n=1}^{H_{\mathrm{P}}} \left\| z_{\mathrm{MPC}_{k+n}} - z_{\mathrm{MPCref},k+n} \right\|_Q^2 + \sum_{n=0}^{H_{\mathrm{c}}-1} \left\| \Delta u_{k+n} \right\|_R^2 \tag{5.67}$$

The first contribution, $\sum_{n=1}^{H_{\mathrm{P}}} \left\| z_{\mathrm{MPC}_{k+n}} - z_{\mathrm{MPCref},k+n} \right\|_Q^2$, relates to the tracking performance of the system (z_{MPCref} is the vector of the reference signals), while the second contribution, $\sum_{n=0}^{H_{\mathrm{c}}-1} \left\| \Delta u_{k+n} \right\|_R^2$, considers the control effort. Similarly to the case of the linear quadratic regulator, the parameters of Q and R can be tuned to define the performance of the model predictive controller, i.e., the variables that need to be tracked with higher precision, and the relative weight between tracking performance and control effort. At each time step, the following finite horizon optimal control problem is solved:

$$\underset{\Delta U}{\arg\min} \; J\left(\xi_t, \Delta U_t\right), \tag{5.68a}$$

s.t.

$$\xi_{k+1,t} = f_{s_{k,t},\mu_{\mathrm{c}k,t}}^{dt}\left(\xi_{k,t}, \Delta u_{k,t}\right) \tag{5.68b}$$

$$z_{MPC_{k,t}} = h\left(\xi_{k,t}\right) \tag{5.68c}$$

$$\mu_{\mathrm{c}k,t} = \mu_{\mathrm{c}t,t}, \;\; s_{k,t} = s_{t,t}, \;\; k = t, \ldots, t + H_{\mathrm{P}} \tag{5.68d}$$

$$\delta_{\min} \le u_{k,t} \le \delta_{\max} \tag{5.68e}$$

$$\Delta\delta_{\min} \le \Delta u_{k,t} \le \Delta\delta_{\max} \tag{5.68f}$$

$$u_{k,t} = u_{k-1,t} + \Delta u_{k,t}, k = t, \ldots, t + H_C - 1 \tag{5.68g}$$

$$\Delta u_{k,t} = 0, k = t + H_C, \ldots, t + H_P \tag{5.68h}$$

$$\xi_{t,t} = \xi(t) \tag{5.68i}$$

The optimization vector at time t is $\Delta U_t = [\Delta u_{t,t} \ldots \Delta u_{t+H_C-1,t}]^T$; H_P and H_C denote the prediction and control horizons, respectively. The solution of problem (5.68) implies a nonlinear optimization, with a very significant computational burden. The optimization of Controller A is solved through the commercial NPSOL software package [78].

Falcone et al. [76] present only experiments at low vehicle speed with the controller based on the nonlinear model in the form of Eq. (5.66a) and the optimization problem (5.68), i.e., Controller A. In fact, as speed increases, larger prediction and control horizons are required "in order to stabilize the vehicle along the path." This implies more evaluations of the objective function and increased size of the optimization problem, which becomes unpractical. As a consequence, Falcone et al. [76] also discuss an alternative formulation, Controller B, based on the linearization of the system at each time step, around the current operating point. This procedure significantly decreases the computational complexity of the optimization problem, even if additional calculations are required for system linearization at each time step. In the case of Controller B, the model output vector is $z_{MPC_k} = [\psi_k \ \dot{\psi}_k \ y_k]^T$. In Controller B, tire slip angle variation is an additional output that is constrained (through a soft constraint and a slack variable) but not tracked.

Controllers A and B were assessed in double lane change tests through simulations and experiments. Controller parameters have a significant effect on the control system performance; therefore, the main parameter values used in [76] are reported for completeness:

- Controller A: $T_s = 0.05$ s, $H_P = 7$, $H_C = 3$, $\delta_{min} = -10$ deg, $\delta_{max} = 10$ deg, $\Delta\delta_{min} = -1.5$ deg, $\Delta\delta_{max} = 1.5$ deg, $\mu_c = 0.3$, $Q = \begin{bmatrix} 500 & 0 \\ 0 & 75 \end{bmatrix}$, $R = 150$;
- Controller B: $T_s = 0.05$ s, $H_P = 25$, $H_C = 10$, $\delta_{min} = -10$ deg, $\delta_{max} = 10$ deg, $\Delta\delta_{min} = -0.85$ deg, $\Delta\delta_{max} = 0.85$ deg, $\mu_c = 0.3$, $Q = \begin{bmatrix} 200 & 0 & 0 \\ 0 & 10 & 0 \\ 0 & 0 & 10 \end{bmatrix}$, $R = 5 * 10^4$.

Finally, Falcone et al. [76] include a simplified version of Controller B, here called Controller C, with $H_C = 1$, allowing a further reduction of the computational load for implementation on actual automotive control hardware (in this case, the set of required calculations at each time step can be predicted a priori).

Table 5.2 Maximum computation times for Controller A and Controller B during lane change tests at different vehicle speeds (from [76])

v[m/s]	Controller A [s]	Controller B [s]
10	0.15 ($H_P = 7, H_C = 2$)	0.03 ($H_P = 7, H_C = 3$)
15	0.35 ($H_P = 10, H_C = 4$)	0.03 ($H_P = 10, H_C = 4$)
17	1.3 ($H_P = 10, H_C = 7$)	0.03 ($H_P = 10, H_C = 10$)

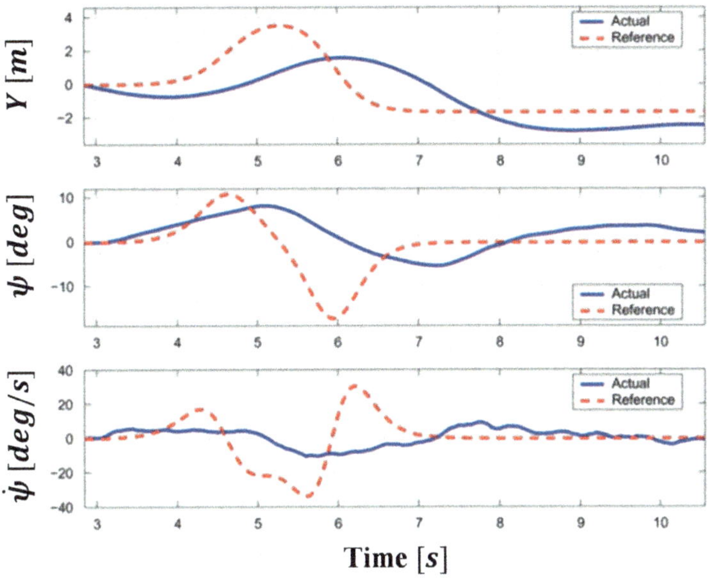

Fig. 5.23 Experimental results with Controller B at 21.5 m/s entry speed. From the top: lateral position, yaw angle, and yaw rate (from [76])

Table 5.2 compares the maximum computation times of Controller A and Controller B during a lane change maneuver in low friction conditions. Clearly, with the control hardware available in [76], it is not possible to run Controller A for vehicle speeds larger than 10 m/s. Figures 5.23 and 5.24 show vehicle performance with Controller B during the double lane change test executed at 21.5 m/s. The control action is characterized by significant chattering. The authors of [76] state that this aspect is not critical, as the vehicle cornering dynamics act as a filter, and, therefore, the vehicle passengers do not perceive the oscillations.

Tables 5.3 and 5.4 report the main tracking performance indicators during the tests, for different vehicle speeds, respectively, for Controller B and Controller C. According to [76], "Controller C performs slightly worse than Controller B;" nevertheless, "it is able to stabilize the vehicle at high speed."

In [79], the same research group presents a model predictive controller (here called Controller D) based on a four-wheel vehicle model including wheel and tire dynamics. As a consequence, the state vector of the model for control system design

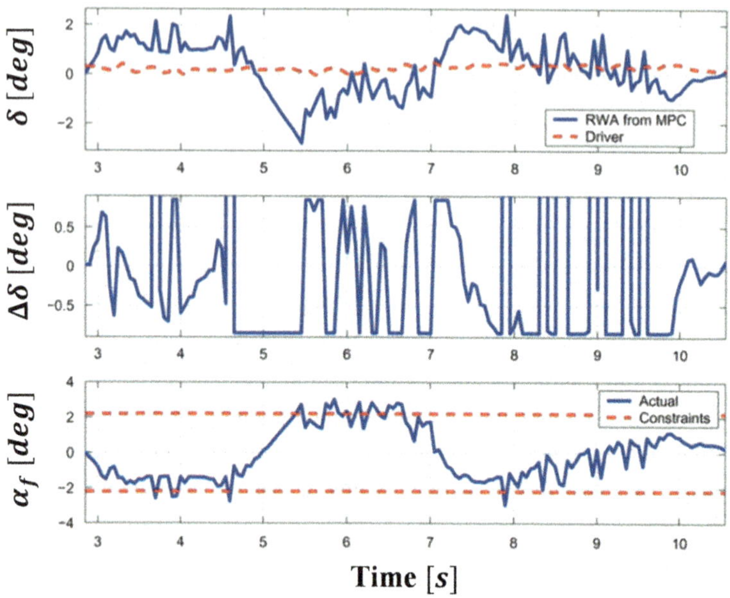

Fig. 5.24 Experimental results with Controller B at 21.5 m/s entry speed. From the top: front steering angle, change in front steering angle, and front tire slip angle (from [76])

Table 5.3 Performance indicators (i.e., tracking errors) for Controller B for double lane change tests at different initial speeds (from [76])

v [m/s]	μ_c	$\Delta\psi_{rms}$ [deg]	ΔY_{rms} [m]	$\Delta\psi_{max}$ [deg]	ΔY_{max} [m]
10	0.3	0.53	1.28×10^{-2}	8.21	0.8
15	0.3	1.172	4.64×10^{-2}	14.71	2.51
19	0.3	1.23	7.51×10^{-2}	16.38	3.10
21.5	0.25	1.81	1.11×10^{-1}	19.02	2.97

Table 5.4 Performance indicators (i.e., tracking errors for Controller C for double lane change tests at different initial speeds (from [76])

v [m/s]	μ_c	$\Delta\psi_{rms}$ [deg]	ΔY_{rms} [m]	$\Delta\psi_{max}$ [deg]	ΔY_{max} [m]
10	0.2	9.52×10^{-1}	5.77×10^{-2}	13.12	3.28
17	0.25	8.28×10^{-1}	2.90×10^{-2}	12.26	1.81
21	0.2	1.037	7.66×10^{-2}	12.49	3.20

is $\xi = \left[\dot{y}\ v_x\ \psi\ \dot{\psi}\ Y\ X\ \omega_{lf}\ \omega_{rf}\ \omega_{lr}\ \omega_{rr}\right]^T$. The tires are modeled through the magic formula, this time including consideration of the interaction between longitudinal and lateral forces. However, the vehicle model considers a constant vertical load on each tire, which is a substantial limitation, as the load transfers induced by longitudinal and lateral accelerations are an important cause of nonlinearity and variation of the understeer characteristic.

The main benefit is that Controller D allows systematic and concurrent control of the steering angle and the individual friction brake torques (and potentially the drivetrain torque as well). This feature is an important point for actual vehicles including stability control systems based on independent caliper pressure control. Therefore, the control output vector for Controller D is $u = [\delta\ T_{\mathrm{b,lf}}\ T_{\mathrm{b,rf}}\ T_{\mathrm{b,lr}}\ T_{\mathrm{b,rr}}]^{\mathrm{T}}$. As the controller was tested for a lane change maneuver only, the traction torque is not considered in u within [79].

Controller D provides better performance in extreme conditions than a controller (here called Controller E) based on a nonlinear single-track vehicle model, in which the control output is represented by $u = [\delta\ M_z]^{\mathrm{T}}$ [79]. In Controller E, heuristics are adopted (within a low-level controller) to calculate the individual friction brake torques required to generate the reference yaw moment, M_z, output by the model predictive controller. The simulation results show that lane change tests can be executed at a higher initial vehicle speed with Controller D than with Controller E. However, the authors admit that the duration of the simulation runs with Controller D was about 15 min, and, therefore, they did not have the time to fine tune the parameters of Controller D. This justifies the simulation results for Controller D (e.g., see Fig. 5.25), showing significant vibrations of the control action, which requires further investigations in the opinion of the authors of this chapter. Falcone et al. [79] also include experimental results with Controller F, which is based on the linearized model used for Controller D (i.e., the model with four vehicle corners), where the linearizations are carried out online, around the current operating point of the vehicle. The resulting controller can, thus, run online with a fixed step size of 50 ms on the control hardware available in [79].

Yin et al. [80] propose a model predictive controller, Controller G, for an autonomous electric vehicle with individually controlled drivetrains, with a formulation very similar to that of Controller F of [79]. The controller is based on a linearized vehicle model including the four vehicle corners, where the reference steering angle and the four slip ratios are the control outputs. This means that a low-level controller is used to calculate the individual wheel torques required to achieve the reference longitudinal slips. In practice, this is very difficult to implement

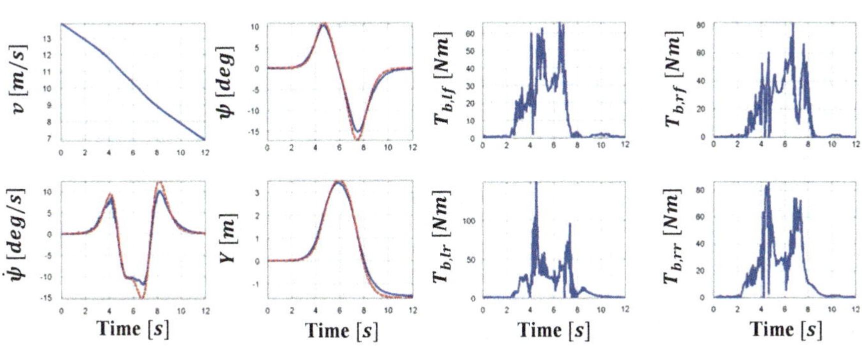

Fig. 5.25 Performance of Controller D at an entry speed of 14 m/s (from [79])

because of the approximations in the slip ratio estimation during normal driving. Slip ratio estimation is much easier in extreme conditions, i.e., when the absolute values of the slip ratios are larger and the conventional traction control and antilock braking systems are usually activated. The benefit of Controller G is that it allows the control of the traction torque during autonomous driving, without the requirement of two separate controllers for steering and longitudinal tracking.

Similarly to [79], Attia et al. [81] present a nonlinear model predictive controller, Controller H, based on a model coupling the longitudinal and lateral vehicle dynamics. The model for control synthesis is a discretized single-track vehicle model, including the degrees of freedom corresponding to the longitudinal and lateral displacements of the center of gravity, the vehicle yaw motion, and the equivalent front and rear wheel dynamics. Therefore, the state vector is $\xi = \left[v_x \; v_y \; \psi \; \dot{\psi} \; \omega_\mathrm{f} \; \omega_\mathrm{r} \; X \; Y\right]^\mathrm{T}$. The main model nonlinearity is represented by the Burckhardt tire model. The model predictive controller is responsible for the steering angle demand only. The longitudinal vehicle dynamics are exclusively used for considering the interaction between longitudinal and lateral tire forces; in fact, the wheel torque demand is controlled by an independent controller based on Lyapunov approach. The paper does not report the details of the numerical aspects related to the algorithm implementation and online optimization, apart that the considered sample time is $T_\mathrm{S} = 10$ ms.

Controller H includes some consideration of the interactions between longitudinal and lateral control. In fact, an excessive level of vehicle speed reference can originate problems in terms of lateral dynamics, as "no active lateral stabilization is considered in the control design." In this respect, the reference vehicle speed of the longitudinal controller can be saturated based on the expected road curvature and the estimated tire-road friction coefficient, according to the following formulations proposed in [82, 83]:

$$v_\mathrm{max} = \sqrt{\frac{g\mu_\mathrm{c}}{\kappa}} \tag{5.69}$$

$$v_\mathrm{max} = \sqrt{\frac{g}{\kappa}\left(\frac{\phi_\mathrm{r} + \mu_\mathrm{c}}{1 - \phi_\mathrm{r}\mu_\mathrm{c}}\right)} \tag{5.70}$$

Also, according to the US National Highway Traffic Safety Administration (NHTSA) [83], the longitudinal acceleration to bring vehicle speed to the maximum value specified in Eq. (5.70) should be limited to:

$$a_{x,\mathrm{max}} = \frac{v^2 - v_\mathrm{max}^2}{2\left(d - t_\mathrm{r} v\right)} \tag{5.71}$$

Criteria (5.69)–(5.71) are easily applicable for reference speed generation; however, safety-critical conditions could happen, for example, caused by an erroneous friction coefficient estimation, thus determining higher reference speed profile than the one compatible with the actual friction limits. In these situations, the stability control system of the vehicle is expected to intervene and overrule the inputs of the autonomous driving controller, as it happens in normal humanly driven passenger cars. Nevertheless, the authors of [81] report a couple of sideslip-related stability criteria (from [84, 85]), mentioned as relevant to automated driving, without clearly specifying how to organically include them within their automated steering controller:

$$\left| \frac{1}{24}\dot{\beta} + \frac{4}{24}\beta \right| \leq 1 \tag{5.72}$$

$$\beta \leq 10\,\text{deg} - 7\,\text{deg}\,\frac{v^2}{(40\ \text{m/s})^2} \tag{5.73}$$

The performance of Controller H was assessed through simulations with a nonlinear vehicle dynamics model developed by the same authors, along a highway exit scenario. The results (Fig. 5.26) show a lateral position error not exceeding 6 cm and a heading angle error not exceeding 4 deg.

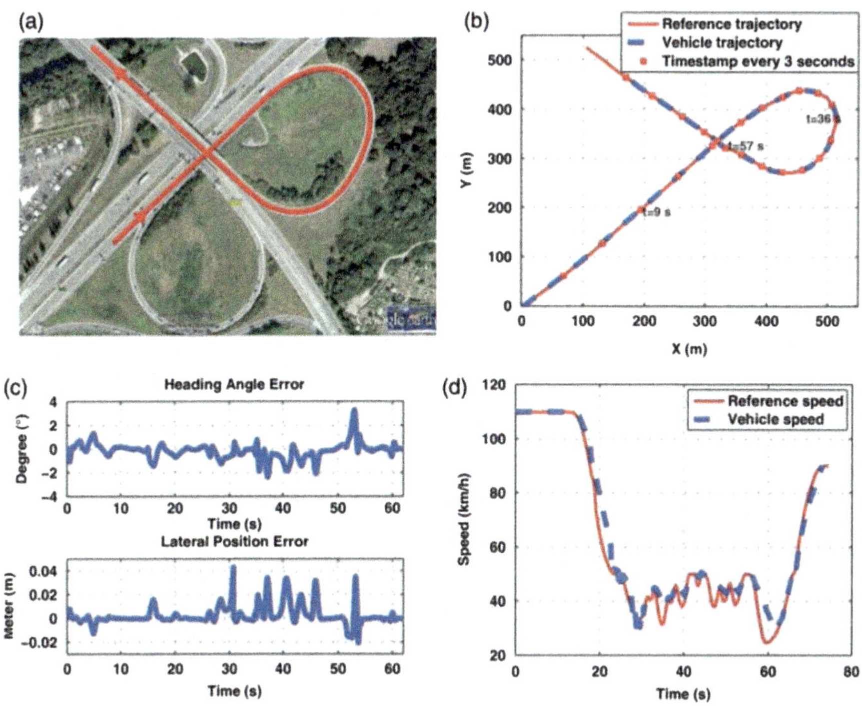

Fig. 5.26 Combined longitudinal and lateral control test with Controller H: (**a**) real-world road; (**b**) reference and vehicle trajectories; (**c**) tracking errors; and (**d**) reference speed tracking (simulation results from [81])

The previous contributions discuss model predictive controllers without any specific feature aimed at providing system robustness. Nevertheless, the presented model predictive controllers guarantee enhanced tracking performance in nominal conditions, i.e., when the model used by the controller provides a good fit with the actual plant. To the purpose of conjugating excellent tracking performance in nominal conditions and robustness, recent contributions are focused on robust model predictive control [65]. For example, Gao et al. [86] discuss a robust tube-based model predictive controller (Controller I, for the theory refer to [87–89]), conceived with the specific purpose of relatively low computational load.

The controller is based on a system model formulation of the form:

$$\xi_{k+1} = A\xi_k + g(\xi_k) + Bu_k + w_k \tag{5.74}$$

with $\xi_k \in \Xi$, $u_k \in \mathcal{U}$, $w_k \in \mathcal{W}$. Equation (5.74) is derived starting from a single-track vehicle model, including the longitudinal force balance equation of the system. The path tracking controller is based on the lateral displacement error at the front center of percussion to eliminate the rear lateral tire forces from the lateral displacement error equation. The state vector is $\xi = \left[\dot{y}_{\mathrm{COP,f}}\ \dot{x}_{\mathrm{COP,f}}\ \dot{\psi}\ \Delta\psi\ \Delta y_{\mathrm{CG}}\ s\right]^{\mathrm{T}}$, and the input vector (i.e., the output of the controller) is $u = [\beta_{x,\mathrm{f}}\ \beta_{y,\mathrm{f}}\ \beta_{\mathrm{r}}]^{\mathrm{T}}$. $\beta_{x,\mathrm{f}}$ and $\beta_{y,\mathrm{f}}$ are the normalized longitudinal and lateral forces on the front wheels, which are controlled through the drivetrain/friction brakes and the steering system, respectively. β_{r} is the normalized longitudinal force on the rear axle. A linear model is used for the lateral force of the rear axle. Essentially, the model in Eq. (5.74) includes a linear term, $A\xi_k + Bu_k$, a small (under the hypotheses of the discussion in [86]) nonlinear term, $g(\xi_k)$, and a disturbance term, w_k.

The control action consists of two contributions: (i) a nominal control input for the nominal system, i.e., the system in Eq. (5.74) under the hypothesis of zero disturbance and (ii) a state feedback controller acting on the error, $e_k = \xi_k - \overline{\xi}_k$, between the actual state of the system and the predicted state of the nominal system. The nominal system is defined as the system with the nominal control input and zero disturbance sequence. As a consequence, the control law has the following shape:

$$u_k = \overline{u}_k + \widehat{u}(e_k) \tag{5.75}$$

where $\overline{u}_k$ is the nominal controller and $\widehat{u}(e_k)$ is the state feedback control action. In the specific case, the stabilizing feedback contribution is based on a linear quadratic regulator: $\widehat{u}(e_k) = K_{\mathrm{LQ}}\left(\xi_k - \overline{\xi}_k\right)$. In practice, in [86], the linear part of the dynamics is separated into two contributions, the first one including the longitudinal dynamics and the second one including the lateral dynamics.

In general, the error dynamics are given by

$$e_{k+1} = Ae_k + B\widehat{u}(e_k) + \left(g(\xi_k) - g\left(\overline{\xi}_k\right)\right) + w_k \tag{5.76}$$

Yu et al. [87] demonstrated that if $\mathcal{Z}$ is a robust positively invariant set of the error system in Eq. (5.76) with control law $\widehat{u}$ and if $\xi_k \in \left\{\overline{\xi}_k\right\} \bigoplus \mathcal{Z}$, then $\xi_{k+i} \in \left\{\overline{\xi}_{k+i}\right\} \bigoplus \mathcal{Z}$ for all $i > 0$ and all admissible disturbance sequences $w_{k+i} \in \mathcal{W}$ (see Appendix for the definitions of invariant set and Minkowski sum, $\bigoplus$). This means that if the system states start close to the nominal state, then the control law in Eq. (5.75) will keep the system trajectory within the robust positively invariant set $\mathcal{Z}$ centered at the predicted nominal states. This statement also suggests that "if a feasible solution can be found for the nominal system subject to the tightened constraints $\overline{\Xi} = \Xi \ominus \mathcal{Z}$ and $\overline{\mathcal{U}} = \mathcal{U} \ominus \widehat{u}(\mathcal{Z})$, then the control law" in Eq. (5.75) "will ensure constraint satisfaction for the controlled uncertain system" (see Appendix for the definition of Pontryagin difference, $\ominus$).

In general, the controller and invariant set pair are very difficult to calculate, unless the nonlinear term in Eq. (5.76) is small, i.e., $\|g(\xi_1) - g(\xi_2)\|_2 \leq L_{\text{Lipschitz}}\|\xi_1 - \xi_2\|_2, \forall \xi_1, \xi_2 \in \Xi$, where $L_{\text{Lipschitz}}$ is the Lipschitz constant of the nonlinear term. Under this hypothesis, the system in Eq. (5.76) can be rewritten as

$$e_{k+1} = Ae_k + B\widehat{u}(e_k) + \widetilde{w}_k \tag{5.77}$$

with $\widetilde{w}_k \in \widetilde{\mathcal{W}} = \mathcal{W} \bigoplus \mathcal{B}(\mathcal{E})$, where

$$\mathcal{B}(\mathcal{E}) = \left\{x \in \mathbb{R}^n \,|\, \|x\|_\infty \leq L_{\text{Lipschitz}}(\Xi) \max_{e \in \mathcal{E}} \|e\|_2\right\} \tag{5.78}$$

For this case, [86] provides an algorithm for the computation of the minimal positively invariant set $\mathcal{Z}$ (see Appendix for the definition) associated with the linear quadratic regulator gain K_{LQ} applied to the system defined by A and B. In the actual calculations of [86], as the system was split into its longitudinal and lateral dynamics contributions, the respective invariant sets were calculated separately.

From a practical viewpoint, the robust model predictive control system design procedure reduces to the following relatively simple steps:

(i) Formulation of the model equations according to the structure of Eq. (5.76).
(ii) Off-line computation of the linear quadratic regulator gain matrix K_{LQ} (for an infinite horizon) based on the linear part of the system model.
(iii) Off-line estimation of the bounded disturbance w. This can be carried out by comparing the one-step prediction of the model in Eq. (5.76), discretized at 50 ms in [86], with the measured vehicle states. For example, in [86] a simple bound (i.e., [0.2 0.2 0.2 0.005 0.05 0.05]) is used "to conservatively estimate the disturbance bound" on the system state prediction.
(iv) Off-line computation of $\mathcal{Z}$ through the algorithm in [86], starting from $\mathcal{W}$.
(v) Online computation of the nominal control action through conventional nonlinear model predictive control methods (e.g., see Eq. (5.68)), e.g., by solving the optimization with the NPSOL tool.
(vi) Online computation of $u_k = \overline{u}_k + \widehat{u}(e_k) = \overline{u}_k + K_{\text{LQ}}e_k$.

The robust formulation in [86] is simply a linear quadratic regulator coupled with an implicit model predictive controller, with an additional algorithm to calculate the invariant set. The calculation of $\mathcal{Z}$ is beneficial to know and consider the expected boundaries of the states, i.e., the projection of the bounds of the robust invariant set, Proj ($\mathcal{Z}$), while the controller is running (hence the concept of tube-based model predictive control).

Gao et al. [86] report simulation results of obstacle avoidance maneuvers, including introduction of random bounded disturbances with uniform distribution into the simulation model, and comparison of the performance of the robust controller, $u_k = \overline{u}_k + \widehat{u}(e_k)$, with the performance of the nominal controller, $\overline{u}_k$ (the difference is evident in Figs. 5.27 and 5.28). Also, Gao et al. [86] include experimental results, such as multiple obstacle avoidance tests carried out on a surface with a tire-road friction coefficient μ_c=0.1, while the controller is set for μ_c=0.3 (Figs. 5.29 and 5.30).

In addition to the tube-based model predictive controller, Carvalho et al. [65] also suggest time-varying stochastic model predictive control for dealing with system

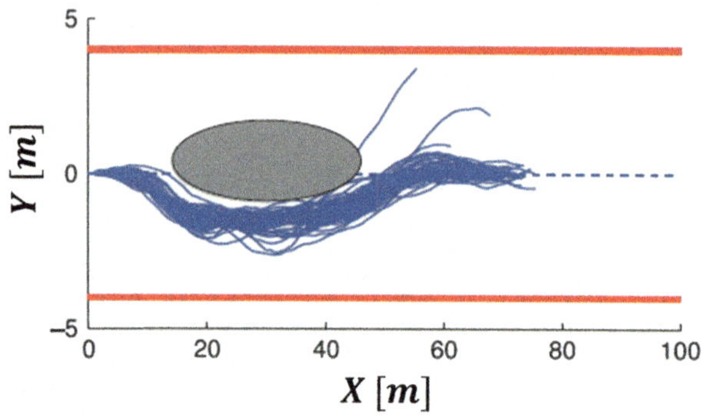

Fig. 5.27 Simulation results: trajectory of the nominal model predictive controller under a random external disturbance (from [86])

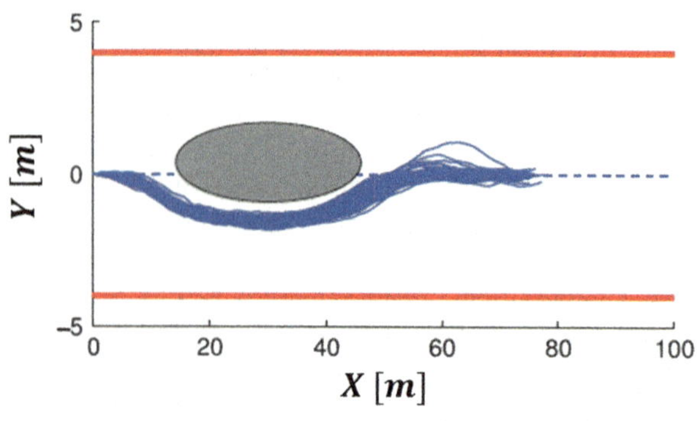

Fig. 5.28 Simulation results: trajectory of the robust model predictive controller under a random external disturbance (from [86])

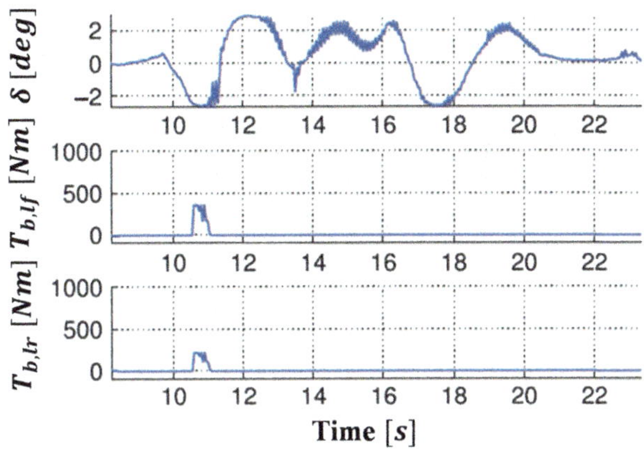

Fig. 5.29 Control commands during an experimental multiple obstacle avoidance test (from [86])

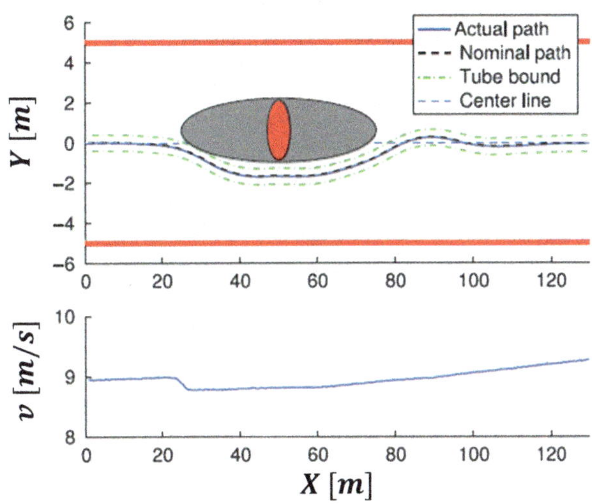

Fig. 5.30 Experimental vehicle trajectory on an ice track during the same test as for Fig. 5.29 (from [86])

uncertainty. The theory and an example of application to the automated driving problem are provided in [90, 91].

Another interesting example of comparison of control structures for path tracking based on model predictive control in uncertain conditions is included in [92], dealing with the problem of obstacle avoidance on a slippery road. The paper supposes that the reference generation layer (see the introduction of this chapter) outputs the reference trajectory, but does not correct it to avoid an obstacle located on the desired path. Therefore, the reference trajectory has to be modified by the control layer, i.e., it has to be replanned to take the obstacle into account. Figures 5.31 and 5.32 show

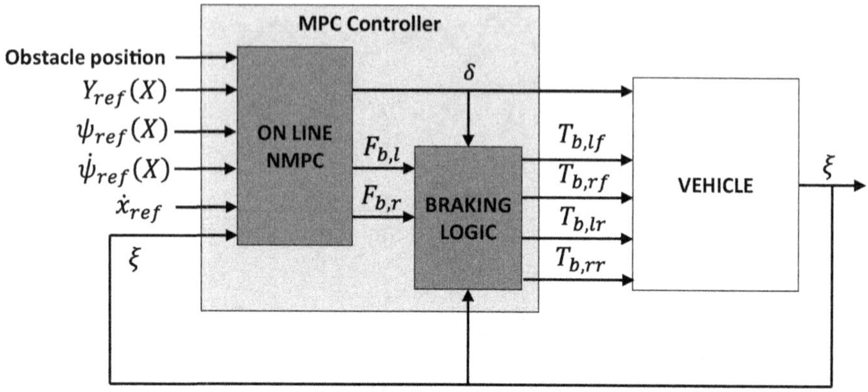

Fig. 5.31 Architecture of the single-level model predictive controller (Controller J) for path tracking with obstacle avoidance (adapted from [92])

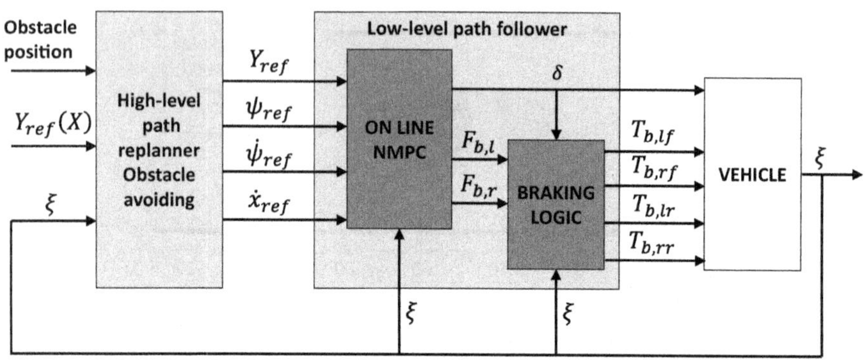

Fig. 5.32 Architecture of the two-level model predictive controller (Controller K) for path tracking with obstacle avoidance (adapted from [92])

the two control layer architectures that are contrasted in [92]:

(i) An architecture (Controller J) consisting of a single-level model predictive controller, based on a four-wheel vehicle model with a nonlinear tire model. The desired trajectory and the obstacle position are fed to the control algorithm. The controller directly calculates the reference steering angle and the reference longitudinal forces for the two sides of the vehicle, $F_{b,l}$ and $F_{b,r}$, in order to avoid the obstacle. These two forces are converted into individual braking torques by the "Braking logic" (see Fig. 5.31);

(ii) An architecture (Controller K) consisting of a two-level model predictive controller. The top-layer (the "High-level path replanner obstacle avoiding" in Fig. 5.32) is based on a simple point-mass model, with the purpose of replanning the reference trajectory starting from the detected position of the obstacle. The lower layer (the "Low-level path follower"), based on the same four-wheel vehicle model as the one used in controller architecture (i), receives the replanned reference trajectory and calculates the same outputs as the single-layer controller.

Controller J and the top layer of Controller K use cost functions with similar structure to the one in Eq. (5.67), with a term referring to the tracking performance and a term referring to the control effort, but they also include an additional term given by:

$$\sum_{k=t}^{t+H_p-1} J_{\mathrm{obs}_{k,t}} = \sum_{k=t}^{t+H_p-1} \frac{K_{\mathrm{obs}}\, v_{k,t}}{d_{\mathrm{min}_{k,t}} + \varepsilon} \tag{5.79}$$

$J_{\mathrm{obs}_{k,t}}$ in Eq. (5.79) is the cost at time k associated to the predicted distance between the vehicle and the obstacle, under the assumption that obstacle position is known for a collection of discretized points $P_{t,j}$. In particular, $d_{\mathrm{min}_{k,t}} = \min_j d_{k,t,j}$. $d_{k,t,j}$ is defined as:

$$d_{k,t,j} = \begin{cases} p_{x_{k,t,j}} - \bar{a} & \text{if } p_{y_{k,t,j}} \in [-\bar{c}, \bar{c}] \text{ and } p_{x_{k,t,j}} > \bar{a} \\ 0 & \text{if } p_{y_{k,t,j}} \in [-\bar{c}, \bar{c}] \text{ and } p_{x_{k,t,j}} \in [-\bar{b}, \bar{a}] \\ & M \text{ otherwise} \end{cases} \quad \forall j = 1, 2, \ldots, N \tag{5.80}$$

where $d_{k,t,j} = 0$ indicates the occurrence of a collision (see also Fig. 5.33).

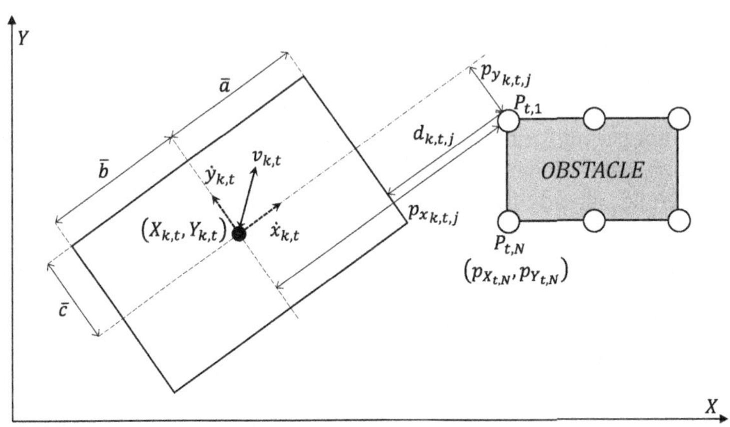

Fig. 5.33 Obstacle avoidance scenario and main geometrical parameters used in the model predictive control formulation (adapted from [92])

The simulation and experimental results demonstrate that the case study obstacle avoidance maneuver can be executed at higher values of vehicle speed with the two-level model predictive controller (Controller K), while the single-level architecture (Controller J) tends to cause stability problems. In general, when the vehicle deviates too far from the reference trajectory, the system becomes uncontrollable. In these conditions, "the vehicle state is outside the region of attraction of the equilibrium trajectory associated to the desired reference." In the single-layer architecture, this phenomenon happens quite often, as it is induced by tire saturation. This situation does not occur with the simple point-mass model of Controller K, "since the path replanner always replans a path starting from the current state of the vehicle, and therefore ensures that the low-level reference is close to current state. This explains why the performance of the two-level approach is better than the one-level approach."

Also, the computational performance of the two-level approach is much better than that of the single-level approach. Hence, the important conclusion is that for effective automated driving through model predictive control, the separation between the reference generation layer, which should replan the trajectory considering the presence of obstacles, and the control layer (see the introduction of this chapter) is not only convenient for control system simplification but also beneficial to control system performance.

Table 5.5 is included as a summary of the main characteristics of the model predictive controllers discussed in this section, with an overview of the adopted models, the possible control action (lateral control or combined longitudinal and lateral control), the involved complexity, and the form of validation presented in the literature, i.e., through simulations and/or experiments.

All the algorithms discussed in this section are based on implicit model predictive control formulations, i.e., the optimizations are run online, which implies a significant computational load for the vehicle control unit. Also, implicit model predictive control does not permit formal analysis of performance, suboptimality and stability [93]. An option that should be assessed in future research is explicit model predictive control, which has already been successfully implemented, including experiments, for concurrent yaw moment and active steering control in [94]. The validation of robust and explicit model predictive control formulations could represent the next step of path tracking control research.

Table 5.5 Comparison of the selected model predictive controllers for path tracking

Controller	Vehicle and tire model	Control action	Controller complexity	Validation
A [76]	Single-track vehicle model with Pacejka tire model (pure cornering) with constant vertical load	Steering control only	Nonlinear model and high computational load	Experiments for vehicle speeds lower than 10 m/s due to computational constraints
B [76]	Single-track vehicle model with Pacejka tire model with constant vertical load	Steering control only	Linearized model for computationally efficient optimization	Experimental lane change tests executed at various speeds
C [76]	Single-track vehicle model with Pacejka tire model with constant vertical load	Steering control only	Further simplified version of Controller B, with $H_C = 1$	Experimental lane change tests executed at various speeds; performance slightly worse than for Controller B
D [79]	Four-wheel vehicle model with Pacejka tire model (combined slip) with constant vertical load	Longitudinal control, steering control, and direct yaw moment control	Nonlinear optimization with high computational load	Simulations of lane change tests executed at 50 km/h and 70 km/h with good tracking performance
E [79]	Single-track vehicle model with Pacejka tire model (combined slip) with constant vertical load	Steering control and direct yaw moment control	Nonlinear (relatively) low complexity model for online optimization	Simulation of lane change tests executed at 50 km/h; controller not able to stabilize the vehicle at higher entry speeds
F [79]	Four-wheel vehicle model with Pacejka tire model (combined slip) with constant vertical load	Steering control and direct yaw moment control	Linearized model based on Controller D for computationally efficient online optimization	Experimental lane change tests at 50 km/h
G [80]	Four-wheel vehicle model and Pacejka tire model (combined slip) with constant vertical load	Longitudinal control, steering control, and direct yaw moment control	Linearized model for computationally efficient optimization	Simulations of a single lane change test

Table 5.5 (continued)

Controller	Vehicle and tire model	Control action	Controller complexity	Validation
H [81]	Single-track vehicle model with Burckhardt tire model with constant vertical load	Only steering control by the model predictive controller; independent Lyapunov approach for longitudinal control	Nonlinear model predictive controller; no details provided regarding the online optimization	Simulations of a highway exit scenario
I [86]	Single-track vehicle model with simplified nonlinear tire model with constant vertical load	Longitudinal control, steering control and direct yaw moment control	Nonlinear model with significant simplifications for computationally efficient online optimization providing system robustness	Experimental obstacle avoidance tests at high speed on a snow track
J [92]	Four-wheel vehicle model with Pacejka tire model (combined slip) with constant vertical load	Longitudinal control, steering control and direct yaw moment control	Single-level nonlinear model predictive controller with significant computational load	Experimental lane change tests in icy road conditions up to 40 km/h
K [92]	Simple point-mass model for the high level controller; the same as Controller J in terms of vehicle model for the low level controller	Longitudinal control, steering control, and direct yaw moment control	Two-level hierarchical nonlinear model predictive controller with significant computational load reduction with respect to Controller J	Experimental lane change tests in icy road conditions up to 55 km/h

5.6 Concluding Remarks

This chapter provided an overview of control structures for path tracking in autonomous vehicles, ranging from basic kinematic controllers to robust model predictive controllers. The presented analysis and results bring the following conclusions:

- Without disturbances (e.g., caused by side wind and the banking of the road), and uncertainties (e.g., caused by the vision systems), the path tracking performance of simple control structures is adequate and was already experimentally demonstrated in research programs in the 1990s [95]. For an assigned vehicle speed, fixed parameter controllers can deal with the range of axle cornering stiffnesses, vehicle masses, and tire-road friction coefficients typical of a real vehicle operation. Apparently, fixed parameter controllers can be effective even in the case of buses and heavy goods vehicles, characterized by significant mass variations during their operation.

- Look-down controllers show significant performance limitations at medium-high speeds. Effective feedback control design for path tracking must be based on the combination of lateral displacement and heading angle control, or lateral displacement evaluated at a look-ahead distance.
- Gain scheduling of the controller parameters, including the preview distance, is recommended as a function of vehicle speed.
- The separation among the reference trajectory level and the control layer is not only convenient for simplifying the automated driving control system design but also allows better system response from the viewpoint of vehicle stability, conjugated with reduced computational effort.
- In order to reduce the effect of disturbances and relax the tuning of the feedback part of the path tracking controller, properly designed feedforward contributions are essential. The main limitation of existing feedforward controllers for path tracking is their reliance on very advanced state estimators, which have to provide smooth outputs. Machine learning techniques, adaptive control and sensor fusion could significantly help with this challenging task.
- An interesting concept is represented by the center of percussion. A path tracking controller based on the position error at the front center of percussion eliminates the effect of the rear axle cornering forces on the lateral position error dynamics, which significantly facilitates the design of a feedback controller for a two-wheel-steering vehicle. A four-wheel-steering path tracking controller based on the lateral position errors at the front and rear centers of percussion simplifies into the design of two decoupled single-input single-output controllers.
- Some of the available experimental studies show that from the vehicle passengers' perspective, vehicle comfort (determined by the frequency and amplitude of the oscillations induced by the control action) is more important than the excellence of the tracking performance. Unluckily, only very limited data are available with respect to the comfort behavior of the most recent and performing path tracking controllers, including analysis of the subjective feedback of the occupants.
- Despite many path tracking formulations have been assessed through experimental tests, an objective comparative assessment of the performance of different control structures for the same vehicle and set of state estimators is missing in the literature. On the one hand, the authors presenting the most advanced control formulations tend to show their benefits without going through a comparison with fine-tuned simple controllers. On the other hand, the authors presenting relatively simple control formulations tend to highlight their performance even at medium-high lateral acceleration levels. Also, the subjective performance of the different path tracking formulations, i.e., in terms of oscillation of the control action and the subsequent vehicle response, should be carefully assessed through

experimental tests, in order to draw clear conclusions on the required level of control system sophistication.

- Current research developments are in the area of automated driving for limit conditions, i.e., at extreme lateral and longitudinal accelerations, and in the area of robust model predictive control, with the main aim of systematically dealing with system uncertainty.
- Future research activities should also systematically cover the interaction between automated steering and direct yaw moment control, which currently represents the main actuation technique for stabilizing the vehicle in extreme transient conditions. Direct yaw moment control can be sporadically actuated through the friction brakes within stability control systems [96] or, in the case of vehicles with torque-vectoring differentials or multiple electric drivetrains, can be continuously actuated during normal vehicle operation [97–100]. Especially in the latter case, further investigations of automated driving during extreme cornering are required, including more detailed and comprehensive experimental demonstrations.

Appendix: Definitions of Invariant Sets, Minkowski Sum ⊕, and Pontryagin Difference ⊖

The following definitions are provided:

(a) Reachable set for systems with external inputs. Consider a system $\xi_{k+1} = f(\xi_k, u_k) + w_k$, with $\xi_k \in \Xi$, $u_k \in \mathcal{U}$, $w_k \in \mathcal{W}$. The one-step robust reachable set from a given set of states $\mathcal{S}$ is $\mathrm{Reach_f}(\mathcal{S}, \mathcal{W}) \triangleq \{\xi \in \mathbb{R}^n | \xi_0 \in \mathcal{S}, \exists u \in \mathcal{U}, \exists w \in \mathcal{W} : \xi = f(\xi_0, u, w)\}$;

(b) Robust positively invariant set. A set $\mathcal{Z} \subseteq \Xi$ is said to be a robust positively invariant set for the autonomous system $\xi_{k+1} = f_a(\xi_k) + w_k$, with $\xi_k \in \Xi$ and $w_k \in \mathcal{W}$, if $\xi_0 \in \mathcal{Z} \Rightarrow \xi_k \in \mathcal{Z}$, $\forall w_k \in \mathcal{W}, \forall k \geq \in \mathbb{N}^+$

(c) Minimal robust positively invariant set. The set $\mathcal{Z}_\infty \subseteq \Xi$ is the minimal robust positively invariant set for the defined autonomous system, if $\mathcal{Z}_\infty$is a robust positively invariant set and $\mathcal{Z}_\infty$ is contained in every closed robust positively invariant set in Ξ (see [89] for the details)

(d) Minkowski sum. The Minkowski sum of two polytopes, $\wp$ and $\mathfrak{H}$, is the polytope $\wp \bigoplus \mathfrak{H} := \{x + h \in \mathbb{R}^n | x \in \wp, h \in \mathfrak{H}\}$;

(e) Pontryagin difference. The Pontryagin difference of two polytopes, $\wp$ and $\mathfrak{H}$, is the polytope $\wp \ominus \mathfrak{H} := \{x \in \mathbb{R}^n | x + h \in \wp, \forall h \in \mathfrak{H}\}$.

References

1. SAE standard J0316, *Taxonomy and Definitions for Terms Related to On-Road Motor Vehicle Automated Driving Systems* (2014)
2. T. J. Gordon, M. Lidberg, Automated driving and autonomous functions on road vehicles. Vehicle Syst. Dyn. **53**(7), 958–994 (2015)
3. C. C. MacAdam, Application of an optimal preview control for simulation of closed-loop automobile driving. IEEE Trans. Syst. Man Cybern. **SMC-11**(6), 393–399 (1981)
4. G. Genta, *Motor Vehicle Dynamics: Modeling and Simulation* (World Scientific, Singapore, 1997)
5. R. S. Sharp, D. Casanova, P. Symonds, A mathematical model for driver steering control, with design, tuning and performance results. Vehicle Syst. Dyn. **33**, 289–326 (2000)
6. R. S. Sharp, Driver steering control and a new perspective on car handling qualities. Proc. Inst. Mech. Eng. C J. Mech. Eng. Sci. **219**(10), 1041–1051 (2005)
7. R. S. Sharp, Rider control of a motorcycle near to its cornering limits. Vehicle Syst. Dyn. **50**(8), 1193–1208 (2012)
8. C. I. Chatzikomis, K. N. Spentzas, A path-following driver model with longitudinal and lateral control of vehicle's motion. Forsch. Ingenieurwes. **73**, 257–266 (2009)
9. G. Markkula, A review of near-collision driver behaviour models. Hum. Factors **54**(6), 1117–1143 (2012)
10. Y. Ohshima et al., Control system for automatic automobile driving. IFAC Tokyo Symposium on Systems Engineering for Control System Design, 1965
11. R. E. Fenton, R. J. Mayhan, Automated highway studies at the Ohio State University – An overview. IEEE Trans. Vehicle Technol. **40**(1), 100–113 (1991)
12. S. Thrun, M. Montemerlo, H. Dahlkamp, D. Stavens, et al., Stanley: The robot that won the DARPA grand challenge. J. Field Robot. **23**(9), 661–692 (2006)
13. C. Urmson, J. Anhalt, D. Bagnell, C. Baker, et al., Autonomous driving in urban environments: Boss and the urban challenge. J. Field Robot. **25**(8), 425–466 (2008)
14. C. Urmson, C. Ragusa, D. Ray, J. Anhalt, et al., A robust approach to high-speed navigation for unhehearsed desert terrain. J. Field Robot. **23**(8), 467–508 (2006)
15. F.S. Campbell, *Steering Control of an Autonomous Ground Vehicle with Application to the DARPA Urban Challenge*. Master's thesis, Massachusetts Institute of Technology, 2007
16. J.M. Snider, *Automatic steering methods for autonomous automobile path tracking*. CMU-RI-TR-09-08, Report, Carnegie Mellon University, 2009
17. A. De Luca, G. Oriolo, C. Samson, Feedback control of a nonholonomic car-like robot, in *Robot Motion Planning and Control* (Springer, London, 1998)
18. J.S. Wit, Vector pursuit path tracking for autonomous ground vehicles. PhD thesis, University of Florida, 2000
19. R. M. Murray, Control of nonholonomic systems using chained forms. Dynamics and control of mechanical systems, the falling cat and related problems. Fields Inst. Commun. **1**, 219–245 (1993)
20. C. Samson, Control of chained systems: Application to path following and time-varying point-stabilization of mobile robots. IEEE Trans. Autom. Control **40**(1), 64–77 (1995)
21. W. F. Milliken, D. L. Milliken, *Race Car Vehicle Dynamics* (SAE International, Warrendale, PA, 1995)
22. M. Guiggiani, *The Science of Vehicle Dynamics: Handling, Braking and Ride of Road and Race Cars* (Springer, Dordrecht, 2014)
23. T. D. Gillespie, *Fundamentals of Vehicle Dynamics* (Society of Automotive Engineers, Warrendale, PA, 1992)
24. A. Broggi, M. Bertozzi, A. Fascioli, G. Conte, *Automatic Vehicle Guidance: The Experience of the ARGO Autonomous Vehicle* (World Scientific, Singapore, 1999)
25. A. Broggi, M. Bertozzi, A. Fascioli, C. G. Lo Bianco, A. Piazzi, The ARGO autonomous vehicle's vision and control systems. Int. J. Intell. Control Syst. **3**(4), 409–441 (1999)

26. P. Furgale, U. Schwesinger, M. Rufli, W. Derendarz et al., Toward automated driving in cities using close-to-market sensors – An overview of the V-Charge project. IEEE Intelligent Vehicles Symposium, 2013
27. S. Tsugawa, An overview on control algorithms for automated highway systems. IEEE/IEE-J/JSAE Conference on Intelligent Transportation Systems, 1999
28. H. Mouri, H. Furusho, Automatic path tracking using linear quadratic control theory. IEEE Conference on Intelligent Transportation Systems, 1997
29. J. Guldner, H. S. Tan, S. Patwardhan, Analysis of automatic steering control for highway vehicles with look-down lateral reference systems. Vehicle Syst. Dyn. **26**(4), 243–269 (1996)
30. J. C. Hsu, M. Tomizuka, Analyses of vision-based lateral control for automated highway system. Vehicle Syst. Dyn. **30**(5), 345–373 (1998)
31. A. Tachibana et al., An automated highway vehicle system using computer vision – A vehicle control method using a lane line detection system. JSAE Autumn Convention **924**, 157–160 (1992)
32. H. Inoue, H. Mouri, H. Sato, A. Asaoka, S. Ueda, Technologies of Nissan's AHS test vehicle. 3rd ITS World Congress, 1996
33. R. Isermann, M. Schorn, U. Stahlin, Anticollision system PRORETA with automatic braking and steering. Vehicle Syst. Dyn. **46**(Supplement), 683–694 (2008)
34. R. Marino, S. Scalzi, M. Netto, Nested PID steering control for lane keeping in autonomous vehicles. Control Eng. Pract. **19**, 1459–1467 (2011)
35. J. Ackermann, J. Guldner, W. Sienel, R. Steinhauser, V. Utkin, Linear and nonlinear controller design for robust automatic steering. IEEE Trans. Control Syst. Technol. **3**(1), 132–143 (1995)
36. J. Ackermann, W. Sienel, Robust control for automatic steering. American Control Conference, 1990
37. J. Guldner, W. Sienel, J. Ackermann, S. Patwardhan, H.S. Tan, T. Bunte, Robust control design for automatic steering based on feedback of front and tail lateral displacement. European Control Conference, 1997
38. V. Cerone, A. Chinu, D. Regruto. Experimental results in vision-based lane keeping for highway vehicles. American Control Conference, 2002
39. M. F. Land, D. N. Lee, Where we look when we steer? Nature **369**, 742–744 (1994)
40. H. S. Tan, B. Bougler, W. B. Zhang, Automatic steering based on roadway markers: From highway driving to precision docking. Vehicle Syst. Dyn. **37**(5), 315–338 (2002)
41. K. Ogata, *Modern Control Engineering* (Prentice Hall, Englewood Cliffs, NJ, 2011)
42. H. Peng, M. Tomizuka, *Optimal preview control for vehicle lateral guidance*. Research Report UCB-ITS-PRR-91-16, California Partners for Advanced Transit and Highways (PATH), 1991
43. J. M. Hedrick, M. Tomizuka, P. Varaiya, Control issues in automated highway systems. IEEE Trans. Control Syst. **14**(6), 21–32 (1994)
44. H. Peng, M. Tomizuka, Preview control for vehicle lateral guidance in highway automation. American Control Conference, 1991
45. Q. Zhou, F. Wang, Robust sliding mode control of 4WS vehicles for automatic path tracking. IEEE Intelligent Vehicles Symposium, 2005
46. T. Hiraoka, O. Nishishara, H. Kumamoto, Automatic path-tracking controller of a four-wheel steering vehicle. Vehicle Syst. Dyn. **47**(10), 1205–1227 (2009)
47. G. Tagne, R. Talj, A. Charara, Design and validation of a robust immersion and invariance controller for the lateral dynamics of intelligent vehicles. Control Eng. Pract. **40**, 81–91 (2015)
48. G. Tagne, R. Talj, A. Charara, Higher-order sliding mode control for lateral dynamics of autonomous vehicles, with experimental validation. IEEE Intelligent Vehicles Symposium, 2013
49. P. Hingwe, M. Tomizuka, Experimental evaluation of a chatter free sliding mode control for lateral control in AHS. American Control Conference, 1997
50. H. Imine, T. Madani, Sliding-mode control for automated lane guidance of heavy vehicle. Int. J. Robust Nonlinear Control **23**(1), 67–76 (2011)

51. C. Hatipolglu, K. Redmill, U. Ozguner, Steering and lane change: A working system. IEEE Conference on Intelligent Transportation System, 1997. ITSC'97, IEEE, 1997
52. A. B. Will, S. Zak, Modelling and control of an automated vehicle. Vehicle Syst. Dyn. **27**, 131–155 (1997)
53. R. T. O'Brien, P. A. Iglesias, T. J. Urban, Vehicle lateral control for automated highway systems. IEEE Trans. Control Syst. Technol. **4**(3), 266–273 (1996)
54. D. McFarlane, K. Glover, *Robust Controller Design Using Normalized Coprime Factor Plant Desciptions* (Springer, Berlin, 1989)
55. S. Skogestad, I. Postlethwaite, *Multivariable Feedback Control: Analysis and Design*, 2nd edn. (Wiley, Chichester, 2005)
56. S. Hima, B. Lusseti, B. Vanholme, S. Glaser, S. Mammar, Trajectory tracking for highly automated passenger vehicles. 18th IFAC World Congress, 2011
57. D. H. Shin, B. Y. Joo, Design of a vision-based autonomous path-tracking control system and experimental validation. Proc. Inst. Mech. Eng. **224**, 849–864 (2010)
58. M. Kristic, I. Kanellakopoulos, P. Kotovic, *Nonlinear and Adaptive Control Design* (John Wiley, New York, 1995)
59. A. El Hajjaji, S. Bentalba, Fuzzy path tracking control for automatic steering of vehicles. Robot. Auton. Syst. **43**(4), 203–213 (2003)
60. L. Cai, A. B. Rad, W. L. Chan, A genetic fuzzy controller for vehicle automatic steering control. IEEE Trans. Vehicle Technol. **56**(2), 529–543 (2007)
61. T. Hessburg, M. Tomizuka, Fuzzy logic control for lateral vehicle guidance. IEEE Trans. Control Syst. **14**(4), 55–63 (1994)
62. J. E. Naranjo, C. Gonzalez, R. Garcia, T. de Pedro, R. E. Haber, Power-steering control architecture for automatic driving. IEEE Trans. Intell. Transp. Syst. **6**(4), 406–415 (2005)
63. S. Chaib, M.S. Netto, S. Mammar, Adaptive, PID and fuzzy control: A comparison of controllers for vehicle lane keeping. IEEE Intelligent Vehicles Symposium, 2004
64. M.S. Netto, S. Chaib, S. Mammar, Lateral adaptive control for vehicle lane keeping. American Control Conference, 2004
65. A. Carvalho, S. Lefevre, G. Schildbach, J. Kong, F. Borrelli, Automated driving: The role of forecast and uncertainty – A control perspective. Eur. J. Control. **24**, 14–32 (2015)
66. C. Voser, R.Y. Hindiyeh, C. Gerdes, Analysis and control of high sideslip maneuvers. Vehicle Syst. Dyn. **48**(Supp.), 317–336 (2010)
67. K. Kritayakirana, J.C. Gerdes, Using the center of percussion to design a steering controller for an autonomous race car. Vehicle Syst. Dyn. **50**(Supp.), 33–51 (2012)
68. K. Kritayakirana, J. C. Gerdes, Autonomous vehicle control at the limits of handling. Int. J. Vehicle Auton. Syst. **10**(4), 271–296 (2012)
69. K. L. R. Talvala, K. Kritayakirana, J. C. Gerdes, Pushing the limits: From lanekeeping to autonomous racing. Annu. Rev. Control. **35**, 137–148 (2011)
70. N. R. Kapanja, J. C. Gerdes, Design of a feedback-feedforward steering controller for accurate path tracking and stability at the limit of handling. Vehicle Syst. Dyn. **53**(12), 1–18 (2015)
71. E.J. Rossetter, *A Potential Field Framework for Active Vehicle Lanekeeping Assistance*. Dissertation, Stanford University, 2003
72. J. M. Maciejowski, *Predictive Control with Constraints* (Pearson Education, London, 2002)
73. E. F. Camacho, C. Bordons, *Model Predictive Control* (Springer, London, 2004)
74. P. Falcone, Nonlinear model predictive control for autonomous vehicles. PhD thesis, Universita' del Sannio, 2007
75. H. Yoshida, S. Shinohara, M. Nagai, Lane change steering maneuver using model predictive control theory. Vehicle Syst. Dyn. **46**(Supplement), 669–681 (2008)
76. P. Falcone, F. Borrelli, J. Asgari, H. E. Tseng, D. Hrovat, Predictive active steering control for autonomous vehicle systems. IEEE Trans. Control Syst. Technol. **15**(3), 566–580 (2007)
77. H. B. Pacejka, *Tire and Vehicle Dynamics*, 3rd edn. (Butterworth-Heinemann, Oxford, 2012)
78. P. Gill, W. Murray, M. Saunders, M. Wright, *NPSOL – Nonlinear Programming Software* (Stanford Business Software Inc., Mountain View, CA, 1998)

79. P. Falcone, H.E. Tseng, F. Borrelli, J. Asgari, D. Hrovat, MPC-based yaw and lateral stabilisation via active front steering and braking. Vehicle Syst. Dyn. **46**(Supp.), 611–628 (2008)
80. G. Yin, J. Li, X. Jin, C. Bian, N. Chen, Integration of motion planning and model-predictive-control-based control system for autonomous electric vehicles. Transport **30**(3), 353–360 (2015)
81. R. Attia, R. Orjuela, M. Basset, Combined longitudinal and lateral control for automated vehicle guidance. Vehicle Syst. Dyn. **52**(2), 261–279 (2014)
82. A. Gallet, M. Spigai, M. Hamidi, Use of vehicle navigation in driver assistance systems. IEEE Symposium on Intelligent Vehicles, 2000
83. D. Pomerlau, T. Jochem, C. Thorpe, P. Batavia et al., Run-off-road collision avoidance using IVHS countermeasures. Technical Report, US Department of Transportation, 1999
84. J. He, D. A. Crolla, M. C. Levesley, W. J. Manning, Coordination of active steering, driveline, and braking for integrated vehicle dynamics control. Inst. Mech. Eng., Part D: J. Automob. Eng. **220**, 1401–1421 (2006)
85. P. Tondel, T.A. Johansen, Control allocation for yaw stabilisation in automotive vehicles using multiparametric nonlinear programming. American Control Conference, 2005
86. Y. Gao, A. Gray, H. E. Tseng, F. Borrelli, A tube-based robust nonlinear predictive control approach to semiautonomous ground vehicles. Vehicle Syst. Dyn. **52**(6), 802–823 (2014)
87. S. Yu, H. Chen, F. Allgower, Tube MPC scheme based on robust control invariant set with application to Lipschitz nonlinear systems. 50th IEEE Conference on Decision and Control and the European Control Conference, 2011
88. D. Q. Mayne, M. M. Seron, S. V. Rakovic, Robust model predictive control of constrained linear systems with bounded disturbances. Automatica **41**(2), 219–224 (2005)
89. K.I. Kouramas, S.V. Rakovic, E.C. Kerrigan, J.C. Allwright, D.Q. Mayne, On the minimal robust positively invariant set for linear difference inclusions. 44th IEEE Conference on Decision and Control and the European Control Conference, 2005
90. B. Kouvaritakis, M. Cannon, S. V. Rakovic, Q. Cheng, Explicit use of probabilistic distributions in linear predictive control. Automatica **46**(10), 1719–1724 (2010)
91. A. Carvalho, Y. Gao, S. Lefevre, F. Borrelli, Stochastic predictive control of autonomous vehicles in uncertain environment. 12th Symposium on Advanced Vehicle Control, 2014
92. Y. Gao, T. Lin, F. Borrelli, E. Tseng, D. Hrovat, Predictive control of autonomous ground vehicles with obstacle avoidance on slippery roads. 3rd Annual ASME Conference on Dynamic Systems and Controls, 2010
93. A. Grancharova, T. A. Johansen, *Explicit Nonlinear Model Predictive Control – Theory and Applications* (Springer, Heidelberg, 2012)
94. S. Di Cairano, H.E. Tseng, D. Bernardini, A. Bemporad, Vehicle yaw stability control by coordinated active front steering and differential braking in the tire sideslip angles domain. IEEE Trans. Control Syst. Technol. **21**(4), 1236–1248 (2013)
95. S. E. Shladover, PATH at 20 – History and major milestones. IEEE Trans. Intell. Transp. Syst. **8**(4), 584–592 (2007)
96. A. van Zanten, Bosch ESP systems: 5 years of experience. SAE 2000-01-1633, 2000
97. L. De Novellis, A. Sorniotti, P. Gruber, J. Orus, J. M. Rodriguez Fortun, J. Theunissen, J. De Smet, Direct yaw moment control actuated through electric drivetrains and friction brakes: Theoretical design and experimental assessment. Mechatronics **26**, 1–15 (2015)
98. L. De Novellis, A. Sorniotti, P. Gruber, Driving modes for designing the cornering response of fully electric vehicles with multiple motors. Mech. Syst. Signal Process. **64–65**, 1–15 (2015)
99. L. De Novellis, A. Sorniotti, P. Gruber, Wheel torque distribution criteria for electric vehicles with torque-vectoring differentials. IEEE Trans. Vehicle Technol. **63**(4), 1593–1602 (2014)
100. T. Goggia, A. Sorniotti, L. De Novellis, A. Ferrara, et al., Integral sliding mode for the torque-vectoring control of fully electric vehicles: Theoretical design and experimental assessment. IEEE Trans. Vehicle Technol. **64**(5), 1701–1715 (2015)

Chapter 6
Vehicle Reference Lane Calculation for Autonomous Vehicle Guidance Control

Werner Kober, Richard Huber, and Ralf Oberfell

6.1 Introduction

Autonomous vehicles need smart sensors, actuators, additional road information, and control devices to achieve their tasks for longitudinal and lateral vehicle guidance control [1]. Recently presented prototype vehicles [2, 3] use radar sensors in combination with actuators for engine and brake control for longitudinal vehicle guidance and camera systems in combination with steering actuators for lateral vehicle guidance. Additional sensors like LIDAR and GPS with detailed map data, which nowadays are installed in prototype vehicles, are used more for research purposes [4]. From all possible sensor technologies for lateral vehicle guidance, the most important sensor information for the actual driving situation is provided by the camera system. The online image processing system of the camera module detects the lane markings of the vehicle lane. The camera software determines the distance of the actual vehicle position on the track to the left and right lane marking, as well as the heading angle of the vehicle to the lane markings on the street. This information is provided in real time via the vehicle bus system to the electronic control unit (ECU) which performs the lane-keeping control task. Besides the information of the actual position of the vehicle on the track or its predicted path, which can be designed according to [5, 6], the lane guidance controller needs also additional information about the desired reference position value of the vehicle. This reference lane is the desired track lane where the vehicle should ideally drive

W. Kober (✉)
Kober Engineering GmbH, Ilz, Austria
e-mail: werner.kober@kober-engineering.com

R. Huber
Kober Engineering Deutschland GmbH, Stuttgart, Germany

R. Oberfell
Daimler AG, Stuttgart, Germany

D. Watzenig, M. Horn (eds.), *Automated Driving*,
DOI 10.1007/978-3-319-31895-0_6

on the track. This value is propagated to the lane-keeping controller by the use of a desired lane offset signal. This desired lane offset signal, or reference lane signal, is usually defined to be zero in the middle of the lane and positive with right-hand side offset values. The controller calculates its controller deviation from this reference value and the current position of the vehicle on the track, trying to minimize the controller deviation by driving the actuators of the system accordingly [7]. It is obvious that the performance of the overall lane-keeping controller is, among others, mainly influenced by the quality of this reference lane input. The reference lane calculation module therefore belongs to the most important controller parts for the lateral vehicle guidance task in autonomous vehicles [8, 9].

One simple approach for reference lane calculation is to keep the vehicle in the middle of the accessible lane, which means to provide the lane-keeping controller with an offset signal of zero. This "mid-lane" approach is widely used by passenger car lane-keeping systems, e.g., seen in [10] and [11]. The "mid-lane" approach shows good results when used in passenger cars without any trailer attached. For vehicles with more complex driving dynamics like vehicles with trailer attached or heavy truck vehicles like truck–semitrailer combinations, the "mid-lane" approach is not sufficient anymore. This is not only because of the bigger dimension of these vehicles but also because of different driving dynamics and especially because of additional risk potential arising, for example, from the tractrix. Drivers of heavy trucks with trailer attached or drivers of truck–semitrailer vehicles therefore often have to choose different vehicle tracks than driver of passenger cars. The task of this paper is to design a reference lane calculation system which meets also the complex requirements of heavy truck and truck–semitrailer driving. Due to the fact that this task is the more general case of reference lane calculation, the easier task for calculating the reference lane for smaller vehicles with simpler driving dynamics like passenger cars will be included in the proposed reference lane calculation method.

6.2 Vehicle Lane-Keeping Boundaries and Requirements

The first step for designing a reference lane calculation system is to gain some overview of the fundamental boundaries of road tracks and the possible diversity of vehicle parameters which have influence on reference lane selection. The environmental boundaries for autonomous vehicles were defined by the geometry of the roads and the associated speed limitation regulations. A short overview of the requirements regarding the road geometry and speed limitation is given in Table 6.1, which shows typical road curvature and road width with the associated design speed regulations.

The columns A, B, and C list different road categories which refer to German guidelines for road construction [12, 13]. All categories are used for national and rural roads; categories B and C are permitted for use within built-up areas. For class C roads, it is furthermore allowed to have adjoining building next to the road. The

Table 6.1 Road characteristics with minimum road radius information [12, 13]

Road category	A	B	C
Lane width (m)	2.75–3.75	2.75–3.75	2.75–3.75
Design speed (km/h)	Minimum road radius for category and speed (m)		
40	–	–	40
50	–	80	70
60	135	125	120
70	200	190	175
80	280	260	–
90	380	–	–
100	500	–	–
120	800	–	–

lane width of a road is chosen in dependence of traffic volume and other factors, but generally it can be supposed that the selected road width decreases from category A to C. Roads with narrow lanes are challenging for lateral vehicle guidance systems because of less safety buffer for keeping the vehicle on track. For vehicles with trailer attached, the additional risk of tractrix hazards increases with reduced lane width and reduced curvature radius. For reference lane calculation development, the environmental boundaries should therefore be selected to cover these use cases best possible. One possible example of an environmental design setup will be a narrow road with small curvature radius, which will typically be used at relatively low vehicle speeds. According to this example, the road type C with small lane width and 50 m curvature radius, used at design speeds of 40 km/h, could be selected. Typical driving scenarios would be, for example, at exits of highways or curvy country roads. The vehicle parameter which influences the reference lane selection the most is the wheelbase of the vehicle as well as the type of the attached trailer or semitrailer. These parameters have influence on the track offset when cornering, which the driver has to take into account when choosing the vehicles lane for avoiding collisions in the tractrix area. The track offset is the difference of turning radius between the first and the last axle of a vehicle when cornering. When surveying vehicles with trailers attached, three main vehicle categories can be separated. These categories are the truck–semitrailer vehicle, the vehicle with rigid drawbar trailer, and the vehicle with pivot plate trailer. Figure 6.1 shows these different vehicle types in cornering maneuvers with the most important vehicle parameters for track offset determination.

As can be found, under Ackermann conditions the track offset calculation equations, as shown in Eq. (6.1), do not differ between truck–semitrailer and truck with rigid drawbar trailer [14]:

$$\Delta = r - \sqrt{r^2 - l_{\mathrm{Tk}}^2 + l_{\mathrm{Co}}^2 - l_{\mathrm{Tl}}^2} \tag{6.1}$$

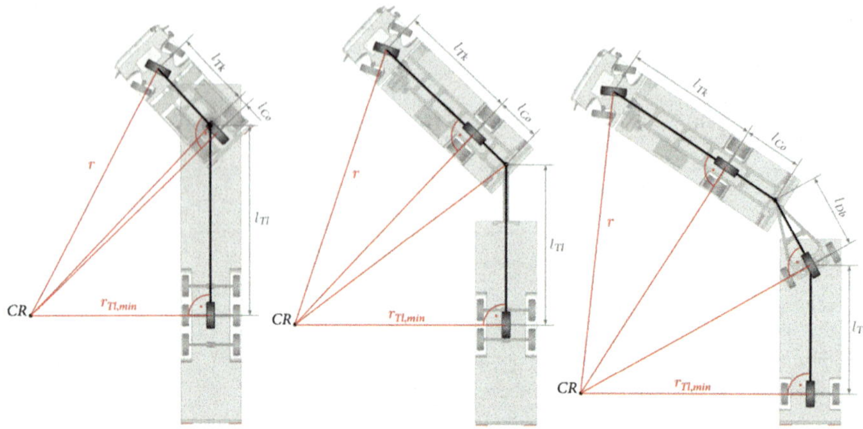

Fig. 6.1 Vehicle types: Truck–semitrailer (*left*), truck with rigid drawbar trailer (*middle*), truck with pivot plate trailer (*right*)

When adapting appropriate vehicle parameters, the track offset calculation relationship is also suitable similarly for passenger cars with trailer, truck with rigid drawbar trailer, and truck–semitrailer applications. The truck with pivot plate trailer has an additional pivot point which leads to a slightly different track offset equation, as shown in Eq. (6.2):

$$\Delta = r - \sqrt{r^2 - l_{\mathrm{Tk}}^2 + l_{\mathrm{Co}}^2 - l_{\mathrm{Db}}^2 - l_{\mathrm{Tl}}^2} \tag{6.2}$$

Both equations require Ackermann conditions, thus applicable in cornering maneuver at low vehicle speed. Considering this driving situation for reference lane calculation, the track offset differences between these three vehicle types have to be examined closer. The results of the gained lane offset Δ, based on typical vehicle parameters of vehicles of each category, are shown in Table 6.2. The comparison of these different vehicle types and their associated vehicle dynamics states under Ackermann conditions on a road of category C at low vehicle speed shows that the truck–semitrailer vehicle typically has the biggest lane offset, whereas the truck with pivot plate trailer has the smallest.

Summarizing the examinations of environmental boundaries and possible vehicle types for most general reference lane calculation design, it can be found that additional risk of tractrix hazards will occur mainly on narrow and curvy roads which are designed for comparatively low vehicle speeds related to the German guidelines for road construction. As can be seen in Table 6.2, the biggest lane offset in these driving situations will be gained by the truck–semitrailer vehicle type ($\Delta = 0.72$ m). For the purpose of designing a reference lane calculation method, which is able to consider the abovementioned boundaries for the most general case, the truck–semitrailer vehicle type has to be chosen.

Table 6.2 Parameters, for example, calculation with Ackermann conditions

Length parameter (m)	Truck with pivot plate trailer	Truck with drawbar trailer	Truck with semitrailer
Curve radius r	50	50	50
Truck wheelbase l_{Tk}	5.5	4.9	3.7
Coupling point (fifth wheel) to rear axle l_{Co}	2.4	1.8	0.65
Drawbar l_{Db}	3	–	–
Kingpin to trailer axle or trailer wheelbase l_{Tl}	5.17	6.43	7.63
Lane offset Δ	0.61	0.62	0.72

6.3 Vehicle Driving Situation Analysis

A reference lane calculation method which is not adapted to the driver's demands will autonomously take steering actions which must or will be actively corrected by the driver. Such a system behavior will be annoying to the driver, which will lead in the end to a deactivation of the assistance system by the driver, if that is possible, or at least will lead to a dishonored system with accordingly bad customer response. For designing a reference lane calculation system, it is therefore essential to analyze the driving behavior of a truck–semitrailer vehicle and the driving style of the users of such vehicles in different driving situations. The first driving situation which has to be analyzed is the simple straightforward driving situation. When a driver of a passenger car will mainly choose the mid of the lane as reference, to get an equal safety buffer to the left and right lane markings, the driver of a truck–semitrailer will choose a reference lane with an offset to the right. This is reasonable because of the bigger dimensions of a truck. Especially the bigger mirrors are in danger to get involved in collisions with the mirror of a truck coming along the opposite lane. Figure 6.2 shows this situation for three different track widths. As track width is declining, also the offset must be chosen smaller. If track width is getting too small to have enough buffer space for choosing a reference lane, the offset is getting zero which leads in the end to mid-lane driving, shown with the truck on the right-hand side of the figure.

When driving vehicles with long wheelbase or vehicles with trailer or semitrailer attached, the driver has to keep attention of the tractrix and track offset of the rear wheels of the vehicle at every cornering situation. Figure 6.3 shows an example of a dangerous situation regarding this issue. The red truck is guided by a "mid-lane" reference lane on a road with 3.75 m width, which is a quite common lane width for highway and larger countryside roads. Nevertheless the rear wheels of the red semitrailer are already leaving the lane and potentially compromising the traffic on the opposite lane.

The green truck–semitrailer uses an adapted (green) reference lane, with an offset to the outside of the turn, taking the track offset of the vehicle in account. This truck–

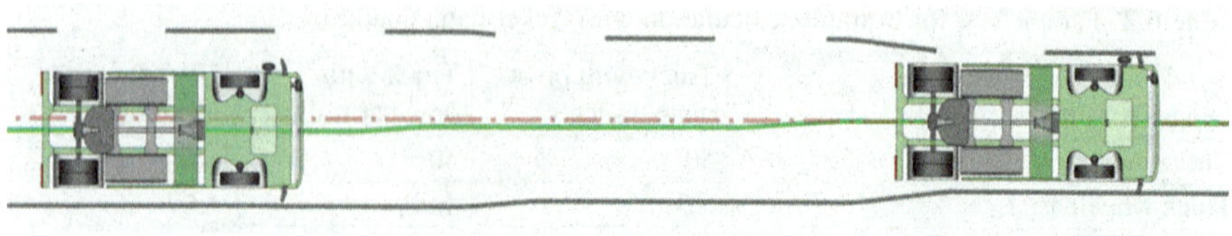

Fig. 6.2 Driving situation: Straightforward driving with lane width change, mid-lane guidance (*chain dotted line*), and adapted lane guidance (*solid line*)

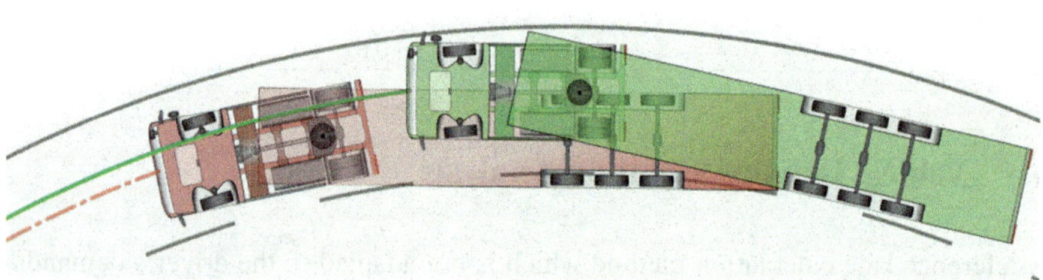

Fig. 6.3 Driving situation: Left-turn driving with mid-lane guidance (*chain dotted line*) and adapted lane guidance (*solid line*)

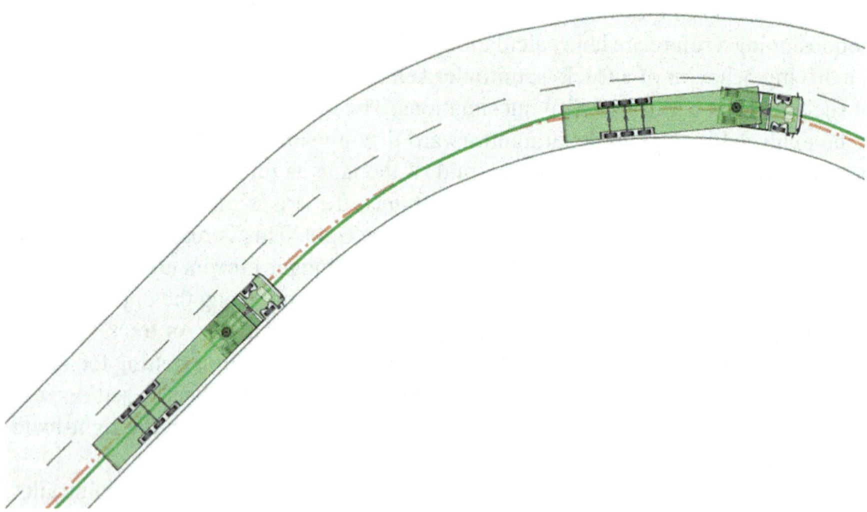

Fig. 6.4 Driving situation: Right-turn driving with mid-lane guidance (*chain dotted line*) and adapted lane guidance (*solid line*)

semitrailer is able to perform the left-hand side cornering safely without leaving the traffic lane. Especially when performing a right-hand side cornering maneuver, the driver of a truck–semitrailer vehicle has often to follow a comparatively sophisticated reference lane, as Fig. 6.4 shows. First the driver of the truck–semitrailer comes from a straight-ahead driving situation, which leads him to drive the vehicle with an offset to the right lane limitation, as mentioned above. To accomplish the cornering without running the risk of departing the traffic lane with the rear wheels of the semitrailer unintentionally, the driver has to adapt his right-hand side reference lane offset to a left-hand lane offset (solid reference lane). If the driver uses the mid-lane reference approach instead (red chain dotted line), the rear wheels of the semitrailer will leave the traffic lane. If the side verge is not able to carry the wheel load of the semitrailer properly, such a situation could easily lead to a rollover accident.

These presented main driving maneuvers cover most of the standard driving situations a driver or an autonomous vehicle has to deal with. Nevertheless there are many more additional driving situations which might occur during a drive. Such additional driving situations, for example, could be obstacles like a stopped vehicle at the side verge. In this situation the driver will take the maximum possible driving offset to the left to have as much safety buffer to this obstacle when passing it. Traffic jams give a further situation where special driving maneuvers like forming an emergency corridor have to be performed. Even if emergency vehicles with flashing blue lights are going to pass, the driver has to react with an adapted reference lane, which will be a maximum offset to the right-hand side in this case.

6.4 Model-Based Reference Lane Calculation Method

Taking the requirements from Sects. 6.2 and 6.3 in consideration, a reference lane calculation algorithm for lane keeping has to comply with the following main tasks:

- **Vehicle Safety Task**: The system has to keep the vehicle in track which also includes preventing the wheels of the vehicle trailer from leaving the road. This means that the tractrix of the vehicle has to be monitored permanently. Therefore the possible reference lane selection within the traffic lane is limited not only by track width but also by tractrix limitations.
- **Driving Comfort Task**: The system has to be adapted to the driver's needs and driving habits at least in the most common driving situations to gain high driver acceptance and high-quality customer response.
- **Connected Emergency Task**: An advanced reference lane calculation system also has to include information from ambient connected vehicles or infrastructure devices in the reference lane calculation process.

To meet the safety requirements, in most general case, the truck–semitrailer vehicle is chosen for designing the reference lane calculation algorithm, because the biggest track offset is gained by this vehicle type as shown in Sect. 6.2. When driving in straightforward driving situations, the possible lane selection isn't affected by the tractrix, so the limitations are in fact the left and right lane markings of the traffic lane the vehicle is actually driving on. When performing a cornering situation, however, the tractrix will take effect and the traffic lane limitations will be lowered due to tractrix limitations. To calculate the tractrix limitations, online in

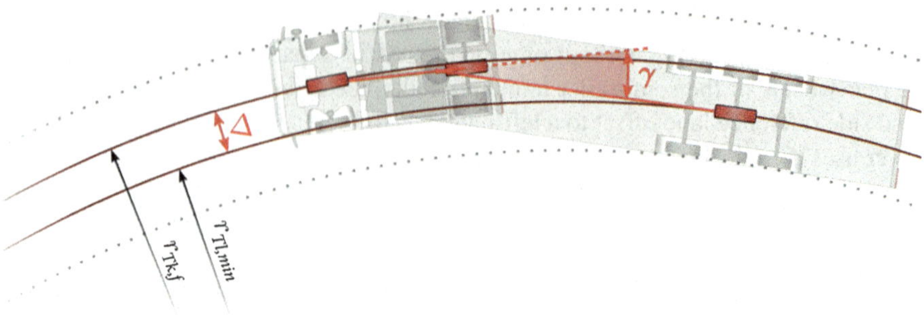

Fig. 6.5 Guidance offset and kink angle at single-track vehicle model

the driving vehicle, a model-based approach based on a single-track model of the truck–semitrailer vehicle, as shown in Fig. 6.5, could be used.

As could be seen in the figure, the vehicle is performing a left-hand side cornering maneuver with kink angle γ and track offset Δ. To assure safe cornering the radius of the vehicle $r_{\mathrm{Tk,f}}$ must be adapted by the track offset Δ so that the turning radius of the trailer $r_{\mathrm{Tl,min}}$ stays within the traffic lane borders. Because online measurement of the track offset is not possible, the offset has to be calculated from available vehicle measurement signals. To determine the track offset of the vehicle, generally the kink angle has to be known. Due to the fact that the vehicle kink angle is not measured by sensors too, a real-time capable kink angle observer has to be designed, which is able to determine the necessary kink angle to pass the road curvatures ahead of the vehicle [15]. The information of the road ahead, which is a necessary input for the kink angle observer, could be gathered by a camera system, which indeed provides the reference lane calculation system with the required information of curvature, vehicle position within the lane and heading angle of the vehicle. By the use of this information, it is possible to set up a track offset calculation module which calculates the necessary track offsets of the road curvature with a beneficial look-ahead functionality. As human drivers do, every lateral vehicle controller needs controller reference values with some look-ahead capacity, to gain the best possible performance at vehicle guidance task. When designing the reference lane calculation module for lane-keeping controllers, this can be assured as mentioned above, by a proper use of the camera look-ahead information available.

For designing the kink angle observer, the single-track model of a truck–semitrailer, as shown in Fig. 6.6, is used. The kink angle γ could generally be determined by numerical integration of kink angle rate $\dot{\gamma}$:

$$\gamma_t = \dot{\gamma} \cdot t_1 + \gamma_{t-t_1} \tag{6.3}$$

The kink angle rate $\dot{\gamma}$ is the difference of the yaw rate of the truck $\dot{\psi}_{Tk}$ and yaw rate of the semitrailer $\dot{\psi}_{\mathrm{Tl}}$:

$$\dot{\gamma} = \dot{\psi}_{\mathrm{Tk}} - \dot{\psi}_{\mathrm{Tl}} \tag{6.4}$$

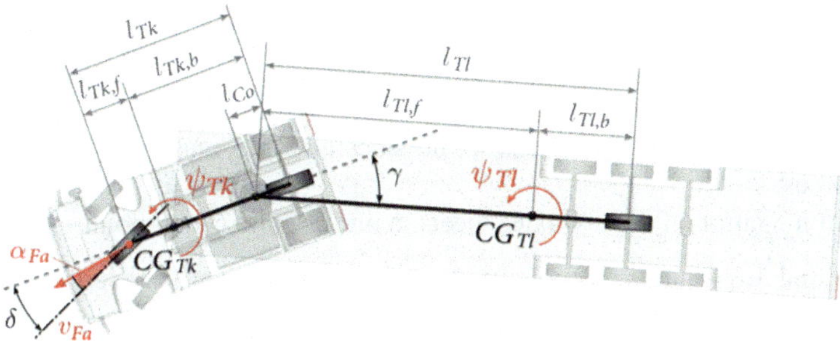

Fig. 6.6 Single-track model of a truck–semitrailer combination

The yaw rate of the truck $\dot{\psi}_{\mathrm{Tk}}$ could be represented by use of the equations of the single-track model as Eq. (6.5) shows. The yaw rate is calculated from vehicle steering angle δ, the front wheel speed v_{Fa}, and the sideslip angle α_{Fa} of the front wheels. Unfortunately the sideslip angle could not be measured online in vehicle when not using expensive measurement equipment; therefore Eq. (6.5) could not directly be used for determining the yaw rate. Alternatively the directly measured yaw rate of the truck by the vehicle's yaw rate sensor, used for the electronic stability program (ESP), could be used:

$$\dot{\psi}_{\mathrm{Tk}} = \frac{v_{\mathrm{Fa}}}{l_{\mathrm{Tk}}} \cdot \sin\delta \cdot \cos\delta - \frac{v_{\mathrm{Fa}} \cdot \cos\delta}{l_{\mathrm{Co}}} \cdot \sin\alpha_{\mathrm{Fa}} \tag{6.5}$$

The yaw rate of the semitrailer could accordingly be represented by using a single-track model as Eq. (6.6) shows. The yaw rate is determined besides geometrical vehicle parameters, from steering wheel angle δ, front wheel speed v_{Fa}, and kink angle γ:

$$\dot{\psi}_{\mathrm{Tl}} = \frac{v_{\mathrm{Fa}} \cdot \cos\delta \cdot \sin\delta}{l_{\mathrm{Tk}}} \cdot \frac{l_{\mathrm{Tk,b}}}{l_{\mathrm{Co}}} \cdot \cos\gamma \tag{6.6}$$

With Eqs. (6.4), (6.5), and (6.6), we get the calculation method for the kink angle rate:

$$\dot{\gamma} = \frac{v_{\mathrm{Fa}}}{l_{\mathrm{Tk}}} \cdot \sin\delta \cdot \cos\delta - \frac{v_{\mathrm{Fa}} \cdot \cos\delta}{l_{\mathrm{Co}}} \cdot \sin\alpha_{\mathrm{Fa}} - \frac{v_{\mathrm{Fa}} \cdot \cos\delta \cdot \sin\delta}{l_{\mathrm{Tk}}} \cdot \frac{l_{\mathrm{Tk,b}}}{l_{\mathrm{Co}}} \cdot \cos\gamma \tag{6.7}$$

The kink angle could be calculated from Eq. (6.3) by the use of sensor input information of vehicle speed, yaw rate of the truck, and necessary steering angle which is gained from the camera curvature information, performing a numerical integration. With the knowledge of kink angle, the track offset has to be determined. When assuming cornering maneuvers under Ackermann conditions, this can be achieved with simple trigonometric equations. The assumed simplifications of the Ackermann conditions are:

- Low vehicle speed with almost no longitudinal tire forces (no slip).
- No influence of lateral acceleration, no sideslip angle, and therefore no lateral tire forces.
- The prolongation of the tire axles meets in one circular center point.

Compared to vehicles without trailer, truck–semitrailer vehicles or vehicles with trailer attached tend to drive cornering maneuvers at relatively low lateral accelerations to avoid rollover risks. Typically values for lateral accelerations during lane change maneuvers, for example, are in a range from 0.2 to 0.5 m/s^2 [16]. So the Ackermann assumptions generally comply quite well with the use cases of the proposed application. Furthermore typically tractrix calculations are essential especially at relatively steep cornering maneuvers which anyway were performed at lower speeds, for example, at highway exits or curvy country roads, as shown in Sect. 6.2. So the Ackermann assumptions are suitable for use in track offset determination task for reference lane calculation.

The trigonometric relationship of the single-track vehicle model under Ackermann assumptions is shown in Fig. 6.7 for a truck–semitrailer vehicle. The steering angle δ can be calculated with Eq. (6.8) from wheelbase of the truck l_{Tk} and the

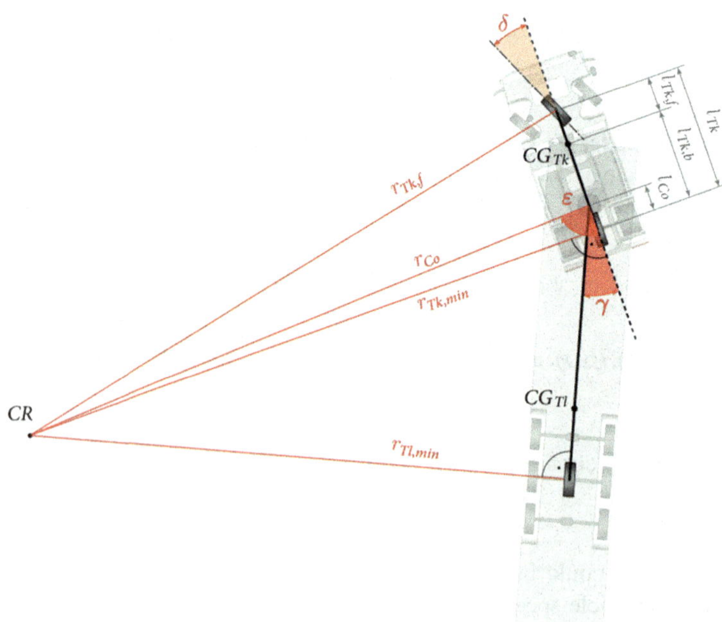

Fig. 6.7 Single-track model of a truck–semitrailer combination

turning radius of the front wheels $r_{\mathrm{Tk,f}}$ [17, 18]:

$$\sin(\delta) = \frac{l_{\mathrm{Tk}}}{r_{\mathrm{Tk,f}}} \tag{6.8}$$

The necessary turning radius of the front wheels $r_{\mathrm{Tk,f}}$ depends on the road curvature c_{r} which the vehicle is driving at. For autonomous vehicles with camera-based lateral vehicle guidance, the curvature information of the road ahead is provided by the lane detection camera. In case of GPS-based lateral vehicle guidance, the curvature information is available from road map data:

$$r_{\mathrm{Tk,f}} = \frac{1}{c_{\mathrm{r}}} \tag{6.9}$$

By the use of Eqs. (6.8) and (6.9), a necessary steering angle δ for driving a road curvature c_{r} can be calculated with Eq. (6.10). This steering angle, depending on the road curvature ahead, can be used as input for the kink angle observer to calculate the kink angle arising when driving the curvature:

$$\delta = \arcsin\left(\frac{l_{\mathrm{Tk}}}{r_{\mathrm{Tk,f}}}\right) = \arcsin\left(l_{\mathrm{Tk}} \cdot c_{\mathrm{r}}\right) \tag{6.10}$$

First step for track offset calculation is the determination of the minimum turning radius $r_{\mathrm{Tk,min}}$ at the rear wheels of the truck. The required cornering radius of the front wheels $r_{\mathrm{Tk,f}}$ could be gained from the camera curvature information. The minimum cornering radius of the rear wheels $r_{\mathrm{Tk,min}}$ could then be calculated with Eq. (6.11) by use of the wheelbase l_{Tk}:

$$r_{\mathrm{Tk,min}} = \sqrt{r_{\mathrm{Tk,f}}^2 - l_{\mathrm{Tk}}^2} \tag{6.11}$$

Next step is the calculation of the turning radius at the semitrailer coupling r_{Co} from minimum rear wheel cornering radius $r_{\mathrm{Tk,min}}$ with Eq. (6.12):

$$r_{\mathrm{Co}} = \sqrt{r_{\mathrm{Tk,min}}^2 + l_{\mathrm{Co}}^2} \tag{6.12}$$

The angle ε which is the angle between turning radius of the semitrailer coupling and the vehicles longitudinal axis could be determined by Eq. (6.13):

$$\varepsilon = \arccos\left(\frac{l_{\mathrm{Co}}}{r_{\mathrm{Co}}}\right) \tag{6.13}$$

By use of ε and the kink angle γ from the kink angle observer module, the minimum turning radius of the trailer $r_{\mathrm{Tl,min}}$ is given in Eq. (6.14):

$$r_{\mathrm{Tl,min}} = r_{\mathrm{Co}} \cdot \sin\left(\varepsilon - \gamma\right) \tag{6.14}$$

Finally the track offset Δ_{tractrix} could be calculated as difference between cornering radius of front wheels $r_{\text{Tk,f}}$ and minimum cornering radius of trailer axle $r_{\text{Tl,min}}$:

$$\Delta_{\text{tractrix}} = r_{\text{Tk,f}} - r_{\text{Tl,min}} \tag{6.15}$$

The track offset Δ_{tractrix} gives the possibility for real-time calculation of required lane space for driving an oncoming traffic lane curvature safely regarding tractrix hazards. Therefore this information is used to generate online safety limitations for possible reference lanes for left-hand side and right-hand side cornering maneuvers. Within this limitation the reference lane could be set freely from safety sight of view. It is recommended to use this potential to add solutions for additional requirements, for example, to gain higher driving comfort. Therefore requirements from driver habits could be taken in consideration to gain the best possible driving comfort. As seen in Sect. 6.3, the driver of the considered truck–semitrailer vehicle will tend to drive this type of vehicles with an offset to the right-hand side. To meet this requirement in the proposed reference lane calculation method, a nonlinear approach of traffic lane width dependent on comfort offset calculation, as shown in Eq. (6.16), is used:

$$\Delta_{\text{comfort}} = \frac{\Delta_{\text{lane,set}}}{w_{\text{l,set}} - w_{\text{veh}}} \cdot \max\left[(w_{\text{l}} - w_{\text{veh}}), 0\right] \tag{6.16}$$

The comfort track offset Δ_{comfort} depends from vehicle width w_{veh} and track width w_{l}. The parameterization of the nonlinear offset function is done by setting a reference offset point $\Delta_{\text{lane,set}}$ at a certain track width $w_{\text{l,set}}$, which can be freely chosen accordingly to the needs of the users. If vehicle width w_{veh} is getting bigger than actual track width w_{l}, the comfort offset is set to zero to perform a mid-lane vehicle guidance approach at this driving scenario. The user of the vehicle could also be enabled to set different comfort parameters, e.g., the reference offset point, of the reference lane calculation software module for adapting vehicle reference lane calculation to his/her needs. Autonomous vehicles also include additional functionalities for special driving situations which have also to be taken in consideration for reference value generation software algorithms. These driving situations include emergency situations like forming emergency corridors when traffic jams occurs or setting the driving lane offset to the maximum possible right direction for enabling emergency vehicles to pass. Another driving situation with high accidental risk is the situation when the vehicle has to pass a stopped vehicle or any other obstacle at the side verge. Again the appropriate reaction would be to set the reference lane offset to the left limit to gain maximum buffer space to the obstacle on the right. To perform these situations, a special emergency track offset is calculated as Eq. (6.17) shows:

$$\Delta_{\text{emergency}} = \frac{\max\left[(w_{\text{l}} - w_{\text{veh}}), 0\right]}{2} \cdot \text{sign}(E_{\text{left,right}}) \tag{6.17}$$

The emergency offset is determined so that the vehicle is guided to the border of the actual lane width of the appropriate side to generate a safety buffer. For activation of the emergency reference lane calculation, additional input and further software modules are used. This module includes also information from other surrounding vehicles and infrastructure devices. The gained information is preprocessed by additional software modules, and the reference lane generation module finally gets the information $E_{\text{left,right}}$ if an emergency lane is needed. The choice of the lane side which is appropriate for the actual emergency situation is defined by the sign of $E_{\text{left,right}}$ in Eq. (6.17). To generate the lane reference output value, the calculated track offsets for vehicle safety, driving comfort, and connected emergency have to be merged together to one target value for the lane-keeping controller of the autonomous vehicle. This task is performed by use of a prioritization logic. The highest priority is to keep the vehicle safely on track. In standard driving situations, Eq. (6.18) is therefore used to limit the determined comfort driving lane with the tractrix limitations. The minimum function $\min_0$ of Eq. (6.18) is used to set the minimum around zero to be able to use left- and right-hand side offsets around the center lane of the track:

$$\Delta_{\text{reference}} = \min_0 \left[\Delta_{\text{tractrix}}, \Delta_{\text{comfort}}\right] \tag{6.18}$$

In cases of emergency situations, the comfort lane is substituted by the appropriate emergency lane, which leads the driving comfort lane having the lowest priority. When using the emergency lane, keeping the vehicle on track has again a higher priority than performing the emergency lane. Equation (6.18) can therefore be used by replacing the driving comfort lane with the emergency lane, shown in Eq. (6.19):

$$\Delta_{\text{reference}} = \min_0 \left[\Delta_{\text{tractrix}}, \Delta_{\text{emergency}}\right] \tag{6.19}$$

By using this approach, the prioritization levels of the calculated track offsets are given by the vehicle safety with highest priority, emergency lane with mid priority, and the driving comfort with lowest priority.

6.5 Functional System Architecture Overview

By use of the developed model-based kink angle observer, it is possible to design a reference lane calculation system which is able to fulfill all the requirements of complex driving dynamics regarding autonomous tractrix observation for truck–semitrailer vehicles. Due to the fact of requirement selection choosing the most general use case in Sect. 6.2, the proposed reference lane calculation method is also valid for vehicle types like passenger car with trailer attached or truck–trailer combinations which generally have less track offset then the truck–semitrailer has. The reference lane calculation method in these cases would only set reference lanes with more safety buffer than a truck–semitrailer would have. Figure 6.8 shows the

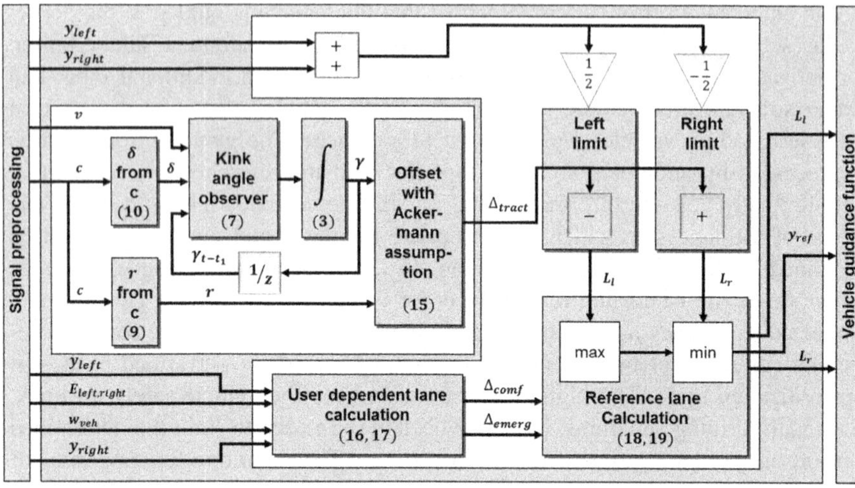

Legend: y_{left}/y_{right} = Left / right spacing to lane mark; v = Vehicle velocity; c = lane curvature; δ = Steering angle; γ = Kink angle; L_l/L_r = Left / right limit; y_{ref} = Reference lane; $E_{left,right}$ = Emergency signals; w_{veh} = Vehicle width; Δ_{tract}= tractrix offset; Δ_{comf}= comfort offset; Δ_{emerg}= emergency track offset

Fig. 6.8 Functional software architecture for reference lane calculation (equations are referenced in brackets)

functional software architecture of the proposed reference lane calculation system with three main parts: the kink angle observer with track offset calculation, the user-specific reference lane generation, and the output of the desired lane after the tractrix limitation of the generated user reference lane. The input stage gathers sensor information from the vehicle bus system, like vehicle speed and steering angle, as well as from the camera system with signals like distance to left/right lane marking and curvature of the lane. This information is used by the subsequent operation modules to calculate the reference lane y_{ref} and the left/right lane limitations (L_l and L_r). The reference lane gives the desired position of the vehicle on the track which is performed and adjusted via the lateral lane controller of the autonomous system. The limitation values L_l and L_r accomplish the driver task of observing the tractrix and ensuring that the rear wheels of the trailer or semitrailer stay on the road in all driving situations. To calculate the tractrix limitations, the kink angle observer, as shown in Sect. 6.4, and the actual lane curvature information from camera system are used to determine the track offset. This track offset is then propagated to the reference lane correction module, to limit the desired lane to possible values regarding tractrix hazards.

Till reference lane correction module limits the desired lane to safe values, a user-specific reference lane can easily be applied at the vehicle without danger of harming any safety restrictions. The user-specific reference generation module implements the algorithm for vehicle-specific reference lanes which could be in the simplest case a "mid-lane" approach for passenger cars with trailer attached. In these cases the tractrix limitation because of trailer usage is permanently assured by the reference lane correction module. For truck-semitrailer combinations, a simple constant lane offset to the right-hand side could be set to meet the requirements of truck drivers straightforward driving desires. For more advanced solutions, the constant offset could be improved by using an offset value which is dependent from actual traffic lane width. If the lane width is equal to vehicle width, the offset will go to zero and the reference lane will then be the same as in the mid-lane approach. In the user-specific reference generation module, also the emergency lane functionalities are implemented. External information received from infrastructure devices or other vehicles are used to set appropriate reference lanes for the vehicle in different emergency situations. Since there where camera input signals used for calculation modules like in the track offset calculation module, a simple look-ahead functionality of the module output can be ensured. This leads to better performance of the subsequent lane controller in the autonomous vehicle guidance system.

6.6 Example Driving Situations and Module Performance

For performance testing and validation of the reference lane calculation method, different driving situations have to be considered to cover all design requirements. As an example the main driving situations, straightforward driving, left-hand side cornering, and right-hand side cornering, as shown in Sect. 6.3, will be subsequently examined with the implemented software module with a measured driving run. The output of the reference lane calculation procedure is given at Fig. 6.9, where the reference lane offset is shown over simulation time. Preprocessed measurement data were used as input for the simulation run; thus the truck–semitrailer is driving with cruise control 50 km/h on a curvy country road during the simulation. To get a short overview of the driving run on which the simulation analysis is based on, the performed driving situations are listed in Table 6.3 in dependence of simulation time.

A reference lane offset of value zero corresponds to the "mid-lane" approach as indicated by the dashed line in Fig. 6.9. The left and right limitation graphs limit the possible reference lane selection area which is safe regarding possible tractrix hazards. The dark red solid line finally shows the calculated reference lane from the proposed reference lane calculation method. In the first section from simulation time from 0 s to approximately 20 s, the vehicle is in a straightforward driving situation, indicated by the given possible reference lane limits of 40 cm which are equally for the left- and the right-hand side. The reference lane of the vehicle, which is given by the green graph, is chosen with a traffic lane dependent offset. As the figure shows, the offset in this road width situation is approximately 25 cm to the right. From 20 to 40 s, the left limitation is lowered which indicates that the vehicle is performing a slight left-hand side cornering maneuver. Due to the fact that the left limitation doesn't affect the actual reference lane, the vehicle is able to continue driving with its previously used right-hand side offset. Shortly after simulation time of 40 s, a

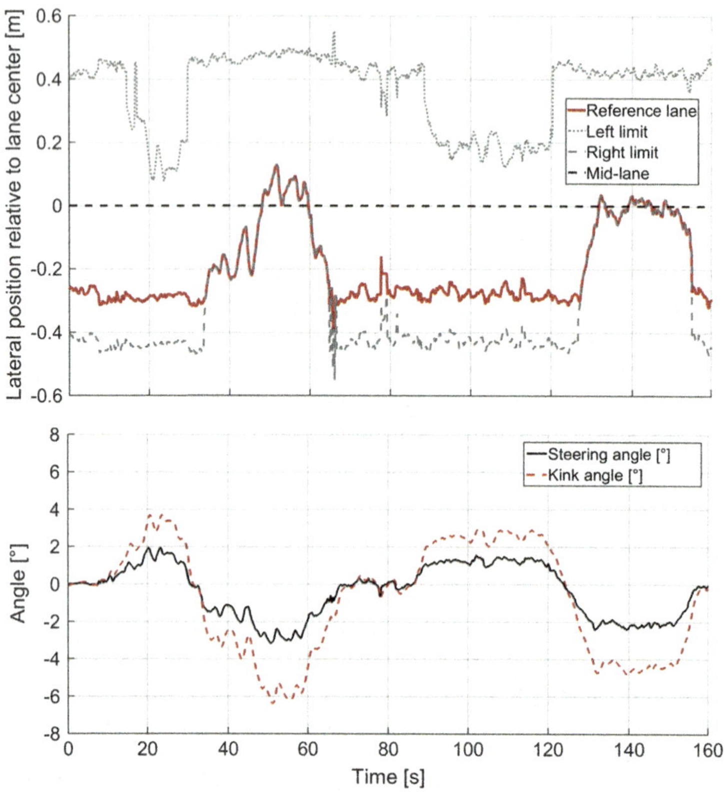

Fig. 6.9 Simulation output reference lane calculation module and angular input signals

Table 6.3 Simulation steps of driving scenario

Simulation time (s)	Driving situation
0–20	Straightforward driving (ref. Fig. 6.2)
20–40	Left-hand cornering (ref. Fig. 6.3)
40–77	Right-hand cornering (ref. Fig. 6.4)
77–98	Straightforward driving (ref. Fig. 6.2)
98–130	Left-hand cornering (ref. Fig. 6.3)
130–165	Right-hand cornering (ref. Fig. 6.4)

steep right-hand side cornering has to be performed. The right limitation calculation uses the information from the camera system and the kink angle observer to detect the oncoming lane curvature. With observer knowledge it is possible to calculate the necessary kink angle to pass this curvature and determine the oncoming track offset in real time.

This information is used to examine the right limitation shown in Fig. 6.9. As can be seen, the right limitation affects the actual vehicle reference lane in that way, that the offset to the right-hand side is lowered to zero in the first stage. When proceeding the cornering maneuver, it is even necessary to set the offset to the left-hand side, to pass safely this steep right-hand side road curvature without having any tractrix hazards. As soon as the road curvature is passed, the reference lane is set back to the lane width dependent offset for straightforward driving situations. Another left-hand cornering maneuver and in comparison to the first a less steep right-hand side cornering situation are following with ongoing simulation time.

6.7 Conclusion

This paper presents the recent developments of reference lane calculation methods necessary for advanced driver assistance systems and autonomous driving vehicles. It could be shown that a simple mid-lane approach as used for reference lane determination for passenger cars is not suitable for vehicles with more complex driving dynamics, e.g., vehicles with trailers or semitrailers attached. For designing an advanced reference lane calculation method, it is essential to determine the requirements and boundaries of road environment as well as the possible variety of vehicles where the system is intended to be applied. The truck–semitrailer vehicle type was chosen for developing the reference lane calculation algorithm, because this vehicle type provides the most general case for this task. This selection approach permits the usage of the proposed reference lane calculation method to be used in principal and also for all other vehicle types. A very important design requirement is also to include considerations of vehicle driver desires in different driving situations in the design specification process to gain a system with good customer response and high driver acceptance. This is equally important for percentage of usage of driver-selectable assistance systems as well as for persistent activated autonomous guidance systems. One of the most important safety hazards when driving vehicles with trailers or semitrailers attached is the risks of tractrix accidents. The proposed reference calculation method includes this safety function of permanent tractrix observation as well as a user-dependent reference lane generation and calculation of special reference lanes for emergency situations to provide a system with all necessary safety features in combination with highest possible driving comfort.

References

1. SAE J3016, Taxonomy and Definitions for Terms Related to On-Road Motor Vehicle Automated Driving Systems, Jan 2014
2. Daimler AG, Mercedes Benz Future Truck 2025, www.mercedes-benz.com/en/mercedes-benz/design/design-of-the-future-the-future-truck-2025/. Accessed 5 Jul 2015
3. Daimler Trucks North America LLC, Freightliner Inspiration Truck, http://www.freightlinerinspiration.com/technology/ . Accessed 5 Jul 2015
4. Google X, Google Self-Driving Car Project, www.google.com/selfdrivingcar/. Accessed 9 August 2015

5. P. Falcone, F. Borrelli, J. Asgari, H.E. Tseng, D. Hrovat, Predictive active steering control for autonomous vehicle systems. IEEE Trans. Control Syst. Technol. **15**(3), 566–580 (2007)
6. P. Falcone, F. Borrelli, H.E. Tseng, J. Asgari, D. Hrovat, A Hierarchical Model Predictive Control Framework for Autonomous Ground Vehicles, in *American Control Conference*, Seattle, 2008
7. A. Nhila, D. Williams, V. Gupta, Integration of Lane Keeping Assistance with Steering, in *SAE 2013 Commercial Vehicle Engineering Congress*, Rosemont, Illinois, 2013
8. B. Mashadi, M. Majidi, Global optimal path planning of an autonomous vehicle for overtaking a moving obstacle. Lat. Am. J. Solids Struct. **11**, 2555–2572 (2014)
9. Y. Gao, T. Lin, F. Borrelli, E. Tseng, D. Hrovat, Predictive Control of Autonomous Ground Vehicles With Obstacle Avoidance on Slippery Roads, in *ASME 2010 Dynamic Systems and Control Conference*, Cambridge, MA, 2010
10. J. Freyer, L. Winkler, M. Warnecke, G.P. Duba, A little bit more attentive audi active lane assist. ATZ Worldw **112**(12), 40–44 (2010)
11. BMW AG, BMW ConnectedDrive Driver Assistance systems in the BMW 7 Series, www.bmw.com/com/en/newvehicles/7series/sedan/2015/showroom/driver_assistance.html? Accessed 9 Aug 2015
12. Forschungsgesellschaft für Straßen- und Verkehrswesen, *Richtlinie für die Anlage von Straßen RAS-Q* (FGSV Verlag, Köln, 1996)
13. Forschungsgesellschaft für Straßen- und Verkehrswesen, *Richtlinie für die Anlage von Straßen RAS-L-1* (FGSV Verlag, Köln, 1984)
14. E. Hoepke, S. Breuer (Hrsg) *Nutzfahrzeugtechnik*, 7. Auflage (Vieweg+Teubner, Wiesbaden, 2013)
15. R. Huber, Model based development of target lane position determination for heavy duty trucks. Thesis, Institut für Verbrennungsmotoren und Kraftfahrwesen, Stuttgart, 2013
16. E.J.-D. Schulze, M. Becke, *Unfallrekonstruktion, LKW Spurwechsel auf mehrspurigen Richtungsfahrbahnen* (Schimmelpfenning und Becke GbR, Münster, 2007)
17. M. Mitschke, H. Wallentowitz, *Dynamik der Kraftfahrzeuge*, 4. Auflage (Springer, Berlin, 2004)
18. B. Heißing, M. Ersoy, *Chassis Handbook* (Vieweg+Teubner, Wiesbaden, 2011)

Part III
Advances in Environment Sensing, Sensor Fusion, and Perception

Chapter 7
The Role of Multisensor Environmental Perception for Automated Driving

Robin Schubert and Marcus Obst

7.1 Introduction

Automated driving is characterized by a computer-based derivation and execution of appropriate driving maneuvers based on the current traffic situation. The basis for evaluating the traffic situation, however, is knowledge about all relevant entities in a vehicle's environment, including traffic participants (vehicles, pedestrians), road infrastructure (lane markings, traffic signs, traffic lights), or obstacles (curbstones, potholes, bushes).

For acquiring such knowledge, numerous sensors are being used that permanently provide relevant data. These data, however, are typically not directly used by the high-level driving functions (such as path planning or maneuver decision-making). Instead, an intermediate processing layer is used that is referred to as the *environmental model* (see Fig. 7.1). This additional layer is used for several reasons:

- **Sensor errors**: Data directly provided by the sensors are subject to different errors. On the one hand, the measured signals can be disturbed by effects such as sensor noise, latencies, or calibration errors. On the other hand (and more severe), sensors may fail to detect objects at all or falsely report objects that are not present in the environment. Those effects are called false-negative or false-positive detections, respectively.

  The environmental model contains algorithms to reduce those errors by temporal filtering and to estimate the magnitude of errors that need to be considered when using its result for taking driving decision. Thus, environment modeling helps to increase the reliability of the perception.

R. Schubert (✉) • M. Obst
BASELABS GmbH, Chemnitz, Germany
e-mail: robin.schubert@baselabs.de; marcus.obst@baselabs.de

D. Watzenig, M. Horn (eds.), *Automated Driving*,
DOI 10.1007/978-3-319-31895-0_7

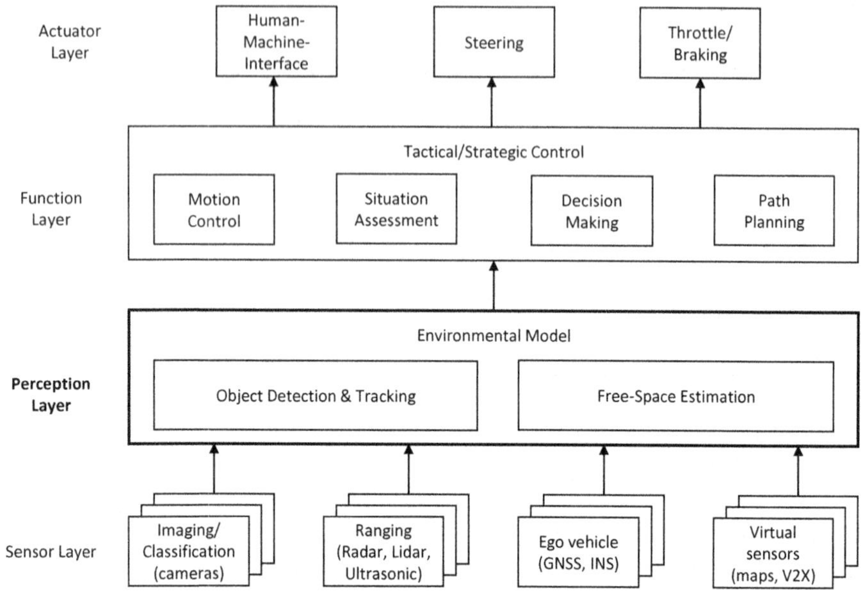

Fig. 7.1 The perception layer constitutes an interface between sensors and high-level functions

- **Sensor abstraction**: The sensor configuration of different vehicles will most likely not be the same. In addition, interfaces and properties of the utilized sensors may change. By providing a separating layer between the sensors and the driving function, the environmental model facilitates the implementation of driving functions that are independent from particular sensors. This has a huge impact on reusability and architectural quality of those functions. The environmental model may, thus, be considered as a *sensor-abstraction layer*.[1]
- **Data fusion**: For many driving functions, information from different sensors are required (e.g., the velocity of an object measured by a radar and its class determined by a camera). The environmental model inherently combines data from all available sensors in order to provide both a higher quantity and a higher quality of available information. While the former is achieved by combining heterogeneous sensors or sensors covering different fields of view, the latter effect is based on sensor redundancy. The environmental model ensures that the driving functions can use a consolidated representation of the environment.
- **Modeling and hypothesis generation**: The required reliability of information representing the environment of the vehicle constitutes an inherent conflict when compared to the quality of the available sensor data. The decisive strategy for bridging this gap is to use *model knowledge*. In fact, the concept of the

[1]To some extent, this can be compared to hardware abstraction layers used in programming to support modular software architectures that facilitate reusability.

environmental model is to form expectations of the environment (often in form of numerous hypotheses) and to confirm or falsify these hypotheses by calculating their fit to the observed sensor data. For instance, lanes may be assumed to be parallel and to follow a clothoidal shape. It is many grades of magnitudes easier to estimate the parameters of a clothoid based on noisy data than to determine a completely arbitrary shape from these data. The same concept holds for other entities such as vehicles, whose stable detection and tracking is only possible because it is assumed that they comply with typical motion patterns. Modeling also facilitates the estimation of quantities that are not directly observable. In fact, this concept forms the basis of the term environmental *model*, as it emphasizes that the main role in this layer is to define the expected structural representation of the environment and to estimate the parameters of this representation based on sensor data.

- **Prediction**: For automated driving, it is not sufficient to describe the current traffic situation. In fact, maneuver decisions or warnings to a driver should be issued *before* a collision or another dangerous situation occurs. This implies the need to predict the traffic situation to the future in order to anticipate the consequences of driving decisions. The environmental model provides this functionality by exploiting model knowledge as described before. For instance, if typical motion patterns of vehicles are modeled and their dynamic parameters (such as their velocities and yaw rates) are estimated, future positions can be determined. As predictions can never be certain, however, also the reliability of such predictions needs to be quantified by the environmental model.
- **Integrity**: Together with the representation of the environment, probably the most relevant role of the environmental model is to provide quantitative indicators that describe its uncertainties. For a single sensor, it is difficult to determine the current error of its measurement. The environmental model, however, can exploit the redundancy between sensors as well as the model knowledge to estimate the current estimation errors. Such an error indicator is considered to fulfill the property of *integrity* if the true error never exceeds the estimated one. As automated driving functions are safety critical, this property is crucial for a safe operation. Only if the driving functions know about the current expected error of the environmental representation, it can take appropriate decisions. The most common way to describe these uncertainties is by means of probabilistic quantities. For this reason, probabilistic inference algorithms are at the heart of environmental modeling algorithms.

7.2 State of Practice

In general, the vehicle environment for automated driving applications comprises dynamic and static entities. In the following sections, typical methods to model and estimate these entities are presented.

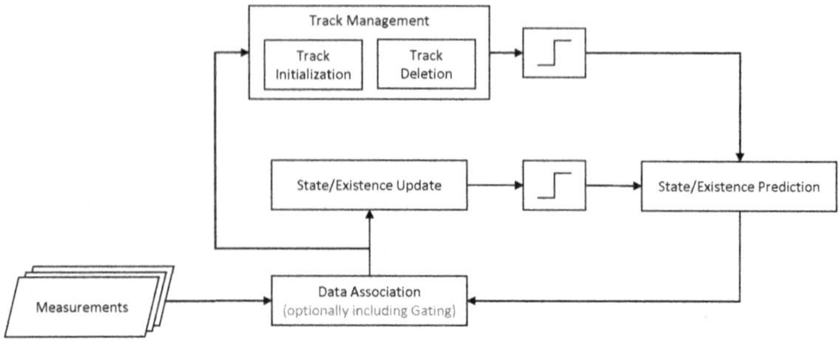

Fig. 7.2 Functional structure of a multiple objects tracking process

7.2.1 Dynamic Environments

Dynamic environments are characterized by numerous entities that can move independently from each other. As described in Sect. 1, it is advantageous to model this motion mathematically in order to improve the robustness of the perception. For this reason, the most common approach for modeling dynamic environments is to apply an *object-based* paradigm by handling each moving object independently.

A commonly applied technique to that end is multiple objects tracking (MOT). Here, object tracking refers to the task of utilizing noisy measurements from heterogeneous perception sensors to derive both the number and the characteristics of relevant objects by filtering the sensor data over time. The characteristics are subsumed in a vector called the *state* and typically include continuous parameters such as position, velocity, and heading angle as well as discrete states like the object class.

Though there is a variety of MOT algorithms, many of them can be generalized to the structure illustrated in Fig. 7.2. The basic assumption of an MOT algorithm is that, due to false positives and false negatives, detections reported by sensors do not necessarily correspond to true objects. This inherent property of perception systems is taken into account by introducing the concept of *tracks* representing *object hypotheses*. A track consists of a state estimate (typically including a position and additional kinematic parameters) and a probability which quantifies the belief in the hypothesis that this track represents a true object (referred to as the *probability of existence*). Both the state estimate and the probability of existence are iteratively updated whenever new sensor data arrive.

The state of a track is modeled independently from the sensor measurements, while the relation between both is modeled by a mathematical function called the sensor model. By using several sensor models, the tracks can be updated by different sensors independent from each other. Thus, all sensors that are related to the state by a sensor model contribute to the estimation of this state as well as the estimation of the track's probability of existence. This concept implicitly facilitates the spatial and temporal alignment of data from different sensors—commonly referred to as (multiple sensor) *data fusion*.

The concept of MOT will be explained in the following subsections based on the structure illustrated in Fig. 7.2.

7.2.1.1 State Estimation by Bayesian Filtering

In general, several approaches to realize multisensor fusion are available [1]. A well-known and successfully applied technique in online multisensor fusion applications is Bayesian state estimation. By that approach, the temporal correlation between consecutive observations from multiple sensors can be continuously exploited and combined with a priori model knowledge. Finally, the result is represented by a probability density function (PDF). The recursive Bayes filter algorithm comprises two main parts:

- **State prediction**: In this step, the knowledge of the system under observation from the last time step t_{k-1} is synchronized to the measurement time t_k of the most recent sensor observations. This is achieved by applying a *system model* (e.g., an appropriate motion model [2] in the case of a vehicle state estimation problem) which describes the expected evolvement of the system over time. As this state synchronization is solely based on model knowledge (which typically does not match reality in every detail), it typically adds uncertainty to the estimate. Due to the fact that the state of the system under observation and the provided sensor measurement are often not defined in the same coordinate system, the predication step also includes an additional transformation process from the synchronized system state to the measurement domain. In order to achieve this transformation, a sensor-specific *measurement model* which considers sensor noise is applied.
- **State update**: After the prediction step, a probabilistic representation of the synchronized system state (also called a state hypothesis) described in the measurement domain and the particular sensor observation sample is available for comparison. Probabilistically speaking, the likelihood of the sensor model given the assumption of the state hypothesis is evaluated at the position of the observed measurement. Based on the result, the predicated state estimate is adjusted and considered the best knowledge of the system at time t_k. The result of this step is a probabilistic combination of the information from the latest sensor measurement and the previously accumulated system state.

Recursive Bayes filtering is an efficient and consistent technique to perform a probabilistic data fusion among multiple sensors for real-time applications. Different measurement principles and the sensor noise are considered by particular measurement models, while the alignment of asynchronous sensor data is implicitly achieved by applying the system model. It is worth mentioning that there are multiple practically relevant implementations of recursive Bayes filters: For linear systems whose uncertainties can be modeled by Gaussian PDFs, the Kalman filter is the optimal estimator. The practically most relevant filters are the extended

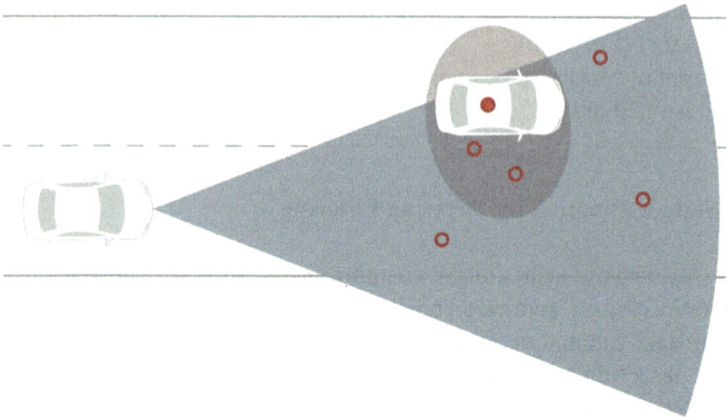

Fig. 7.3 Position uncertainties and false-positive detections are causing ambiguities

Kalman filter and the unscented Kalman filter for nonlinear systems that are subject to unimodal disturbances which can be approximated by Gaussian PDFs. For multimodal PDFs or in case of severe nonlinearities, particle filters can be applied [3].

7.2.1.2 Data Association

In order to update a track's state estimate, sensor measurements are used. In practice, however, this process is subject to severe ambiguities as illustrated in Fig. 7.3. The output of a sensor is a list of detections within the field of view that are not related in any way to the object hypotheses that are handled by the MOT. For updating the state of a track, it is crucial to know which of these detections shall be used for updating a particular track. One part of the ambiguity related to this issue stems from the uncertainty of the state estimate itself. As the track's position is not exactly known, it is difficult to distinguish between a detection that displays a different position than the track due to measurement noise and a detection coming from an adjacent object. The other part of the ambiguity is caused by the fact that sensor detections can be erroneous. In Fig. 7.3, some of the sensor detections may be related to the vehicle, they may be false positives, or they may represent a vehicle which is not yet tracked. Solving this ambiguity is referred to as data association (or, more precisely, measurement to track association) and can be considered the key challenge in multiple objects tracking.

The basis of data association is some kind of distance metric between the tracks and the measurements. Instead of using a geometrical measure such as the Euclidean distance, MOT requires to apply a statistical distance measure that, on the one hand, accounts for the uncertainties of the tracks and the sensor measurements and, on the other hand, facilitates the combination of different state variables such as position and velocity into a single metric. The metric of choice for this application is the *Mahalanobis distance*, which is a statistical measure describing the distance between a multi-dimensional probability density function and a point [4]. If the track's state is represented by a Gaussian PDF, the metric can be considered as a track's likelihood evaluated at the position of the measurement.

Based on the Mahalanobis distance between each track and each measurement, the measurement to track association takes place in several steps.

The first step is called *gating*. In this phase, detections that exceed a predefined distance to the closest track are excluded from the further steps. Those detections are considered as *non-associable*, which means it is considered very unlikely that those detection represent objects which are already being tracked. Instead, those detections are assumed to be potential objects that are not yet tracked—hence, they are used to create new object hypotheses (i.e., new tracks). In addition to using those non-associable detections as a source for new tracks, another rationale for gating in combination with decision-based association techniques is to avoid unreasonable association decisions.

Regarding the second step after the gating, there are two groups of techniques: The first one reduces complexity by conducting a unique n-1-pre-association in the sense that every detection is associated unambiguously to the closest track (which implies that several detection can be pre-associated to each track). Those approaches can be considered as *local* association techniques. The second group does not make use of this simplification, which makes it more performant in ambiguous situations but also more computationally complex.

In the final association step, again two different categories may be distinguished:

- **Decision-based data association**: In the first category, a unique association decision is derived in the sense that not more than one detection is associated to a track and vice versa. This can be achieved by minimizing the distance between tracks and detections. The local algorithm in this category is the *local nearest neighbor* association which determines the closest detection for each track. The *global nearest neighbor* association, however, enumerates overall association hypotheses and chooses the one with the smallest sum of distances.
- **Probabilistic data association (PDA)**: In this category, no hard decisions are made. Instead, all detections (possibly pre-associated to a track) are considered valid association hypotheses, and a probability is derived for each hypotheses which is higher for detections closer to certain tracks than for detections that are farther away. Depending on whether a pre-association has been applied, either a local variant (PDA) or a global variant (joint PDA—JPDA) of this approach can be used.

PDA and JPDA can be further distinguished into several subclasses of association algorithms depending on certain assumptions and models:

- **PDA/JPDA vs. IPDA/JIPDA**: The basic PDA/JPDA is using the association weights for the state update but not for updating the existence probability. An extension is the *integrated* PDA (or the corresponding joint integrated PDA) that updates also the existence probability accordingly [5].

- **GPDA**: The basic PDA assumes that each object cannot generate more than one detection. This assumption limits the number of hypotheses and, thus, reduces complexity. However, it also implies that if several measurements are received for one object, all but one are per definition considered false positives. In some scenarios (like tracking of large trucks with a radar sensor), this assumption does not hold. The generalized PDA (that is also available in the integrated and joint variant) relaxes this assumption at the costs of a higher computational complexity [6].

As an interim conclusion, data association is a complex topic that can be solved by numerous algorithms that are summarized in the Tables 7.1 and 7.2. The choice of the right algorithm depends on the sensors as well as on the scenarios and the available computation power.

Table 7.1 Systematization of data association techniques

Data association				
Gating	Detections too far away from existing tracks are considered *non-associable* and are used to create new tracks[a]			
Pre-association (per detection)	n-1-association ("local"): Detections are associated to not more than one track		n-m-association ("global"): Detections are associable to any track (no pre-association)	
Deterministic vs. probabilistic association (per track)	**Local nearest neighbor** (hard 1-1 decision)	**Probabilistic data association (PDA)** (no decision)	**Global nearest neighbor** (hard 1-1 decision)	**Joint probabilistic data association (JPDA)** (no decision)

[a]This approach is called *data-driven* track creation. While it is the practically most relevant approach, the theoretically consistent way of creating tracks is to draw them randomly from a birth distribution

Table 7.2 Further systematization of probabilistic data association techniques

(Joint) probabilistic data association		
Impact on existence estimation	Classical PDA/JPDA: No influence between existence and association probabilities	**Integrated PDA/JPDA (IPDA/JIPDA):** Mutual influence between both probabilities
Assumed number of possible true detections from an object	Assumption of classical PDA/JPDA/IPDA/JIPDA: The sensor never receives more than one true detection from an object	**Generalized PDA/JPDA/IPDA/JIPDA**: A sensor can receive several true detections per object

7.2.1.3 Existence Estimation

For the estimation of a track's probability of existence, Bayesian filtering can be applied as well. However, in contrast to the estimation of a continuous state, the existence of a track is a binary variable. This simplifies the estimation, as it allows the usage of a discrete Bayes filter.

The belief that the track under consideration represents a true object is represented by the conditional probability $P(\exists x_k|\mathbf{Z}_k)$ where the binary variable $\exists x_k$ denotes the existence of an object x at time k and $\mathbf{Z}_k$ denotes the set of all measurements from time 0 to k. This conditional probability can be calculated using Bayes rule (using η as a normalizing constant), which leads to

$$P(\exists x_k|\mathbf{Z}_k) = P(\exists x_k|z_k, \mathbf{Z}_{k-1})$$

$$= \eta \cdot P(z_k|\exists x_k, \mathbf{Z}_{k-1}) \cdot P(\exists x_k|\mathbf{Z}_{k-1})$$

$$= \eta \cdot P(z_k|\exists x_k) \cdot P(\exists x_k|\mathbf{Z}_{k-1})$$

This equation contains two relevant terms: $P(\exists x_k|\mathbf{Z}_{k-1})$ denotes the probability of existence conditioned on all measurements except the last one. It can be obtained by predicting the probability of existence from the last time step to the current time using the Chapman-Kolmogorov equation:

$$P(\exists x_k|\mathbf{Z}_{k-1}) = \sum_{\exists x_{k-1}} P(\exists x_k|\exists x_{k-1}) \cdot P(\exists x_{k-1}|\mathbf{Z}_{k-1})$$

The term $P(\exists x_{k-1}|\mathbf{Z}_{k-1})$ corresponds to the probability of existence from the previous time step, which illustrates that this value can be calculated iteratively from each time step to the next. The conditional probability $P(\exists x_k|\exists x_{k-1})$ describes how likely it is that an existing object will persist or a new object will appear and is referred to as birth/persistence model. The far more relevant term, however, is the existence likelihood $P(z_k|\exists x_k)$. This term describes the expected sensor measurements for both for the case of an existing as well as a nonexisting object. Its usage depends on the selected data association strategy.

For a nearest neighbor association, the only relevant information is if a measurement could be associated to a track or not, that is, a binary variable. This reduces the existence likelihood to a probability distribution consisting of four values (of which only two are required due to the fact that probabilities sum up to unity). The required likelihoods are:

- $P(z_k = 1|\exists x_k = 1)$: This is the true positive rate of the sensor, that is, the probability that the sensor will detect an existing object.

- $P(z_k = 1 | \exists x_k = 0)$: This is the false-positive rate of the sensor, that is, the probability that the sensor will provide detections from locations without existing objects.

Both quantities can be obtained by characterizing the sensor. If the true positive rate is considered zero outside the field of view, the fusion of several sensors with partially overlapping observation areas can be implicitly handled by updating the existence probability independently for each sensor. Both the true and the false-positive rate may vary over the field of view (for instance, the detection performance of a radar may depend on the azimuth angle due to its antenna characteristics).

In a system using PDA, the existence likelihood also depends on the true and false-positive rate of the sensors, but also on the distance between the measurement and the track. Further details can be found in [6].

7.2.1.4 Track Management

In MOT, a dedicated, application-specific strategy to inject and discard track hypotheses is required. A typical choice for the track initialization step is to apply a data-driven regime where new track hypotheses are proposed based on not associated sensor measurements after each sensor's duty cycle. However, this approach assumes that the information from the observed sensor data is sufficient to perform the full inverse transformation from the measurement domain to the targeted system representation. If this requirement is not fulfilled (e.g., it is generally not possible to derive information about the absolute velocity of a vehicle from a radar measurement which only includes angle, range, and radial velocity) further assumptions have to be made.

The same applies to the discarding strategy of already confirmed object hypothesis. Again, it depends on the targeted application and particular use case which hypotheses are considered relevant and thus have to be kept and which can be safely removed (e.g., in order to save computational resources).

7.2.2 *Static Environments by Occupancy Grid Mapping*

Whereas the modeling of dynamic environments as introduced in Sect. 2.1 aims to represent individual entities, occupancy grid mapping is a complementary approach more suitable for arbitrary static environments which does not assume any type of feature. The concept of grid mapping is to model the environment as an evenly spaced two-dimensional grid[2] with a fine-grained cell size. Each cell inside of the grid comprises a binary random variable that represents the presence of an

[2]There are also other variants of grid implementations which consider more dimensions such as the 4D grid [8].

obstacle at that particular location. In order to estimate this occupancy from sensor measurements, the individual cells are continuously updated by applying a discrete Bayes filter algorithm similar to the existence estimation step in MOT of Sect. 2.1. Thus, the occupancy grid can be considered a storage for spatially and temporally distributed measurement data.

In contrast to dynamic environment modeling with MOT, which provides confirmed object hypotheses, the result of an occupancy grid can be efficiently used to infer information about free space areas. This is in particular important for decision-making and path planning of automated maneuvers.

Due to the lack of dynamic quantities in the estimate, occupancy grids are typically not well suited for dynamic environments. There are however, hybrid approaches such as particle grids that are attempting to overcome this limitation and to constitute an estimation approach for both static and dynamic environments [7].

7.3 Challenges in Data Fusion

In the previous section, typical techniques for modeling dynamic and static environments for automated vehicles have been presented. The following section introduces selected challenges which practitioners may encounter during the implementation of data fusion systems for environmental perception.

7.3.1 Sensor Characterization

Sensor observations are a crucial input of every perception system. However, in general, the sensors which are used to observe the environment have weaknesses. First, not every measurement necessarily belongs to a real existing object. Those false-positive measurements (whose number is time varying) have to be eliminated during the tracking process. Vice versa, not every existing entity generates a measurement at sensor cycle and thus leads to a false negatives. In addition, the measurements are disturbed by noise. Finally, sensors have a limited detection range and finite resolution.

For the successful implementation and application of a multisensor data fusion system such as MOT, a correct and realistic modeling of the previously mentioned sensor properties is inevitable. In the following text, common techniques to derive and specify these sensor properties are stated. Moreover, examples of sensor mismodeling and their influence to the environmental model will be given:

- **Measurement noise**: Random fluctuations in measured signals are referred to as measurement noise. They can be characterized by statistical parameters that describe the average deviation of a sensor observation from the true value over time. For example, for a radar system that measures angle, range, and range rate, the measurement noise is usually given independently for each dimension and described by a zero-mean Gaussian distribution with appropriate variances. Finally, the realization of a sensor observation and the knowledge of its behavior is used by statistical filtering such as Bayesian estimation in order to reduce the uncertainty over time.

Usually, these parameters are either extracted from the sensor manual or derived via empirical testing under controlled conditions. Although initial values from a sensor manual are usually a good starting point, practice reveals that these parameters need to be tested and validated carefully within the target scenario and the utilized data fusion algorithm. In MOT, for example, the measurement noise directly influences the performance of the data association step and the accuracy of the estimated objects. If the measurement noise is set too small, the tracking algorithm will not be able to generate stable object hypotheses (i.e., object with constant track IDs). On the other hand, assuming a too conservative value for the sensor noise limits the performance of the environmental model to represent high-dynamic maneuvers.

- **Field of view**: As the concept of MOT is based on object hypotheses (see Sect. 2.1), the perception system naturally aims to predict whether and how a particular object hypotheses will be measured by a sensor system. Generally, speaking, a sensor observation is only expected inside the sensor's field of view (FOV). Inside the FOV, a sensor observation might either correspond to a real object (and thus confirms its existence) or is considered clutter and should be ignored. If a sensor observation is missing for a predicted object hypothesis, this usually indicates that the hypothesis inside of the MOT is incorrect and should be removed. Of course, this decision is not made at one particular time step but derived over time. The time span necessary to make this decision directly depends on the detection behavior of a sensor.

  A typical configuration in environmental perception for automated vehicles is to have multiple perception sensors realizing a partially overlapping FOV in order to increase robustness and ensure proper handover. It is worth mentioning that the consistent performance of MOT is rather sensitive to the correct modeling of the overlapping area. If sensor FOVs are modeled too optimistic (i.e., sensor range is modeled to large), the MOT algorithm will consider nonexistent sensor observations at the FOV border as potentially missed detection and thus propose to delete the object hypothesis. This may lead to the unintuitive situation where two sensors work against each other instead of confirming a common object hypothesis.
- **Detection behavior**: Although the measurement noise already gives an indication of the quality of a sensor observation in terms of accuracy, it does not include any information about the reliability of a sensor system. Similarly, while the sensor's FOV already gives a binary indication whether a sensor observation can be expected at a particular position around the host vehicle or not, it does not quantify the probability of detection in a statistical way. A straightforward

approach is to quantify the probability whether a sensor will detect an object given its existence inside of the FOV with a constant detection probability.

Given an exemplary high-reliable sensor which usually always detects objects inside the FOV, already one missing detection might indicate that the object hypotheses inside the perception is wrong and should be removed.

Although the assumption of a constant detection probability yields reasonable results for simple setups and scenarios, practical evaluation have shown that most perception systems require a more sophisticated modeling of the detection probability. A typical evolution in MOT modeling is to make the detection probability dependent on the position inside of the sensor FOV. Again, empirical evaluations may be used to derive these values.

- **Occlusion**: In general, a particular sensor is able to detect an object inside its FOV with a given detection probability, may it be constant or dynamically derived from other parameters such as object position or weather conditions. However, for a certain class of perception sensors such as lidar, radar or imagery the FOV may be temporarily impaired due to occlusion of other objects. In order to consistently distinguish between the systematic outage caused by temporarily occlusions and missing detections according to the behavior described by the detection probability, the sensor model needs to explicitly consider the position and spatial extension of other track hypotheses. If an MOT implementation does not consider occlusion, objects which are going to be occluded will be typically instantly discarded from the track list and are consequently no longer part of the environmental model. Proper occlusion handling is especially important for advanced scenarios where communication-based information such as vehicle-to-vehicle data is considered inside of the environmental perception.

7.3.2 Extended Objects

As previously mentioned in Sect. 2.1, MOT aims to estimate the time-varying number and the characteristics of the dynamic road entities in the surveillance area in an integrated manner. Current practical MOT implementations already consider the fact that the sensor observations are in general not perfect and thus are subject to *noise* and *false positives*. Moreover, temporary outages—so called *false negatives*—are typically treated by modeling with an appropriate sensor-specific detection probability.

In addition, the core part of almost every MOT implementation, the data association algorithm, typically assumes that each object generates not more than one physical sensor detection. Basically this implies that the object to be tracked is considered a *point source* that has one unique and fixed reflection center. The rationale behind that choice is to ensure an efficient implementation with a computational complexity that can be handled under real-time conditions.

However, not in all situations, this strong assumption is appropriate. For example, Fig. 7.4 illustrates a scenario where a remote vehicle generates three true

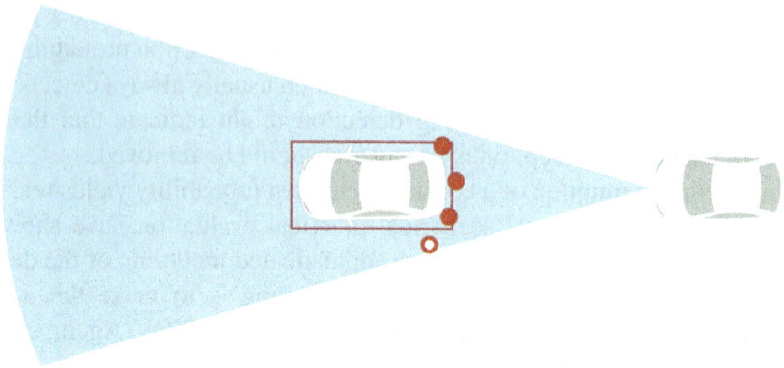

Fig. 7.4 In extended object tracking, an entity may generate more than one detection

observations on the rear bumper which is directly observable by the front-looking perception sensor of the host vehicle. Apparently, whether an object generates more than one true sensor observations directly depends on the sensor resolution and the dimensions of the object under observation. By taking into account that the performance of modern perception sensors constantly increases, the issue of multiple detections per object becomes more relevant for future applications. Moreover, automated systems are going to target denser environments such as urban areas where the average distance to other road users is lower, and thus the number of detections per object increases as well.

A straightforward approach in practical systems to address the challenge of multiple detections per object is to introduce a dedicated detector stage between the sensor interface and the perception layer. Typically a *clustering* approach is chosen which refines multiple observations of an extended object from one epoch to one particular object measurement. By that approach, the point source assumption is kept, and the usual data association algorithms can be applied. However, one major drawback of non-statistical methods such as clustering for object extraction is the still unsolved question of the consistent handling of measurement uncertainties and false-positive phenomena. In addition, the introduction of an explicit detection stage makes it nearly impossible to reverse that decisions once they are made using the shown approaches.

7.3.3 Track Initialization

The basic concept of MOT is to improve the reliability of the state estimate as well as the confidence on the existence of a track by iteratively updating it using measurements from several subsequent time steps. As described in Sect. 3, the main challenge for steady-state tracks is data association. However, another significant step that can have a crucial impact on the tracking performance is *track initialization*. Track initialization is required if new object hypotheses (tracks) are created based on sensor data that is unlikely to be generated from existing tracks. Initialization of a track means to derive a first-state estimate from a single-sensor detection. If the quality of this initialization is insufficient, the track will most likely not be associated to measurements from subsequent time steps. This, in turn, will prevent a steady-state tracking and lead to a repeated creation and deletion of tracks representing the same physical object. In particular for applications where the time to confirm an object is particularly short (e.g., automatic emergency braking), such an effect can lead to severe functional failures.

The main challenge in track initialization is to find initial values for nonobservable quantities, mainly velocity and heading. The difficulty of this step substantially depends on the sensors and the scenarios in which the perception system shall be applied. This shall be discussed on two examples.

Cameras and lidars are capable of providing position measurements of relevant objects. They are not capable of directly measuring velocities. A radar, however, can measure velocities but only in radial direction. This means that the velocities of objects in front of the ego vehicle can be directly measured by the radar, while the longitudinal velocities of objects moving on one side of the ego vehicle cannot.

The heading angle of an object cannot directly be measured.[3] The variability of this quantity, however, depends on the scenario. In a highway scenario, it can be appropriate to initialize objects to travel in the same direction as the host vehicle. In urban environments like intersections, however, this assumption is not valid. Instead, objects may move in arbitrary directions.

A common way to handle initialization ambiguities is to use multiple initialization hypotheses. In later time steps, those hypotheses can be evaluated using their fit to the additional sensor data available at that time. However, this approach requires a rigorous handling of existence probabilities in order to avoid violations of the system's integrity in terms of correctly determining the confidence on the existence of an object.

7.3.4 Asynchronous Sensors and Out-of-Sequence Processing

One challenge in data fusion is the combination of independent measurements from multiple sensors to get a more accurate and reliable estimate of the environment. In general, sensors provide measurements in fixed time intervals. However, they are usually not synchronized with a common clock, and thus it is rather likely that the measurements are asynchronously arriving at the data fusion. Especially in dynamic environments, a synchronization strategy is needed as otherwise the true state of an object might have changed between the sensor observations. By using Bayesian filtering, as proposed in Sect. 2, measurement synchronization is naturally included by applying the prediction step of the system model.

[3] Save by high-resolution sensors such as lidars in close distances as a result of a clustering process.

Furthermore, the data fusion of several heterogeneous sensor systems raises the problem of correctly reordering the individual observations according to their time of validity. This problem commonly referred to out-of-sequence measurements (OOSMs) arises mainly due to different inherent sensor latencies [11]. In general, sensor latencies comprise acquisition time due to the measurement principle, processing time inside of the sensor, as well as the time required for communication toward the perception system. The latencies can also be classified to be either deterministic or nondeterministic. If these latencies are not properly handled inside of the multisensor fusion, the original order of the measurements is lost. Exemplarily sensor latencies of radar-based perception sensors are in the range of 80–200 ms [8].

However, a typical Bayes filter as used in MOT relies on the arrival of data in the correct chronological order and will therefore generate corrupted estimation results under an OOSM scenario without proper handling. Especially scenarios with high dynamics such as vehicle tracking (e.g., automatic emergency braking) are vulnerable to this problem.

Although there are several potential approaches with different levels of complexity to handle OOSMs, it is usually not obvious which technique should be implemented in a particular MOT application. A straightforward solution which gives reasonable results when the variance of the latencies among the perception sensors is rather low is to completely ignore the latency while slightly increasing the modeled measurement noise of the sensors. The rationale behind that approach is to transform the error in the time domain into the measurement domain. Other, more complex approaches, either aim to perform an extrapolation of the sensor measurements [9] or directly try to apply dedicated OOSM strategies such as retrodiction to include the measurements inside of the multisensor data fusion according to their correct time of validity [10].

7.4 Implementation Workflows and Paradigms for Perception

In the previous sections, the core algorithms as well as selected challenges of environmental perception systems have been presented. In this section, these contents shall be mapped on state-of-practice design paradigms and tool-based workflows that support the implementation process. This includes the implementation and configuration of perception software modules and their validation.

7.4.1 Design Paradigms for Perception Software

The straightforward approach for implementing a software component for environmental perception is to apply one of the commonly used models for system design in the automotive area (such as the V-model) and to start the software implementation based on previously defined requirements. This implies using a software development environment that is suitable for embedded systems as the final software is supposed to be executed in an electronic control unit (ECU) of a vehicle.

While this approach is well suited for other parts of automotive software, it displays significant disadvantages for the implementation of environmental perception software. The reasons for this can be found in the challenges described in the previous section: Due to the variability of environmental perception with respect to sensors and scenarios, it is rather challenging to specify the utilized algorithms in advance. Starting the implementation immediately from the specification bears the risk of either failing to comply with those requirements or using an algorithmic configuration that is significantly more complex than required and, thus, wasting computational resources in the final system. Due to the lack of applicability of modern software design patterns to embedded software, also the reusability of such software for future systems is limited.

For these reasons, the most commonly used design paradigm is an iterative approach that is based on an early *prototype* of the system which is continuously tested and improved to refine the system requirements. The advantages of this approach are the higher development speed (due to the usage of a non-embedded environment) as well as the better extendibility and reusability of the implemented software. This paradigm, however, can be further divided: One common option is to implement the software prototype by manual coding. The sole advantage of this approach is the usage of a non-embedded programming language. A second approach is to use model-based paradigms building on predefined libraries and architectural models. This approach yields a significant acceleration of the development process, as it provides a wide range of pre-implemented algorithms and models. The transition to the series code can be done using automatic code generators, which makes the manual coding of embedded software obsolete. In addition to the advantages in development speed, automatic code generation is also widely accepted to provide safer code as typical manual coding errors can be avoided. The disadvantage is that (given sufficient efforts) manual embedded code typically can be more optimized for particular platforms and applications than auto-generated code.

Model-based design using auto-generated code is a well-established design paradigm for many automotive software components (for instance, engine ECU or controller design). The basis of this paradigm is the assumption that the software follows a domain-specific architecture and, thus, can be modeled in a unified way using a predefined set of building blocks. While this is true for many applications, the process of defining a standardized model for environmental perception is still ongoing. As of today, a major part of environmental perception algorithms can be modeled in a unified way; however, there are other parts that need to be adapted to the sensor configuration or the scenario, respectively.

In addition, classical model-based design tools are designed for *signal-oriented* systems and allow for handling and processing a set of scalar signals. The input of

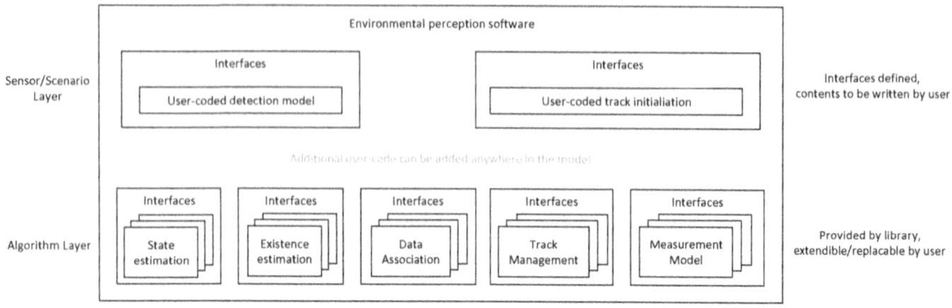

Fig. 7.5 Hybrid design paradigm combining model-based and coding-based aspects

Table 7.3 Comparison of design paradigms for perception software

Design paradigm	Transition to series system	Development time	Flexibility
Direct ECU implementation	Not necessary	Long	None
Coding-based prototyping	Manual re-implementation	Long	Low
Model-based prototyping	Template-based code generation	Short	Medium
Hybrid prototyping	Generic code generation	Short	High

perception systems, however, is typically more complex. Data types to be processed include object lists of variable length, point clouds, or images. The usage of signal-oriented tools for the processing of such data makes the design process particularly complex.

Based on these two limitations of current model-based design paradigms, an additional workflow is increasingly being used. This paradigm is referred to as *hybrid prototyping* in this chapter, as it combines characteristics of model-based and coding-based prototyping. The concept of hybrid prototyping is illustrated in Fig. 7.5: On the one hand, this paradigm assumes that the key algorithms described in Sect. 2 are provided by a perception library in the numerous available variants. By defining a generic architecture of environmental perception, such a library can be used to configure the algorithmic layer.

In a second layer (called the sensor and scenario layer), the variants are too numerous to be covered by one generic model. Thus, standardized interfaces can be defined that allow the user to implement important parts of the perception, such as the detection model and the track initialization, by adding his own code. Finally, the user may add additional code anywhere in the model. This may be necessary to handle special scenario conditions or to implement particular processing prioritization strategies.

The application of this paradigm requires a code generator that is not restricted to a limited set of models and templates, but facilitates a generic code generation from both library-provided models and user code. Table 7.3 summarizes the different design paradigms discussed in this section.

7.4.2 *Software Environments for Testing and Validation*

Testing and validating software typically requires feeding defined input data, executing the software, and comparing the results to the expected output. For environmental perception systems, the dimensionality of the input data is significant. In addition, data for particular scenarios and use cases are required for testing and validation. Finally, the complexity of the system makes it particularly challenging to identify error sources without a domain-specific data representation and the calculation of dedicated key performance indicators (KPIs). These characteristics motivate the usage of domain-specific environments for implementing, debugging, and testing environmental perception software systems. In the following, key properties of such environments are discussed in order to provide guidelines for the reader to select an appropriate environment:

- **Sensor interfaces**: Testing perception software implies feeding it with data from automotive perception sensors such as radars, cameras, or lidars. Multisensor frameworks shall provide ready-to-use interfaces to typical sensors and automotive interfaces (including access to automotive bus systems). To facilitate the usage of proprietary or prototypical sensors, those sensor interfaces should be extendible by the user.
- **Data synchronization**: One key characteristic of perception systems is the usage of asynchronous sensors with different update rates. While some perception algorithms support an implicit data synchronization by predicting the state estimate to the time of the data to be processes, other techniques require a manual data synchronization. A testing environment should provide different configurable ways to synchronize data and active processing components.
- **Data recording/replay**: While tests and demonstrations in a vehicle in live (real-time) conditions are useful, debugging and manual validation requires different modes of feeding data into the perception software. State-of-practice multisensor environments facilitate the convenient recording of complex, timestamped data. This facilitates a synchronous replay including features such as pausing, single-step processing, or the selection of particular scenes.
- **Interfaces to simulation software**: Testing perception software requires sensor data. However, sensors and test vehicles are not always available at the beginning of a development process. Furthermore, not all scenarios can be safely tested in a real environment. For this reason, simulation software[4] is available to allow for defining scenarios and run simulations. Most simulation tools facilitate closed-loop test in which the result of a sensor data processing is fed back to modify the behavior of the simulation. A testing environment for perception systems should provide interfaces to common simulation tools in order to facilitate virtual validation.

[4]For instance, PreScan by TASS International or Carmaker by IPG

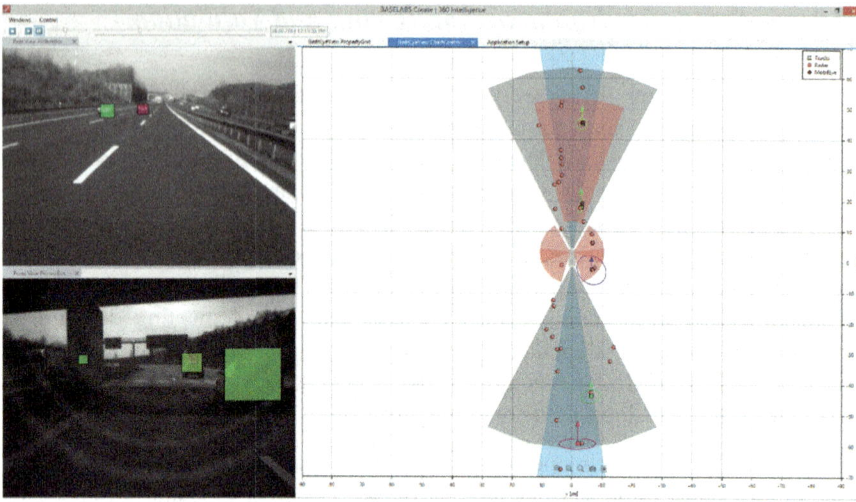

Fig. 7.6 Domain-specific data visualization supports the debugging of perception software

- **Visualization**: Due to the complexity of perception systems, a domain-specific data representation is required to reasonably debug the system and identify errors. For manual debugging, a rich visualization is an invaluable tool for analyzing the perception performance. Typical requirements to that end include overlaying camera images with sensor data or perception results and visualizing the complete traffic scene in a top-down perspective. Figure 7.6 illustrates this notion on the example of a multisensor 360° perception.
- **Interfaces to other domain-specific software**: Parts of software for automated vehicles are written in specific development environments, such as MATLAB Simulink for controller design. Testing tools for perception software should support interfaces to such environments in order to facilitate a unified testing of perception and controller software.
- **Test automation**: In addition to manual debugging, testing perception systems typically requires the automated testing. This implies the automated processing of huge amounts of recorded sensor data as well as the calculation of specific KPIs. Testing environments should, thus, provide interfaces for external test management tools to facilitate selection of data to be processed, parameterization of components, and automated control of applications under test.

These requirements illustrate that the test and validation of environmental perception software requires domain-specific tool support to facilitate an efficient process. Experiences from practice indicate that the tools available for this purpose can contribute significantly to the implementation of reliable perception software.

7.5 Conclusions

The environmental model has been identified as a key component of automated vehicles in this chapter. Typical algorithms for static and dynamic environments, in particular multiple objects tracking, have been presented and discussed. In addition, additional information from a practitioner's perspective have been given:

On the one hand, it has been argued that there is no generic perception for all sensor configurations and use cases, but rather a need for configurable perception software. This configuration needs to be based on the sensor configuration and the targeted use cases (which, in turn, influence the choice of appropriate algorithms). Thus, one conclusion of this chapter is that, for automated driving, *the* environmental model does not exist. Instead, we will most likely see a variety of implementations.

On the other hand, it has been shown that a configurable perception requires new design paradigms. Though classical manual coding or pure model-based design are possible implementation approaches, a much higher design efficiency can be achieved using state-of-the-art methodologies as presented in this chapter.

On the other hand, it has been shown that due to state-of-practice design paradigms and tool-based workflows, the efficiency of the design process for this software could be significantly increased over the last years.

References

1. B. Khaleghi et al., Multisensor data fusion: A review of the state-of-the-art. Inf. Fusion **14**(1), 28–44 (2013)
2. R. Schubert et al., Empirical evaluation of vehicular models for ego motion estimation, in *Intelligent Vehicles Symposium (IV), 2011 IEEE*. IEEE (2011)
3. S. Thrun et al., *Probabilistic Robotics* (The MIT Press, Cambridge, 2005)
4. P.C. Mahalanobis, On the generalised distance in statistics. Proc. Natl. Inst. Sci. India **2**(1), 49–55 (1936)
5. D. Musicki, R. Evans, *Joint Integrated Probabilistic Data Association—JIPDA*, in Information Fusion, 2002. Proceedings of the Fifth International Conference on, vol. 2, pp. 1120–1125, 8–11 Jul 2002
6. C. Adam, R. Schubert, G. Wanielik, *Radar-based extended object tracking under clutter using generalized probabilistic data association,* in Intelligent Transportation Systems—(ITSC), 2013 16th International IEEE Conference on, pp. 1408–1415, 6–9 Oct 2013. doi: 10.1109/ICIF.2002.1020938
7. R. Danescu, F. Oniga, S. Nedevschi, Modeling and tracking the driving environment with a particle-based occupancy grid. Intell. Transp. Syst. IEEE Trans. **12**(4), 1331–1342 (2011). doi:10.1109/TITS.2011.2158097
8. C. Coué et al., Bayesian occupancy filtering for multitarget tracking: An automotive application. Int. J. Robot. Res. **25**(1), 19–30 (2006)
9. M.M. Muntzinger et al., Reliable automotive pre-crash system with out-of-sequence measurement processing, in *Intelligent Vehicles Symposium (IV), 2010 IEEE*. IEEE (2010)

10. A. Rauch et al., Car2x-based perception in a high-level fusion architecture for cooperative perception systems. *Intelligent Vehicles Symposium (IV), 2012 IEEE*. IEEE (2012)
11. Y. Bar-Shalom, Update with out-of-sequence measurements in tracking: Exact solution. Aerosp. Electron. Syst. IEEE Trans. **38**(3), 769–777 (2002)

Chapter 8
Galileo-Based Advanced Driver Assistance Systems: Key Components and Development

Dirk Abel, Bassam Alrifaee, Frank-Josef Heßeler, Matthias Hoppe, and Matthias Reiter

8.1 Introduction

Accompanying the introduction of the European navigation satellite system "Galileo," possibilities of Global Navigation Satellite Systems (GNSS) for advanced applications in the field of driver assistance systems are currently discussed.

GNSS-based applications have gained increasing relevance in the field of modern Advanced Driver Assistance Systems (ADAS), as they contribute to higher safety and comfort in road traffic. While classic driver assistance systems such as antilock braking systems (ABS) and electronic stability control (ESC) use onboard sensors, future ADAS developments take the environment into account. In this context, the effective detection of stationary or moving vehicles in the environment is essential.

Classic driver assistance systems actively support the driver. Nevertheless, current systems are restricted by the limited detection range of the sensors employed and are subject to varying environmental conditions. In this context, a significant improvement can be achieved by the additional use of GNSS in combination with vehicle-to-vehicle (V2V) communication, to enable long range detection of the environment.

The European navigation satellite system "Galileo" will have better integrity and availability than current systems. In addition, it will offer improved position accuracy. In this respect, GNSS-based ADAS are subject of current research at the Institute of Automatic Control of RWTH Aachen University.

This chapter is structured as follows: Section 8.2 gives an overview of the test environment in use. Section 8.3 presents an introduction into sensor data fusion of Galileo and onboard sensors using Kalman filters. Subsequently, Sect. 8.4 presents

D. Abel (✉) • B. Alrifaee • F.-J. Heßeler • M. Hoppe • M. Reiter
Institute of Automatic Control RWTH Aachen University, Steinbachstrasse 54, 52074 Aachen, Germany
e-mail: D.Abel@irt.rwth-aachen.de

D. Watzenig, M. Horn (eds.), *Automated Driving*,
DOI 10.1007/978-3-319-31895-0_8

two applications of Galileo in ADAS, a Cooperative Adaptive Cruise Control and a Collision Avoidance System.

8.2 Test Environment Aldenhoven Testing Center and automotiveGATE

For the testing of safety-relevant advanced driver assistance systems, which can influence the driving behavior of the vehicle, a dedicated automotive testing center is a very important development component. Due to the fact that the automotive manufacturers themselves operate most of the European automotive testing centers, it is very difficult for small and medium enterprises or research facilities to get access to these testing centers. Furthermore, there exists no possibility to test GNSS-based systems in real vehicles under controllable conditions. For these reasons, RWTH Aachen University developed the idea of a manufacturer-neutral automotive testing center in combination with a Galileo testing environment.

The automotiveGATE consists of six terrestrial transmitting antennas, so-called pseudolites. These pseudolites simulate the signals of the European satellite navigation system Galileo within the area of the Aldenhoven Testing Center. The precision of position measurement is up to 0.8 m. The automotiveGATE offers the possibility to test Galileo-based applications independently of the real Galileo satellites. The signals of the automotiveGATE can be manipulated. It is, for example, possible to investigate the influence of different levels of position accuracy on the newly developed Galileo-based driver assistance systems. One additional advantage of this center is that it allows for testing applications under preassigned and reproducible conditions.

In Fig. 8.1, the different track elements of the Aldenhoven Testing Center (ATC) and the positions of the pseudolites of the automotiveGATE are depicted. The ATC provides all necessary track elements for automotive tests. In detail, these are the oval circuit, the handling track, the braking test track, the vehicle dynamics area, the rough road, the hill section, and the highway. For example, the oval circuit has a length of 2 km and the vehicle dynamics area has a diameter of 210 m. Especially, the vehicle dynamics area is ideal for the testing of advanced driver assistance systems, as it is possible to set up arbitrary traffic scenarios.

The combination of the ATC and the automotiveGATE is unique and gives the possibility to test not only standard driver assistance systems but also advanced driver assistance systems, which use position, velocity, and time information (PVT) of different road users under controllable conditions.

The automotiveGATE makes it possible to develop Galileo-based applications before the satellite system is in full operation. The Aldenhoven Testing Center in combination with the automotiveGATE provides ideal conditions for the development of Galileo-based driver assistance systems, which are described in the following sections.

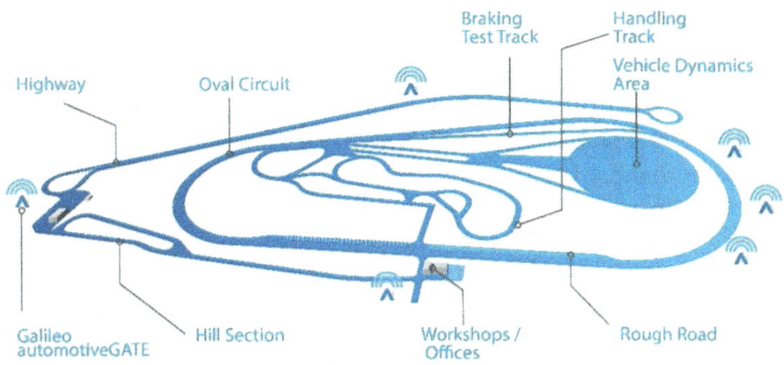

Fig. 8.1 The Aldenhoven Testing Center and automotiveGATE

8.3 Galileo-Based Sensor Fusion

In this chapter, certain fundamental characteristics of Galileo-based (or more generally GNSS-based) driver assistance systems are described. Also, an introduction is given on how these characteristics need to be considered when implementing a GNSS-based control system. More specifically, aspects that will be addressed are the update rates of typical GNSS sensors, delayed input data, and the need for sensor fusion.

8.3.1 GNSS Characteristics

One aspect that almost inevitably needs to be addressed when using GNSS data in control systems is the need for sensor fusion. Unlike other sensors used in automotive control systems, a GNSS sensor needs to be treated as unreliable. Although the position of highly specialized GNSS sensors can reach the order of centimeters or even less, the achievable accuracy is highly situation-dependent. The accuracy can degrade, for example, due to an insufficient number of satellites in view or multipath effects. Also, situations in which no or only imprecise GNSS information is available are manifold, such as tunnel or car-park driving. Obstacles along the road, especially high buildings ("urban canyon"), can also dramatically reduce accuracy. Even in best conditions, a GNSS sensor is subject to a startup acquisition time; therefore, one cannot rely on the signal to be readily available at system start. Once in proper operations, two other aspects are very common for GNSS sensors. For one, data is typically output at a relatively low sample rate of approximately 5–10 Hz. Also, GNSS sensors need a certain computation time in order to calculate a position from the acquired satellite signals. This time delay is very often not negligible and can be on the order of tenths of a second.

8.3.2 Sensor Fusion

Sensor fusion is a very common and useful means in order to overcome restrictions arising from the characteristics mentioned above. For this, the GNSS signal is augmented using onboard sensors, such as accelerometers, gyroscopes, or wheel speed sensors. These sensors are highly reliable and provide information at a high update rate. However, this information is only incremental, such that no absolute position can be calculated from onboard sensors alone. Also, due to error integration, navigation results based on onboard sensors alone are prone to drift. In a typical sensor fusion implementation for navigation applications, GNSS data is used to provide an absolute position measurement whereas the onboard sensor data is used for "interpolation." That way, higher update rates can be achieved. Also, the aforementioned sensor delay can be accounted for. Furthermore, it is possible to provide valid position information for short outages of the GNSS sensor.

From an algorithmic point of view, a predictor–corrector structure is well suited to implement sensor fusion tasks. For this, a dynamic model, e.g., in state space form, can be used. In the following, the indices k and $k-1$ are used to denote the time step. For example, x_k and x_{k-1} denote the system states at time steps k and $k-1$. In analogy, u_{k-1}denotes the input at time step $k-1$ and y_k the (measurement) output at time step k. A_{k-1}, B_{k-1},and C_k describe the system dynamics as well as input and output behavior; no feedthrough is considered:

$$\begin{aligned} x_k &= A_{k-1}\, x_{k-1} + B_{k-1} u_{k-1} \\ y_k &= C_k\, x_k \end{aligned} \tag{8.1}$$

The functionality of the predictor–corrector structure can be outlined as follows. First, the model is used to predict the current state and outputs of the system based on the state of the system at the last time step as well as on known inputs (prediction step).

$$\begin{aligned} \widehat{x}_k^- &= A_{k-1}\, \widehat{x}_{k-1}^+ + B_{k-1} u_{k-1} \\ \widehat{y}_k^- &= C_k\, \widehat{x}_k^- \end{aligned} \tag{8.2}$$

Then, the outputs of the system model are compared to measured system outputs, feeding back the difference in order to correct the estimated model state (correction step):

$$\widehat{x}_k^+ = \widehat{x}_k^- + K_k \left(\widetilde{y}_k - \widehat{y}_k^- \right) \tag{8.3}$$

Here, $\widetilde{y}_k$ denotes the measured system outputs and K_k the feedback gain for the correction. Please note that the nomenclature does not use the "actual" state x_k. Instead, an "estimated" state $\widehat{x}_k$ is used. Furthermore, there is a distinction between the estimated state $\widehat{x}_k^-$ and output $\widehat{y}_k^-$ before ("a-priori") and the estimated state $\widehat{x}_k^+$ and output $\widehat{y}_k^+$ after ("a-posteriori") the most current measurement values have been used to correct it.

8.3.3 Kalman Filter, Extended Kalman Filter

Many methods and variations exist on how the actual prediction and correction steps are implemented. Among the most popular methods are the Kalman Filter [1] and an extension of it, the Extended Kalman Filter (EKF).

A Kalman filter does not only estimate the system state itself but also keeps track of the uncertainty of these estimates in form of an (estimated) covariance matrix P_k of the system state:

$$P_K = E\left[(x_k - \widehat{x}_k)(x_k - \widehat{x}_k)^T\right] \tag{8.4}$$

P_K describes the expected probability distribution of the estimation error. This covariance increases every time a prediction is performed (as more insecurity is introduced through the prediction) and drops every time a new measurement value is used to correct the system state. A thorough introduction into Kalman filtering can be found in [2].

For the design of a Kalman Filter, a system as well as a measurement model of the form

$$\begin{aligned} x_k &= A_{k-1}\, x_{k-1} + B_{k-1} u_{k-1} + w_{k-1} \\ y_k &= C_k\, x_k + v_k \end{aligned} \tag{8.5}$$

is used. Here, w_{k-1} and v_k are white noise disturbances that are assumed to be acting on the process and its measurement output. Their noise levels are quantified using the *Process Noise Covariance* Matrix Q_k and the *Measurement Noise Covariance* Matrix R_k:

$$E\left(w_i, w_j^T\right) = \begin{cases} Q_k, & i = j \\ 0, & i \neq j \end{cases} \tag{8.6}$$

$$E\left(v_i, v_j^T\right) = \begin{cases} R_k, & i = j \\ 0, & i \neq j \end{cases} \tag{8.7}$$

The Kalman Filter Gain K_k to be used in the correction step of the filter according to Eq. (8.3) is then given as

$$K_k = P_k^- C_k^T \left(C_k P_k^- C_k^T + R_k\right)^{-1} \tag{8.8}$$

The update of the covariance estimation is performed as

$$P_k^- = A_{k-1}\, P_{k-1}^+ A_{k-1}^T + Q_{k-1} \tag{8.9}$$

and the correction as

$$P_k^+ = (I - K_k C_k)\, P_k^- \tag{8.10}$$

Given the assumptions of a linear system and that the process and measurement noise are correctly quantified using the white noise assumption, the Kalman filter is an optimal iterative estimator. That means it is not possible to find an iterative filter that can provide a better approximation of the system state [3].

Of course, it is seldom possible to perfectly describe a practical system using a linear model as described above. Furthermore, often neither the white noise assumption holds nor is it always possible to obtain a precise quantification of the disturbances. Still, this type of filter is very powerful and therefore used in many applications. Many extensions and variants of the Kalman Filter and other observer algorithms exist that use the predictor–corrector structure. For example, the EKF uses a nonlinear model within the prediction step and a linearization of it within the correction step [4], allowing to consider certain nonlinear systems. Other filters, such as the Sigma Point filter [5], use a nonlinear model and a sampled probability distribution to improve the covariance update when the system is subject to nonlinearities.

8.3.4 Example: Simple 2D Case

In the following, an example of a model for the implementation of a simple filter is used. The filter is based on measurements of the yaw rate, wheel speeds, and a GNSS sensor. Despite being relatively simple, this filter can already be a substantial improvement over using a raw GNSS signal.

A model widely used in vehicle dynamics is the two-track model (see Fig. 8.2). Here, the vehicle is modeled as a rigid body with the four wheels attached to it.

The forces $F_{X,xx}$ and $F_{Y,xx}$ acting on each tire are modeled through lateral and longitudinal slip that arises from non-holonomous movement. This model is able to consider effects in lateral dynamics such as over- and understeer as the vehicle heading is uncoupled from the direction of movement of the center of gravity. For applications that do not need to consider lateral dynamics in such detail, a simpler model, such as the simplified Kinematic Single-Track Model (see Fig. 8.3), can be sufficient. Here, perfect holonomous movement is assumed; therefore kinetics as well as tire slip are completely neglected. If the reference point (and therefore ideally the mounting point of the GNSS receiver) is selected as the center of the rear axle, its movement can be described as a circular motion.

The turning radius is determined by the vehicle length L and the steering angle δ:

$$R = \frac{L}{\tan(\delta)} \tag{8.11}$$

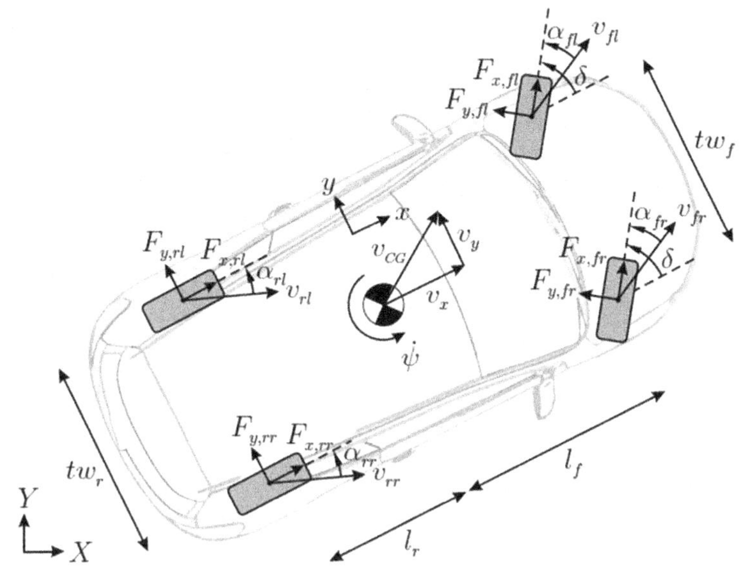

Fig. 8.2 Two-track model

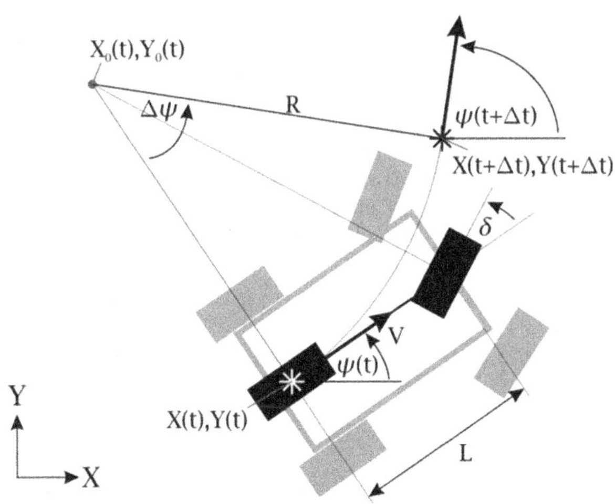

Fig. 8.3 Simplified kinematic single-track model

The yaw rate $\dot{\psi}$ then results as

$$\dot{\psi} = \frac{V}{\mathrm{R}} \tag{8.12}$$

If a small enough time step T is assumed, even a stepwise linear movement can be used to finally describe the model in a state space form:

$$
\begin{aligned}
x_{k+1} &= \begin{bmatrix} X \\ Y \\ \psi \end{bmatrix}_{k+1} = \begin{bmatrix} 1 & 0 & 0 \\ 0 & 1 & 0 \\ 0 & 0 & 1 \end{bmatrix} x_k + \begin{bmatrix} \cos(\psi) \cdot T & 0 \\ \sin(\psi) \cdot T & 0 \\ 0 & T \end{bmatrix} u_k \\
y_k &= \begin{bmatrix} 1 \\ 1 \\ 1 \end{bmatrix} x_k \text{ with } x = \begin{bmatrix} X \\ Y \\ \psi \end{bmatrix} \text{ and } u_k = \begin{bmatrix} V \\ \dot{\psi} \end{bmatrix}
\end{aligned}
\tag{8.13}
$$

The velocity V of the reference point can be taken as the mean speed of the rear wheels whereas a measurement of the yaw rate can be obtained from a (bias-compensated) inertial sensor.

Even if the application does not have high demands on the accuracy, it is advisable to consider the time delay of the GNSS measurements. Although for GNSS sensors, the time stamp of a measurement is known very precisely, it is only available after a (varying) processing time, which can be on the order of tenths of a second. If the measurements are simply used for state correction with no further processing, the system state will be corrected using an up-to-date estimate of the output and a measurement that is several time steps old. An intuitive but effective possibility to handle the delay is to use a ring buffer to save the old system state estimates as calculated by the filter. Then, when a new measurement becomes available, the state estimate corresponding to the actual time of measurement can be obtained from the buffer. More sophisticated approaches to handle the delay are available, for example, as described in [6], where closer attention is paid to not only consider the delayed measurement but also to correctly propagate the covariance estimates.

8.3.5 Example: 3D Case

In case of highly dynamic maneuvers, a simple single-track representation does not capture the important dynamics of the vehicle. Thus, a more complex representation has to be chosen.

For a complete description of the vehicle, the position, the velocity, and the attitude of the vehicle have to be estimated in all three dimensions. An estimate can be calculated using the signals provided by an inertial measurement unit (IMU). This IMU uses three accelerometers and three gyroscopes to measure the accelerations in x-, y-, and z-direction and the roll, pitch, and yaw angular rates. By integration of these three accelerations and three rotational speeds, it is possible to predict position, velocity, and orientation. This integration is able to provide calculation results at a high data rate. However, the prediction is only accurate for a short time being prone to drift.

A GNSS receiver provides long-time stable position and velocity information but at a low data rate. This information can be used to estimate the errors $\delta \boldsymbol{x}$ resulting from the integration of the IMU sensor data. Therefore, a Kalman filter in error state space formulation is used. This filter estimates the error covariance matrix in the prediction step and computes the estimation errors $\delta \boldsymbol{x}$ once a new measurement from the GNSS is available. Besides the estimation of the error of position $\delta \boldsymbol{p}$, velocity $\delta \boldsymbol{v}$, and attitude $\delta \boldsymbol{\epsilon}$, the bias of the accelerometers $\delta \boldsymbol{b_a}$ and gyroscopes $\delta \boldsymbol{b_\omega}$ can be estimated in order to correct them online. All in all, the resulting model consists of the 15 state variables

$$\delta \boldsymbol{x} = \Big[\underbrace{\delta x_n\ \delta x_e\ \delta x_d}_{\text{position error } \delta \boldsymbol{p}^T} \quad \underbrace{\delta v_n\ \delta v_e\ \delta v_d}_{\text{velocity error } \delta \boldsymbol{v}^T} \quad \underbrace{\delta \phi\ \delta \theta\ \delta \psi}_{\text{attitude error } \delta \boldsymbol{\epsilon}^T} \quad \underbrace{\delta b_{a,x}\ \delta b_{a,y}\ \delta b_{a,z}}_{\text{accelerometer bias error } \delta \boldsymbol{b}_a^T} \quad \underbrace{\delta b_{\omega,x}\ \delta b_{\omega,y}\ \delta b_{\omega,z}}_{\text{gyroscope bias error } \delta \boldsymbol{b}_\omega^T} \Big]^T \tag{8.14}$$

Linearization of the model leads to a state space model in the form:

$$\frac{d}{dt} \begin{bmatrix} \delta \boldsymbol{p} \\ \delta \boldsymbol{v} \\ \delta \boldsymbol{\epsilon} \\ \delta \boldsymbol{b_a} \\ \delta \boldsymbol{b_w} \end{bmatrix} = \boldsymbol{F} \begin{bmatrix} \delta \boldsymbol{p} \\ \delta \boldsymbol{v} \\ \delta \boldsymbol{\epsilon} \\ \delta \boldsymbol{b_a} \\ \delta \boldsymbol{b_w} \end{bmatrix} + \boldsymbol{G} \begin{bmatrix} \boldsymbol{n}_a \\ \boldsymbol{n}_\omega \\ \boldsymbol{n}_{b_a} \\ \boldsymbol{n}_{b_\omega} \end{bmatrix} \tag{8.15}$$

with the process noise originating from accelerometer and gyroscope measurements ($\boldsymbol{n}_a$ and $\boldsymbol{n}_\omega$) and sensor biases ($\boldsymbol{n}_{\mathrm{b}_a}$ and $\boldsymbol{n}_{\mathrm{b}_\omega}$), respectively. The sensor biases are assumed to follow a random walk process, leading to a zero-mean Gaussian white noise distribution. The complete description of the linearized state space model is omitted due to its complexity. For the complete mathematical description and the derivation, the interested reader is referred to [7] or [6].

8.4 Applications Examples

In the following, two examples for systems using of the filters and filter models described in Sect. 8.3 are presented. Both examples describe systems that use fused data originating from a GNSS sensor and onboard sensors. Whereas in the first example (Cooperative Adaptive Cruise Control), the sensor fusion is based on the simpler 2D case, the system in the second example (Collision Avoidance System) is based on the 3D case.

8.4.1 Application 1: Cooperative Adaptive Cruise Control

A possible application of GNSS in ADAS is the extension of Adaptive Cruise Control (ACC) systems for road vehicles to situations where the vehicles cannot locate each other using only onboard sensors. This can, for example, be driving through tight corners and on hilly roads, where radar- or lidar-based sensors reach their geometric limitations or situations where vehicles located far ahead have to be accounted for (e.g., at the end of a traffic jam). This section describes an extension of an ACC system to a GNSS- and Map-Based Cooperative Adaptive Cruise Control (CACC) system for road vehicles. The CACC system implements a distance control based on position data acquired from a GNSS sensor fused with onboard sensors, a digital road map, and vehicle-to-vehicle (V2V) communication. The system is validated in experiments.

Figure 8.4 describes the principle structure of the CACC system. First, each vehicle has to be located on a digital road map. Then, the leading vehicle provides its position data to the following vehicle using V2V communication, enabling it to determine the distance between both vehicles. Additionally, the CACC system uses vehicle data from both the following and the leading vehicle. In the following, we give an overview of the main components **distance determination** and **controller** and show experimental results.

8.4.2 Distance Determination Between Two Vehicles

In order to keep a reference distance between two vehicles, the actual distance between them should be determined. However, the Euclidean distance between two vehicles, which is easy to calculate, does not represent the real route distance in between them. The route distance is the actual distance to be travelled on a road network. Therefore, it is the relevant distance concerning the CACC system.

In order to use the map data, the vehicles first need to find the logical equivalent of their measured, physical position on the map. It has to be determined on which map segment and where within this segment the vehicle is located. This process is called map matching. In [9], a map matching algorithm for application in GNSS-based ADAS has been developed that is also applied here.

In the first step of **distance determination,** the map is converted to a directed line graph $G = (V, E)$. The vertices of the directed line graph represent the map segments. Its edges contain the connection information of the map segments. Two vertices are connected by an edge if their respective map segments are connected and if it is possible to drive from the first segment to the second. It is **not** possible to drive between two segments if the second segment represents a one-way road in the opposite direction of travel. Each vertex has a cost value, which is assumed here to be its length. The graph is saved in an adjacency matrix.

The shortest route distance between two vertices (map segments) can be determined by the generated directed line graph. This is done using the Dijkstra graph

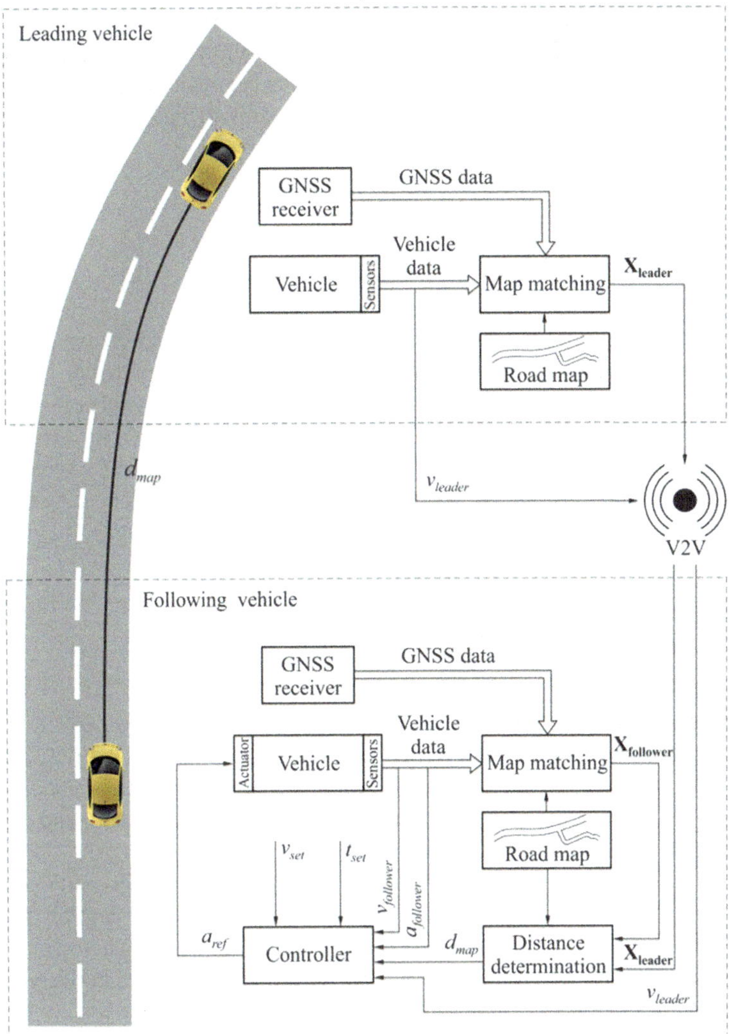

Fig. 8.4 Structure and main components of the CACC system [8]

search algorithm that solves the shortest path problem [10]. In addition, the Dijkstra algorithm is modified to apply to directed line graphs and to fulfill the real time requirements of the controller device. The returned cost from the Dijkstra algorithm of the shortest path is the sum of all vertex costs on it. The route distance between two vehicles is determined as

$$d_{\text{map}} = d_{\text{dijkstra}} - d_{\text{follower}} - d_{\text{leader}} - \frac{l_{\text{follower}}}{2} - \frac{l_{\text{leader}}}{2} \tag{8.16}$$

where d_{dijkstra} is the route distance computed by Dijkstra algorithm, d_{follower} is the route distance between the current position of the following vehicle and the start point of its map segment, d_{leader} is the route distance between the current position of the leading vehicle and the end point of its map segment, and l_{follower} l_{leader} are the following vehicle and leading vehicle length, respectively.

8.4.3 Design of the Distance Controller

The distance controller computes a reference acceleration a_{ref} using a cascaded controller [11]. The inner loop controller is a proportional controller controlling the velocity, and the outer loop controller is a proportional-integral controller with disturbance feedback controlling the route distance. As the velocity error acts as ramp disturbance on the controlled output d_{map}, the integral controller is required to achieve steady-state offset-free tracking in the closed loop setup. The CACC system keeps a reference route distance d_{ref} according to a constant time gap [12].

8.4.4 Experimental Results

The results are based on measurement data that has been recorded on the test track (Aldenhoven Testing Center) using a Volkswagen Passat CC as the following vehicle and a BMW 7 Series as the leading vehicle. The performance of the proposed CACC system is evaluated using a sensor data fusion of radar and a camera as a reference system in the following vehicle.

The distance determination is evaluated by comparing the GNSS- and map-based route distance with the radar-based distance from the reference system on a straight line (1st plot in Fig. 8.5). Thereby, the radar-based distance and relative velocity are 0 if no target object is selected. The 2nd plot shows the difference between several measurement methods of the relative velocity $\Delta v = v_{\text{leader}} - v_{\text{follower}}$. Thereby, Δv_{wheel} (the difference of the raw measurements of the mean wheel velocities) delivers the best availability. Hence, it is used in the CACC system although its accuracy is lower than Δv_{GNSS} (the difference of the raw values of the velocities measured using GNSS).

To validate the CACC system, the following test scenario is chosen. The leading vehicle initially drives with approximately constant velocity (as far as possible for the driver), since the leading vehicle has no cruise control system. The driver of the following vehicle tries to follow it as possible with the same initial velocity and an initial route distance $d_{\text{map}} > d_{\text{ref}}$ without control. The CACC system is enabled at $t = 9$ s (5th plot). Subsequently, the following vehicle accelerates and joins up to the leading vehicle keeping the reference route distance (3rd and 4th plot). The reference route distance is computed using a time gap of 1.8 s and an additional safety distance of 5 m. The experimental results show the applicability of

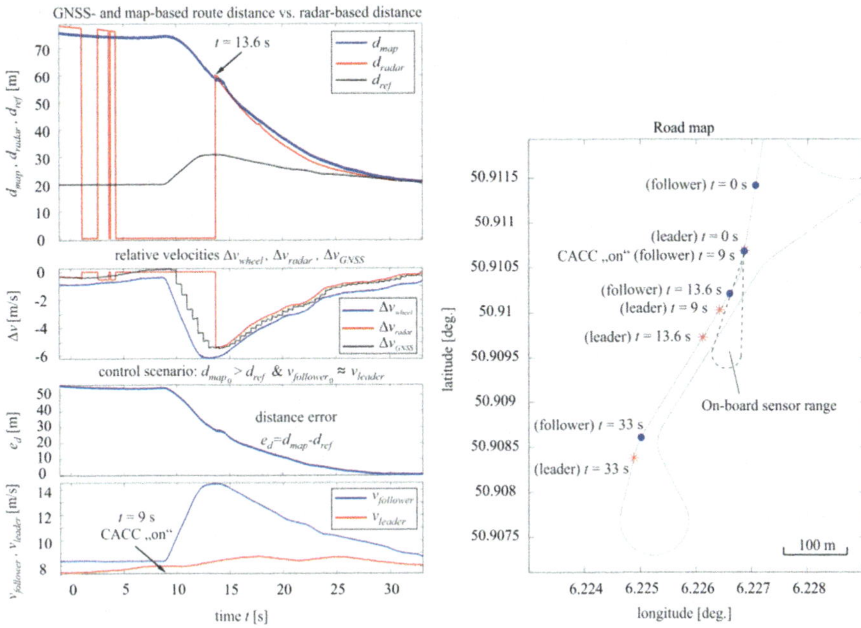

Fig. 8.5 Experimental results [8]

the proposed CACC system. Note that the control could also be conducted at times where geometric limitations did not allow for valid distance measurement using the onboard sensors.

8.4.5 Application 2: Collision Avoidance System

The task of a Collision Avoidance System (CAS) is to observe the surroundings of the vehicle and to perform an autonomous emergency braking or evasion maneuver in case of an imminent collision in order to avoid the collision or mitigate collision damage. The emergency maneuver is started once it is clear that the driver does not react appropriately in time. This leads to an intervention of the system at the last possible moment. Thus, the algorithm must be able to guide the vehicle around the obstacle while driving at the vehicle handling limits.

A collision avoidance system can generally be divided into several different components:

- Sensor fusion: In order to realize reliable collision avoidance, the position, the velocity, and attitude of the ego-vehicle is needed. Since the evasion maneuver will be highly dynamical, a 3D approach as described in Sect. 8.3.5 is needed.

- Environment recognition and collision detection: The CAS needs proper information about the position of possible collision targets in order to avoid them reliably. One possibility to get that information is the use of environmental sensors such as camera, LiDAR, or radar sensors. Another possibility is the combination of GNSS systems with digital road maps and vehicle-to-vehicle (V2V) communication; see [13, 14]. The usage of such a combination can lead to a significant improvement of detection ranges.
- Maneuver coordination: Once a collision is imminent, the CAS needs to decide on a possible maneuver to avoid the collision. This decision includes the choice between braking and evading and the best time of a driver warning.
- Path/trajectory planning: For an evasion maneuver, a feasible and collision-free evasion path has to be found.
- Vehicle Control: Once the evasion path is known, the task of the controller is the longitudinal and lateral guidance along the evasion path.

In the following, it is assumed that a collision is imminent. It is further assumed that an appropriate evasion path is computed in order to concentrate on the vehicle control.

Since the evasion path is computed directly and is thus known, a model predictive control scheme is chosen [15]. The main advantage of a model predictive controller is that limitations from the actuator dynamics and vehicle stability can directly be taken into account. A model predictive controller uses a mathematical plant model to predict the future outputs $\boldsymbol{y}\left(k+i\middle|k\right),\ i=1,\dots,H_p$ of the plant over a finite prediction horizon H_p. The control inputs $\boldsymbol{u}\left(k+j\middle|k\right),\ j=1,\dots,H_u$ are then optimized over a finite control horizon H_u such that the deviation of the outputs from a specified reference trajectory $\boldsymbol{r}\left(k+i\middle|k\right),\ i=1,\dots,H_p$ is minimized.

For the optimization, a quadratic cost function of the form

$$J(k)=\sum_{i=0}^{H_p-1}\left\|\boldsymbol{y}\left(k+i|k\right)-\boldsymbol{r}\left(k+i\middle|k\right)\right\|^2_{Q(i)}+\sum_{i=0}^{H_u-1}\left\|\Delta\boldsymbol{u}\left(k+i\middle|k\right)\right\|^2_{R(i)} \tag{8.17}$$

is used, where $Q(i)$ and $R(i)$ penalize deviations of the control outputs from the reference and changes in the control input, respectively. In this regard, the notation $(k+i|k)$ indicates that the future value of a variable is predicted for time $k+i$ at time k. The minimization of the cost function gives a sequence of optimal input steps $\Delta\boldsymbol{u}_{\mathrm{opt}}\left(k\middle|k\right),\dots,\Delta\boldsymbol{u}_{\mathrm{opt}}\left(k+H_u-1\middle|k\right)$. The control input $\boldsymbol{u}_k=\boldsymbol{u}_{k-1}+\Delta\boldsymbol{u}_{\mathrm{opt}}\left(k\middle|k\right)$ is applied to the plant, and the optimization is repeated with a shifted prediction horizon. This principle is called receding horizon.

In order to apply the predictive control scheme, an appropriate plant model for the vehicle dynamics and the relative kinematics of the vehicle and the evasion path is needed. Figure 8.6 shows a principle sketch of the single-track model which is used.

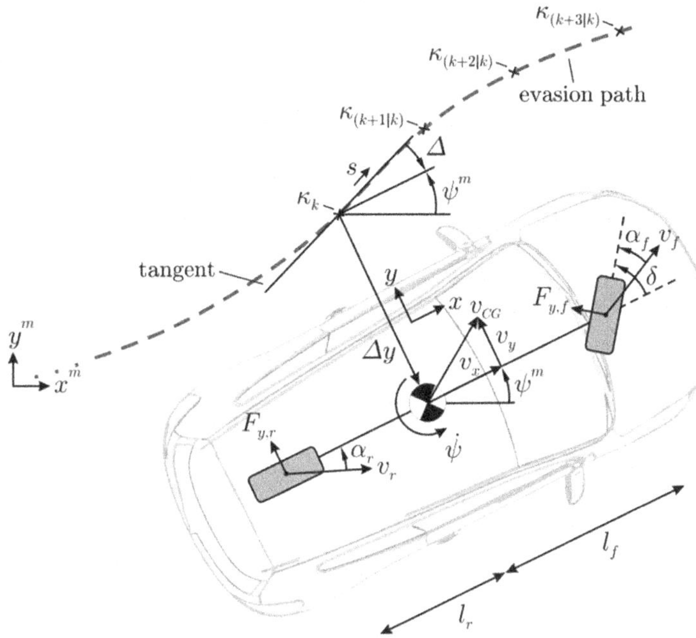

Fig. 8.6 Free-body diagram of the employed prediction model [7]

All in all, the resulting nonlinear prediction model can be written in state space representation as

$$\begin{aligned}\dot{\boldsymbol{x}}(t) &= \boldsymbol{f}\left(\boldsymbol{x}(t), u(t), z(t), \theta(t)\right) \\ y(t) &= g\left(\boldsymbol{x}(t)\right) = \Delta y\end{aligned} \tag{8.18}$$

In that model, the state vector

$$x^T = \left[\, v_x \; v_y \; \dot{\psi} \; \delta \; \Delta\psi \; \Delta y \; d_{\Delta\dot{y}} \,\right] \tag{8.19}$$

consists of the lateral and longitudinal velocity v_x and v_y, the yaw rate $\dot{\psi}$, the actual steering angle δ, the relative yaw angle between the vehicle's center of gravity and the evasion path $\Delta\psi$, the lateral displacement Δy from the evasion path, and the lateral velocity disturbance $d_{\Delta\dot{y}}$. The control input u is the demanded steering angle δ_{ref}. The path's curvature k is assumed as a known disturbance value. Additionally, an adaption parameter $\theta(t)$ is added to account for unknown changes in tire-road contact.

Figure 8.7 shows the control results for a double lane change maneuver. This corresponds to the evasion of a standing obstacle in a scenario with oncoming traffic. The initial velocity of the vehicle is approximately 15.5 m/s.

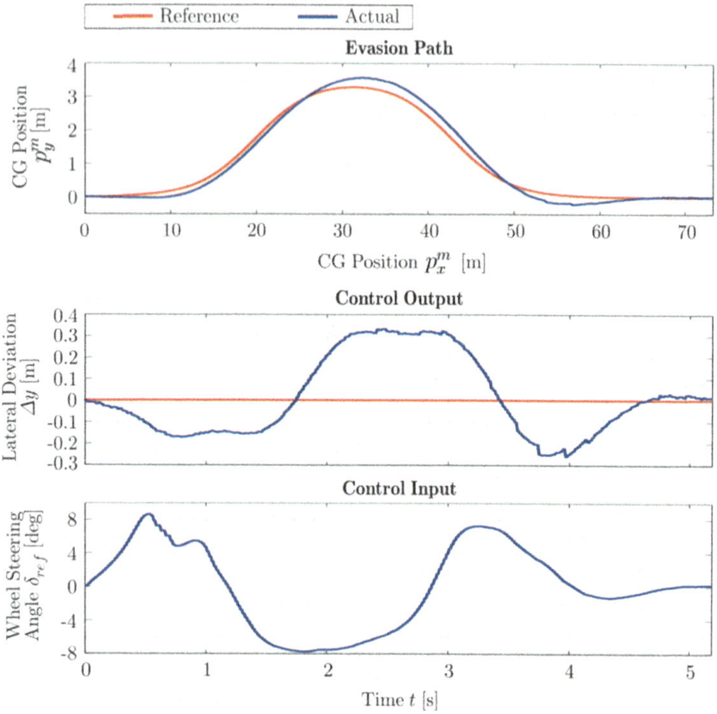

Fig. 8.7 Experimental results for MPC (based on [7])

It is visible that the controller follows the evasion path well with a maximum absolute lateral deviation of 0.33 m. Best control results were achieved with a prediction horizon of about 1 s and a control horizon of about 0.5 s. Furthermore, a sample time of 0.02 s is chosen. Simulations have shown that shorter horizons lead to a degraded control performance which may even lead to stability problems for very short horizons.

8.5 Conclusion

This contribution presented a test infrastructure that can be used to develop Galileo-based Advanced Driver Assistance Systems. The infrastructure consists of the Aldenhoven Testing Center (ATC) in combination with a pseudolite system that provides a **G**alileo **T**est and **D**evelopment **E**nvironment (automotive**GATE**). This infrastructure allows company-independent research on Galileo-based control systems. An overview is given on characteristics that are imminent to GNSS-based control systems and how some of the arising issues, such as measurement delay and low update rates, can be addressed using sensor fusion techniques.

Then, implementations of two GNSS-based ADAS applications developed using the presented infrastructure were presented. The first system uses navigation data in combination with a digital road map as well as V2V communication in order to extend the range and functionality of an Adaptive Cruise Control system. The second system performs an emergency Collision Avoidance maneuver based on GNSS and inertial data.

Acknowledgement The presented research work has been performed within the projects "Galileo above" and "GALILEO-basierte Assistenzsysteme."

The project "Galileo above" is sponsored by the Space Agency of the German Aerospace Centre (DLR) with funding by the Federal Ministry of Economics and Technology, in compliance with a resolution of the German Parliament (project/grant no. 50 NA 0902).

The project "GALILEO-basierte Assistenzsysteme" is funded by the Ministry of Economics, Energy, Industry, and Small Business of the State of North Rhine-Westphalia and the European Union with Financing of the European Regional Development Fund (ERDF).

The idea of a manufacturer neutral testing center (Aldenhoven Testing Center) was realized with the help of the county of Düren and fundings by the State Government of North Rhine-Westphalia and the European Union. The idea of the Galileo testing center for automotive applications (automotiveGATE) was implemented within the aforementioned project "Galileo above" by the funding of the DLR Space Administration with funds of the Federal Ministry for Economic Affairs and Energy. Both facilities explicitly guarantee the possibility to be used especially by small and medium enterprises and research facilities. All other customers are also welcome.

References

1. R.E. Kalman, A new approach to linear filtering and prediction problems. J. Basic Eng. **82**(1), 34–45 (1960)
2. M.S. Grewal, A.P. Andrews, *Kalman filtering: Theory and Practice with MATLAB* (Wiley, Hoboken, NJ, 2014)
3. P.S. Maybeck, The Kalman filter: An introduction to concepts, in *Autonomous Robot Vehicles*, ed. by I.J. Cox, G.T. Wilfong (Springer, New York, 1990), pp. 194–204
4. H.W. Sorenson (ed.), *Kalman Filtering: Theory and Application* (IEEE Press, New York, 1985)
5. R. van Der Merwe, E.A. Wan, Sigma-Point Kalman Filters for Integrated Navigation, in *Proceedings of the 60th Annual Meeting of the Institute of Navigation (ION)*, pp. 641–654, 2004
6. J. Wendel, *Integrierte Navigationssysteme* (Oldenbourg Verlag, Munchen, 2011)
7. A. Katriniok, Optimal Vehicle Dynamics Control and State Estimation for a Low-Cost GNSS-based Collision Avoidance System, Ph.D. thesis, RWTH Aachen University, Aachen, Germany, 2014
8. B. Alrifaee, M. Reiter, J. P. Maschuw, F. Christen, L. Eckstein, D. Abel. Satellite- and Map-based Long Range Cooperative Adaptive Cruise Control System for Road Vehicles, in *AAC 2013, 7th IFAC Symposium on Advances in Automotive Control*, pp. 732–737, Tokyo, Japan, Sept 2013
9. M. Reiter, D. Abel, D. Gulyas, C. Schmidt. Konzept, Implementierung und Test eines echtzeitfähigen Map-Matching-Algorithmus für Anwendungen in GNSS-basierten Fahrerassistenzsystemen, in *AUTOREG*, 5, pp. 389–400, Baden-Baden, Nov 2011

10. E.W. Dijkstra, A note on two problems in connexion with graphs. Numerische Mathematik **1**, 269–271 (1959)
11. R.C. Dorf, R.H. Bishop, *Modern Control Systems*, 9th edn. (Prentice-Hall, Upper Saddle River, NJ, 2000)
12. R. Rajamani (ed.), *Vehicle Dynamics and Control* (Springer, New York, 2005)
13. R. Schubert, M. Schlingelhof, H. Cramer, G. Wanielik, Accurate Positioning for Vehicular Safety Applications—The SAFESPOT Approach, in *IEEE Vehicular Technology Conference*, pp. 2506–2510, 2007
14. P. Papadimitratos, A. La Fortelle, K. Evenssen, R. Brignolo, S. Cosenza, Vehicular communication systems: Enabling technologies, applications, and future outlook on intelligent transportation. IEEE Commun. Mag. **47**(11), 84–95 (2009)
15. J. Maciejowski, *Predictive Control with Constraints* (Prentice Hall, Harlow, 2002)

Chapter 9
Digital Maps for Driving Assistance Systems and Autonomous Driving

Alexandre Armand, Javier Ibanez-Guzman, and Clément Zinoune

9.1 Introduction

The use of digital maps in the form of on board vehicle navigation systems or in mobile telephones has been widely deployed for in past years. Their use has become part of current driving tasks, particularly for drivers unfamiliar with the roads leading to their destination. Digital maps store information on the road and its attributes; further currently, vehicle connectivity means that geolocalized information can be cascaded from multiple sources to drivers enhancing the information in them. Digital maps allow drivers to anticipate what will encounter as they drive, to form better mental models with regard to their expectations and ultimate shape their intentions. For example, it is possible to anticipate the arrival to a complex intersection or the crossing of a sharp bend or a sudden change in vehicle speed limits.

After a slow start like with the adoption of radars for adaptive cruise control (ACC) systems, the deployment of exteroceptive sensors on board passenger vehicles has rapidly expanded over the past years. Vision-based ADAS systems form part of the current offer in particular for the detection of lane markings, vehicles and pedestrians. In Europe, the independent assessment of the safety level available in passenger cars, Euro NCAP, has encouraged safety improvements in new car designs and thus propelled the use of sensor-based safety systems on board of European vehicles as proactive safety systems. Most of the current offer relies on the use of video cameras, whose outputs are processed by purpose built processors achieving results that were only found in research laboratories years back. The output of perception systems are classified objects with their position and time orientation is estimated with respect to the vehicle navigation frame. Whilst this

A. Armand (✉) • J. Ibanez-Guzman • C. Zinoune
Renault S.A.S., Technocentre de Guyancourt, Research Department, France
e-mail: alexandre.armand@renault.com

D. Watzenig, M. Horn (eds.), *Automated Driving*,
DOI 10.1007/978-3-319-31895-0_9

is a major progress, due to the complexities found in using machine vision systems in real time, challenges remain.

Perception systems are far from perfect. The presence of false positives and false negatives leading to hazardous situations remain. This is being addressed through the use of multi-sensor approaches and the fusion of their acquired information to provide a better perceived world as well as increased interest on the use of a new generation of active sensors like laser rangers or the application of convolution network techniques for the classification of video data. Nevertheless, once the perceived world is classified and registered, the machines using this information for decision-making (e.g. stop the vehicle or perform a collision avoidance manoeuvre) need to make sense of this information.

Situation understanding is a major function prior to decision-making and the lateral/longitudinal motion control of the vehicle. It is necessary to understand the spatio-temporal relationship between the perceived objects with respect to the road network and the subject vehicle. However, making sense of the perceived object without context is a difficult task for a machine. For example, as the subject vehicle arrives to an intersection, it needs to scan for vehicles likely to arrive perpendicular to its path or those in front that might turn. Without contextual information, the presence of vehicles around the intersection could be confusing to the machine and likely lead to hazardous situations. By projecting the perceived objects on top of a digital map, it can be inferred, e.g. how far a vehicle is from an intersection, or who has priority at such an intersection by reading the road attributes associated with it. Similarly, the presence of a pedestrian next to a road intersection or next to a pedestrian crossing would lead to an interpretation that the pedestrian is likely to cross in front of the vehicle, whilst without such information, the machine will only know that there is a pedestrian close to its path.

The perceived objects are projected into segments of the stored maps in order to facilitate the understanding of spatio-temporal relationships between these objects, road infrastructure and the subject vehicle. This information is then used to provide a situation understanding and subsequently to infer decisions. Therefore, a local world model is first constructed by mapping the vehicle position on the digital navigation map. This results in the extraction of a segment where the perceived objects can be projected on. This provides an instantaneous representation of the world.

The first part of this chapter introduces a methodology to infer situation understanding and thus to identify the likely interactions between perceived road users and road features in order to estimate their impact on vehicle navigation. Therefore, a framework for the structured representation of the data in terms of a purpose built ontology is proposed. This is then used to assess risk as applied to road intersections, information that is used to infer decisions on the action to take at a road intersection. The principle is shown in Fig. 9.1. A major feature of this approach is that the application is inferred from existing production sensors and that the output can be either used to provide anticipatory information to drivers in the case of ADAS systems or to the decision-making mechanisms in the case of highly automated vehicles. The chapter includes results of the application of the approach in an experimental vehicle evolving in public roads.

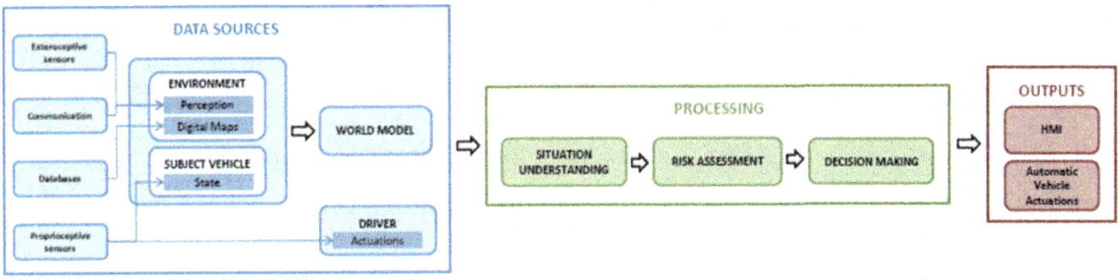

Fig. 9.1 Framework for inferring spatio-temporal situation understanding applicable to ADAS systems and highly automated vehicles

The digital map is the underlying component in this process. It provides the stored information that is exploited for contextualization and to infer situation understanding. The source for these maps originates from map suppliers, the most known being HERE (previously known as Navteq) and TOM TOM. Digital maps have been built over the years from different sources including the use of purpose built surveying cars that will collect road information from laser rangers, video cameras and vehicle position estimates. That is, digital maps can be regarded as a patchwork of maps originating from various sources. Their precision and resolution varies and thus not only road attribute errors but also geometric errors exist. This can represent several metres in relative terms between the links and nodes but larger errors in absolute terms, that is with respect to the world reference frame like the WGS84. Another error could reside on the road structure itself, that is maps are not up-to-date as the road structure would have changed, for example, the incorporation of a roundabout replacing a road intersection. If the maps are wrong, the approach proposed in the first part of this chapter will not be applicable, the inference of situations will be made based on the wrong assumptions which shall lead to errors, and eventually hazardous situations. The second part of this chapter proposes a framework that enables the detection of faults in the geometry of navigation maps through the use of standard vehicle proprioceptive sensors plus GNSS receivers. A particular feature of the described approach is that it can solve the ambiguity that could arise due to the presence of errors in the localization systems. That is when an spacial error occurs, this can be either from errors in the map geometry but it can also originate due to errors in the position estimates. Whilst in the first run, the framework can indicate that an error exists, its origins on whether this originates at the map or the position estimate is not know, however, once the vehicle crosses again the same area, the framework can discern that the error is a map geometric error or a position estimate. This approach can be possible as most driving journeys are repetitive, and thus vehicles will drive over similar paths often as when commuting from home and the work place in a daily manner. The different algorithms applied to implement the fault detection are included as well as results of trials performed using data collected from public roads.

9.2 Ontology-Based Situation Understanding

9.2.1 Ontologies

Ontologies are an essential concept for semantic-based approaches and are therefore introduced first.

The term *ontology* was first used by philosophers to designate the study of being of existence. It has been adopted by researchers in artificial intelligence from the beginning to design computational models for automated reasoning [11]. The most famous definition describes an ontology as 'an explicit *specification of conceptualization*' [9]. The conceptualization of a domain is the manner how a domain is perceived and understood, and the specification of this conceptualization is actually a formal description of this conceptualization.

More concretely, an ontology is a description of the concepts and relationships that are relevant to model a domain of interest. It specifies the vocabulary that is necessary to make assertions, and which may be inputs/outputs of knowledge agents (e.g. software, etc.). Moreover, it provides the language for communication between agents [10]. Figure 9.2a illustrates this definition.

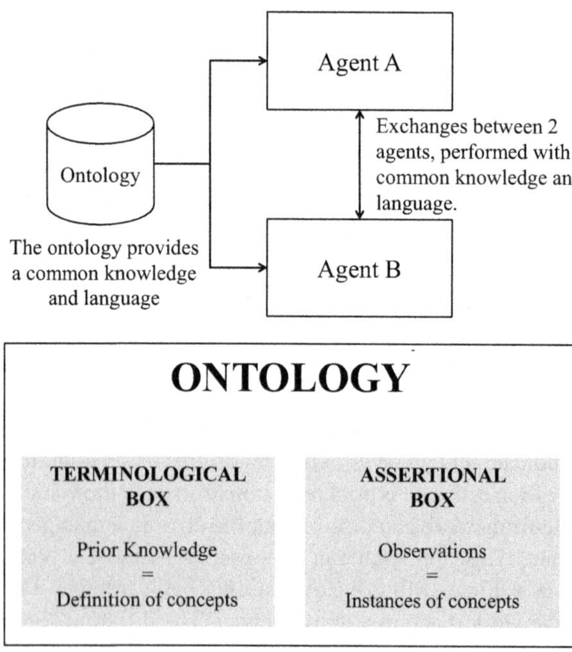

Fig. 9.2 Ontology. (**a**) Ontology used by agents (e.g. software) for communication. (**b**) Structure of an ontology-based on Description Logic

Ontologies are based on Description Logics (DL) which is a formal language for knowledge representation [3]. A DL enables to model Concepts, Roles and Individuals through its two functional parts, the Terminological Box (TBox) and the Assertional Box (ABox). Figure 9.2b illustrates this structure.

The TBox consists of the definition of all the concepts that the ontology aims to describe. An analogy can be done between the TBox and the knowledge that human have. The knowledge that humans acquire along their life is used to understand and to interpret the world. The ontology TBox represents prior knowledge, and the definition of it is performed through the definition of Concepts, Roles and Relations. The following definitions were established after [13]:

- *Concepts* (or classes) are concrete representations of the concepts of the domain that the ontology aims to describe. These concepts can be organized into a superclass–subclass hierarchy, which is generally called *Taxonomy*.
- *Roles* are properties which can be defined and assigned to concepts. Roles can be classified into two groups:
  - *Object Properties* aim to define axioms in the form of *Triples*. In other words, they are binary relationships between two concepts in the form *Concept1—Object Property—Concept2*. Characteristics may be attributed to object properties, such as symmetry or transitivity with respect to other object properties.
  - *Data Properties* are used to assign properties to single classes or instances of classes in the form *Concept1—Data Property —Property Value*.
- *Relations* between concepts are defined with taxonomic relations (hierarchical relations), axioms (classes linked by object properties) and rules. The definition of rules can be done using basic description logic axioms which only enables the definition of basic class equivalence. More sophisticated languages enable to define more complex and expressive rules. Among these languages, the Semantic Web Rule Language (SWRL) is one of the most common [14].

The ABox consists of the definition of instances of classes previously defined in the TBox. These instances, commonly called *Individuals*, represent real life data that the ontology aims to interpret. Again, an analogy may be done with humans as the ABox can represent objects that humans observe and understand because of prior knowledge (TBox). Further, in the same way as properties can be attributed to concepts defined in the TBox, Object and Data Properties can be attributed to individuals defined in the ABox.

The main advantage of ontologies is the possibility to reason on knowledge using *Reasoners*. Reasoners are pieces of software able to infer logical consequences from a set of asserted facts or axioms [1]. In other words, they aim to exploit information stored in the TBox in order to infer new information and knowledge which is not specifically expressed about *Individuals* defined in the ABox.

The reasoning methodology refers to the algorithm that is used for the reasoning task (e.g. tableau-based or hypertableau algorithms [23, 35]). The methodology has a significant influence on computational efforts required for reasoning. Performance of a reasoner also includes the soundness and completeness which is the capability to correctly perform all inferences which should be done in theory.

9.2.2 Situation Understanding

Three terms are generally used when talking about situation understanding: *Scene*, *Situation* and *Scenario*. The meaning of each of these terms in the context of intelligent vehicles is rather vague and definitions are often contradicting. Geyer et al. formally defined the meaning of these terms in the context of assisted and automated driving guidance [7] as shown in Fig. 9.3. The definitions are rather abstract and have been interpreted with respect to the problems addressed in this article:

- A ***scene*** is a snapshot of a collection of cohabiting road entities, including the subject vehicle and the surrounding static and dynamic entities. Each entity is defined by its type and state. A *scene* can therefore be represented by all data returned by the data sources.
- A ***situation*** is the *scene* as it has to be understood by a particular dynamic entity of this scene (i.e. the subject vehicle). This consists in understanding how the interactions between all entities of the Scene are propagated to this entity and constrain it in its navigation. Therefore, Situation understanding consists in giving sense to a Scene.
- A ***scenario*** is understood as a sequence of Scenes which is the consequence of the Situation of all interacting road entities present in a primary Scene. In the context of this thesis, predicting a Scenario consists in predicting the future state of the participating entities in order to estimate the risks lead by the Situation.

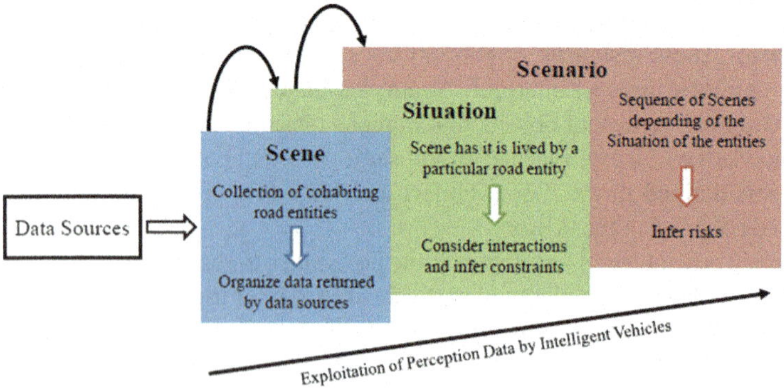

Fig. 9.3 Relation between scene, situation and scenario

9.2.2.1 Methods for Situation Understanding

Situation understanding consists of reasoning on the scene representation, i.e. adding meaning to the scene information. From the subject vehicles point of view it means to infer how it is constrained by the surrounding environment. Two main categories for situation understanding were identified: probabilistic approaches and semantic-based approaches.

Probabilistic Approach

Road situations are often highly complex due to the variability of situations, especially in urban environments. Vehicles are able to perceive plenty of surrounding entities, however, most of these entities are not relevant. Platho et al. proposed to decompose situations into 'parts of situations' or 'configurations' [25]. This simplifies the scene by considering only relevant entities. The notion of relationship between road entities is introduced by, for example, considering that the behaviour of a perceived vehicle can be affected by a red traffic light or by another vehicle which has stopped. The recognition of configurations is performed through a Bayesian Network. The approach was tested in simulated environments, and was used to predict the velocity profiles of other road users in intersection situations [26]. However, the approach only allows to consider direct relationships between entities but no chain reactions (e.g. if the subject vehicle follows a vehicle that approaches to a pedestrian, the interaction between the lead vehicle and the pedestrian is not taken into account).

Probabilities allow to take into account uncertainties on perception data, but so far they can only be used to model basic situations based on scenes represented with conventional methods. The amount of possible situations which may occur makes it difficult to define a generic probabilistic model which would be well suited for all situations and which would be capable of considering interactions between road entities. This may explain why literature does not propose other probabilistic frameworks for situation understanding than the one presented above.

Semantic-Based Approaches

Semantic-based approaches do not only use semantics for description, but also apply reasoning. Most of these approaches either use First-order Logics [36] or Description Logics [3] to describe concepts in ontologies. First, semantic-based approaches were used to model and understand the road network from the point of view of the subject vehicle. Later, they were also used to model and understand the whole interaction between road network and dynamic entities.

One of the first works exploiting DL for situation understanding was done by Hummel et al. [17]. The proposed ontology introduces the concepts of road networks (roads, lanes, dividers, road markings and junctions) and is used as a complement of vision sensors and digital maps to retrieve relevant information about intersections. For example, if the precision of the localization sensors is not able to determine the current lane the vehicle is navigating on, this information can be inferred by the ontology from map and camera data. Whilst this formalism does not take into account cohabiting road entities (vehicles, pedestrians, etc.), it shows that ontologies can be used to reason, at least partially, on road situations.

The representation of road intersection networks through ontologies was introduced by Regele [32]. It was used to solve the traffic coordination problem of autonomous vehicles, i.e. to handle conflicts between vehicles reaching the same intersection or cohabiting in the same area. This work inspired Hülsen et al. who proposed a generic description of road intersections for situation understanding [16]. This ontology enables to infer conflicts and thus potential risk situations for vehicles reaching the same intersection. Figure 9.4 illustrates a sample representation of a relationships. The framework was tested on several intersections and its efficiency was proved even for very complex intersections. A real time implementation of the framework in simulated environments was successfully performed [15]. In these ontologies, vehicles and other road entities are not formally represented.

Vacek et al. [37] introduced an ontology-based representation of other road entities. In that way, semantic information about road entities (i.e. types, etc.) were defined in an ontology which is used within a case-based reasoning framework to perform scene understanding. The tenet was to recover similar or resembling situations in order to infer the behaviour which seems to be the most appropriate to negotiate the situation. Whilst the ontology allows to represent different types of

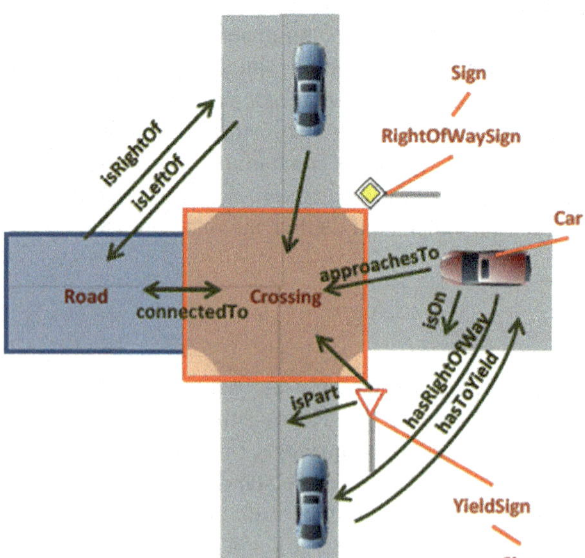

Fig. 9.4 Example of semantic representation of an intersection. After [16]

road entities, relationships between them are ignored, which finally prevents from understanding the situation as a whole.

First-order probabilistic languages (FOPL) was used by Schamm et al. [34] to perform situation assessment. A FOPL knowledge base was used to model driving situations and interactions. This a priori knowledge is thus used with sensor information to automatically create a probabilistic network and then to infer the situation in a structured manner. Since all entities are conceptually of the same type, interactions between entities of different types are not considered. Moreover, first-order logic suffers from poor expressivity, which therefore prevents from editing complex rules in the knowledge base. Further, it seems difficult to extend such a system for more complex situations than those presented in [34].

Zhao et al. built a knowledge base which contains information about maps and traffic regulation. It is used within safety ADAS to take decision at road intersections in case of over speed [39]. Three ontologies were defined for this purpose. The first one aims to describe information which may be stored in a digital map, the second aims to describe control strategies and the last one aims to describe vehicles. Interactions between road entities are considered only between vehicles reaching the same intersection, for the generation of collision warning.

An ontology that models traffic scenes in order to establish the state space of the subject vehicle with respect to other vehicles and the road network was proposed by Kohlhaas et al. [19]. Two categories of objects are considered, namely the environment objects (related to the road network) and the dynamic objects (related to vehicles). The interactions between the vehicles and the road network as well as the lateral and longitudinal interactions between vehicles are formally stated. Further, the ontology contains information about traffic rules through defined conditions.

Finally, Pollard et al. proposed an ontology that represents features of the road network, environmental conditions, sensors states, subject vehicle state and presence of moving obstacles [27]. This ontology enables vehicles to perform self-assessment on their automated driving capabilities, with the aim to decide on the most appropriate automation level (from fully manual to fully automated).

Semantic-based approaches enable to model road scenes and interactions between entities as well as to reason in a straightforward manner on situations. Their main limitation is their inability to take data uncertainties into account. FOPL enables to fill this gap, however, it does not enable to model complex situations because of the low expressivity of the language.

9.2.3 Framework for Ontology-Based Situation Understanding

The developed ontology is part of an overall framework for situation understanding shown in Fig. 9.5.

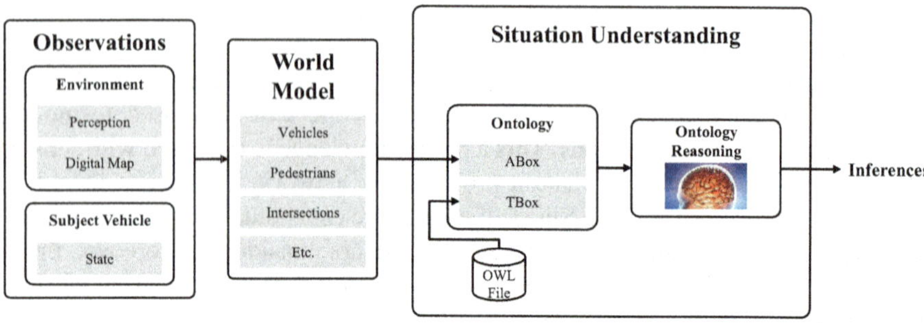

Fig. 9.5 Exploitation of the ontology within a framework

The ontology presented here is not exhaustive and must be considered as a draft that is used to confirm the coherence of the approach. Further extensions and optimizations are required for a real-world validation.

9.2.3.1 Observations

The first prerequisite for the framework is information about surrounding entities. This step is represented by the Observations box in Fig. 9.5. Two types of data sources are considered for the awareness of the environment. Modern vehicle sensors such as smart cameras, radars or lasers allow for the real time perception of moving entities. Most of them are able to perform classification on the perceived entities and provide an estimation of their state with respect to the subject vehicle on which they are embedded. Further, digital maps can store and provide contextual information about the road network features. For instance, this a priori information can contain information about the coming road intersections, about coming pedestrian crossing, etc. Figure 9.6 shows a sample situation as it may be perceived by a vehicle.

9.2.3.2 World Model Principles

Information is provided in a piecemeal manner from different mostly independent data sources. It is therefore necessary to organize this data as a list of surrounding entities. This structured and organized list is called the *World Model*. Figure 9.7 provides a sample world model for the situation shown in Fig. 9.6.

The creation of this world model may require some preliminary processing on perception data. One entity may be perceived by several sensors at the same time or sensors may not be synchronized. For these purposes, data and sensor fusion techniques have to be employed [20]. These problems are complex to solve and today they remain a meaningful challenge for the data fusion community. Here,

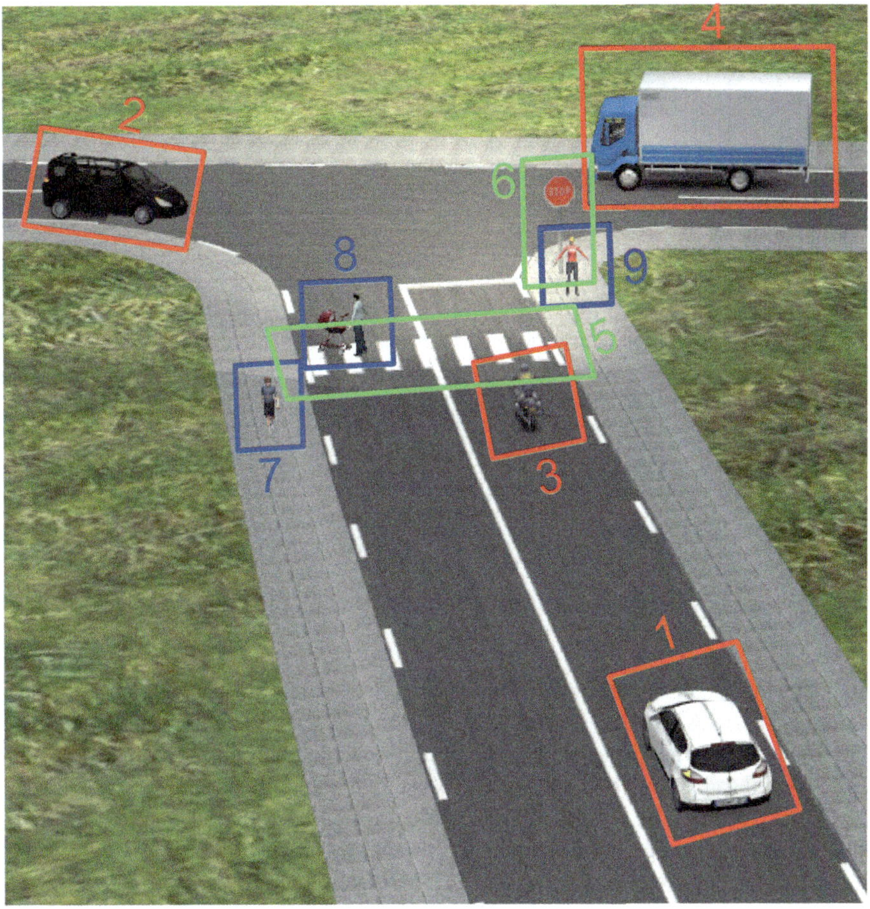

Fig. 9.6 Example of perceived world. The perceived entities are given IDs

perception is considered as a black box performing sensor and data fusion on the data returned by a set of perception sensors.

9.2.3.3 Situation Understanding

Situation understanding is based on the use of an ontology (see Fig. 9.5) consisting of two fundamental parts, the TBox and the ABox.

The TBox consists of a conceptual description of the entities and contextual objects which can be met by a vehicle in a navigable space. In other words, it enables to define the types of entities which can be met and the relationships and interactions which are likely to exist between them. An analogy can be done with the knowledge that drivers acquire when they learn driving at driving school, which is fundamental

#	Type	Information
1	Car	Subject Vehicle
2	Car	Coming from the left Turning right
3	Motor Bike	Same lane x meters ahead
4	Truck	Coming from the right x meters before intersection
5	Pedestrian Crossing	x meters ahead
6	Stop Intersection	x meters ahead
7	Pedestrian	x meters ahead On left pavement
8	Pedestrian	x meters ahead On the road
9	Pedestrian	x meters ahead On right pavement

Fig. 9.7 World presented in Fig. 9.6 in the form of World Model table

to makes them able to understand situations. The TBox is the permanent part of the ontology and was developed with respect to the Description Logic specifications. The focus of this ontology is only on situations which can be represented in one-dimensional space (i.e. road entities on the same navigation lane).

The ABox can be considered as the conversion of the *World Model* into the ontology language. For each entity in the *World Model* an instance of the corresponding concept is created. The ABox is the changing part of the ontology and is updated at each update of the *World Model*. After each update of the ABox, reasoning can be performed on the whole ontology in order to give more sense to the data. More precisely, it means to take into consideration the interactions which are likely to exist between the entities and also chain reactions which may happen as a consequence of these interactions. The purpose is to infer a high level interpretation of the perceived situation in order to select the risk assessment algorithms which suit the situation best.

TBox

Figure 9.8 shows the taxonomy which defines the ontology. It is described in the following:

- The *Context entity* aims to list and classify the road entities which may be met in a driving space. Road entities were classified into two sub-concepts, *Mobile Entity* and *Static Entity*. Information about a mobile entity (i.e. pedestrians and vehicles)

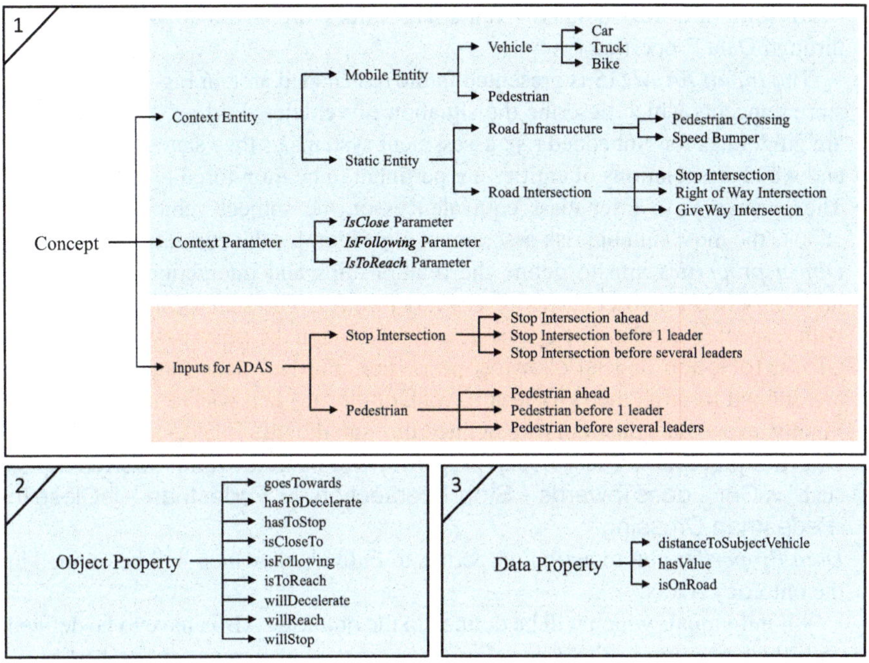

Fig. 9.8 Ontology concepts, object properties and data properties

cannot be a-priori known. This information has to be obtained in real time from perception sensors. Static entities are assumed to be part of the road network. Their presence is perfectly predictable and can be stored in digital maps. The presented ontology represents two types of static entities: *Road Infrastructures* which effect the behaviour of vehicles such as speed bumpers and pedestrian crossings and *Road Intersections* (classified into three categories: "Stop", "Right of Way" and "Giveway Intersections").

The *Context Parameter* aims to define spatio-temporal thresholds which allow to decide whether interactions between two entities are likely to exist. To illustrate the IsFollowing Parameter, lets imagine two vehicles (the leader and the follower) navigating at the same speed, on the same road, and in the same direction. If the two vehicles are separated by 90 m, the interaction between them depends on their speed. If they are moving at 30 km/h, the leader is 6 s ahead of the follower, so it may be considered that there is no interaction between them. However, if they are moving at 90 km/h, the leader is only 2 s ahead. It can therefore be considered that interaction between the two vehicles is established. The IsFollowing Parameter allows to set the threshold in the form of time duration which enables to consider if a vehicle is following another one. Following the same logic, the IsClose Parameter and the IsToReach

Parameter are also defined. Numerical values are given to these concepts through Data Properties.

The *Inputs for ADAS* is presented in the red shaded area in Fig. 9.8. It aims to store concepts which describe the situation of vehicles. Further, these concepts are guidelines for embedded risk assessment systems as they state which entities and which associations of entities are pertinent to be monitored to ensure safety. The purpose is to infer class equivalences on the subject vehicle in order to choose the most suitable risk assessment algorithm for the current situation.

- *Object properties* aim to define the relationships and interactions which may happen between two concepts of *Context Entities*. The state of a mobile entity with respect to another one can be described through the goesTowards, isCloseTo, isToReach and isFollowing properties. Further, near future behaviours are defined through the isToReach, willDecelerate and willReach properties. Finally, expected behaviours are defined through the hasToStop and hasToDecelerate properties. These object properties will be used within inferred triples such as Car - goesTowards - Stop Intersection, or Pedestrian - isCloseTo -Pedestrian Crossing.
- *Data Properties* aim to assign properties to individuals which will be defined in the ontology ABox.

  All individuals which will be defined in the ontology ABox have to be defined with their position in the scene. For this purpose, a reference frame had to be chosen. As most observations are performed with respect to the subject vehicle, it was chosen as the reference frame. Further, since the world is represented in one dimension in the ontology, the positions of entities with respect to the subject vehicle are defined as curvilinear abscissas along the road (in the same manner as the position of static entities are defined in the Electronic Horizon). In that way, the property distanceToSubjectVehicle was created, which expects arguments as numerical values.

  Further, some entities such as pedestrians can be either on the road or on the pavement. For a vehicle this information is important in order to decide if the entity has to be considered. Therefore, the boolean parameter isOnRoad enables to define in the ontology whether a pedestrian is on the road.

  Finally, the Context Parameter concepts require to be set. For this purpose the numerical data parameter hasValue was created.
- *Relations* aim to provide a priori knowledge about road entity concepts and their potential interactions and to extract the most relevant features of the situation. Relations consist of axioms aiming to affect object properties to the individuals which are stored in the ontology ABox.

Relations were created in two steps. In the first step, axioms which enable to infer the likely interactions between the road entities stored in the ABox are defined. Most of this axioms require an expressiveness which cannot be provided by a Description Logic language. Therefore SWRL rules have been chosen for this purpose. The second step adds additional axioms to exploit the interactions which were inferred during the first step and thus to extract for all vehicles the most relevant features

Table 9.1 Example of 3 SWRL rules edited in the ontology

#	SWRL rules	Meaning
1	vehicle(?v1) ∧ vehicle(?v2) ∧ distanceToSubjectVehicle(?v1,?d1) ∧ distanceToSubjectVehicle(?v2,?d2) ∧ subtract(?sub,?d2,?d1) ∧ isFollowingParameter(?fParam) ∧ hasValue(?f,?fParam) ∧ lessThan(?sub,?f) → isFollowing(?v2,?v1)	The position *d1* and *d2* of the vehicles *v1* and *v2* are known thanks to the *distanceToSubjectVehicle* parameter. By performing a subtraction (line 4), it is possible to determine the distance *sub* between both vehicles. By comparing this distance with the threshold of the *isFollowingParameter* (line 7), it is determined whether one vehicle is following the other one (line 8).
2	vehicle(?v1) ∧ StopIntersection(?stop1) ∧ willReach(?v1,?stop1) → willStop(?v1,?stop1)	The vehicle *v1* will reach the stop intersection *stop1*. This condition means that *v1* will probably stop at *stop1* (line 4).
3	vehicle(?v1) ∧ StopIntersection(?stop1) ∧ isToReach(?v1,?stop1) → hasToStop(?v1,?stop1)	The vehicle *v1* is about to reach the stop intersection *stop1*. This condition means that *v1* has to stop at *stop1* (line 4).

of the situation. For this purpose, it was possible to use the DL language as the corresponding axioms are simple. Note that SWRL could have been used, however, reasoning on SWRL rules is more expensive than reasoning on DL axioms.

For the first part, 14 SWRL rules were defined. Table 9.1 presents 3 of these rules. These rules aim to make it possible to infer spatio-temporal relationships between entities, near future behaviours of mobile entities and expectation about mobile entities manoeuvres. Rule 1 in Table 9.1 is one of five rules dealing with spatio-temporal relationships. Rule 2 is one of three rules dealing with near future behaviours of the mobile entities. Finally, rule 3 is one of six rules dealing with expected manoeuvres of the mobile entities. Some of these rules were defined to take into consideration possible chain reactions (e.g. vehicle that is following another vehicle that has to stop has also to stop in order to avoid a collision).

For the second part, one basic DL axiom was created for each Output For ADAS concept. In total, six axioms were edited for the ontology, two of them are presented in Table 9.2 in order to help the reader understand the principles. Axiom 1 aims to define that if a single vehicle is expected to stop at a stop intersection, it is pertinent to run an ADAS that makes sure that the driver is aware of the stop intersection. Further, axiom 2 aims to define that if a vehicle following another vehicle expected to stop at a stop intersection, it is pertinent to run an ADAS that makes sure that the driver is aware that the lead vehicle will stop soon.

Table 9.2 Example of 2 Description Logic Axioms edited in the ontology.

#	DL axioms	Meaning
1	StopIntersection $\doteq$ Vehicle $\sqcap$ $\exists$hasToStop · StopIntersection	If an instance of concept *Vehicle* is linked to an instance of concept *StopIntersection* through the object property *hasToStop*, then the instance of concept *Vehicle* is also an instance of the *StopIntersectionAhead* concept.
2	StopIntersectionBefore1Leader $\doteq$ Vehicle $\sqcap$isFollowing · StopIntersectionAhead	If an instance of concept *Vehicle* is linked to an instance of concept *StopIntersectionAhead* through the object property *isFollowing*, then the instance of concept *Vehicle* is also an instance of the *StopIntersectionBefore1Leader* concept.

The ABox

The ontology ABox contains two types of individuals. There are those which are mandatory and created independently from the World Model, and those which are created according to the World Model.

- Mandatory Individuals

  Even if the World Model does not include any information about surrounding road entities, the ontology ABox requires four individuals to be defined. These individuals enable the ontology to work properly and are defined as follows.

  The World Model will always contain information about the subject vehicle that perceives its surrounding environment. Therefore, the ontology ABox has to store one instance of the *Vehicle* concept, representing the subject vehicle. This entity is taken as the origin of the frame, so the distanceToSubjectVehicle data property is affected to this individual and is set at 0.

  The three other individuals refer to the three concepts included in the Context Parameter major concept. These individual aim to activate the context parameters in the ontology and thus to assign a value to the three of them. In that way, one instance of the isCloseParameter concept has to be created. This individual is given the hasValue data property which sets the maximum distance between a pedestrian and a static entity to consider them close enough to interact. Further, one instance of the isFollowingParameter concept has to be created with the hasValue data property. The value of this property sets the distance between two vehicles from which it is considered that the following vehicle is no longer following the leader. This value depends on the vehicle speeds in stabilized conditions. Finally, one instance of the isToReachParameter concept has to be created, again with the hasValue data property. This parameter sets the distance of a vehicle to a static entity from which it is considered that the vehicle is about to reach the static entity. This parameter also depends on the vehicle speed in stabilized conditions.

- World Model Dependent Individuals

  These individuals can be considered as the conversion of the World Model contents into the ontology language. Thus, each entity stored in the World Model has its equivalent in the ontology ABox. For each entity, one instance of the corresponding concept is created and is affected the distanceToSubjectVehicle data property. The value of this property is the position of the concerned entity, with respect to the subject vehicle. Note that uncertainties on the position of entities are not considered by the ontology. Finally, the isOnRoad data property has to be attributed to all instances of the Pedestrian concept to declare whether the corresponding pedestrians are on the road or on the pavement.

9.2.4 Implementation and Experimental Evaluation

The ontology's ability to infer pertinent information about a road situation has been tested using test data and real data. It was edited in the Protégé software (TBox and ABox), version 4.3, developed by the Stanford Center for Biomedical Informatics Research [29, 30]. This ontology editor enables the edition of SWRL rules.

9.2.4.1 Case Study Using Manual Data

This first case study was performed with an ontology whose ABox was filled manually, Fig. 9.9a describes the situation that was manually edited in the ontology. This situation consists of three vehicles (called *Subject Vehicle*, *Vehicle 2* and *Vehicle 3*) going towards a stop intersection (called *Stop 1*). *Vehicle 3* is the closest to the intersection and just passed a pedestrian crossing (called *Pedestrian Crossing 1*). *Vehicle 2* goes towards *Pedestrian Crossing 1*, and *Subject Vehicle* follows *Vehicle 2*. Finally, a pedestrian (called *Pedestrian 1*) is walking next to *Pedestrian Crossing 1*.

As mentioned before, the ABox contains four mandatory individuals, one for the subject vehicle and three others for context parameters. In this case study, the highest allowed speed is 50 km/h, therefore the context parameters are set according to this speed. In that way, it was set that a vehicle is following another one if the following time is lower than 3 s. Therefore, an individual of the isFollowingParameter concept was created with the hasValue data property set to 42 m (distance travelled in 3 s at 50 km/h). Further, it was set that a mobile entity is about to reach a static entity if it is reaching it within 5 s at constant speed. Therefore an individual of the isToReach concept was created with the hasValue data property set to 70 m. Finally, an instance of the isCloseParameter concept was created with the hasValue property set to 3 m.

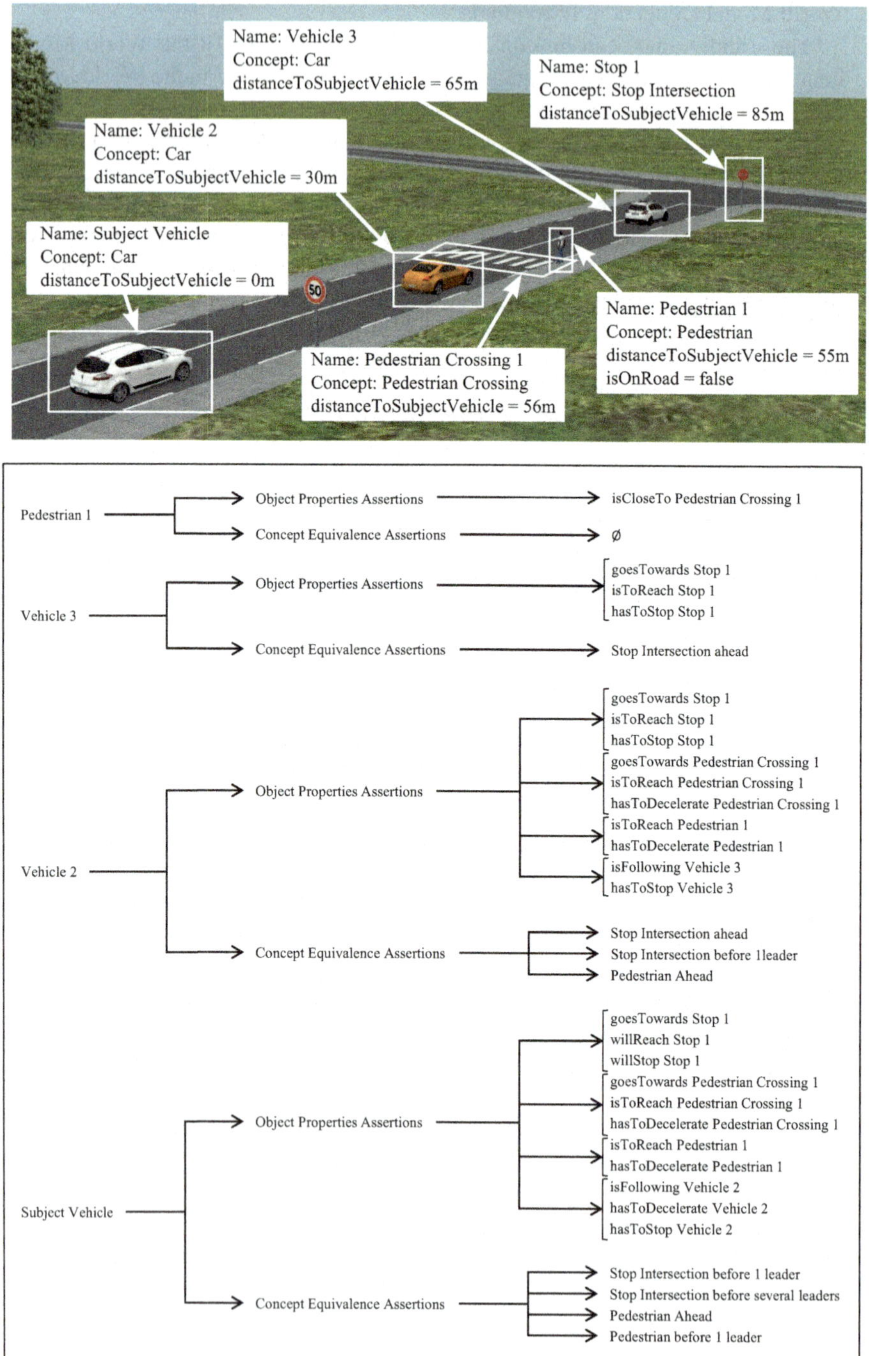

Fig. 9.9 Case study. (**a**) On the *left*, an illustrative picture of the case study (scale is not respected). In the boxes on the *right*, the World Model Dependant Individuals stored in the ABox. (**b**) Object properties and concept equivalence assertions after reasoning

Results

Reasoning was performed through the Protégé software and the Pellet reasoner as it is compatible with SWRL and offers good performances [6]. Figure 9.9b shows the object properties and concept equivalence assertions performed by Pellet for the chosen case study. Some of these inferences are detailed bellow.

Pedestrian 1 is inferred to be close to *Pedestrian Crossing 1*. The reasoner computes the distance between these two entities according to the distanceToSubjectVehicle data parameter set on the two corresponding individuals. This distance is of 1 m and satisfies the condition (that depends on the isClose context parameter) that was set in the ontology to claim that a pedestrian is close to a pedestrian crossing. It implies that the pedestrian is likely to have the intention to cross the road, therefore it means that a vehicle that would approach to the pedestrian would have to take care of the pedestrian. No concept equivalence is asserted on *Pedestrian 1* because there is no axiom for concept equivalence defined in the TBox for the Pedestrian concept.

Three object property assertions are inferred for *Vehicle 3*. These assertions concern interactions between this vehicle and the stop intersection *Stop 1*. Thanks to the position of these two entities, is was inferred that *Vehicle 3* passed all the static entities except *Stop 1*. Therefore, it is inferred that *Vehicle 3* goes towards *Stop 1*. Further, the distance between these two entities is low enough to consider that *Vehicle 3* is about to reach *Stop 1*. Moreover, since it was defined in the ontology that all vehicles about to reach a stop intersection have to stop at the intersection, it is inferred that *Vehicle 3* has to stop at *Stop 1*. Finally, it is inferred that *Vehicle 3* is an instance of the Stop Intersection Ahead concept. This is performed using the first DL axiom presented in Table 9.2.

Ten object property assertions are inferred for *Vehicle 2*. The ontology infers that this vehicle passed neither *Pedestrian Crossing 1* and *Stop 1*, therefore it is inferred that it goes towards these two static entities. Moreover, *Vehicle 2* is close enough to *Pedestrian Crossing 1* and *Stop 1* to say that it is about to reach them. Since all vehicles have to stop at stop intersections, it is inferred that *Vehicle 2* has to stop at *Stop 1*. This assertion implies *Vehicle 2* to be an instance of concept Stop Intersection Ahead. Moreover, all vehicle have to decelerate before reaching a pedestrian crossing, therefore *Vehicle 2* has to decelerate for *Pedestrian Crossing 1*. Further, as it was inferred that *Pedestrian 1* is close to *Pedestrian Crossing 1*, and since *Vehicle 2* is about to reach *Pedestrian 1*, it is inferred that it has to decelerate for the pedestrian. This assertion implies *Vehicle 2* to be an instance of concept Pedestrian Ahead. Finally, *Vehicle 2* is close enough to *Vehicle 3* to claim that it is following this latter. However, it was inferred that *Vehicle 3* has to stop at *Stop 1*, and since *Vehicle 2* is following *Vehicle 3*, *Vehicle 2* has to top behind *Vehicle 3*. This chain reaction implies *Vehicle 2* to be an instance of concept Stop Intersection before 1 leader.

Eleven object property assertions are inferred for *Subject Vehicle*. Like *Vehicle 2*, *Subject Vehicle* passed neither *Pedestrian Crossing 1* and *Stop 1*. It is therefore inferred that it goes towards these two entities. Moreover, *Subject Vehicle* is too far from *Stop 1* to consider that it is about to reach it. However it is inferred that *Subject Vehicle* will reach *Stop 1*, and therefore that it will stop at *Stop 1*. Further, as it is close enough to *Pedestrian Crossing 1*, *Subject Vehicle* is about to reach it, and thus has to decelerate. In addition, as it is for *Vehicle 2*, the chain reaction with *Pedestrian 1* and *Pedestrian Crossing 1* implies that *Subject Vehicle* has to decelerate for *Pedestrian 1*. This implies Subject Vehicle to be an instance of concept Pedestrian Ahead. Further, *Subject Vehicle* is close enough to *Vehicle 2* to claim that it is following it. This implies several chain reactions with the other context entities. First, *Subject Vehicle* is following *Vehicle 2* that is an instance of Stop Intersection ahead. This implies *Subject Vehicle* to be an instance of concept Stop Intersection before 1 leader. In addition, *Vehicle 2* is also an instance of concept Stop Intersection before 1 leader, therefore is also implies that *Subject Vehicle* is an instance of concept Stop Intersection before several leaders. Finally, *Vehicle 2* is an instance of concept Pedestrian Ahead, it therefore implies that *Subject Vehicle* is an instance of concept Pedestrian before 1 leader.

These results show that the proposed ontology enables to perform coherent reasoning on global road situations. It shows that interactions between road entities can be understood and considered to anticipate the behaviours of the mobile entities.

9.2.4.2 Case Study Using Recorded Data

The next step is to test the ontology with data recorded from sensors embedded on an experimental vehicle. Figure 9.10 shows a representation of the case study that was chosen for the evaluation of the ontology in real time conditions. It consists of the subject vehicle that is following a lead vehicle. Both vehicles are navigating towards a pedestrian crossing that precedes a stop intersection. Ten meters separate the pedestrian crossing and the intersection. Additionally, a pedestrian is located next to the pedestrian crossing.

Figure 9.11 presents the framework that was used for the real time exploitation of the ontology.

The framework requires several data sources. A priori information about the position of the pedestrian crossing and of the stop intersection were stored in a digital map in the Open Street Map format. In addition to this map, the localization data returned by the GPS receiver was used by the Navigation System in order to generate the Electronic Horizon in real time. Real time information about leading vehicles and pedestrians are provided by the Lidar sensor. Finally, the ontology TBox was stored in a Ontology Web Language (OWL) file [22]. This file format is the reference for the storage of ontologies.

Three pieces of software were necessary to exploit the ontology in real time. The first one is the navigation system that exploits the OSM digital map and that returns the Electronic Horizon at each new measurement of vehicle location. That is, it provides information about the static entities, i.e. the distance of the subject vehicle to the pedestrian crossing and to the stop intersection.

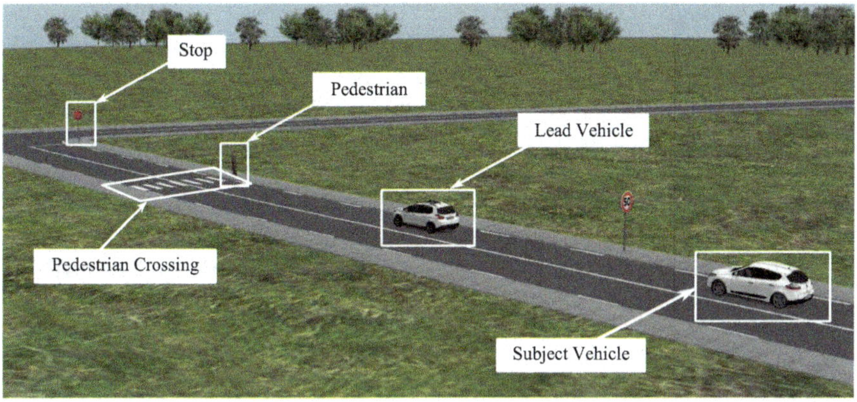

Fig. 9.10 Case study for real time evaluation of the ontology

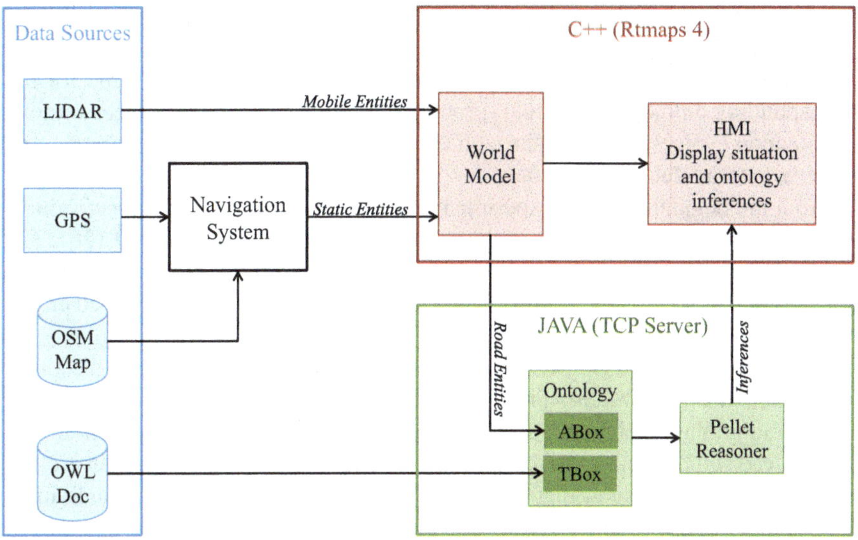

Fig. 9.11 Framework for real time situation understanding

The second piece of software was developed in the C++ programming language for the RTMaps 4 middle-ware. This software allows to get information about the mobile and static entities as they are returned by the data sources. An RTMaps component was developed to feed the World Model structure according to the information about road entities.

The last piece of software was developed in the Java programming language. It enables to exploit the ontology and thus to reason about the World Model. Even if the Java language is not the best language for real time functions, it was chosen to use it as it was the only programming language that proposes accessible libraries for ontology handling. For this purpose, the OWL API library was used [12]. Moreover, the software was developed as an ontology server, that is it communicates with clients which need to reason about World Models. The communication between the server and RTMaps is performed through the TCP protocol. The World Model structure is exchanged after having been serialized using the Protobuf library [31]. After reception of the World Model structure by the server, ontology individuals are created, completing the core ontology that was preliminary loaded from the OWL file. Reasoning is then performed through the Pellet reasoner and inferences are sent back to the TCP client. The inferences can therefore be exploited by an ADAS, which is, in this Chapter, an HMI that displays the ontology inferences.

Results

Figure 9.12 presents the results of the experimental evaluation of the ontology. Figure 9.12a shows the state of the lead vehicle and the inferred class equivalences over time for the corresponding ontology individual. Further, Fig. 9.12b shows the state of the subject vehicle and the inferred class equivalences over time for the corresponding ontology individual. From the point of view of the subject vehicle, the situation evolves over time through eight main events happening at times t_1–t_8. These events are detailed hereafter.

From the beginning of the experiment, the distance between the subject vehicle and the lead vehicle is lower than the isFollowing threshold (see Fig. 9.12b). The ontology therefore considers that the subject vehicle is following the lead vehicle. It means that as soon as the lead vehicle interacts with at least one other road entity, this interaction is propagated to the subject vehicle.

At time t_1, the distance between the lead vehicle and the pedestrian becomes lower than the isToReach threshold (see Fig. 9.12a). Therefore, the ontology considers that there is interaction between the lead vehicle and the pedestrian and that the lead vehicle is about to reach the pedestrian. However, the pedestrian is close to the pedestrian crossing, therefore it is inferred that the lead vehicle individual becomes an instance of the Pedestrian Ahead concept (see Fig. 9.12a). Moreover, since the subject vehicle is following the lead vehicle, the interaction between the lead vehicle and the pedestrian is propagated to it. The subject vehicle individual therefore becomes an instance of the Pedestrian Before 1 Leader concept (see Fig. 9.12b).

At time t_2, the distance between the lead vehicle and the stop intersection becomes lower than the isToReach threshold (see Fig. 9.12a). Thus, the ontology considers that the lead vehicle is about to reach the stop intersection. The lead vehicle individual therefore becomes an instance of the Stop Intersection Ahead concept. Further, since the subject vehicle is still following the lead vehicle, the subject vehicle individual becomes an instance of the Stop Intersection Before 1 Leader concept (see Fig. 9.12b).

At time t_3, the distance between the subject vehicle and the pedestrian becomes lower than the isToReach threshold (see Fig. 9.12b). Since the pedestrian is still

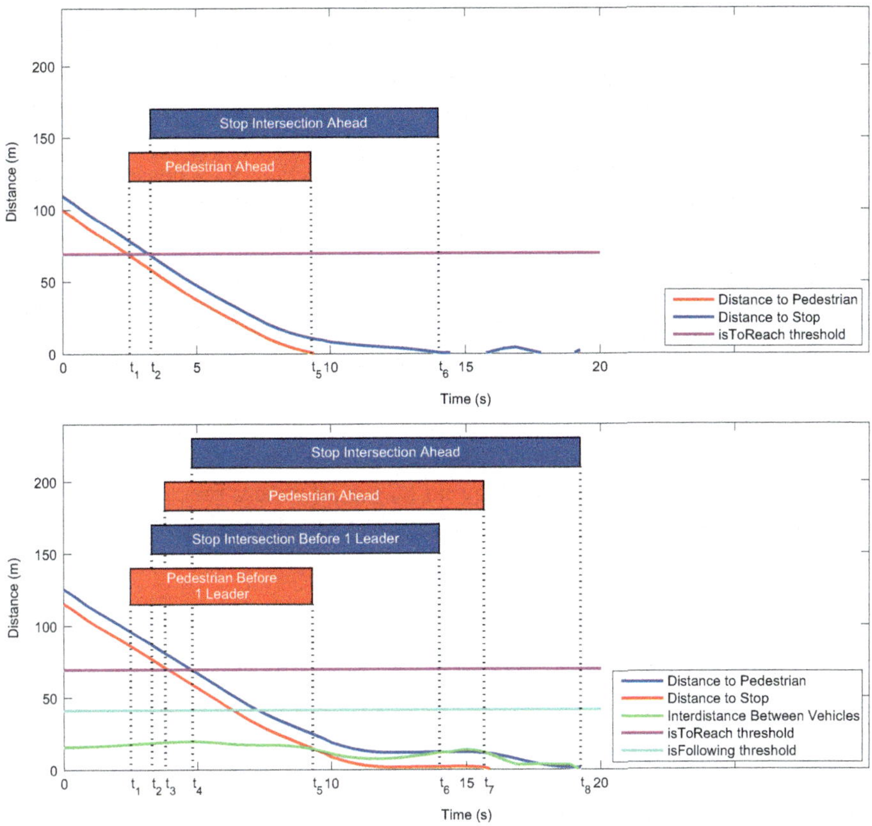

Fig. 9.12 Results of the experimental evaluation. (**a**) Situation of the lead vehicle with respect to the static entities and ontology inferences. (**b**) Situation of the subject vehicle with respect to the static entities and ontology inferences

close to the pedestrian crossing, the subject vehicle starts to interact with him and therefore the subject vehicle individual becomes an instance of the Pedestrian Ahead concept. Note that at this time the lead vehicle did not pass the pedestrian, therefore the subject vehicle individual is still an instance of the Pedestrian Before 1 Leader concept.

At time t_4, the distance between the subject vehicle and the stop intersection becomes lower than the isToReach threshold (see Fig. 9.12b). Therefore, the ontology considers that the subject vehicle starts to interact with the intersection and thus the subject vehicle individual becomes an instance of the Stop Intersection Ahead concept. Note that at this time the lead vehicle did not pass the stop intersection, therefore the subject vehicle individual is still an instance of the Stop Intersection Before 1 Leader concept.

At time t_5, the lead vehicle passes the pedestrian (see Fig. 9.12b). As a consequence, the lead vehicle individual is no longer an instance of the Pedestrian

Ahead concept. Further, that implies that the subject vehicle is no longer following a vehicle that is about to reach a pedestrian. Therefore, the subject vehicle individual is no longer an instance of the Pedestrian Before 1 Leader concept (see Fig. 9.12b). It means that the subject vehicle no longer indirectly interacts with the pedestrian.

At time t_6, the lead vehicle passes the stop intersection. As a consequence, the lead vehicle individual is no longer an instance of the Stop Intersection Ahead concept (see Fig. 9.12b). Therefore, the subject vehicle is no longer an instance of the Stop Intersection Before 1 Leader concept (see Fig. 9.12b).

At time t_7, the subject vehicle passes the pedestrian. Therefore, the subject vehicle individual is no longer an instance of the Pedestrian Ahead concept (see Fig. 9.12b). The stop intersection therefore becomes the only pertinent road entity for the subject vehicle.

Finally, at time t_8, the subject vehicle reaches the stop intersection. Therefore, the subject vehicle individual is no longer an instance of the Stop Intersection Ahead concept (see Fig. 9.12b). The ontology no longer infers any concept equivalence, therefore there is no more pertinent perceived surrounding entity that have to be monitored by the subject vehicle.

9.2.5 *Discussion*

It was shown that the proposed ontology enables to reason on road environments as they can be perceived by a vehicle. Reasoning on road environments can be performed with respect to the types of the entities which are concerned, while considering the interactions which are likely to happen between entities. The ontology enables to consider chain reactions in a straightforward manner, that is, the interaction between two entities can have consequences on the behaviour of another entity. In comparison, most conventional ADAS would have considered each perceived entity independently from the others and would have monitored the closest entity only.

The proposed ontology cannot be exploited to reason on every road context. Only situations compatible with it can be understood, that is, situations which only meet entities which have been described in the ontology TBox. It means that if the World Model contains an entity that is not formally described in the ontology, the latter will not be able to reason about this entity. If in real life this entity has influence on other entities known by the ontology, a great part of the reasoning will not be representative of reality and thus will not be consistent. Further, for the experiments which were presented, the values of the Context parameters were set in an ad hoc manner. This was because no studies aiming to define conditions for which entities can be considered as interacting were found in the literature. It would therefore be pertinent to carry out studies to fill this gap.

In addition to the quality of the knowledge that is stored in the ontology, the consistency of the inferred information depends on the quality of the information stored in the World Model. If a pertinent entity of the situation misses in the World Model, reasoning about the situation cannot be coherent. Moreover, if the World Model contains incoherent information about the situation, reasoning will neither be coherent. For instance, let's consider a situation for which the World Model contains an intersection and a lead vehicle in addition to the subject vehicle. If the distance between the subject vehicle and the intersection is under-estimated, the ontology may understand that the lead vehicle already passed the intersection while it did not. The consequence would therefore be that no interaction between both entities is inferred, therefore the situation would be misunderstood by the subject vehicle.

In the current state of research, one weak point of ontologies is their inability to take uncertainties into account. Again, this means that the precision of the data stored in the World Model is of great importance. This implies that all perception and localization sensors must provide precise and accurate measurements and that navigation maps are precise and up to date. In addition, this lack of uncertainty implies that it has to be assumed that drivers comply with rules, and that it is not considered that rules can be violated. Finally, neither uncertainties on interactions between entities, neither uncertainties on concept equivalence assertions can be estimated. Such uncertainties could be of great interest, especially for the risk assessment systems which may have to exploit the ontology inferences.

Finally, the time necessary to reason on an ontology is significant and has to be considered. For the experimental evaluation presented in Sect. 9.2.4, the average processing time necessary for reasoning was 71 ms on a 4 GB RAM laptop with a dual core 1.9 GHz processor. This processing time depends both on the complexity of the ontology TBox (number of axioms and especially the number and complexity of the SWRL rules) and on the number and types of road entity individuals stored in the ontology ABox. If the ontology has to be extended, an effort would have to be made in order to limit the number of axioms and rules and thus to limit the complexity of the reasoning step. For real time applications, if the processing time is too high in comparison with the frequency at which the World Model returns data, it would be conceivable to reason on the ontology asynchronously with the rest of the system as it was done in [15].

9.3 Map Error Detection

Intelligent and autonomous vehicles applications require the information provided by the navigation function in the form of the Electronic Horizon (EH). Erroneous data in the EH may result in undesirable behaviour of client systems and generate hazardous situations. These can be the consequence of faults that arise at any step of the EH generation. Localization system used in combination with the map may be perturbed and provide a position estimate that contains large errors. This may result in large errors on the estimation of the vehicle position onto the map. A map is a complex entity that represents the environment that is constantly evolving, it then necessarily contains Faults. The scope of this section is on faults originating in the navigation map.

This section first defines the concepts of fault and error which are essential here. Next the pathology of faults found in navigation maps and their effects on the vehicle applications is described.

9.3.1 Definitions

The terms of fault, error, failure and integrity may have different meanings according to the application domain. The following definitions are used in the context of this research and are based on those given in [28]:

- Fault: Error generative process. The presence of a fault may not lead to an error.
- Error: A discrepancy between a computed, observed or measured value and the true, specified or theoretically correct value.
- Failure: Instance in time when a required function exceeds the acceptable limits or is terminated.
- Integrity: Reliability of the confidence indicator associated with an information with respect to specifications of the client application.

There is a loss of integrity of EH data if the error is greater than the estimation.

Figure 9.13 shows the geometry of the segment of a road network. The shaded shape represents admissible error associated with it. In this illustration, a fault occurs if the EH data is outside this envelop. Figure 9.14 shows the loss of integrity for the position estimate. The true vehicle position is indeed outside the confidence domain of the error associated with the position estimate.

The accuracy of the vehicle position estimate provided by the navigation function in the EH is closely related to both the localization and the navigation map.

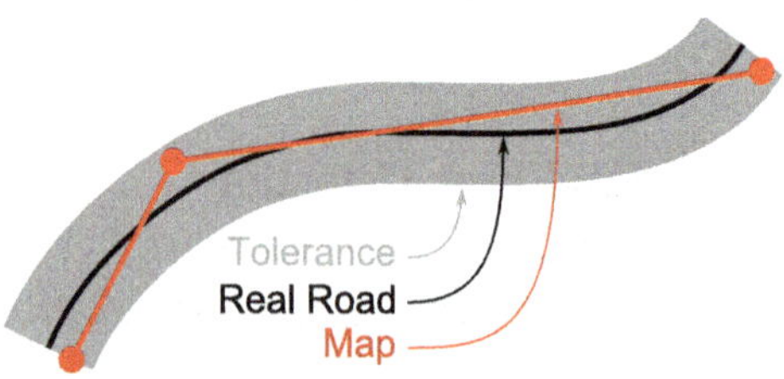

Fig. 9.13 Navigation estimates where integrity is kept

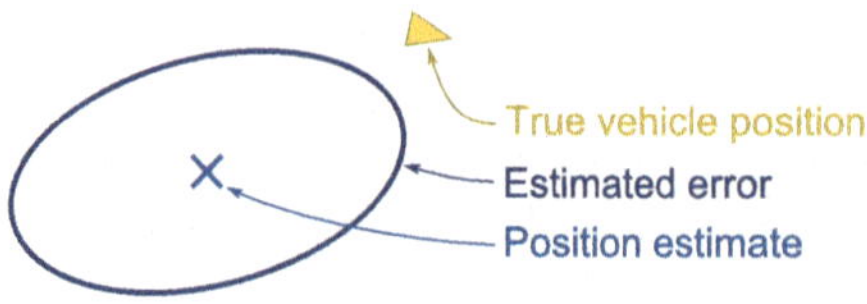

Fig. 9.14 Loss of integrity for position estimate

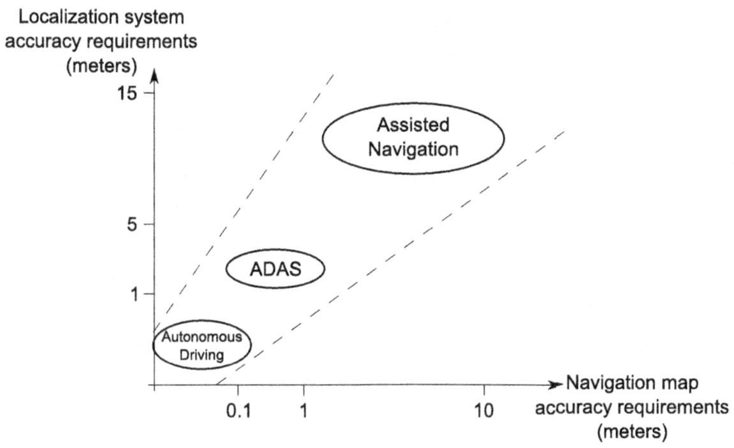

Fig. 9.15 Needs in localization and map accuracy for the development of driving assistance functions (after [8])

Figure 9.15 highlights this dependency and the consequences on the development of driving assistance functions, from the assisted navigation to autonomous driving [8].

Passenger vehicles are mainly used either for commuting or to travel on known roads. Except for occasional journeys like going on vacations, the vehicle is therefore driven on roads on which it has already been driven. It must be noticed that the dysfunctions due to map faults are mainly systematic. Whenever the vehicle crosses again the erroneous road, the EH provides the same false data to ADAS which, themselves, will have the same uncomfortable behaviour. The quality perceived by the driver decreases significantly due to the frustration of repetitive illogical warnings or reactions of the vehicle.

9.3.2 *Pathology*

The road network is constantly evolving. Map providers consider that 15 % of the road network of a mature country change each year. This value increases in developing countries. With the exception of few countries like Japan, road changes are not monitored by a central administration. It is therefore difficult for cartographers to keep the geographic databases up-to-date.

Map creation is a complex process that takes a long time. When loaded in the vehicle, the navigation map data is therefore already several months old and partially outdated. Mapmakers adopt two approaches to pursue fast map updates. The first is to facilitate mapping surveys by deploying a large fleet of vehicles. These are equipped with a specific set of sensors (e.g. lidar, inertial measurement systems, differential GNSS) for rapid mapping of the road network. Priorities are given to main roads and areas where changes have been reported. This strategy allows precise mapping but is costly in material and human resources. The second is to rely on user data. The records of the journeys travelled with the navigation devices are automatically uploaded to the map provider server. Data mining algorithms are then applied to update the central database. This approach is less expensive and allows to update the road geometry or the mean travel times. However other structural elements such as traffic signs, speed limits still require field surveys.

The typical faults are summarized in a pathology. It provides an overview of the issues involved by each fault and associated scientific challenges. The distinction is made here between fault in the navigation map structure, geometry and attributes. Structural faults are related to the manner in which the elements of the map are connected or are identified in the map. Geometric faults are related to the shape or the geographic placement of these entities. This distinction is made because they require different approaches to be detected and corrected. These are detailed in the following paragraphs.

9.3.2.1 Structural Faults

Road Connectivity

The correct representation of road junction in the navigation map is essential for optimal path planning. The missing connection between two road may cause the path planner choose a suboptimal path and bother the driver. Reciprocally, a connection between two roads in the navigation map that does not exist in reality may result in impossible path. In this case, the driver may be misled by the navigation assistant and cause hazardous driving situations.

Type of Intersection

The significant information at road junctions is the right way of one road with respect to the others. Hazardous situations could occur if this information is missing. In navigation maps, the type of intersection can be associated with junctions in order to describe implicitly the order of priorities. Over the past years, a particular type of intersection that is roundabouts (or traffic circles) is preferred to others types and is built in many places. They indeed reduce fatalities and increase traffic exchange between roads. Their construction means that the network structure is changed [33, 38]. The manner in which the vehicle crosses them is different to a classic crossroad and the reachability to the intersection branches is different. The inclusion of a roundabout must be registered in the navigation map in order to provide convenient guidance information to the driver and adapt the approaching manoeuvre (e.g. speed reduction, lane placement).

9.3.2.2 Geometric Fault

Road Shape and Road Offset

For optimization purposes, the on board navigation maps are compressed and compiled. In this process, some road shape points are removed. The road curvature estimated using the shape points is therefore perverted. In the GIS domain, absolute and relative accuracies are distinguished [5]. The former describes the accuracy of the geographic feature with respect to a global reference coordinate system whilst the later describes accuracy relative to other features. Roads with low relative accuracy result in poor shape definition. The curve warning related ADAS (i.e. Contextual ACC, LKA and PFL) are directly affected by low relative accuracy of the road. In case of low absolute and high relative accuracies (road offset), the map-matching algorithm may not choose the right road candidate, especially in dense road network area. Moreover, Fig. 9.16 shows that the offset on junctions induces malfunctions of intersection warning systems for vehicles running on the other roads [2].

Missing Roads

Road networks change over time and some roads are created or closed. Until the next cartographic survey, every contextual ADAS application cannot operate properly. The case in which the vehicle is driven on a new road is split into two situations. First, if the new road is far from the roads stored in the navigation map, the map-matching function will switch to failure mode and provide the output *off road*. The contextual ADAS applications could then adopt a suitable strategy and limit the

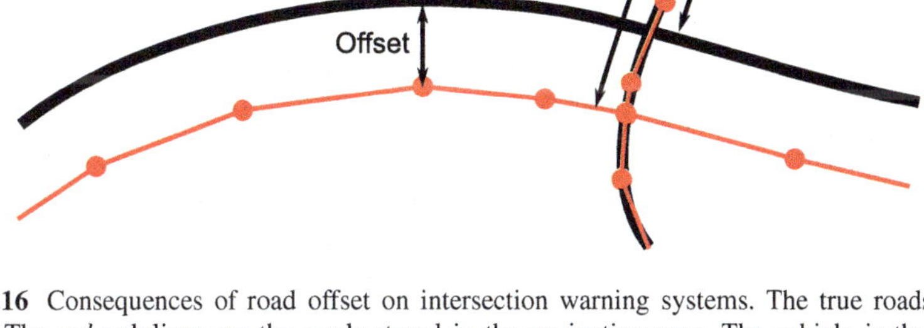

Fig. 9.16 Consequences of road offset on intersection warning systems. The true roads are in *black*. The *red* polylines are the roads stored in the navigation map. The vehicle is the *yellow* triangle

consequences. Second, if the new road goes along a road stored in the navigation map, the vehicle position is likely to be matched on the wrong road with a high confidence level. The contextual ADAS could operate based on inconsistent data and cause hazardous situations.

9.3.2.3 Attributes Faults

Speed Limits

The speed limit can be displayed to the driver in Over-Speed Prevention (OSP) system or used to set the vehicle cruise speed in Contextual ACC applications. The navigation map stores the speed limit as an attribute for each road. This attribute originates from direct surveys or is inferred based on the road class, the number of lanes and the area (inside or outside built-up area). Smart cameras are nowadays capable of detecting speed signs in real time [18, 24], however, the challenge resides in determining whether the detected sign is applicable to the vehicle. Speed limits indicated by traffic signs may be dedicated to one particular lane (e.g. for a motorway exit), to a class of vehicles (e.g. trucks, vehicles towing caravans, buses) or to special whether conditions. An inappropriate estimation of the speed limit by the vehicle would bother the driver and decrease the perceived vehicle quality.

Driving Directions and Vehicle Restrictions

In navigation maps, roads are assumed to be drivable in both direction unless a dedicated attribute is associated with the road. Similarly, attributes are defined in order to establish some traffic restrictions (e.g. trucks, pedestrian and maximum height). The path planner uses these attributes to exclude wrong-way roads and roads that does not comply with the vehicle type. Faults in these attribute may cause the path planner to choose a suboptimal route or ask the driver to take a forbidden route. This may have severe consequences especially for large goods vehicles that cannot manoeuvre easily.

As a conclusion on this fault pathology, it must be noticed that the variety of the information stored in the navigation map induces a variety in the faults in the map. The methods employed to address the detection and correction of these faults are necessarily diverse in terms of sensors and formalisms. Some of the formalisms that could permit to address this problem require an a priori knowledge on the correctness of the navigation map. However, to the author's knowledge, no trustworthy study on the navigation map overall reliability has been done. Here, the navigation map is considered to be globally correct but potentially locally erroneous.

9.3.3 Page's Trend Test

The quality of the road representation within the navigation map in terms of geometry has a direct impact on the performance of intelligent vehicles navigation systems. The knowledge of the road shape in front of the vehicle is used in existing intelligent vehicles to improve sensor tracking (e.g. lane markings for lane keeping functions or leading vehicle for adaptive cruise control applications) and anticipate hazardous situation by adapting the vehicle speed. The navigation map road geometric description is also essential for autonomous driving like for path planning, decision-making and control functions. To avoid dysfunction of these systems, the quality of the geometric description of the road in the navigation map must be monitored.

In passenger vehicles, no access is permitted to internal variables or data of navigation function which is therefore considered as a black box in the approach presented here. When a failure occurs on the application functions it is difficult to identify their origin, to correct them or reduce their effect. A method to detect error in the map geometry is to compare the vehicle position estimate provided by the navigation function with the positioning information given by the available sensors as illustrated by Fig. 9.17. The challenge here is that the level of performance of vehicle sensors is limited considering measurement applications. The low accuracy and high level of noise introduced by vehicle sensors make probabilistic approach appropriate.

Statistical tests are appropriate to evaluate parameters of a probability law based on a set of outcomes. In our application, we aim at detecting a change of the mean of the probability density function (PDF) of a set of observed data while the standard deviation of this PDF is in the same order of magnitude than the expected mean gap. Page's trend test works sequentially and is especially efficient for stream data. The problem is therefore formulated as the detection of a change of the mean of a

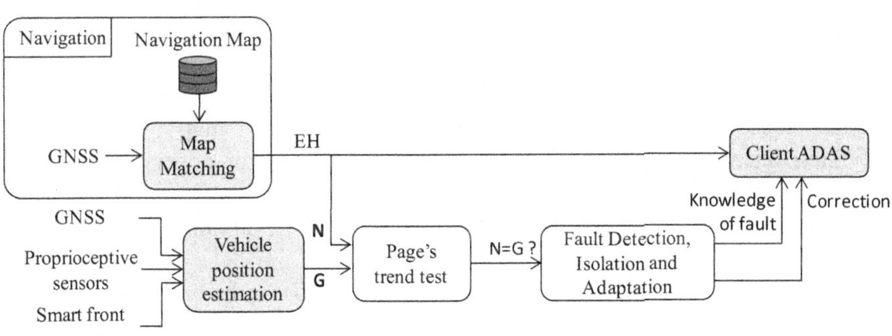

Fig. 9.17 Page's trend test for fault detection in navigation integrity monitoring context

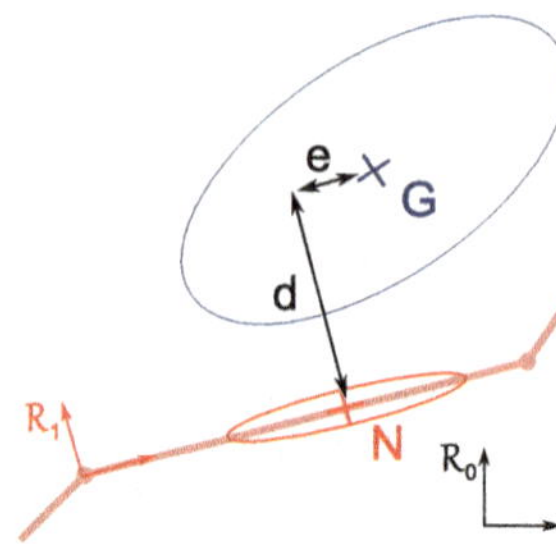

Fig. 9.18 The distance between the estimate from sensors G and the estimate from navigation N taken as random variables. d and e are the lateral and longitudinal offsets between the estimates from navigation and from sensor, respectively. $\mathcal{R}_0$ is the East–North plane locally tangent to the Earth surface and $\mathcal{R}_1$ is the frame aligned with the road segment on which the position is matched by the navigation function.

random variable that represents the distance between the estimates from sensors and from navigation as illustrated by Fig. 9.18.

9.3.3.1 Signal Generation

This section details how the distance signal is generated and described in terms of mean and standard deviation. Let us consider the estimate N from the navigation as the result of a random process based on the true vehicle position P in a frame $\mathcal{R}_1$ aligned with the road:

$$N = P + \alpha \tag{9.1}$$

$$\Sigma_\alpha = \begin{bmatrix} \sigma_a^2 & 0 \\ 0 & 0 \end{bmatrix}_{\mathcal{R}_1} \tag{9.2}$$

where α is a noise supposed zero-mean with a diagonal covariance matrix Σ_α. Indeed, since the roads are represented in the navigation map by zero width polylines, the variance of the navigation map-matched error normal to the road segment is by definition null. However, a map-matched position error along the road segment exists and σ_a denotes the longitudinal standard deviation of the navigation estimate.

The estimate of the vehicle position from sensors G can be encoded as a two-dimensional point $G = (x, y)^T$ in the East–North plane $\mathcal{R}_0$ locally tangent to Earth with the covariance matrix Σ_β of the estimation error β:

$$G = P + \beta \tag{9.3}$$

$$\Sigma_\beta = \begin{bmatrix} \sigma_x^2 & \sigma_{xy}^2 \\ \sigma_{xy}^2 & \sigma_y^2 \end{bmatrix}_{\mathcal{R}_0} \tag{9.4}$$

In order to make the distance signal independent of the road direction, an isotropic approach is chosen and it consists in using the outer circle of the ellipsoid. Its radius is $\eta = \max(\eta_i)$, η_i being the eigenvalues of Σ_β. So, the covariance matrix expressed in $\mathcal{R}_1$ is $\eta \cdot I$ (with I being the identity matrix).

In $\mathcal{R}_1$, the difference between the map-matched and estimated positions is given by a vector L which has two independent components.

$$L = \begin{bmatrix} d \\ e \end{bmatrix} = N - G = \alpha - \beta \tag{9.5}$$

Under the hypothesis of independent errors, the lateral d and longitudinal e signals have the following variances:

$$\begin{aligned} \sigma_d^2 &= \eta \\ \sigma_e^2 &= \eta + \sigma_a^2 \end{aligned} \tag{9.6}$$

The relevant information in terms of the application is the lateral position of the roads in the navigation map. The fault detection is therefore made by detecting mean changes of the signal d.

9.3.3.2 Formulation of the Test

Page's test consists in statistically detecting a change in the mean of a random variable [4]. Let us consider q samples d_i of a random variable D. The likelihood of two hypotheses H_0 and H_1 are compared. The first hypothesis states that D has a constant mean μ_0 among the q samples. The second one assumes that, given $0 < r \leq q$, the mean of D is μ_0 for the first $r - 1$ samples and μ_1 for samples r to q:

$$\begin{aligned} &H_0 : \ d_i = \mu_0 + b_i\,,\ i = 1, \ldots, q \\ &H_1 : \begin{cases} d_i = \mu_0 + b_i\,,\ i = 1, \ldots, r-1 \\ d_i = \mu_1 + b_i\,,\ i = r, \ldots, q \end{cases} \end{aligned} \tag{9.7}$$

where b is a zero-mean noise of standard deviation σ. The generalized likelihood ratio of both hypotheses is given by (9.8).

$$\Lambda(D) = \frac{\prod_{i=1}^{q} p\left(d_i, r|H_1\right)}{\prod_{i=1}^{q} p\left(d_i|H_0\right)} \tag{9.8}$$

Since the likelihood of the alternative hypothesis H_1 depends of an unknown parameter r, its maximum likelihood estimation is considered.

$$\Lambda\left(D\right) = \frac{\sup_r \left(\prod_{i=1}^{r-1} p\left(d_i|H_1\right) \prod_{i=r}^{q} p\left(d_i|H_1\right)\right)}{\prod_{i=1}^{q} p\left(d_i|H_0\right)} \tag{9.9}$$

As the likelihood of the null hypothesis H_0 does not depend on r and having $\prod_{i=1}^{r-1} p\left(d_i|H_1\right) = \prod_{i=1}^{r-1} p\left(d_i|H_0\right)$, the likelihood ratio can be simplified as follows:

$$\Lambda(D) = \sup_r \left(\prod_{i=r}^{q} \frac{p\left(d_i|H_1\right)}{p\left(d_i|H_0\right)}\right) \tag{9.10}$$

Let δ denote the mean gap ($\delta = \mu_1 - \mu_0$). Under Gaussian assumption, one can get [21]:

$$\ln\left(\Lambda(D)\right) = \frac{\delta}{\sigma^2} \sup_r \left(\sum_{i=r}^{q} \left(d_i - \mu_0 - \frac{\delta}{2}\right)\right) \tag{9.11}$$

The decision of choosing either H_0 or H_1 is made by comparing the likelihood ratio with a threshold λ_Λ:

$$\begin{cases} H_0 \ : \ \ln\left(\Lambda(D)\right) < \ln(\lambda_\Lambda) \\ H_1 \ : \ \ln\left(\Lambda(D)\right) > \ln(\lambda_\Lambda) \end{cases} \tag{9.12}$$

For a real time implementation, it is especially convenient to formulate the test sequentially. Let us then define the cumulative sum as in (9.13):

$$S_r^q\left(\mu_0\,,\,\delta\right) = \delta \sum_{i=r}^{q} \left(d_i - \mu_0 - \frac{\delta}{2}\right) \tag{9.13}$$

which can be re-written as:

$$S_1^q(\mu_0, \delta) = S_1^{r-1}(\mu_0, \delta) + S_r^q(\mu_0, \delta) \tag{9.14}$$

The decision rule then becomes

$$\begin{cases} H_0 : S_1^q(\mu_0, \delta) - \inf_r \left(S_1^{r-1}(\mu_0, \delta)\right) < \gamma \\ H_1 : S_1^q(\mu_0, \delta) - \inf_r \left(S_1^{r-1}(\mu_0, \delta)\right) > \gamma \end{cases} \tag{9.15}$$

where $\gamma = \sigma^2 \ln(\lambda_\Lambda)$.

Let δ_m be the minimum value of δ which must be detected. The test is split into two sub-tests running in parallel, the first aiming at detecting a mean growth and the other a decrease of the mean.

Finally, at the current time k, a mean growth is detected as soon as (9.16) is true:

$$U_k - m_k > \gamma \tag{9.16}$$

where

$$U_k = U_{k-1} + d_k - \mu_0 - \frac{\delta}{2} \tag{9.17}$$

$$m_k = \min(m_{k-1}, U_k) \tag{9.18}$$

Conversely, a mean decrease is detected when:

$$M_k - T_k < \gamma \tag{9.19}$$

where

$$T_k = T_{k-1} + d_k - \mu_0 + \frac{\delta}{2} \tag{9.20}$$

$$M_k = \max(M_{k-1}, T_k) \tag{9.21}$$

As soon as threshold γ is reached, the cumulative sums are reset to zero. The actual mean change happened at the last time m (resp. M) has reached its minimum (resp. maximum) before crossing γ. Another formulation for the localization of the mean change which is used in this work is to find the last time the decision variable ($U_k - m_k$ for increase detection or $M_k - T_k$ for decrease detection) is null. This test is then very efficient in terms of required computational load. Indeed, at every new time step, it only requires to make additions and comparisons of scalar variables. The mean change cannot be detected at the samples prior to the last one at which the decision variable is null. Those samples are useless and can be removed from the computer memory.

The choice of γ has consequences on the false alarm probability. It is not possible to express it formally because the PDF of hypothesis H_1 depends of an unknown parameter r. However, it can be set based on the number h of estimated parameters in the PDF and on the number n_σ of standard deviations: $\gamma = 2 \cdot h \cdot n_\sigma \cdot \sigma/\delta_m$ [21]. Here the mean is the only estimated parameter of the PDF and n_σ is set to 2 as nominal tuning which is a good compromise between false alarm rate and time to detection. Then $\gamma = 4 \cdot \sigma/\delta_m$.

Let us consider the example shown in Fig. 9.19. The signal plotted in the upper graph of the figure is generated using a Gaussian function with a constant standard deviation of 2. The mean of this signal is 0 from sample 1 to sample 40 and from sample 61 to sample 100. The mean is set to 5 for sample indexes 41–60. Page's trend test is run using this signal as an input and the cumulative sum U_k as well as the decision variable $U_k - m_k$ are plotted in order to illustrate the behaviour of the test. A larger threshold than indicated previously is chosen in order to make the explanation more understandable; here $n_\sigma = 6$ thus $\gamma = 4.8$. When a sample of the signal deviates from zero and get close to 5 (at samples 5, 18, 21, for instance) the cumulative sum increases and thus the decision variable increases. In those situations a mean change of the signal is expected but not confirmed. Since the actual

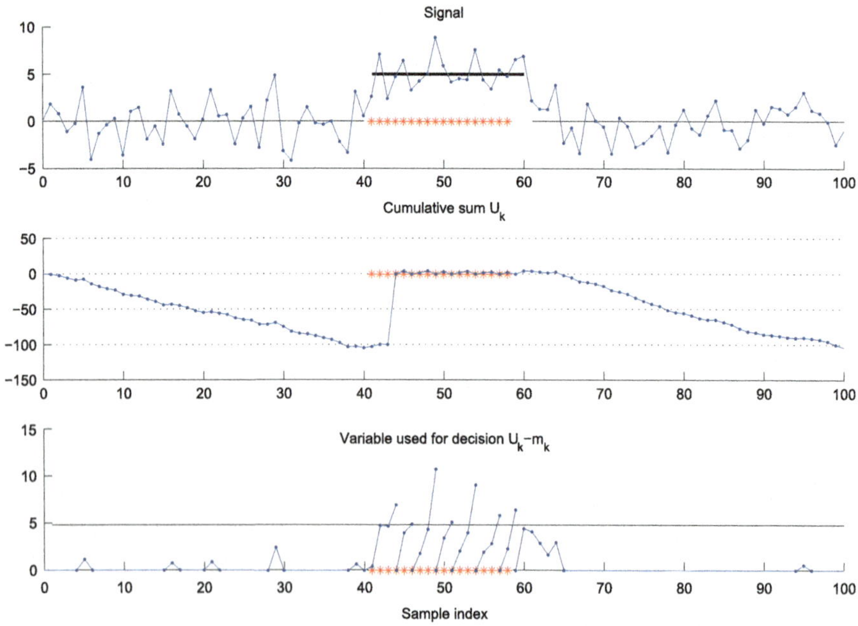

Fig. 9.19 Illustrative example of Page's test. The signal to consider is plotted on the *top graph*. The corresponding cumulative sum and decision variable are beyond. The *red stars* denote the sample indexes for which the test detected a mean change

mean of the signal is 0 from sample 1 to 40, the cumulative sum keeps decreasing and the decision variable goes back to zero before crossing the threshold. The expectation of a mean change is therefore rejected. At the sample indexes 41–43, the cumulative sum and the decision variable increase like before but the decision variable finally crosses the threshold at index 44 which confirms the mean change. Samples 41–44 are therefore marked with red stars on the graph. The cumulative sum and the decision variable are set to zero which is indicated by discontinuities of the plots. The Page's test starts again: it detects at the sample 46 that a mean change occurred since the sample 44, and so on. In this example the test requires 3 or 4 samples to detect a mean change and missed the sample 59 and 60 because the decision variable was close to the threshold but did not reach it.

9.3.3.3 Experimental Evaluation of Fault Detection

Analysis Methodology

The critical issues of the fault detection are the reactivity and the ability to detect the fault when it occurs. It was shown in the previous section that the sequential formulation of Page's trend test resets depending on the value of the samples. It therefore has the particularity to constantly adapt the size of the sliding window. The reactivity of this test is not known before hand. In order to measure this, the test is evaluated in terms of distance-to-alert, distance-to-recovery and accuracy of map error localization, as shown in Fig. 9.20. The distance-to-alert and the distance-to-recovery are derived from the usual time-to-alert and time-to-recovery and adapted in the context of this work in which the distance travelled by the vehicle is the reference. The distance-to-alert $\delta_{H_0 \to H_1}$ is the distance travelled by the vehicle before detecting a fault. Reciprocally, $\delta_{H_1 \to H_0}$ is the distance-to-recovery, that is, the distance after which the test detects the end of a fault. Since the test may detect a fault a few samples in the past, the a posteriori accuracy of the test is measured by $e_{H_1|H_0}$ and $e_{H_0|H_1}$. $e_{H_1|H_0}$ denotes the length of the road that has been identified as faulty while being actually fault-free. This can be named false alarm distance. Reciprocally, $e_{H_0|H_1}$ stands for the length of road that actually contains a fault that has not been detected by the algorithm which can be seen as a missed detection distance.

The performance of Page's trend test is compared to two other usual methods based on fixed length sliding window. These both aim to discriminate between H_0 and H_1:

$$\begin{cases} H_0 \, : \, d_i = b_i \, , \, i = 1, \ldots, q \\ H_1 \, : \, d_i = \delta_m + b_i \, , \, i = 1, \ldots, q \end{cases} \tag{9.22}$$

where $b_i \sim \mathcal{N}\left(0 \, , \, \sigma^2\right)$.

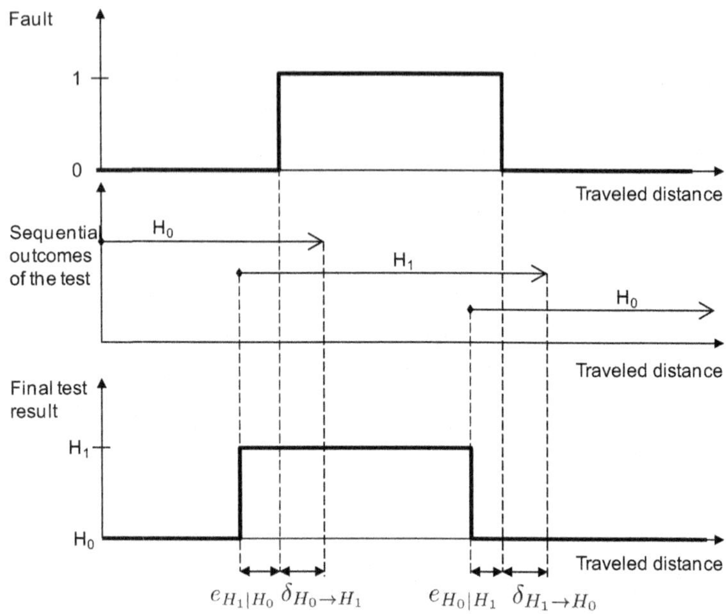

Fig. 9.20 Definition of the metrics used for tests assessment based on a simple example. The fault to detect is represented on the *upper part* of the figure. As the test may go back in the past to localize a detected fault, the a posteriori outcomes of the test appear on the *second line*. The fault finally detected and localized appears on the *bottom graph*

On the one hand, a simple decision rule based on the empirical mean of the sliding window was implemented:

$$\frac{1}{q}\sum_{i=1}^{q} d_i \underset{H_0}{\overset{H_1}{\gtrless}} \delta_m \tag{9.23}$$

On the other hand, the Neyman Pearson probabilistic decision rule was used for comparison. This is based on the generalized likelihood ratio of the hypotheses. Under Gaussian noise assumption. The choice follows the rule (9.24) [4, 21]:

$$\sum_{i=1}^{q} d_i \underset{H_0}{\overset{H_1}{\gtrless}} \sigma\sqrt{2 \cdot q \cdot \log(\Phi)} \tag{9.24}$$

Where the threshold Φ arises from a compromise between desired false alarm (type I error) and missed detection probabilities (type II error) of the decision rule. The false alarm probability has been set to its usual value (0.1 %) for an appropriate comparison with the proposed method.

The size q of the sliding window must be large enough to be statistically representative and short enough to detect map errors as fast as possible. Moreover, the samples D must be reinitialized as soon as the vehicle leaves one road for another which happens frequently in urban environment. It has then been set to $q = 20$ which is equivalent to approximately 200 m of travelled distance.

Experimental Evaluation

The tests were run on a set of roads that were recently modified due to the construction of a new motorway in Normandy, France (see Fig. 9.21). This area is representative of typical geometric errors that a map may hold and which may cause severe malfunctions in driving assistance systems. In area 1 and 2, sharp bends were added to the road which was previously straight. This would makes a curve warning system inefficient. In area 3 a new carriage way was added to the old single track road. This induces a constant lateral offset of the new road which would make obsolete intersection warning systems on crossing roads. On the fourth area, the lateral road offset decreases while the vehicle goes. This situation is very useful to highlight the distance to recovery of the tests.

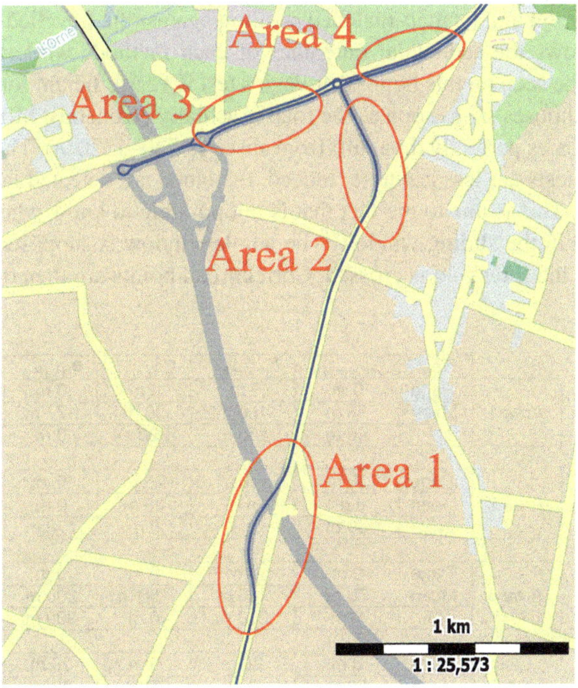

Fig. 9.21 Global view of the test areas. The out-of-date map being assessed is shown by *yellow lines*. A map of the actual road network is at the background in *grey*. The vehicle trajectory is in blue (starting from the *bottom* of the figure). The four test zones are *circled in red*

Since the experiment is done in a rural environment and also for explanation convenience, it is assumed here that a fault always originates from an error in the navigation map.

Tests have been run on the lateral euclidean distance d. The most restrictive constraint comes from intersection warning systems which require a longitudinal precision of 10 m for the placement of an intersection on a road link. Indeed, a lateral offset of a road link induces a longitudinal misplacement of intersection on crossing roads. This value has then been chosen as the mean change to detect $\delta_m = 10\,\text{m}$. Finally, the cumulative sum threshold is set dynamically to be $\gamma = 4 \cdot \sigma/\delta_m$ which represents a good compromise between false alarm rate and time to alert.

The results summarized in Fig. 9.22 show that Page's trend test is appropriate for detection and localization of faults. Indeed, faults are detected less than 20 m after the beginning of the fault and well localized. The two other methods provide less suitable false alarm and missed detection rates. This is mainly due to the fact that fixed length sliding windows are used. The fault detection has then undesirable collateral effects on the tail of the window. These methods show bad results when the road error is small with respect to the size of the sliding window.

Let us focus on area 1, to better understand the strength and weakness of the three methods. The upper part of Fig. 9.23 shows the lateral error between the vehicle's estimated position and its map-matched position against the travelled distance. The lower part shows the sequential outcomes of each methods while the vehicle is driven. It can be seen on this figure that Page's test is very efficient for detecting the fault since it chooses H_1 as soon as the road is actually erroneous. Moreover, in this example, it locates perfectly the fault (from abscissa 520–1520 m). The outcomes of the two other tests are less accurate. Indeed, the faults are detected later and locate it very poorly. This is due to the fact that is not possible to know where the change happened within the sliding window. The whole window is supposed to belong to H_1 as soon as the threshold is crossed. Correct road points are then declared faulty

	Test	$\delta_{H_0 \to H_1}$	$\delta_{H_1 \to H_0}$	$e_{H_0 \mid H_1}$	$e_{H_1 \mid H_0}$
Area 1	Page	0 m	0 m	0 m	0 m
	Mean	73 m	150 m	90 m	177 m
	N.P.	46 m	255 m	220 m	0 m
Area 2	Page	20 m	n.a.[3]	20 m	0 m
	Mean	n.d.[4]	n.a.	35 m	0 m
	N.P.	80 m	n.a.	0 m	170 m
Area 3	Page	0 m	0 m	0 m	0 m
	Mean	50 m	100 m	150 m	200 m
	N.P.	0 m	300 m	0 m	300 m
Area 4	Page	0 m	20 m	0 m	20 m
	Mean	50 m	0 m	250 m	40 m
	N.P.	0 m	250 m	0 m	210 m

[3] not applicable
[4] not detected

Fig. 9.22 Comparative results with nominal tuning of each test

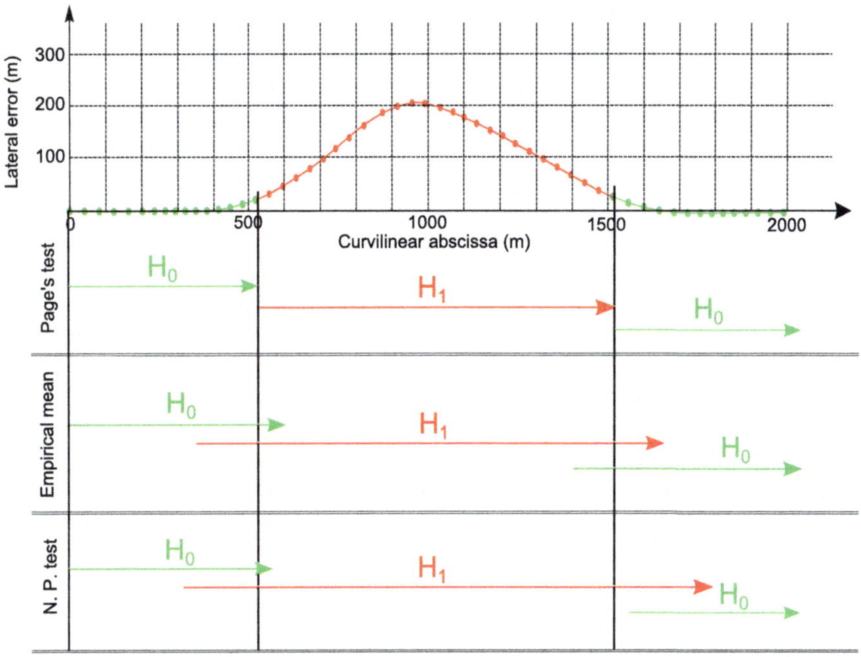

Fig. 9.23 Sequential outcomes of three trend tests on hypotheses H_0 (correct road segment) and H_1 (erroneous road segment). The lateral distance to the map-matched position is denoted on the *upper part*. The plot colour shows the true state of the road: correct in *green*, with fault in *red*

while they are not and *vice versa*. This example illustrates why these methods induce false alarms, missed detections and inaccuracies in fault localization.

9.3.4 Discussion

The mathematical formulation of this test and its application to the comparison of vehicle position estimates was detailed. Given this, its ability to detect a discrepancy between the estimate from sensors and the estimate from navigation was evaluated using real vehicle data. Since the metrics usually employed to measure the performance of this test are not relevant in the intelligent vehicle context, the evaluation was done using metrics introduced in this chapter. The test showed convincing performance since it detected quickly (short distance-to-detection and distance-to-recovery) and accurately (short false alarm and missed detection distances) the discrepancies between the estimates.

Page's trend test allows to detect with robustness when the two estimates of the vehicle position are significantly different and to conclude that a fault affects at least one of them. However, it does not permit to determine which one is affected by the fault if it is assumed that vehicle sensors can also be responsible for the fault. In this sense, this test performs fault detection but does not perform isolation. In order to solve this ambiguity, an approach consists in taking benefit of the repeated vehicle journeys on the same roads. The repeatability of the map geometrical faults permits to isolate and correct faults after a few number of journeys [40].

9.4 Summary

The chapter examined a fundamental issue in the decision-making process for vehicle navigation under computer control, namely situation understanding of the spatial-temporal relationship between entities sharing the same segment of a road network. A solution is proposed through the use of ontologies that establish this relationship in an 'ordered' manner to structure this relationship. During this process, a fundamental source of information and knowledge resides within navigation maps. They provide contextual information that it is used to facilitate the situation understanding process. Whilst much progress has been attained on the deployment of digital navigation maps for vehicle guidance applications, their geometric descriptions is far from perfect, there are modifications to maps structures, there are geometric errors, etc. Thus the second part of this chapter presented a novel approach for the detection of faults in the geometric descriptions of the roads, a feature of this being the use of close to production components that are used in other ADAS functions. The theoretical developments in this chapter have been implemented in passenger vehicles and different trials performed in standard road networks.

The overall results have shown that further work is needed, in particular on the information needed to infer understanding of the situation. That is, how to infer information from the close environment to improve the spatio-temporal relationships amongst the relevant entities. For this purpose an area to explore is semantic road segmentation. It should facilitate classification as well as the definition of the driving or navigation space. Maps remain a challenge, the concepts portrayed in this chapter are being extended to the notion of learning maps by leveraging on the repeated trajectories that often drivers, e.g. daily commuting. The success of these endeavours shall be demonstrated when applied to autonomous driving as the machine has to interpret data, detect errors and interact with its environment in a safe manner as it drives to its destination.

References

1. S. Abburu, A survey on ontology reasoners and comparison. Int. J. Comput. Appl. **57**(17), 33–39 (2012)
2. A. Armand, D. Filliat, J. Ibañez-Guzmán, Modelling stop intersection approaches using gaussian processes, in *Proceedings of the 16th International IEEE Conference on Intelligent Transportation Systems-ITSC* (2013)

3. F. Baader, *The Description Logic Handbook: Theory, Implementation, and Applications* (Cambridge University Press, Cambridge, 2003)
4. M. Basseville, I.V. Nikiforov, *Detection of Abrupt Changes: Theory and Application* (Prentice-Hall, Englewood Cliffs, NJ, 1993)
5. D.J. Buckley, The GIS primer an introduction to geographic information systems. Technical Report, Innovative, 1997
6. K. Dentler, R. Cornet, A.T. Teije, N. De Keizer, Comparison of reasoners for large ontologies in the OWL 2 EL profile. Semantic Web **2**(2), 71–87 (2011)
7. S. Geyer, M. Baltzer, B. Franz, S. Hakuli, M. Kauer, M. Kienle, S. Meier, T. Weißgerber, K. Bengler, R. Bruder et al., Concept and development of a unified ontology for generating test and use-case catalogues for assisted and automated vehicle guidance. IET Intell. Transp. Syst. **8**(3), 183–189 (2013)
8. P.-Y. Gilliéron, H. Gontran, B. Merminod, Cartographie routière précise pour les systèmes d'assistance à la conduite, in *Proceedings of the GIS-SIT Conference*. TOPO-CONF-2006-015 (2006)
9. T.R. Gruber, Toward principles for the design of ontologies used for knowledge sharing? Int. J. Hum. Comput. Stud. **43**(5), 907–928 (1995)
10. T. Gruber, Ontology, in *The Encyclopedia of Database Systems*, ed. by L. Liu, M.T. Özsu (Springer, Berlin, 2009)
11. P.J. Hayes, The second naive physics manifesto, *Formal Theories of the Commonsense World* (1985), pp. 1–36
12. M. Horridge, S. Bechhofer, The OWL API: a java API for OWL ontologies. Semantic Web **2**(1), 11–21 (2011)
13. M. Horridge, S. Jupp, G. Moulton, A. Rector, R. Stevens, C. Wroe, A practical guide to building OWL ontologies using protégé 4 and CO-ODE tools edition 1. 2. The University of Manchester (2009)
14. I. Horrocks, P.F. Patel-Schneider, H. Boley, S. Tabet, B Grosof, M. Dean et al., Swrl: a semantic web rule language combining OWL and RuleML. W3C Member Submission **21**, 79 (2004)
15. M. Hulsen, J.M. Zollner, N. Haeberlen, C. Weiss, Asynchronous real-time framework for knowledge-based intersection assistance, in *2011 14th International IEEE Conference on Intelligent Transportation Systems (ITSC)* (IEEE, New York, 2011), pp. 1680–1685
16. M. Hulsen, J.M. Zollner, C. Weiss, Traffic intersection situation description ontology for advanced driver assistance, in *2011 IEEE Intelligent Vehicles Symposium (IV)* (IEEE, New York, 2011), pp. 993–999
17. Hummel, W. Thiemann, I. Lulcheva, Scene understanding of urban road intersections with description logic, in *Dagstuhl Seminar Proceedings* (Schloss Dagstuhl-Leibniz-Zentrum fr Informatik, Dagstuhl, 2008)
18. C.G. Keller, C. Sprunk, C. Bahlmann, J. Giebel, G. Baratoff, Real-time recognition of U.S. speed signs, in *2008 IEEE Intelligent Vehicles Symposium* (2008), pp. 518–523
19. R. Kohlhaas, T. Bittner, T. Schamm, J.M. Zollner, Semantic state space for high-level maneuver planning in structured traffic scenes, in *2014 IEEE 17th International Conference on Intelligent Transportation Systems (ITSC)* (IEEE, New York, 2014), pp. 1060–1065
20. L. Lamard, R. Chapuis, J.-P. Boyer, Multi target tracking with CPHD filter based on asynchronous sensors, in *2013 16th International Conference on Information Fusion (FUSION)* (IEEE, New York, 2013), pp. 892–898
21. D. Maquin, J. Ragot, Diagnostic des systèmes linéaires (Lavoisier, Paris, 2000)
22. D.L McGuinness, F. Van Harmelen et al., Owl web ontology language overview. W3C Recommendation **10**(10), 2004 (2004)
23. B. Motik, R. Shearer, I. Horrocks, Optimized reasoning in description logics using hypertableaux, in *Automated Deduction–CADE-21* (Springer, Berlin, 2007), pp. 67–83
24. F. Moutarde, A. Bargeton, A. Herbin, L. Chanussot, Robust on-vehicle real-time visual detection of American and European speed limit signs, with a modular traffic signs recognition system, in *2007 IEEE Intelligent Vehicles Symposium* (2007), pp. 1122–1126

25. M. Platho, J. Eggert, Deciding what to inspect first: incremental situation assessment based on information gain, in *2012 15th International IEEE Conference on Intelligent Transportation Systems (ITSC)* (IEEE, New York, 2012), pp. 888–893
26. M. Platho, H.-M. Gros, J. Eggert, Predicting velocity profiles of road users at intersections using configurations, in *2013 IEEE Intelligent Vehicles Symposium (IV)* (IEEE, New York, 2013), pp. 945–951
27. E. Pollard, P. Morignot, F. Nashashibi, An ontology-based model to determine the automation level of an automated vehicle for co-driving, in *2013 16th International Conference on Information Fusion (FUSION)* (IEEE, New York, 2013), pp. 596–603
28. V. Popovic, B. Vasic, Review of hazard analysis methods and their basic characteristics. FME Trans. **36**(4), 181–187 (2008)
29. Protégé website, http://protege.stanford.edu/. Accessed 29 April 2015
30. Protégé Wiki website, http://protegewiki.stanford.edu/wiki/webprotege. Accessed 29 April 2015
31. Protobuf Website, https://developers.google.com/protocol-buffers/. Accessed 29 April 2015
32. R. Regele, Using ontology-based traffic models for more efficient decision making of autonomous vehicles, in *Fourth International Conference on Autonomic and Autonomous Systems, 2008. ICAS 2008* (IEEE, New York, 2008), pp. 94–99
33. R.A. Retting, B.N. Persaud, P.E. Garder, D. Lord, Crash and injury reduction following installation of roundabouts in the united states. Am. J. Public Health **91**(4), 628–631 (2001)
34. T. Schamm, J.M. Zollner, A model-based approach to probabilistic situation assessment for driver assistance systems, in *2011 14th International IEEE Conference on Intelligent Transportation Systems (ITSC)* (IEEE, New York, 2011), pp. 1404–1409
35. L. Serafini, A. Tamilin, Local tableaux for reasoning in distributed description logics, in *Proceedings of the International Workshop on Description Logics, DL*, vol. 4 (2004), pp. 100–109
36. R.M. Smullyan, *First-Order Logic* (Courier Corporation, North Chelmsford, MA, 1995)
37. S. Vacek, T. Gindele, J.M. Zollner, R. Dillmann, Situation classification for cognitive automobiles using case-based reasoning, in *2007 IEEE Intelligent Vehicles Symposium* (IEEE, New York, 2007), pp. 704–709
38. R. Wang, H.J. Ruskin, R. Wang, H.J. Ruskin, Modeling traffic flow at a single lane urban roundabout. Comput. Phys. Commun. **147**, 570–576 (2002). Proceedings of the Europhysics Conference on Computational Physics Computational Modeling and Simulation of Complex Systems
39. L. Zhao, R. Ichise, S. Mita, Y. Sasaki, An ontology-based intelligent speed adaptation system for autonomous cars, in *Semantic Technology* (Springer, Berlin, 2014), pp. 397–413
40. C. Zinoune, P. Bonnifait, J. Ibanez-Guzman, Sequential FDIA for autonomous integrity monitoring of navigation maps on board vehicles. IEEE Trans. Intell. Transp. Syst. **17**(1), 143–155 (2016)

Chapter 10
Radar Sensors in Cars

Holger H. Meinel and Wolfgang Bösch

10.1 Introduction

Advanced Driver Assistance Systems (ADAS) and autonomous driving, also based on mm-wave radar, are in discussion everywhere these days. It is already several years back—about 50 really—since the first attempts of implementing radar sensors into cars have been started. In those days, the car was entirely a mechanical device, even the injection control was done mechanically; the ignition coil—besides lightning—then was the only electric device in the car.

Already 50 years ago, there were some people envisioning the changes towards the electronic content in a car that have become reality today. However, they were laughed at then. Diagram 10.1 shows the increase of electric/electronic parts in a car as thought of in 1990s.

More electronics needed higher and—most importantly—lower cost integration by the utilization of automated assembly lines and the use of integrated circuits RFICs and MMICs. It was only in the early 1990s that in a MELECON Conference in Ancona, Prof. Alberto Sangiovanni-Vincentelli (today at the University of LA, Cal., USA) postulated to have Si wafers as large as 15 in. at the end of the century (being said in a time with production wafers having a diameter of 4–6 in. only)—half of the audience left the auditorium laughing.

Recently, Infineon has sold more than ten million radar chips for cars (Microwave Journal—Microwave flash—July 2015).

H.H. Meinel (✉)
Daimler AG, Ulm, Germany
e-mail: holger.h.meinel@gmx.de

W. Bösch
TU Graz, Graz, Austria

D. Watzenig, M. Horn (eds.), *Automated Driving*,
DOI 10.1007/978-3-319-31895-0_10

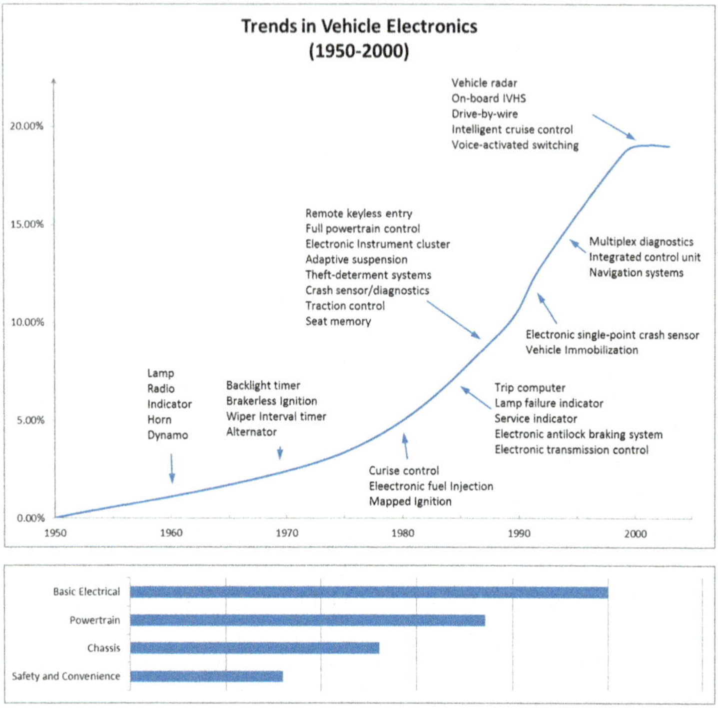

Diagram 10.1 The increase of electric/electronic parts in a standard sedan—as predicted in 1990s. *Source*: DASA, Germany—business development

In the early 1970s, we had the oil crisis with ***car free sundays*** all over Europe (Fig. 10.1),—the development of more comfortable and safer cars was not at all in focus—the objective was to lower the fuel consumption.

> Was that really the right time to develop microwave radar systems for cars, or the Automotive Collision Avoidance System (ACAS)—as it was dubbed then?!

Still it was done and beginning in 1972, the German government sponsored a research program investigating radar-based Automotive Anti Collision Systems—named NTÖ 49—which was launched at AEG-Telefunken in Ulm, Germany, employing 35 GHz technology. Earlier attempts implementing car radar systems were carried out at X-Band and Ku-Band, around 10 GHz and 16 GHz, shown in Figs. 10.2 and 10.3, respectively. However, the radar units were quite too large to fit into a standard sedan.

Fig. 10.1 Autobahn near Frankfurt on 25th November 1973—the first car free Sunday in Germany. *Source*: SWP, Ulm

Fig. 10.2 10 GHz automotive radar system built by VDO, early 1970s. *Source*: private collection of the author

10.2 Forward Looking Radar (FLR)

Within the research project, NTÖ AEG-Telefunken thoroughly investigated automotive radar design strategies and developed the first radar system operating at 35 GHz in the millimeter wave frequencies range. Many innovative design ideas were implemented, such as narrow beam width antennas with 2.5 by 3.5° pencil beams. The 2.5° in azimuth was illuminating just one lane 100 m in front, while 3.5° in elevation was good enough to look over hills and under the bridges. The utilization of miniaturized semiconductor two terminal devices, namely

Fig. 10.3 16 GHz automotive radar system built by SEL (Standard Electric Lorenz) 1975. *Source*: private collection of the author

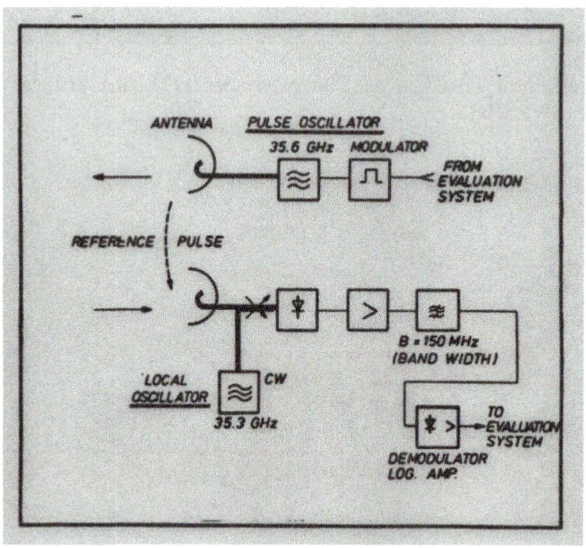

Fig. 10.4 Block diagram of a 35 GHz collision avoidance radar built by AEG-Telefunken in 1973. *Source*: AEG-Telefunken, Ulm, Germany

IMPATT-, GUNN-, and Schottky diodes and hybrid assembly technology, made it possible to significantly reduce the size of such a radar unit and fit it into the front of a standard sedan. IMPATT diodes were used for the radar pulse generation in the transmitter, while GUNN diodes served as a local oscillator (LO) in the receiver unit, and discrete beam-lead-type Schottky diodes were used as the mixing element in the receiver.

Figure 10.4 shows the block diagram of this first 35 GHz collision avoidance radar designed by AEG-Telefunken. The transmit (TX) and the receive (RX) units

Fig. 10.5 35 GHz automotive radar system built by AEG-Telefunken 1974 (the driver being the 1st author of this paper). *Source*: private collection of the author

Fig. 10.6 MAN truck equipped with a 35 GHz sensor. *Source*: private collection of the author

were realized as separate modules with their own transmit and receive antenna, as no adequate and technically sufficient circulators at 35 GHz were available then. The achievable pulse output power was about 100 mW with 20 ns of pulse width.

Already then, at the very beginning, the idea of using an automotive radar as a means to reduce the accident rates on our streets was one of the major driving forces. Therefore various cars, buses, and trucks were equipped with radar sensors (Figs. 10.2, 10.3, 10.5, and 10.6) and tested worldwide.

In other countries, mainly in the USA and Japan, research on automotive radar systems was initiated as well, however, at very different frequencies such as 35, 47, 60, and 94 GHz, the later being a "leftover" from military applications and

Fig. 10.7 Antenna configuration schematic of "Distronic Plus"—combining LRR and SRR for urban employment (*yellow*: 77 GHz LRR-Sensor, *green*: 24 GHz SRR-Sensor). *Source*: Daimler AG, Stuttgart, Germany

thus components were readily available, e.g., from the DARPA sponsored MMIC program in the USA.

Nevertheless, these first mm-wave units operating at 35 GHz showed promising performance. Several ten units had been built and were tested on several millions of highway test kilometers in Germany, jointly together with Bosch, one of the AEG-Telefunken partners within the NTÖ 49 research program.

Even trucks have been equipped and tested with the first 35GHz radar sensors. Based on these results, similar and slightly improved 35 GHz radar units were forwarded to SEL company working together with Mercedes-Benz within the regime of the sponsored NTÖ 49 program.

The Japanese 60 GHz approach was a political and very pragmatic one; mm-wave radar and future communication systems could be set up and investigated, employing the entirely same semiconductor devices and mm-wave components being developed at 60 GHz then, Fig. 10.13.

Only the introduction of 77 GHz radar sensing, being started in Germany in the early 1980s, as a worldwide standard for long range automotive radar (LRR) (see also Fig. 10.7) within WARC 89 cleared this turmoil of different frequencies.

10.3 Blind Spot Detection Radar

A first Blind Spot Detection (BSD) sensor, i.e., a Short Range Radar sensor (SSR) for the recognition of vehicles, being in the optical blind spot of the standard rear mirror, was presented in the early 1970s by Dunlop & Assoc. and Bendix (Fig. 10.8) in the USA, employing slotted array-type antennas at 16 GHz. "*The antenna patterns intersect adjacent* (street) *lines to illuminate the blind spot areas*

Fig. 10.8 Slotted array-type antennas with waveguide feeds at 16 GHz. *Source*: Harokopus, W.P., IEEE G-MTT, May 19, 1971 [6]

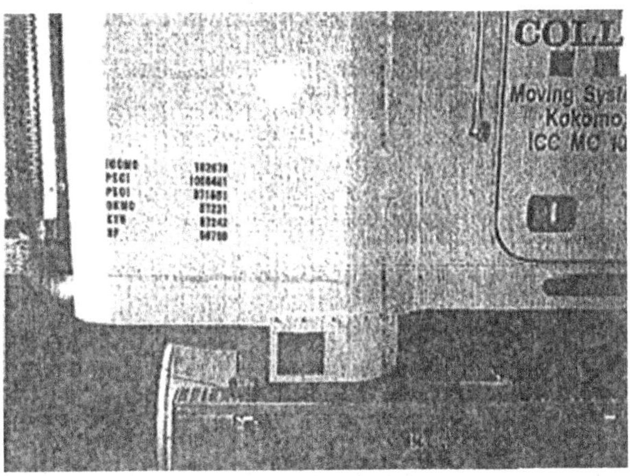

Fig. 10.9 24 GHz SDS system built by HE Microwave, 1995. *Source*: HE Microwave publication, IEEE MTT-S, 1995, San Francisco, USA [7]

and to warn of the presence of approaching automobiles with a light or audible signal."

More than 20 years later, in 1995, HE Microwave Corp. (Hughes Electronics) in Tucson, AZ, USA, already proposed 24 GHz for their SDS (side detection sensor) system for trucks (Fig. 10.9). "*The SDS was conceived to assist the driver in accessing the viability of a planned lane change*," as well a BSD sensor.

Today, the 24 GHz range is the general frequency approach being taken for BSD sensing (Fig. 10.10).

Narrow-Band (NB) systems, operating in the ISM-Band (24.05–24.25 GHz)—standing for Industrial, Scientific and Medical—and Ultra-Wide-Band (UWB)

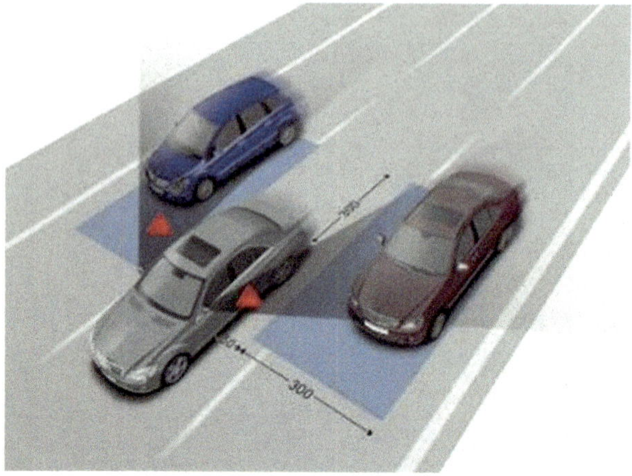

Fig. 10.10 Blind Spot Detection (BSD) incorporating two separate 24/26 GHz SRR sensors. *Source*: Daimler AG, Stuttgart, Germany

systems operating between 21.65 and 26.65 GHz with various advantages and disadvantages, are on the market today. Companies such as Valeo or Hella, just to mention two companies in Germany, are already producing more than two million of these NB sensors per year (2014) each. Autoliv is manufacturing UWB Blind Spot Detection systems for Mercedes-Benz, e.g., the so-called CPA (Collison Prevention System).

Nowadays, basically all long range automotive radar (LRR) sensors with "pencil" beams are operating in the 77 GHz frequency range that has been allocated by the ETSI standardization organization, whereas the SRRs for urban traffic applications with wider azimuth antenna values are operating in the 24 GHz range—these days (more on this subject in the Sect. 10.5).

The development of new markets and the worldwide deployment of a technically innovative product like automotive radar is typically a reaction to social developments and the upcoming of specific needs. The worldwide trend towards huge megacities accompanied by the democratization of mobility in highly populated countries like the BRIC States (Brasil, Russia, India, and China—together amounting to more than 40 % of all car sales in 2012) has led to a dramatic increase in traffic density and thus accident rates on the streets, prompting the need for enhanced vehicle safety and more driver assistance.

Due to their unique physical performance, automotive radar sensors are the backbone of modern vehicle safety and driver assistance systems, though not being the only means; laser scanner and video camera systems are mostly complementary and sometimes competing technologies. However, based on the described trends, it is quite simple to forecast that the density of radar systems will explode over the next decades, along with the worldwide hugely increasing number of cars (Diagram 10.2).

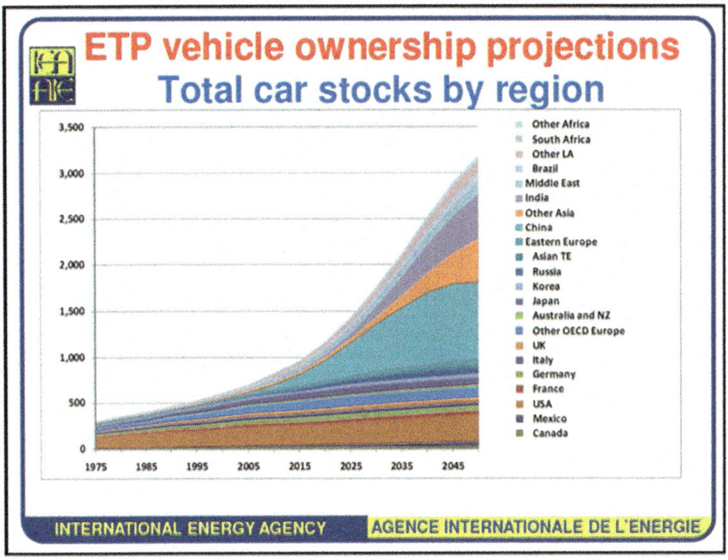

Diagram 10.2 Total car stocks by region. *Source*: IEA—International Energy Agency

The higher the automotive radar density operated on the road will grow the more important it will be to guarantee the interoperability of such sensors, especially as we are on the move from simple comfort systems to highly sophisticated safety applications, e.g., for autonomous driving utilizing automotive radar systems. TÜV certification based on further developed ASIL definitions (i.e., setup of sensor structures) within DIN 26262 will become necessary.

10.4 Early Systems and Their Results

1992 As early as 1992, the **EATON VORAD** CWS (collision warning system) operating at 24 GHz was installed in more than 4000 buses and trucks in the USA, from Greyhound buses (Fig. 10.11) to rental trucks providing an acoustic warning for the driver only.

Being driven on more than 900 million road km, the study confirmed that the amount of accidents per km travelled was reduced by more than 50 %; more than that, the resulting severity of accidents still occurring was significantly reduced (Fig 10.12).

However, these radar system units had to be de-installed on the long run due to heavy protests of the US-driver unions. The CWS radar system tracked the driving hours on the road as well and hence produced "transparent" drivers to their employers. Obviously, the drivers did not like that at all. The time was not yet ready for such a system.

Fig. 10.11 24 GHz CWS system built by EATON-VORAD, 1996. *Source*: private collection of Wolfgang Menzel

Fleet Type	Number of Trucks	Km Driven With CWS (Millions)	Accidents per Million Km without CWS	Accidents per Million Km with CWS	Accident Reduction Percentage
Private Carrier	20	17.0	1.3	0.0	100%
For Hire Carrier	170	37.8	3.7	2.4	34%
Truckload Carrier I	350	55.8	0.5	0.0	100%
Truckload Carrier II	850	115.2	0.8	0.3	71%
Leasing Company I	9	0.5	19.7	0.0	100%
Leasing Company II	8	0.32	16.3	0.0	100%
Beverage Distributor	58	2.2	29.6	0.0	100%
Totals	**1,465**	**224.8**	**1.0**	**0.3**	**69%**

Fig. 10.12 Accident reduction results. *Source*: Woll, J. D., M&RF Conf. 1997, London, UK [8]

1996 Four years later in 1996, the "Automated Highway System" was installed and tested in Japan on about 100 km of the newly built Jo-Shin-Etsu highway. The interoperability of several systems was tested. A "leakage coaxial cable" (LCX) was used for vehicle2roadside (v2x) communication. Magnetic nails in the road surface and corresponding sensors in the cars provided lane control during driving, and a mm-wave radar at 60 GHz was utilized for distance control. In addition, optical lane markers were detected by a CCD camera installed on the car (Fig. 10.13). The overall system was called ACAS (Automotive Collision Avoidance System) and demonstrated the first driverless cars. The "Electronic Information Systems Research Laboratory" of Nissan Motor Co. LTD. was one of the protagonists of this system.

The achieved and demonstrated results were very promising; however, the required road-side installations—LCX, magnetic nails, etc.—and the sensor systems in the car were thought to be too expensive at the time.

However, these days NISSAN is one of the technology leaders towards autonomous driving.

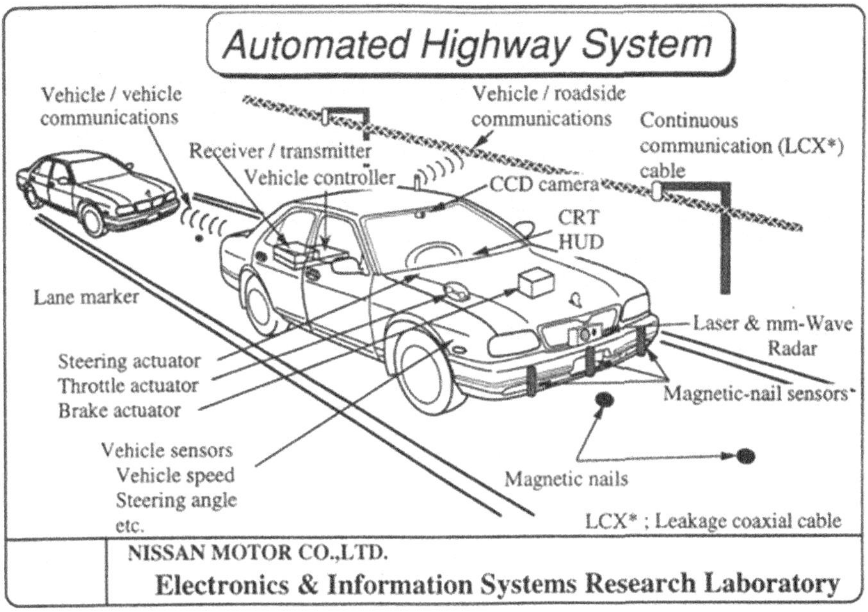

Fig. 10.13 Automated Highway System built by NISSAN Motor Co. Ltd. *Source*: Nissan Motor Co. Ltd

1998 Two more years later, when the *DISTRONIC* system was introduced by Mercedes-Benz in 1998, the acceptance rate was not at all favorable and the timely spread from the premium S-Class cars to other classes was quite slow—however, since the introduction of *DISTRONIC PLUS* (1x LRR plus 4x SRR), another 8 years later (2006), this has changed significantly. With "*Brake Assist Plus*" and "*PRE-SAFE Brake*" having become part of the system package (Fig. 10.14), such a system became directly driver recognizable and was well perceived by the public.

2011 The paradigm change from **comfort systems to safety systems** started to take place, as well as the "**democratization**" process with the introduction of the new Mercedes-Benz B-class and A-class in 2011 and 2012, respectively. "*DISTRONIC PLUS*" became available as a special equipment (Sonderausstattung) product fitted also in smaller cars. *CPA—Collision Prevention Assist* was already a series product although in the beginning only as an acoustical warning signal.

2014 With the introduction of the new Mercedes-Benz C-class in spring 2014, the further developed CPAplus (Collision Prevention Assist plus) and thus automatic brake assist became a standard series equipment in Mercedes-Benz cars.

The "*PRE-SAFE BRAKE*" system of Mercedes-Benz or the "*Intelligent Brake Assist*" from NISSAN are *the* prerequisite for advanced braking systems, that significantly reduce the number and the severity of road accidents.

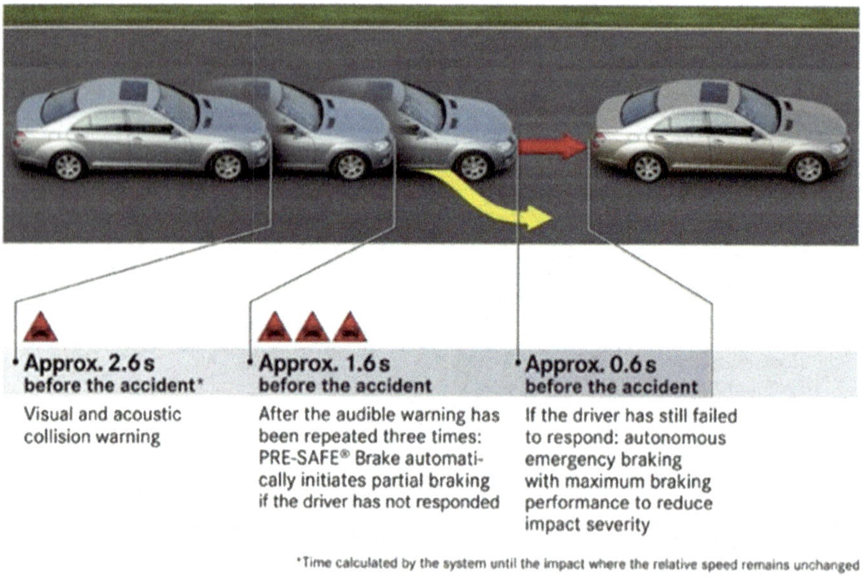

Fig. 10.14 Pre-safe brake—developed and confirmed for sedans—now also in trucks. *Source*: Daimler AG, Stuttgart, Germany

Figure 10.15 shows the reduction of accidents (in number as well as in severity) due to the introduction of "*Distronic PLUS*," a radar-based electronic braking system.

A study from KPMG—August 2015—came to the result that ADAS could reduce accidents by 80 %.

Today, we have roughly one accident per 280,000 miles driven; in 2040 this could be only one accident per 1,600,000 miles driven.

10.5 Trends

In the future, as described above, many more sensors will be operated on the road simultaneously. Therefore, it is quite obvious for any OEM investigating and planning the next generation of radar; it will not be sufficient to take a sensor that has demonstrated to be insensitive against others. The sensor itself has also to be proven not to interfere or to blind other systems. Testing the interoperability of radar sensors for OEMs, as well as for the corresponding supplier, it would be very convenient to utilize a standard interferer [10] as a qualification device. Such a test device should be able to generate a large variety of modulation schemes and bandwidths that could be tested against the sensor under question. Another objective has to be keeping the opportunities open to meet future demands in

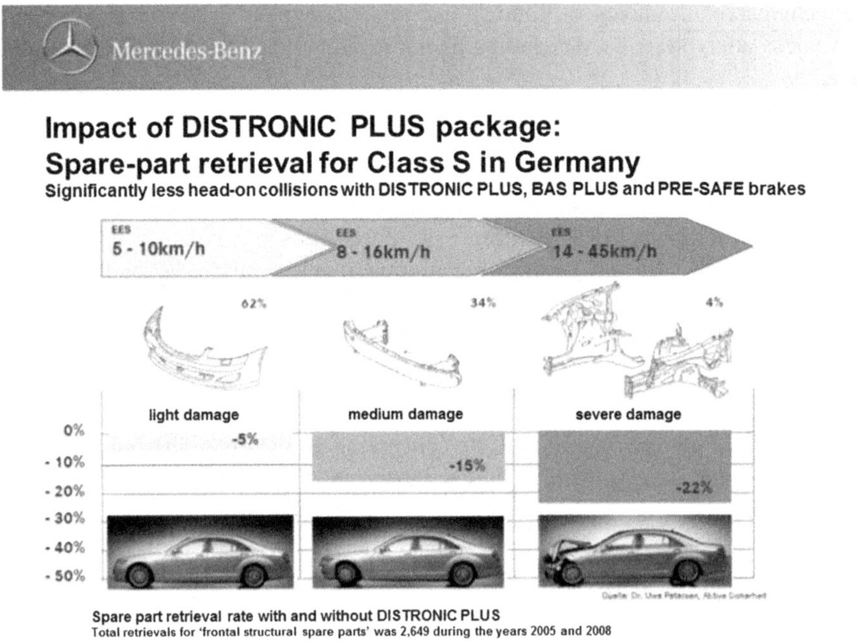

Fig. 10.15 Reduction of accidents based on "Distronic PLUS." *Source*: Daimler AG, Stuttgart, Germany

the process towards interoperability. One prominent issue of this concern is the corresponding frequency band as well as the bandwidth taken per sensor to allow, for example, frequency hopping. Future frequency regulation strategies should aim to get worldwide access to higher frequency bands, as well as using larger bandwidths per sensor. A minimum bandwidth of 1–2 GHz is very likely to be the requirement for future traffic scenarios, like pedestrian recognition (micro-Doppler) or side-impact-alert.

As a consequence, all this demands a common effort between radar manufacturers, backed by their corresponding OEMs, in order to identify and to agree upon a *common design rule book*. This guideline has to quote commonly agreed countermeasures against interference, while keeping opportunities open for each manufacturer to develop his individual radar. The EU-funded project MOSARIM (MOre Safety for All by Radar Interference Mitigation) was a first step in this right direction. MOSARIM was openly described and broadly discussed within the microwave community during the EuMW 2012 in Amsterdam; the EuMC/EuRAD WS23 being the official and final event of that program.

Today 24/26 GHz is the general frequency approach, being taken for BSD sensing, as stated above. However, the time for changing the operating frequency of future radar sensors from 24/26 GHz to 76–81 GHz is very likely to come soon.

There are four main reasons:

1. Technical requirements of future (and more complex) driver assistance and vehicle safety systems, demanding higher resolution and accuracy in space and time.
2. Vehicle integration and sensor packaging demands minimization, while enhancing sensor performances.
3. Cost reduction based on "economies of scale." Shifting to the 76–81 GHz range allows to develop radar modules being able to be used for all automotive radar types from LRR via MRR to SRR.
4. Interoperability, since the market penetration of automotive radars will explode, interference mitigation has to become the key for further market growth. A large bandwidth enabling frequency hopping or other efficient frequency separation procedures has to be mandatory and is worldwide available only between 76 and 81 GHz.

However, more and ongoing efforts concerning automotive radar performance are already under development, as they will be necessary for the future.

10.6 Future Directions

Taking off from today's already available and market introduced radar sensors, injury free driving and a better and more efficient vehicle management, resulting in lower consumption, will be a near future issue (Fig 10.16).

Applications like RCTA (Rear Cross Traffic Alert), supporting a driver while backing-up out of a parking lot or RPC (Rear-Pre-Crash) and detecting and calculating critical objects approaching from the rear, are already being implemented, e.g., in the **Mercedes-Benz S-Class**, having been launched in July 2013 (Fig. 10.17).

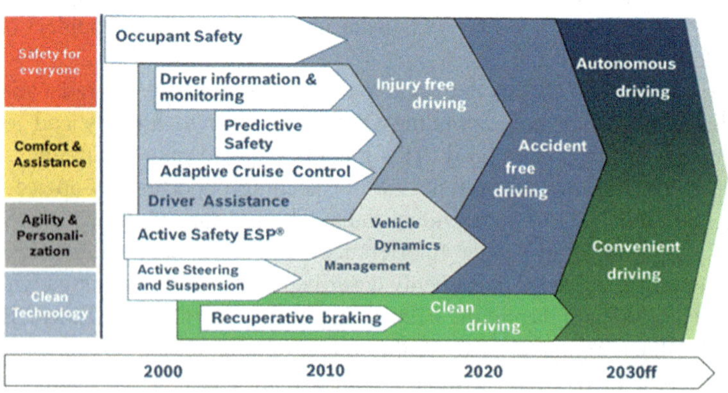

Fig. 10.16 Visions for vehicle motion and safety. *Source*: Bosch GmbH, Stuttgart, Germany, from a EuRAD 2012, Amsterdam, paper

Fig. 10.17 The Mercedes-Benz S-Class of 2013, featuring a RPC (rear-Pre-Crash) sensor. *Source*: Daimler AG, Stuttgart, Germany

In 2013, Valeo had introduced a combined BSD/RTCA sensor, employing DBF (Digital Beam Forming) antenna technology.

The "All-around collision free" car from **NISSAN** utilizes the "safety shield," a concept being already introduced in 2005. The "Safety Shield" is an effort to proactively achieve active safety. The car is able to help to provide assistance to the driver for safer driving depending on the actual situation. By combining aspects of active and passive safety, NISSAN was able to help reduce the number of fatalities and serious injuries.

"Distance Control Assist," based on ACC, or "Lane Departure Warning" and even more "Lane Departure Prevention" are parts of NISSAN's "safety shield."

TOYOTA announced details of its entry into the autonomous vehicle race at the annual "Consumer Electronics Show" (CES) 2013 in Las Vegas, USA, in January. The "Advanced Safety Research Vehicle," a **Lexus RS** (Fig. 10.18), utilized a variety of technologies, like GPS, mm-wave radar, laser tracking, and stereo cameras, to achieve its autonomy.

In December 2013, the **Volvo** Car Corporation (VCC) in Gothenburg, Sweden, has announced the "*Drive Me—Self-driving cars for sustainable mobility*" project. "*Drive Me*" *is* a world unique pilot project with up to 100 self-driving cars on public roads before 2017. If everything is running as planned and smoothly, series production is envisioned to start in 2020 (Fig. 10.19).

Recently, in September 2015, the **Dutch government** has announced an ambitious program to have automotive vehicles on the roads before 2025. Necessary changes are going to be implemented in existing laws to open the door for such vehicles on public roads. In parallel, the national consortium DAVI (Dutch

Fig. 10.18 The Lexus Advanced Safety Research Vehicle, as shown in Las Vegas, CES 2013. *Source*: Toyota, Japan, from the Internet

Fig. 10.19 *DRIVE-ME* test car. *Source*: Volvo Cars Corporation (VCC)

Autonomous Vehicle Initiative) was organized, with the TU Delft as part of the consortium.

Advanced signal processing of today—incorporating "big data" at very high data rates—has made new solutions and applications possible for car radar employment. *Micro-Doppler detection for pedestrian recognition* or the *distinguished wheel Doppler detection* of a car are such recent examples. Super computing with Deep learning capabilities thus will be the backbone for future and advanced signal processing—bringing Automated Driving nearer to reality.

Accident free driving and subsequently *autonomous driving* have come into reach technically.

The following chapters will describe this situation in detail.

Further Reading

1. H. Meinel, B. Rembold, Commercial and scientific applications of millimetric and sub-millimetric waves. Radio Electron. Eng. **49**(7/8), 351–360 (1979)
2. H. Meinel, Millimeter-Wave Systems and Applications in Europe, Invited Paper, IEEE MTT-S, New York, 1988, pp. 649–653
3. H. Meinel, Millimeter Wave System Design and Application Trends in Europe, Invited Paper, Alta Frequenza LVIII(5–6), 441–456 (1989)
4. H. Meinel, Automotive Radar—Present Status and Future Trends in a 'Market to Be' for Millimeterwaves, Invited Key-Note Talk, in *20th International Conference on IR & Millimeterwaves*, Lake Buena Vista, FL, USA, Dec 1995
5. H. Meinel, J. Wenger, G. Rollmann, H. Dominik, M. Chinn, M. Kunert, 24 GHz UWB Technology for Automotive Radar Applications, in *IWUWBT2005, 2005 International Workshop on UWB Technologies*, Yokosuka Research Park, Yokosuka, Kanagawa, Japan, Paper: TA4-1, 8–10 Dec 2005, pp. 79–82
6. W.P. Harokopus, Application of Radar to Automobile Control and Sensing, IEEE G-MTT, 19 May 1971
7. J.C. Reed, Side Zone Automotive Radar Criteria for Target Classification, in *Intelligent Vehicle Technology Conference*, VTC 1995, Detroit, 25/26 Sept 1995, pp. 361–363
8. J.D. Woll, Monopulse Doppler Radar for Vehicle Applications, in *M&RF Conference 1997*, London, UK, Sept 30–Oct 2, 28 pages
9. C. Gauss et al., Comparative Test of Advanced Emergency Braking Systems, ADAC Internal Test Report, ADAC Technik Zentrum, Landsberg, Germany, 30 Aug 2012
10. H. Meinel, J. Dickmann, Automotive radar: From its origins to future directions. Microw. J. **56**(9), 24–40 (2013). Cover Feature/Invited Paper
11. TOYOTA Deutschland GmbH—Public Relations & Press, TOYOTA entwickelt automatisierte Fahrsysteme—neue Abstandsregelung und Fahrspursteuerung vorgestellt, 11 Oktober 2013
12. H. Furushou, Nissan Research Center: *Private communication*, and Nissan Motor Company, Kanagawa, Japan: *Technological development activities*, Internet adv., Dec 2013
13. Volvo Car Corporation, Gothenburg, Sweden, Press release on behalf of *"DRIVE-ME"*, Dec 2013
14. H. Yan, *Drivers Stay Behind the Wheel Awhile*, China Daily, Oct 2015, p. 18

Part IV
In-Vehicle Architectures and Dependable Power Computing

Chapter 11
System Architecture and Safety Requirements for Automated Driving

Jan Becker, Michael Helmle, and Oliver Pink

11.1 Toward Automated Driving

Assisted driving functions already support the driver today by taking over either the longitudinal or the lateral driving task in specific situations. Examples are Adaptive Cruise Control or Lane Keeping Support. While these functions support the driver with regard to the longitudinal or lateral vehicle guidance within defined situations, handover of control back to the driver is required in case functional system boundaries are reached or a critical fault is detected. Therefore, the driver has to remain available all the time and provide fallback and recovery by means of human intervention.

Automated driving introduces the first driving functions that will carry out both longitudinal and lateral control tasks simultaneously and allow the driver to be absent from the active driving task for a limited amount of time. The driver will be responsible for permanently supervising partially automated functions, which is not required for highly and fully automated functions.

Different approaches exist on how to reach fully automated driving. One strategy is to incrementally increase the amount of automation for the respective function. For example, adaptive cruise control on highways is followed by combined longitudinal and lateral control with the driver constantly supervising the function. This is followed by highly automated driving where the driver may perform side tasks while driving, followed ultimately by fully automated driving on highways. Another approach is to jump directly to fully automated driving while significantly limiting

J. Becker (✉)
Robert Bosch LLC, Palo Alto, CA, USA
e-mail: Jan.Becker@us.bosch.com

M. Helmle • O. Pink
Robert Bosch GmbH, Gerlingen-Schillerhöhe, Germany

D. Watzenig, M. Horn (eds.), *Automated Driving*,
DOI 10.1007/978-3-319-31895-0_11

velocity, e.g., to 25 km/h. For a detailed discussion on the different strategies and their implications the reader is referred to [1].

Below, we will adopt the incremental view and introduce a set of driving functions with increasing degree of automation to set the stage for the technical discussion. The functions referred to serve as a set of example specifications which are currently not standardized and represent a subset of possible future automated functions. We expect wide-scale introduction of automated driving in well-defined situations and restricted environments first. For example, a freeway provides an environment with unidirectional traffic flow, whereas urban driving scenarios include cross-traffic situations, pedestrians, and bicyclists. Consequently, the requirements on perception, situation recognition, and decision making are considerably higher in a more complex environment.

Figure 11.1 shows our expectation of the evolution of automated driving. For a detailed taxonomy of automated driving functions, we refer to [2]. Our vision for automated driving can be seen in [3] and, more specifically, how a user may experience a highway pilot system is shown in [4].

Therefore, we are expecting a function such as the traffic jam pilot to be the first highly automated driving function deployed to the market, followed by the highway pilot. Both functions will be restricted to highway-like environments. Below, we present a brief functional description of these two exemplary driving functions. The function definitions given here are tentative and serve as examples to derive requirements regarding the onboard network but are far from being standardized and, therefore, are subject to changes in definition.

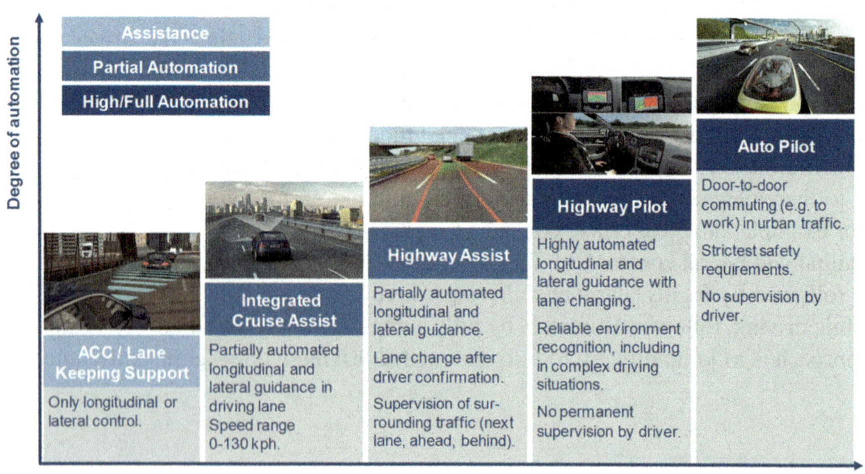

Fig. 11.1 Examples for automated highway driving functions according to the level of automation: Adaptive Cruise Control and Lane Keeping Support function are examples for assisted functions (level 1); Integrated Cruise Assist and Highway Assist are examples for partial automation (level 2). The highway pilot and autopilot resemble high and full automation, respectively (level 3 and 4)

11.1.1 Traffic Jam Pilot

The traffic jam pilot is designed to provide automated guidance of the vehicle in situations with traffic congestions on highway-like road environments. This requires combined lateral and longitudinal guidance of the vehicle at velocities typically less than 60 km/h on roads with more than one lane per driving direction, wide lanes, and low curvature. Lateral guidance aims at keeping the vehicle in the current lane; automated lane changes are not supported. Longitudinal guidance aims at keeping a safe distance to the preceding vehicle. In case a system boundary is reached, the driver is requested to take over control of the vehicle. If the driver is not responding accordingly within a defined time limit, the system will start switching to the safe state.

11.1.2 Highway Pilot

The highway pilot will extend the traffic jam pilot to higher velocities of up to 130 km/h and to situations without surrounding vehicles. In addition, automated lane change maneuvers and finer lateral guidance within the ego lane are provided, resulting in a more comfortable distance to adjacent vehicles. Another evolution of the highway pilot is represented by the exit to exit function. It implements functional features such as transitions from one highway to another highway, including on-ramps and off-ramps. This allows the driver to enter a city area as target destination in the vehicle navigation system, and the pilot function automatically selects relevant combinations of highways to reach that destination. We refer to [4] for a potential implementation of such a highway pilot system.

11.2 System Architecture

The use cases in the previous section enable us to define requirements for individual components of an automated driving system. While an automated driving function does not necessarily have to cover all situations, it is desirable to cover as many as possible in order to obtain higher availability of the function and less frequent takeover requests to the driver. Furthermore, in case of the occurrence of any unforeseen situation, the function must be able to handle the situation until the driver has taken back control.

Similarly, in case of hardware failures, the system needs to stay operational, at least with reduced functionality, until the driver has taken back control. This imposes additional requirements on the sensor set, electronic control units (ECUs), communication network, power supply, and actuators [5].

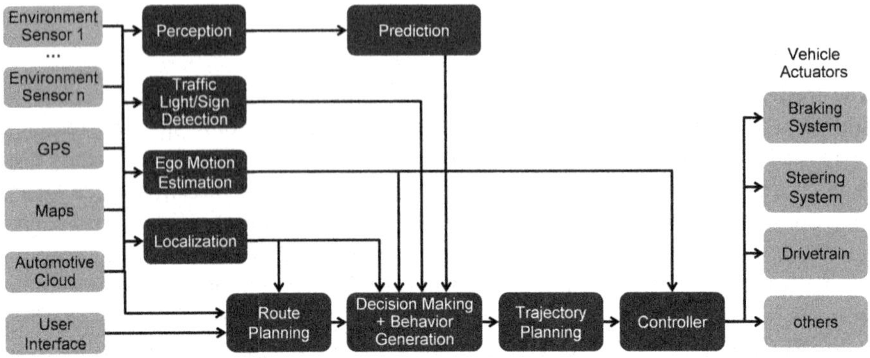

Fig. 11.2 Simplified example for the functional architecture of a highly automated driving vehicle. Redundant sensors and other system input are on the *left*. Algorithmic components in the *center* (*dark gray boxes*). Redundant actuators on the *right*

The use cases to be covered by a respective function typically have an impact on all components. For example, snowfall can lead to limited sight of the sensors, will affect a surround-sensor-based localization system due to different appearance of the surroundings, may require decision making to drive more slowly, and causes a low friction road surface which requires appropriate motion control capabilities [6].

Technical solutions for automated driving have been developed during the 2007 DARPA Urban Challenge [7] and have been improved since [8–10]. These systems need to be further extended in order to be capable of handling all situations that occur in real traffic. In the following sections, the impact of handling the aforementioned use cases will be detailed for some of the key components of an automated driving system.

Figure 11.2 shows a simplified functional architecture for highly automated driving. Functional redundancy is employed on the sensor and actuator level.

11.2.1 Surround Sensors

The surround sensor set for automated highway driving must be capable of reliably detecting all relevant obstacles in any situation the vehicle may encounter. In addition to physical redundancy for handling, e.g., hardware failures, this requires a diverse sensor set with different sensing technologies. Even under adverse circumstances, relevant obstacles have to be detected by at least one sensor. Sensors have to fulfill different requirements in different areas around the vehicle. The detection range and reliability in each area are defined by the respective most challenging use case.

For the front sensor set, this could be a comfortable stop behind a standing vehicle, given that there is no oncoming traffic on highways, for example, the rear

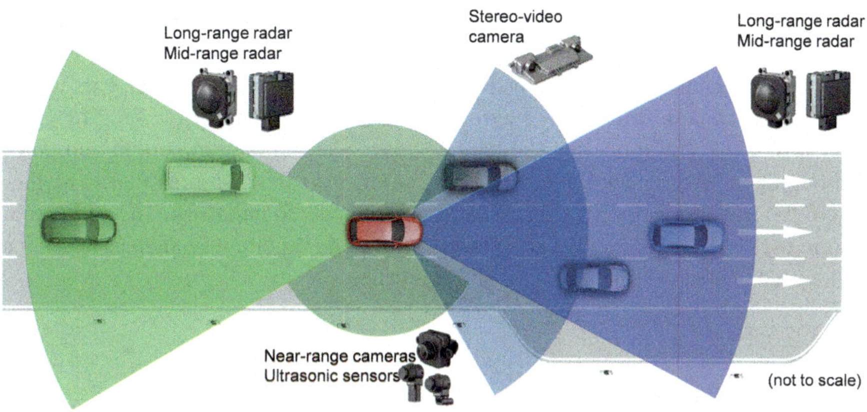

Fig. 11.3 Example of sensor set and associated field of view for highly automated driving

end of a traffic jam. The object detection reliability is dependent on the distance from a vehicle ahead. Late detection of a slow vehicle ahead will lead to a harder and potentially uncomfortable braking maneuver. However, any potentially harmful object has to be detected early enough so that an emergency maneuver can be triggered. Consequently, the area guaranteed to be free of potentially harmful obstacles needs to be determined, and the vehicle must be able to perform a potential emergency maneuver within this area.

For the rearwards facing sensors, the most challenging use case encountered on highways is typically a fast approaching vehicle from behind with an approaching speed of up to 250 km/h.

For the side sensor set, the detection range is defined by the lane change use case and has to cover an area around the vehicle large enough to ensure that the neighboring lane is free. Consequently, the detection range has to cover at least two adjacent lanes to the left and two adjacent lanes to the right of the vehicle.

Figure 11.3 shows an exemplary sensor set for highly automated driving.

11.2.2 Perception

The main goal of the perception system is to combine all sensor measurements into a consistent representation of the surrounding world. There are many fusion algorithms proposed to this end. They can be categorized into object-level fusion, where each individual sensor delivers object hypotheses which are combined in a subsequent fusion step, e.g., [11], and feature-level fusion, where lower-level sensor data is directly used to update a world model, e.g., a grid or particle representation [12]. In general, the output of any fusion system is an obstacle representation, for example, an occupancy grid or an object list or a combination of both.

In addition to a representation of all known obstacles, perception also needs to provide a notion of the unknown for many use cases. For example, if the field of view is limited due to fog or obstructions, the system needs to adapt its speed accordingly, because there may be undetected obstacles outside the field of view. Instead of explicitly modeling the "knownness," our perception system computes and outputs a "free space," known to be free of obstacles. An area that is neither part of the free space nor part of an obstacle is considered unknown. The perception system uses measurements from the road surface and other clues, such as positions of detected obstacles, to determine the free space.

11.2.3 Localization

Localization is the process of estimating the vehicle position and orientation with respect to a given map. We distinguish three levels of localization precision:

- Road precision: On which road am I?
- Lane precision: On which lane am I?
- Sub-lane precision: Where within the lane am I?

The different levels are used at different steps during the decision making process. Localization on road level is required for function activation, e.g., if the function is limited to a certain set of roads, e.g., highways, and in order to determine the route from the current position to the desired destination. Lane-level localization is mainly required for lane change use cases, e.g., whether a lane change is possible or whether we have to change lanes in order to reach our navigation goal. Sub-lane level localization is required for maintaining the proper position within the lane. Unavailability of any of these localization levels may lead to degradation of the functional performance or even to a driver takeover request. For example, if the activation relies on road-level localization, it has to be available to offer the function to the driver. If lane localization is unavailable, lane changes may not be possible during this time. Special situations, such as driving in a tunnel or under a bridge, may lead to temporary unavailability of any of the aforementioned levels, even when using state-of-the-art satellite-based localization systems. To recover in these situations, we use surround-sensor information in addition to Global Navigation Satellite System (GNSS)-based information for localization. Such surround-sensor-based localization systems have been proposed for several sensor technologies (e.g., camera [13] or Lidar based [14]). Similar to the perception system, combining multiple sensing technologies increases the overall availability of the localization system. In order to guarantee a high availability, we are using a combination of all available sensing technologies for localization.

11.2.4 Decision Making

It is essential to adapt the behavior of an automated vehicle to the current driving situation, especially in challenging situations. This includes determining a safe maximum vehicle speed and safety distances for the current situation, for example, based on the current sensor viewing ranges, road surface condition, or the speed of other vehicles. Using the free space representation from the previous section, the maximum vehicle speed could be selected so that we can perform an emergency maneuver within the current free space.

Before executing any given maneuver, decision making also has to determine whether this maneuver can be performed safely in the respective situation. For example, lane changes will not be performed when fast vehicles are approaching from behind or if the vehicle is driving in a narrow curve where the field of view is limited.

Decision making is also responsible for following traffic rules. Essentially, the entire subset of local traffic rules [15, 16] that applies to highway driving needs to be implemented in the planning system. This differs significantly from country to country, which is illustrated by some of the rules that are implemented in our planning system for lane changes in the USA and Germany, respectively:

USA:

- Perform a lane change to another lane (left lane preferred but not mandatory) if the vehicle ahead is driving significantly slower than our desired speed.
- If approaching a highway split or exit, change onto a lane that continues toward our navigation goal.

Germany:

- Rightmost lane: Perform a lane change to the left if the vehicle ahead is driving significantly slower than our desired speed. Since passing other vehicles on the right-hand side is illegal, the same rule applies to any other vehicles ahead on our left-hand side.
- Middle lane(s): Perform a lane change to the left if the vehicle ahead in our lane (or in any lane left to us) is driving significantly slower than our desired speed. Change to the right if there is no vehicle in sight (the law states that continuous driving in the middle lane(s) is allowed if there is a vehicle in the rightmost lane "occasionally" present).
- Leftmost lane: Immediately change lanes back to the right if we can continue there with our desired speed for at least several seconds.
- If we approach a highway split or exit, the above rules apply to all lanes that continue toward our navigation goal. That is, if no other traffic is present, change onto the rightmost lane that continues toward our navigation goal.
- At speeds below 80 km/h, passing on the right-hand side with moderately higher speed is permitted.

– In traffic jams, only change lanes if required (e.g., due to a lane ending), the rule of driving in the rightmost lane does not apply.

There are several additional rules related to lane changes that are not listed, for example, based on the distances and speeds of vehicles in the target lane. Our implementation makes use of a hierarchical state machine, which is a common approach for automated driving systems [17, 18]. Different driving behaviors, such as lane following or lane changing, are modeled as states. Maneuver decisions, including the aforementioned traffic rules, safety considerations, and interactions with other vehicles, are modeled as state transitions.

11.3 Functional Safety Concept

As one of the first activities in the safety lifecycle according to ISO 26262 [19], the hazard analysis and risk assessment (H&R) shall be conducted. As a result, the safety goals and the related ASIL ratings shall be defined.

For the traffic jam pilot, the hazard analysis and risk assessment yields among others the following safety goal: "Avoid insufficient vehicle deceleration when traffic jam pilot is active." Evaluation of the severity of the possible harm (S3), exposure (E3), and controllability by the driver and other involved traffic participants (C3) leads to assignment of ASIL C for this safety goal. The definition of the safe states is based on this safety analysis. According to ISO 26262-1, the safe state is defined as "operating mode of an item without an unreasonable level of risk" (cf. [19]). In the context of automated driving, there are usually two types of safe states including different criteria (see Table 11.1).

In general, zero collision risk would imply to have a prediction horizon to infinite future times. We refer to the discussion in [20], where the authors conclude that all of the classic planning methods are arguably unsafe. However, in the context of a certain world model, it can be argued according to [20] that an automated vehicle does not actively harm and that it can be designed to be as safe as humanly possible. In simple terms, the safe state translates to bringing and keeping the vehicle in a

Table 11.1 Potential safe states for automated driving functionalities

Safe state	Criteria for the safe state
The driver takes over control of the vehicle	• Maximum period of time allowed for driver takeover • Functionality to be maintained while waiting for driver takeover
The automated driving functionality switched to a degraded operation mode and finally into the safe state	• Maximum estimated time to reach the safe state • Minimum functionality to be maintained while transitioning to the safe state

standstill condition in a safe location while sending out warning signals to other traffic participants, e.g., via hazard messages or activated warning lights.

For the traffic jam pilot, a possible strategy is to moderately decelerate the vehicle within the current lane and keep it in standstill there, i.e., parking the vehicle. At first, a request for driver takeover is issued; however, if the driver does not respond or a critical failure is detected, the vehicle slows down automatically. The deceleration is accompanied by warning signals for the surrounding traffic participants and possibly by an emergency call, depending on the driver's state.

For vehicle velocities lower than 60 km/h and an assumed deceleration of at least -3 m/s^2, the time required to reach standstill is approximately 6 s. The vehicle travels a maximum distance of approximately 50 m during this time within the current lane. This emergency trajectory has to be planned ahead and stored securely, since the traffic jam pilot will rely on it as the fall back reaction in case of a relevant system failure. Consequently, a valid emergency trajectory has to be available before the automated driving function can be activated.

This definition is insufficient for the highway pilot due to the higher relative velocity range and potentially more complex traffic situations compared to a traffic jam situation. The safe state has to be determined situation dependent and could be one of the following states:

- Driver takes over control of the vehicle.
- Stopping the vehicle in the current lane.
- Decelerate and pull over to the rightmost lane (for right-hand traffic).
- Decelerate and pull over to the emergency lane.
- Continue driving at reduced speed to a safe location.

These examples show that safe states for automated functionalities are very complex and can take quite some time to reach. The time to reach the safe state is defined as emergency operation interval in ISO 26262-1 (cf. [19]).

The maximum transition time to the safe state can be split in the following two contributions:

- The maximum time span to wait for driver takeover.
- The time span required to take the vehicle to the safe state without support by the driver.

Both add up to the fault reaction time, which is defined in Fig. 11.4. During both time durations, the system completely controls the vehicle. Therefore, the corresponding system function has to be provided with high reliability. For most cases, this corresponds to a safety goal rated ASIL D according to ISO 26262.

Table 11.2 summarizes preliminary concepts regarding maximum allowance for driver takeover and corresponding safe state for the traffic jam pilot and highway pilot. We want to emphasize that these numbers are given as an indication only. The exact values depend on many parameters such as

- Type of automated functions
- Definition of the safe state(s)

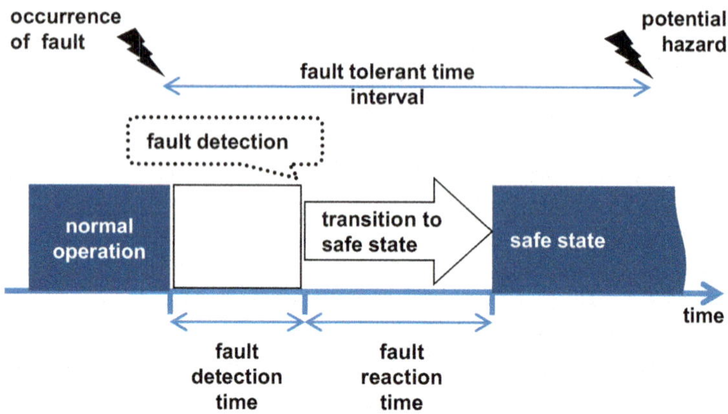

Fig. 11.4 Definition of fault reaction time and fault-tolerant time interval adapted from ISO 26262-1, 1.44 (cf. [19])

Table 11.2 Estimation of maximum tolerable times

Automated driving function	Estimated time for driver takeover (s)	Possible safe state	Criteria for the safe state
Traffic jam pilot	10–15	Standstill in current lane	Up to 5 s
Highway pilot	10–15	Standstill on rightmost lane	Approximately 30 s
Highway pilot	10–15	Standstill on emergency lane	Approximately 60 s
Highway pilot	10–15	Standstill in breakdown bay or parking lot	Up to 30 min

– Driving conditions
– Vehicle type

Final assessment of these parameters will require in-depth HMI and user experience studies. From the safety perspective, critical procedures are activation and deactivation of the automated function. At these particular events, the responsibility for controlling the vehicle is transferred from the human driver to the system and vice versa. The HMI is responsible for providing transparent state information to the driver, i.e., whether the automated function can be activated and subsequently control the vehicle so that the driver can release control. The same argument applies for deactivation where handover of control to the driver is required. In addition, warning signals and indications of the planned driving maneuver in fallback mode need to be issued reliably to other traffic participants.

In the fallback mode of a highly automated vehicle, the system cannot rely on driver actions any more. The mechanical fallback solution that is offered by many brake systems in use today has no benefit in automated mode. Instead, the safety concept requires a redundant "drive-by-wire" functionality in the automated mode.

Legal aspects have to be considered as well, for instance, the Official Journal of the European Union, ECE_R13-H for the braking system [21] and ECE_R79 for the steering system [22], specifies explicit design and test criteria. At least two independent channels are required for the operating braking system as well as an additional fail-operational parking brake system. For the steering system, there is currently a test catalog defined in [22] which demands a degraded steering operation for an extended time interval of 40 min. We refer to [23] for a detailed discussion of the legal situation in the context of tele-operated and automated vehicles with respect to braking and steering systems.

Contemporary surround sensors have a limited ability to reliably detect and classify objects. Therefore, functional redundancy is required in a way that at least in critical regions around the vehicle surround sensors using different technologies can provide an overlapping field of view. This is in the functional safety concept addressed as functional redundancy (see Fig. 11.2). For the traffic jam pilot, a minimal set of surround and localization sensors is required to safely navigate to the safe state. For the highway pilot system, the sensor cluster becomes even more important since feed-forward control on a pre-computed fallback trajectory is not sufficient for the longer operation times in fallback mode, instead the fallback trajectory needs to be updated permanently.

11.4 Technical Safety Concept

In the following, we discuss exemplary architectural issues regarding the technical safety concept. Due to space restriction of our presentation, we will not be able to discuss software- and hardware-related safety measures but rather restrict ourselves to some aspects of the onboard communication and power network.

The technical safety concept referenced in the ISO 26262 has to specify technical safety requirements covering system external interfaces, environmental and functional constraints, and safety mechanisms related to detection and control of faults.

For highly safety relevant systems, it is recommended to carry out the safety analysis both in a deductive and an inductive way. However, a detailed safety analysis is beyond the scope of this article. We will restrict the discussion here to the most prominent components from the perspective of the safety concept. Firstly, these are the braking and steering systems and the control units. Depending on the definition of the safe state, also the power train can be of concern (not considered here).

Maintaining a *fail-operational* or *fail-degraded* system operation, while transitioning into the safe state, requires high availability of the actor control and in particular the actors. From the safety perspective, this can be achieved by a corresponding design or by employing redundancy concepts. As far as braking systems are concerned, redundancy concepts can be established based on technology already available today, for an example configuration see Sect. 11.4. As a prerequisite for the braking system redundancy, the power supply system must reach a maximal availability (see Table 11.2). This results in requirements for the power supply system, which are beyond today's implementations.

11.5 Requirements Imposed on the Onboard Network by Automated Driving Functions

In Sect. 11.3, we discussed the different automated driving functions together with the corresponding safe states and transition times to the safe state. In this section, we derive resulting requirements for the onboard electrical systems, i.e., the electric power network and the communication network.

11.5.1 Requirements for the Electric Power Supply

The electrical power net has to provide two independent energy storage capabilities that are able to supply the components required for transitioning to safe state. The electrical storage has to provide the energy capacity under all environmental and climatic conditions for which automated driving has been released. Monitoring of the energy content has to be implemented, and the energy content has to be checked reliably before the automated driving mode is offered to the driver. Moreover, the diagnostic system has to guarantee that automated driving mode can only be activated if the energy storage will be capable of supplying the critical components for at least a defined time that is needed to reach a safe state. This is a more delicate task than just monitoring voltage levels.

The power supply topology has to be designed in a way that defined minimum functionality can be provided in single-point failure situations. Resulting critical failures are rated ASIL D, and, therefore, the related failure rate must be sufficiently low. Wiring of components has to be organized in a way that a minimal set of required components in the fallback scenario will stay operational.

11.5.2 Requirements for the Communication Network

Our presentation of the communication network in Fig. 11.5 is kept very abstract, for instance, we have drawn all communication channels as point-to-point connections. Bus topologies can be used as well; however, the single-point failure tolerance has to be carefully evaluated.

A challenging requirement is the maximum tolerable latency time, which is imposed on the communication network. Maximum latency times can be derived

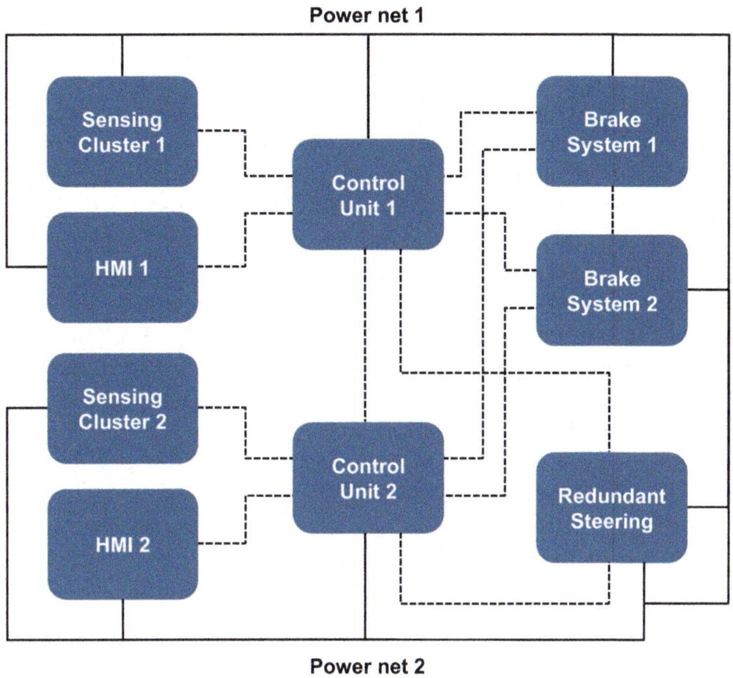

Fig. 11.5 System architecture for a potential traffic jam pilot/highway pilot system. *Solid lines*: power net 1 and 2. *Dotted lines*: communication lines. The sensing clusters contain environment and localization sensors. While not necessarily required for the traffic jam pilot system, the highway pilot demands a redundant electrical supply for the steering system

from the specification of the fault reaction time defined for each safety goal by subtracting the time required by the control units, sensors, and actors to process the request. Thereby, existing gateways and switches have to be considered as well. A guaranteed latency time with respect to the communication channel is easily achieved using time-triggered protocols but more difficult to assure for event-triggered communication.

Regarding the detection of errors that lead to missing or falsified data content, standard measures like end-to-end safeguarding have to be applied; the concepts for this are, for instance, described in [24].

11.6 Requirement Implications

Considering the requirements derived above, we present a coarse architecture solution, which addresses part of the constraints. In Fig. 11.5, we sketch the coarse E/E architecture of a hypothetical traffic jam pilot system. We have chosen to display a distributed architecture variant with two separate control units. Using different control units supports the argumentation for freedom of interference of the redundant units. An alternative architectural variant could be based on a central ECU hosting both units sufficiently decoupled, i.e., with independent electrical supply, independent communication channels, and sufficiently low thermal coupling. The latter variant will save construction space and supports high data bandwidth between both units but requires a careful design for the assertion of freedom of interference.

Control unit 1 is responsible to compute the vehicle's future collision-free trajectory. Therefore, control unit 1 receives input data from the environment sensors and the location sensors (sensing cluster 1) in order to judge the relative positions and velocities of objects in relation to the vehicle. The sensing clusters in Fig. 11.5 can be realized in different ways: as physically redundant sensors, as sensor units with redundant electrical and communication interfaces, or as sensors with different diverse data processing interfaces, provided that common cause faults are sufficiently controlled. We will not elaborate this issue further but rather focus on the onboard network. Since the correctness of this computation is rated safety critical, there has to exist a redundant path composed from sensing cluster 2 and control unit 2. The redundancy has to be static or dynamic in hot standby mode (see definition in [25], Fig. 11.6) in order to meet the timing requirements.

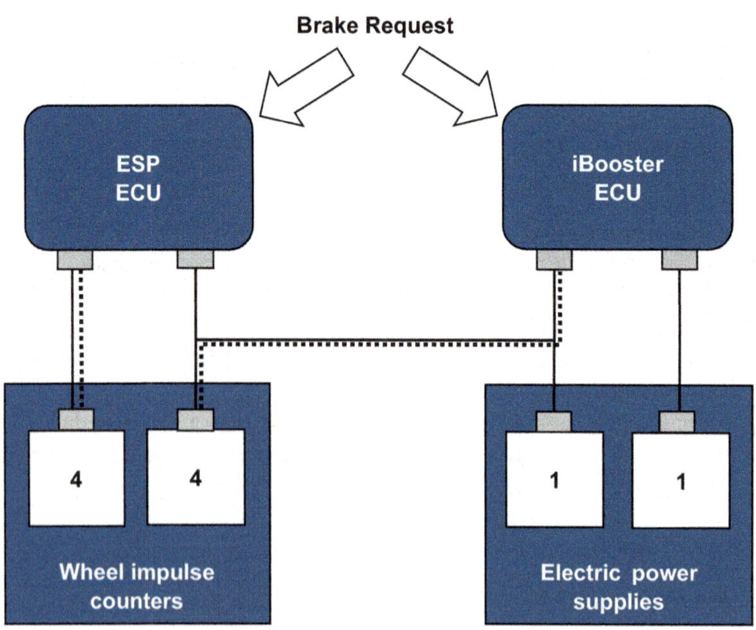

Fig. 11.6 Redundancy concept for fail-degraded brake system. Each brake system has to be independently electrically supplied and connected to one of the two sets of wheel impulse counters (four for each set)

Depending on the interpretation of the situation, the redundancy can be asymmetric, i.e., the function executed on control unit 2 is not necessarily required to understand the situation in the same level of detail as the function hosted on control unit 1. This offers a good opportunity to implement diversity in the development of the function running on control unit 2. The same argument is applicable to the sensing clusters, where the use of sensors with diverse principles of physical measurement is usually required to cope with their weaknesses. The emergency trajectory has to be stored in both braking systems in order to be available with high reliability, even in degraded mode.

For the traffic jam pilot, we assumed in Sect. 11.1 that the safe state is defined as decelerating the vehicle to standstill in its current lane and keep it parked. Transition to the safe state is primarily a task for the braking system, which in case of a nonfunctional steering system can even provide small lateral corrections by asymmetric brake interventions required to keep the vehicle in the current lane. In fact, this is a steering control redundancy with limited performance.

Therefore, the capability to provide controlled braking of the vehicle is crucial and has to be maintained with high reliability. Taking into account legal requirements, which mandate two independent channels for a braking system (cf. [21]), this requirement translates into a fail-operational braking system where each unit has to be independently electrically supplied. In addition, a fail-degraded HMI is required to reliably communicate the system status to the driver in the context of activation and takeover.

For the highway pilot, this system architecture has to be complemented with a fail-operational steering system. The steering system thus needs a second electrical feed. This requirement is imposed by the possibly higher vehicle velocity leading to a longer distance to reach the safe state and the less constrained motion of other vehicles. Therefore, lateral control is necessary in the fallback mode for a certain period of time and a minimum of perception capabilities in order to avoid collisions with other traffic participants. Depending on the definition of velocity range and safe state, a reliable limp home functionality has to be supported. This would require a fail-degraded power train architecture and would represent an extension to the architecture given in Fig. 11.5.

11.7 Safety Architecture Solutions

In the previous section, the exemplary safety goal "avoid insufficient vehicle deceleration when traffic jam pilot is active" was derived. Below, the consequences with respect to the system architecture of the braking system will be elaborated.

In order to reach the traffic jam pilot safe state (vehicle is decelerated and parked within the driving lane), the following functionalities of the braking system have to be fail degraded:

- Active pressure buildup
- Avoidance of locked wheels at the rear axle in order to maintain stability of the vehicle

Control of the front-wheel slip in low μ situations to provide the lateral forces required for small corrective steering maneuvers in order to follow the pre-calculated emergency trajectory.

A possible concept for a fail-degraded brake system realizing this safe state is shown in Fig. 11.6. This figure focuses on specific parts of the redundancy concept and does not show all components in detail.

This exemplary setup contains two independent systems, each of them capable to fulfill the requirements regarding emergency operation to reach the safe state. During normal operation (both systems are free from errors), the ESP takes the tasks of vehicle stability control incl. the processing of vehicle deceleration requests (issued by the automated driving functionality). In case of a fault forcing the ESP into degraded mode (or even fail safe mode), the automated driving functionality is switched into degraded mode. In this scenario, the iBooster performs the task of bringing the vehicle into the safe state. This concept is called dynamic hot standby redundancy (cf. [25]).

The mechanical push-through that is available in today's braking systems has no benefit in the automated driving mode. This is due to the fact that the driver is out of the loop in these automated driving situations. The same argument applies to the steering system; therefore, a fail-operational electrical steering is required. Redundant steering systems are likely to be designed as one-box solutions, embedding inherent redundancy rather than using two redundant systems. Such steering systems then require dual electrical feed from independent power sources as sketched in Fig. 11.5.

Our overall concept for redundant actuation is shown in Fig. 11.7. The redundant braking is realized by combining ESP and iBooster, while a redundant steering system is realized by combining electronic power steering with ESP, which can issue a yaw momentum through braking of individual wheels.

This paper has not yet discussed a detailed architecture of the environment sensor set. It will be a formidable future task to design an electrical and communication topology for the sensor clusters which will support all functional use cases, all safety goals, and at the same time manage to achieve target costs.

Finally, we would like to point out that the coexistence of other driving functionalities with an automated driving system, for instance, the functional behavior at defined system boundaries, shall also be considered for the design of the onboard network.

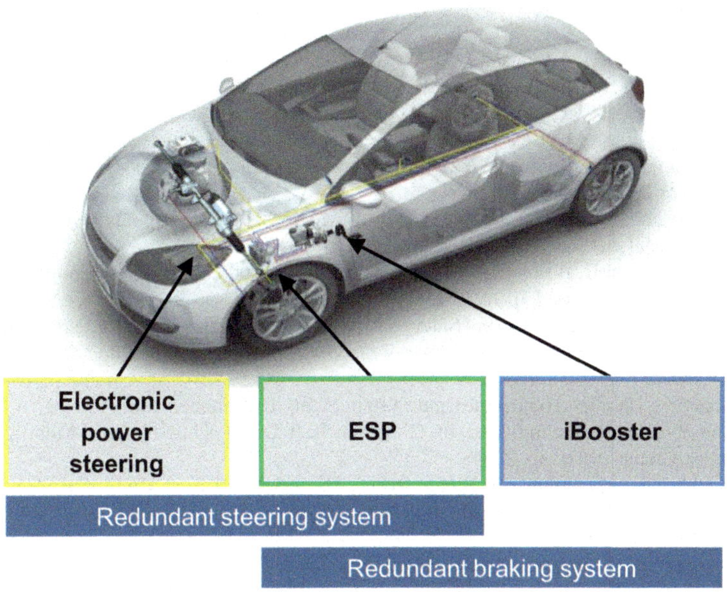

Fig. 11.7 Overall concept for redundant actuation. Redundant braking is realized by combining ESP and iBooster, redundant steering is realized by combining electronic power steering with ESP, which can issue a yaw momentum through braking of individual wheels

11.8 Conclusion

In this paper, we provided a high-level overview of our development of highly automated driving systems. We illustrated challenging situations and use cases and outlined their impact on system design, key technologies, and their technical realization. The paper also exemplifies how certain aspects of the system design as well as the implementation are country and use case specific. We are convinced that automated driving is becoming a reality, offering benefits for safe, relaxed, and economical driving, and we expect a stepwise introduction of automated driving starting with increased levels of automation on the highway. The first highly automated driving function will be a traffic jam pilot. The trend toward automated driving is generating new technical challenges for the sensors, algorithms, actuators, as well as for the E/E architecture of future vehicles. We discussed the implications of the new functionalities on the onboard supply and communication network in this paper. Derived from safety aspects and legal constraints, we argued that redundant electric onboard network and redundant communication network is required even for closer-to-market highly automated driving functions such as the traffic jam pilot. For the highway pilot, the standard concept of one safe state per safety goal has to be revisited in favor of a cascade of safe reactions. The corresponding transition times are likely to be on the order of minutes. The onboard net has to provide the required energy capacity during this time safely.

References

1. International Organization for Standardization, Road Vehicles—Functional Safety. ISO 26262-10, 2012 (2012)
2. L. Lutz, T. Tang, M. Lienkamp, Analyse der rechtlichen Situation von teleoperierten (und autonomen) Fahrzeugen (2012), http://www.ftm.mw.tum.de/uploads/media/07_Lutz.pdf. Accessed 28 Mar 2014
3. Robert Bosch GmbH, Bosch Automated Driving (2013), http://youtu.be/0D0ZN2tPihQ. Accessed 10 Jul 2015
4. Robert Bosch GmbH, Bosch User Experience for Automated Driving (2015), https://youtu.be/2i-t0C7RQWM. Accessed 10 Jul 2015
5. M. Helmle et al., Transient System Safety and Architectural Requirement for Partly and Highly Automated Driving Functions, in *Electric & Electronic Systems in Hybrid and Electric Vehicles and Electrical Energy Management* (2014)
6. J. Becker, M. Helmle, Architecture and System Safety Requirements for Automated Driving, in *Road Vehicle Automation 2*, ed. by G. Meyer, S. Beiker. Lecture Notes in Mobility, vol. 2 (Springer, Cham, 2015), pp. 37–48
7. M. Buehler, K. Iagnemma, S. Singh (eds.), *The DARPA Urban Challenge. Autonomous Vehicles in City Traffic*. Springer Tracts in Advanced Robotics, vol. 56 (Springer, Berlin, 2009)
8. Google Inc, Google Self-Driving Car Project, Monthly Report, May 2015, http://static.googleusercontent.com/media/www.google.com/us//selfdrivingcar/files/reports/report-0515.pdf. Accessed 8 Jun 2015
9. H. Lategahn, C. Stiller, Vision-only localization. IEEE Trans. Intell. Transp. Syst. **15**(3), 1246–1257 (2014)
10. C. Stiller, J. Ziegler, Situation Assessment and Trajectory Planning for AnnieWAY, in *IEEE/RSJ International Conference on Intelligent Robots and Systems. IROS Workshop on Perception and Navigation for Autonomous Vehicles in Human Environment*, San Francisco, 30 Sep 2011
11. M. Aeberhard et al., Object Existence Probability Fusion Using Dempster-Shafer Theory in a High-Level Sensor Data Fusion Architecture, in *Intelligent Vehicles Symposium (IV)*, 2011 IEEE, Baden-Baden, 5–9 Jun 2011
12. T. Gindele, S. Brechtel, J. Schroeder et al., Bayesian Occupancy Grid Filter for Dynamic Environments Using Prior Map Knowledge, in *2009 IEEE Intelligent Vehicles Symposium (IV)*, Xi'an, 3–5 Jun 2009
13. T. Jiang, S. Petrovic, U. Ayyer et al., Self-driving cars: Disruptive or incremental? in *Applied Innovation Review*, vol. 1 (Sutardja Center for Entrepreneurship & Technology, University of California, Berkeley, 2015), pp. 3–22
14. J. Levinson, S. Thrun, Robust Vehicle Localization in Urban Environments Using Probabilistic Maps, in *2010 IEEE International Conference on Robotics and Automation (ICRA)*, Anchorage, 3–8 May 2010
15. California DMV, California Driver Handbook (2015), https://apps.dmv.ca.gov/pubs/dl600.pdf. Accessed 10 Jul 2015
16. D. Sales, D. Correa, L. Fernandes et al., Adaptive finite state machine based visual autonomous navigation system. Eng. Appl. Artif. Intell. **29**, 152–162 (2014)
17. SAE International, On-Road Automated Vehicle Standards Committee, Taxonomy and Definitions for Terms Related to On-Road Motor Vehicle Automated Driving Systems (2014), http://standards.sae.org/j3016_201401/. Accessed 30 Oct 2014
18. Federal Ministry of Justice and Consumer Protection, Straßenverkehrs-Ordnung StVO. Teil I (2013), http://www.gesetze-im-internet.de/stvo_2013/index.html. Accessed 1 Mar 2013
19. International Organization for Standardization, Road Vehicles—Functional Safety. ISO 26262-1, 2011 (2011)

20. R. Benenson, T. Fraichard, M. Parent, Achievable Safety of Driverless Ground Vehicles, in *10th International Conference on Control, Automation, Robotics and Vision*, 2008, Hanoi, 17–20 Dec 2008
21. United Nations Economic Commission for Europe, Regulation No 13-H of the Economic Commission for Europe of the United Nations (UN/ECE). Uniform provisions concerning the approval of passenger cars with regard to braking. Off. J. Eur. Union L **230**, 1–80 (2010)
22. United Nations Economic Commission for Europe, Regulation No 79, Addendum 78, Revision 2, of the Economic Commission for Europe of the United Nations (UN/ECE). Uniform provisions concerning the approval of vehicles with regard to steering equipment. Off. J. Eur. Union L **137**, 25–51 (2008)
23. J. Levinson et al., Towards Fully Autonomous Driving: Systems and Algorithms, in *Intelligent Vehicles Symposium (IV)*, 2011 IEEE, Baden-Baden, 5–9 Jun 2011
24. AUTOSAR R4.0, Rev 2, Technical Safety Concept report Status V1.1.0
25. R. Isermann, R. Schwarz, S. Stölzl, Fault-tolerant drive-by-wire systems. IEEE Control Syst. Mag. **22**(5), 64–81 (2002)

Chapter 12
Advanced System-Level Design for Automated Driving

Jan Micha Borrmann, Sebastian Ottlik, Alexander Viehl, Oliver Bringmann, and Wolfgang Rosenstiel

12.1 Motivation

Automated driving (A.D.) requires concurrent execution of multiple complex driving functions on automotive embedded platforms. In general, such systems can be partitioned into early stages including sensor processing, individual perception, and cognition functions and into later, more centralized stages that perform data fusion, planning, and decision making. In this chapter, we exemplarily concentrate on automotive embedded processing systems for perception and cognition problems, however, we expect similar problems also on later stages such as data fusion. For perception and cognition, one can observe a wide gap between required processing power and the achievable embedded realizations which have to fulfill non-functional requirements such as low power and small cost. Furthermore, these systems must perform all processing under strict safety requirements that guarantee deadlines and provide high system robustness.

Regarding their development, there is a lack of efficiency when applying established methodologies to modern heterogeneous architectures that employ hardware acceleration to meet performance requirements. Finally, there is a lack of hardware architectures that can scale towards the ever-growing processing requirements. In this chapter, we first identify these gaps based on a case study of a single perception and cognition function. Afterwards we present novel solutions that tackle these gaps

J.M. Borrmann (✉) • S. Ottlik • A. Viehl
FZI Research Center for Information Technology, Haid-und-Neu-Str. 10–14, 76131 Karlsruhe, Germany
e-mail: borrmann@fzi.de; ottlik@fzi.de; viehl@fzi.de

O. Bringmann • W. Rosenstiel
Wilhelm-Schickard-Institute for Computer Science, University of Tübingen, Sand 13, 72076 Tübingen, Germany
e-mail: oliver.bringmann@uni-tuebingen.de; wolfgang.rosenstiel@uni-tuebingen.de

D. Watzenig, M. Horn (eds.), *Automated Driving*,
DOI 10.1007/978-3-319-31895-0_12

individually. Finally, we introduce a powerful framework for modern automotive platform design based on these solutions.

12.2 State-of-the-Art

This section consists of two parts: First, we give an overview on current heterogeneous automotive processing hardware and discuss state-of-practice for automotive timing analysis and multitasking. Then we discuss embedded prototyping. Focusing on perception and cognition, we then present a case study based on our previous work [7] to identify current deficiencies of embedded automotive processing architectures and their development.

12.2.1 Overview on Current Embedded System Design

We first give an overview on current heterogeneous embedded platforms for advanced driver assistance systems (ADAS) and automotive processing before we summarize current automotive software development with regards to timing analysis, automotive multitasking, and prototyping. The shortcomings of current design methods are then identified based on the case study.

12.2.1.1 Current Embedded Automotive Hardware

Currently, automotive embedded architectures rapidly change in favor of complex heterogeneous system-on-chips (SoCs) that do not only integrate multiple embedded CPU cores but also hardware accelerators. Table 12.1 gives a summary of

Table 12.1 Overview on current high-performance embedded ADAS system-on-chips (selection)

Product	Core types	Accelerators
TI TDA2x [40]	2x ARM Cortex-A15, 2x ARM Cortex-M4, 2x TI C66x DSP	4x TI EVE Vision accelerator
TI TDA3 [41]	2x ARM Cortex-M4, 2x TI C66x DSP	1x TI EVE Vision accelerator
nVidia Tegra K1 [25]	4+1 ARM bigLittle Cortex-A15	6x12 nVidia Kepler Shaders
nVidia Tegra X1 [26]	4+4 ARM bigLittle Cortex-A57/A53	2x16x8 nVidia Maxwell Shaders
Mobileye EyeQ3 [38]	4x MIPS32 1004K [31]	4x Vector Microcode Processors
Mobileye EyeQ4 [23]	4x MIPS InterAptive, 1x MIPS M5150 peripheral core	6x Vector Microcode Processors, 2x Multithreaded Processing Cluster, 2x Programmable Macro Array

prominent architectures on the market and their architectural features. Examples for these complex SoCs come from vendors such as nVidia or Texas Instruments but also from more specialized companies such as Mobileye [21]. Heterogeneity differs in type and number of cores or accelerators. Accelerators include arrays of GPU shaders as well as fixed-function and programmable accelerators. Regarding safety qualification, in contrast to nVidia's chips, EyeQ4 is designed to provide ISO 26262 ASIL-B(D) [23] while Texas Instruments' TDA2x provides AEC-Q100 qualification [2].

12.2.1.2 Timing Analysis for Automotive Systems

Timing analysis and verification is a crucial task in automotive system development. Such a timing analysis is typically already integrated in model-driven and component-based development workflows on highest abstraction level. The AUTOSAR timing extensions [4] allow for specification of timing properties on component level but also the specification of end-to-end timing requirements over component boundaries. By modeling dependencies and annotating periodicity, early schedulability analyses can already be executed on such high-level models. One prominent automotive tool is SymTA/S [39] which integrates low-level analysis results into a model- and component-based flow and can be coupled to ECU software development tools such as Ascet [12]. Early timing behavior can be validated using provided simulation models for generic schedulers, bus systems, and RTOS, such as the AUTOSAR OS. In model-based flows, for early analysis, manual estimation of timing properties is still state-of-the art (even for space missions [14]), while timing models can be refined iteratively based on updated results from low-level static analysis or timing measurements (e.g., execution on ECU hardware during a test drive to extract timing data) on generated code.

For tighter, lower level, estimates, timing analysis that considers application knowledge can be done manually (see analysis of the pipeline of Fig. 12.1 in [8]), or more generalized based on timing-annotated graphs [13]. Manual analysis is cumbersome and not generalizable. Here, guaranteed worst-case execution times can only be given when largely overestimating execution time although theoretical worst-cases practically never occur on real executions (cf. Sect. 12.3.2.2).

Graph-based modeling such as scenario-aware-dataflow graphs (SADFG) [13] considers data dependencies and control flow dynamics on a more generic level, e.g., by modeling them as scenarios. Both, model-based analysis and graph-based

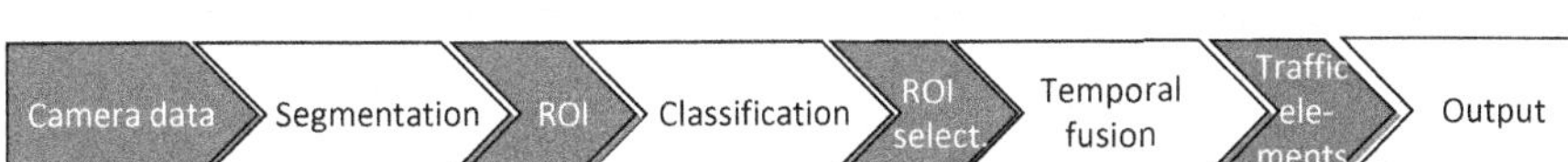

Fig. 12.1 Structure of the software pipeline for camera-based assistance systems (adapted from [24])

analysis, share the fact that they require the annotation of timing properties which are (iteratively) extracted from lower level tools from implementation results: For example, SymTA/S allows the import of timing properties from the Absint aiT [1] static analysis tool or from recorded trace data from execution on real hardware.

Measurement-based extraction of timing data introduces big uncertainty [3] due to the small coverage of real-world scenarios. For automotive manufacturers, this manifests in the state-of-practice of many required test drive kilometers. On the other hand, commercially available static timing analysis tools such as Absint aiT [1] are not available for complex microarchitectures with features such as out-of-order execution or SMT, both of which can be found in current, commercially available, high-performance automotive platforms (cf. Sect. 12.2.1.1).

Furthermore, the trends towards automotive multicores and hardware acceleration introduce side-effects that complicate such analyses. Examples are common access to shared resources such as shared Level 2 caches but also the side-effects of DMA transfers from other masters in the system, such as accelerators [3]. Schmidt et al. [33] showed that such effects already occur on current automotive multicores. Also, low-level static analyses usually require application knowledge, such as bounds on loop iteration counts, which can be hard to establish for complex A.D. algorithms while limiting overestimation to reasonable margins (cf. Sect 12.3.2.2).

12.2.1.3 Automotive Multitasking

Multitasking systems for automotive processing have to consider strong timing requirements. As apparent from the previous section, those requirements can be annotated early when using a component-based or model-based workflow. Then, tools can provide schedulability analyses or simulation of real-time schedulers such as *earliest deadline first (EDF)*. Static (cyclic) scheduling is also common when using AUTOSAR [22]. Here within a scheduling hyperperiod, multiple AUTOSAR runnables are scheduled at fixed offsets within smaller time-slots [22]. Such static scheduling can especially be observed for near-sensor or data fusion driving applications which are typically tightly synchronized to input data capturing rates. Here, the length of scheduling slots is usually coupled to the input sensor update rates, such as camera frame rates for perception problems. Scheduling within Mobileye EyeQ1 is an example for such static scheduling [37]. In EyeQ1, timeliness of execution times against the static slot periods was verified using simulation on large amounts of recorded data from test drives. While static scheduling provides highest predictability, it limits concurrent utilization of processing resources, especially with growing number of concurrent pipelined applications (cf. Sect. 12.2.3.2).

12.2.1.4 Current Embedded Prototyping

Prototyping by utilizing simulations of a full system, so-called virtual prototypes, has emerged as a potential solution for many challenges in general embedded system development. However, the approach is yet to be adopted by many automotive software developers and system designers. The key issue of virtual prototyping is approximating system properties such as accurate timing behavior of target software at a simulation performance that makes a practical application viable. The needed trade-off between performance and accuracy can vary over designs and development progress. For example, exact timing provides little advantage in early stages of development when a system and thus its timing will still undergo large changes. Here we give a brief overview of simulation abstraction levels during an idealized development process.

During pre-development or early development, abstract prototypes can be constructed using tools such as Simulink [43] or usual desktop programming. These prototypes enable activities such as algorithmic development or training of models for machine learning. However, due to the high level of abstraction, an accurate prediction of most product properties is impossible.

For automotive embedded real-time systems, timing behavior is as crucial as functional behavior. Therefore a target hardware platform has to be selected considering its timing but also its cost. Virtual prototypes that provide timed simulations enable a systematic exploration of the design space, such as potential platforms and hardware/software partitions. SoC vendors rely on detailed hardware models (e.g., of the CPU cores) that closely approximate hardware at this stage [9]. However, this approach is unrealistic for application development—for example, by an automotive OEM—due to two reasons: Firstly, the abstraction of these models is typically orders of magnitude too low to allow a fast timed simulation of real applications. Secondly, if the sole purpose is selecting a suited target platform, the high licensing cost of such detailed architectural models is not justified, as a large library of many of those models would be required.

Pressler et al. [29] propose a solution based on the estimation of platform independent source code and the computational cost of operations on a given target. While this approach, due to the higher level of abstraction, certainly has a lower accuracy than more detailed simulations, it is less subjective and much faster than alternatives (e.g., selecting based on benchmark results).

System-level simulations approximate a full system by co-simulating hard- and software models that have been compiled for the simulation host. They can be applied during most development stages and offer a wide range of possible trade-offs between simulation accuracy and performance. Software is simulated based on either source or binary code. Source-code-based approaches (source-level simulations, SLS) rely on heuristics [9, 17, 35, 36] to accurately reflect non-functional properties such as timing of target binary code. This step can introduce an additional error compared to approaches based on binary code, the so-called binary-level simulation (BLS). Hardware is usually modeled in SystemC [15], a standard language for system-level simulation. Abstraction of hardware can range from techniques such as TLM+ [11], where bus transfers are simulated for whole application-level object (e.g., a full video frame), to a detailed simulation of internal hardware signals.

12.2.2 Challenges in Hardware/Software Co-design for Automated Driving: A Case Study

In this section, we exemplarily demonstrate the deficiencies of the current design of automotive processing architectures, with a bias towards perception and cognition, based on a case study (also see our previous work [7]) that covers the development of a heterogeneous embedded realization of a complex driver assistance function, using a state-of-the-art approach. We then discuss the shortcomings of such classic development and propose novel solutions.

12.2.2.1 A Vision-Based Traffic Light Detection

Our case study is concerned with an efficient implementation of a camera-based traffic light recognition (TLR) system. The targeted algorithm [24] consists of a multi-level image segmentation for finding light points, a search for the surrounding traffic light boxes and a subsequent classification of found candidates by a support vector machine (SVM). Candidates are tracked over time and classification results are fused temporally over multiple frames. The application's output are the detected light positions and their states. Such a result can be used by subsequent applications such as an intersection assistance which is a key component in automated driving. On 1280×720 color video data it could achieve about 25 frames per second when executed on an Intel Core i7-2820QM quad-core general-purpose development PC, outperforming state-of-the-art approaches in both, detection robustness and speed (when executed on a PC platform). Figure 12.1 shows the generic algorithmic pipeline structure that is used by the TLR and also applies to other perception and cognition applications.

12.2.2.2 Systematic Hardware/Software Co-design

As the algorithm evaluates the results from a history of several frames before deciding on a recognition result, a high average frame rate, ideally near 24 fps, is targeted. Our workflow for realizing the recognition under embedded constraints is depicted in Fig. 12.2. Based on past experience together with the performance results from executing the application on the high-performant developer platform, we concluded that we most probably would not be able to realize the application at required performance on an embedded platform, when executed in software only.

Therefore, the first design decision was choosing the Xilinx Zynq SoC [47] as target platform. It does not only contain a powerful ARM Cortex-A9 dual core CPU but it also allows for custom hardware accelerators by providing an on-die FPGA logic fabric. Also, for maximum possible software performance and minimized overhead, we additionally decided to use no operating system but only a small hardware abstraction layer that contains basic library and I/O support (bare metal programming). Thus we save the overhead from device drivers, virtual memory, and also keep memory footprint low.

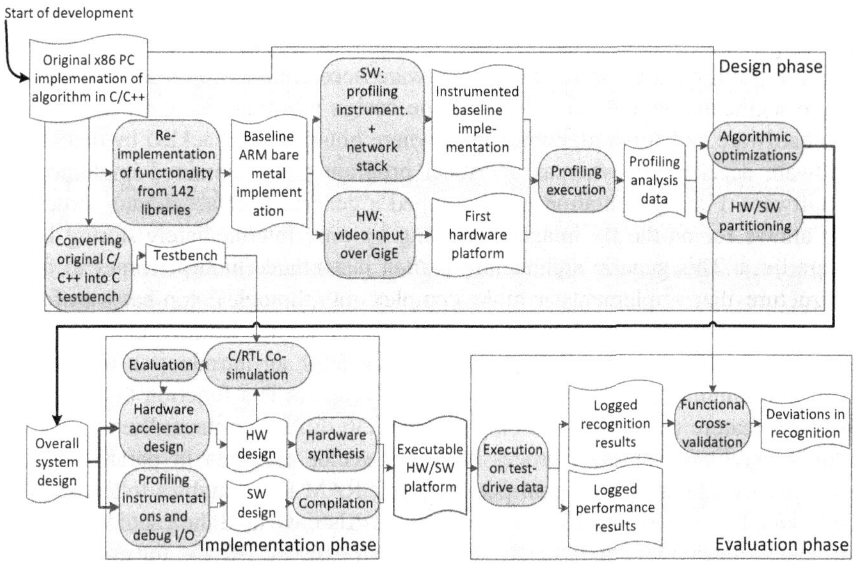

Fig. 12.2 Workflow towards an embedded realization of the TLR algorithm of [24] (based on [7])

Starting point was the x86 architecture implementation from [24] of the assistance function on a PC. First, an initial port to the target ARM CPU architecture had to be done, which included replacing a large number of libraries by custom implementations as they were relying on an operating system or were not available for the ARM architecture. Then, preparations for profiling this software port were done. For being able to receive test drive video data from a host PC, a hardware interface for GigE had to be provided together with adding a UDP software stack.

Additionally, the software port was instrumented by adding calls to the ARM CPU timers around individual functional blocks to allow for timing measurements. This port of the software was then executed for profiling, where input video data was fed over gigabit Ethernet.

While memory consumption was fair (about 22 MiB), the performance profiling results of this solution were orders below the targeted goal: Processing one single video frame could take multiple seconds, resulting in a average-case throughput of 0.5 frames per second. As a major bottleneck we identified the classification step which uses a computationally high-demanding SVM. Also, it could occur that several tens of traffic light box candidates would be selected for classification by the segmentation stage. The selection itself was performed by solving a complex optimization problem which discarded candidates based on iterative scoring. We replaced this approach by a simpler plausibility check that processes a list of rules that includes color and geometry checks and still showed robust detection while lowering complexity of decision making at similar detection rates.

We also performed a manual data type analysis for the SVM's feature vector elements, testing different data resolutions against the introduced error. It was

Static scheduling as introduced in Sect. 12.2.1.3 gives highest predictability, but due to the inherent dynamics of input data and subsequent execution time, big over-provisioning has to be applied to the common static slot period, leaving slots highly underutilized in the average case. Again, this dilemma grows with growing number of concurrently executed applications.

12.2.3.3 Development Gap

Prevailing approaches to developing automotive embedded systems, usually electronic control units (ECUs), rely on target hardware to evaluate implementation properties such as performance. This has numerous drawbacks: The effort required for a preliminary implementation restricts the number of evaluated platforms. In the case of our case study, development took several months due to the system complexity. A preselection can often only be based on imprecise factors such as benchmark results or developer experience.

Furthermore, observability in modern SoCs is restricted, as classical debugging aids such as in-circuit emulation are not feasible anymore.

12.2.3.4 Scalability Gap

While current automotive systems employ multiple embedded CPU cores and accelerators, their symmetric multiprocessing together with currently used interconnect will eventually hit a scalability gap. Regarding interconnect, buses, crossbars (as currently found on automotive SoC, cf. Sect. 12.2.1.1), but also ring interconnects do not scale well towards large numbers of processing elements (PE) such as CPU or accelerators [16, 32]. As authors of [32] point out: Buses are limiting due to the serialization of concurrent requests. Crossbars, while limiting serialization, suffer from quadratical area and power demand wrt. number of PEs, while average hop count for rings grows proportional to the number of PEs. The current crossbars and shared memory architectures of current SoC will eventually not be sufficient anymore to fulfill the increasing demand of concurrent driving functions.

12.3 Concepts for Efficient Future Automated Driving

In this section we present solutions to individually tackle the challenges discussed in Sect. 12.2.3: overcoming the performance gap, the utilization gap, as well as the development gap and the scalability gap.

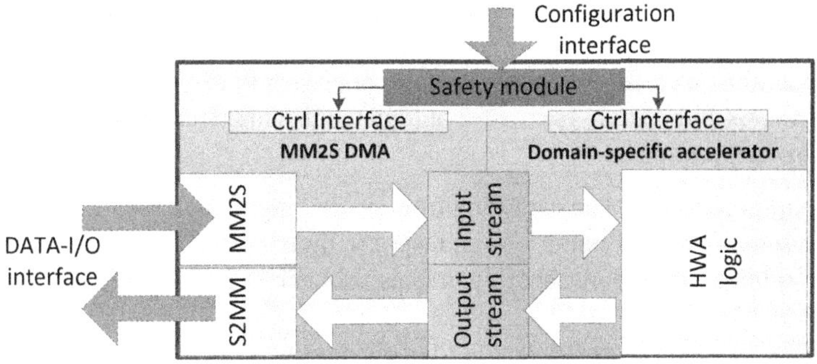

Fig. 12.4 Generic hardware accelerator, separating data I/O from its configuration. Accelerator-local DMAs are used to read/write data and to perform memory-mapped to streaming translations

12.3.1 Bridging the Performance Gap with Heterogeneity

As successfully implemented in our case study, for bridging the gap between demanded and available processing power in embedded realizations, we propose hardware acceleration. For not equally enlarging the utilization gap, these accelerators should not be application-specific but should allow for reconfiguration towards usage by a large number of concurrent applications. For example, for our RBF Kernel accelerator this means that the application-specific lookup tables (cf. Fig. 12.3) can be reloaded at runtime to contain different trained classification models.

For design reuse and scalability towards multiple instantiations, hardware accelerators (HWA) should share a common interface separating configuration from input/output data transfers. Thus, only the communication channels have to be adapted when changing from a classic bus interconnect to a modern network-on-chip (NOC). This also allows for simple addition of novel accelerator types.

Furthermore, hardware safety measures are employed: Configuration of the DMA's memory addresses is only possible via privileged access, not allowing the DMAs to be accidentally configured to overwrite foreign memory areas. The accelerator structure is depicted in Fig. 12.4.

12.3.2 Bridging the Utilization Gap

Hardware acceleration for demanding computational tasks can only be cost-efficient when high utilization is ensured. Otherwise additional chip area cost may not be justified (cf. Sect. 12.2.3.2). However, sharing resources in the context of usually safety-relevant systems has to be carefully designed and verified towards safety requirements. We now present means on how such high utilization can be achieved.

12.3.2.1 Runtime Resource Management

In our previous work [8], we give a twofold approach to ease the utilization gap in heterogeneous embedded systems for automated driving. Our approach consists of two main contributions:

1. Decoupling of input data capturing from processing to increase utilization while at the same time shortening system response time.
2. Inter-application resource sharing that also considers safety aspects.

Decoupling of Input Data Capturing from Processing Our approach targets applications that decide on a vehicle reaction based on observing the environment over a certain period of time. Practically, this means that such a decision is based on a temporal fusion of results from past application iterations (i.e., multiple executions of the full application's pipeline, see Fig. 12.5). Many driver assistance applications follow such a scheme.

In our approach, applications follow a common software pipeline model similar to that of Fig. 12.1. Adhering to this pipeline scheme, we use two camera-based applications, a TLR (also see Sect. 12.2.2 for details) and a traffic sign recognition (TSR) which performs a circle-detection-based segmentation into Region-of-Interest, which are then classified by an SVM against a trained model using a radial-basis-function kernel.

Both algorithms further track results over time and can robustly decide on a reaction based on the last 15 iterations. Deadlines of the system's response times are derived from functional requirements (see [8]). Our first contribution now lies in the fact, that by decoupling input data capture from processing, in contrast to state-of-the-art automotive static scheduling (cf. Sect. 12.2.1.3), we do not have leave resources idle until the next capturing period is started but can immediately start processing of a fresher frame. By capturing at higher rates than classic automotive designs we thus can minimize the per-application idle time and shorten reaction time by earlier availability of the results from 15 iterations. The scheme is depicted in Fig. 12.5 for a two-application setting executed on a dual core processor.

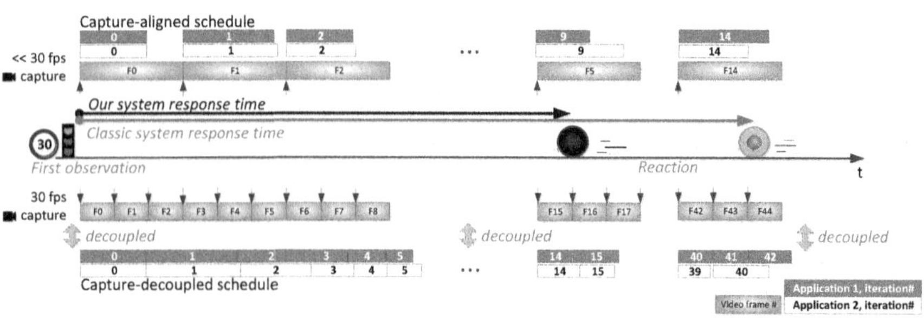

Fig. 12.5 Decoupling data acquisition from processing allows for better utilization of processing resources at reduced system response time

Sharing of Common Processing Hardware Under Safety Requirements Our second measure for increased efficiency is the definition of sharable clusters of processing resources of a common type. A cluster may consist of multiple CPUs or (programmable) hardware accelerators. We define *tasks* as being sub-computations within each application pipeline that can be offloaded to the processing resources located within such a cluster of shared resources.

Access to shared resources is restricted: Applications cannot directly issue tasks to such a cluster but only indirectly to a unit we name *strategy controller*, controlling resource access. Following this scheme, utilization of these resources can be granted in a way, so that application deadlines are fulfilled. Each cluster is equipped with such a strategy controller which in our implementation is a small synthesizable Xilinx MicroBlaze CPU core. The decision on using a programmable CPU was made for flexibility in evaluating different scheduling algorithms. As soon as the scheduling algorithms are decided on, in a series production, one would rather replace that CPU by custom hardware for chip area efficiency. The strategy controller does not only arbitrate requests to shared resources, it also collects status information at runtime from each executed application using a heartbeat scheme: At each completion of a software pipeline stage, the controller is accordingly notified. Also, individual resources within a cluster notify their controller when a task has been successfully processed. Different scheduling strategies such as *EDF* can be implemented that consider application deadline requirements over hardware/software boundaries. For an evaluation of scheduling algorithms, please see [8]. A strategy controller can additionally act as safety manager when heartbeat rates of an application drop below a threshold, indicating a malfunction.

12.3.2.2 A Runtime Fail-Operational Mechanism

For individual driving functions, response-time requirements can be derived from functional requirements and parameters such as camera optics or supported vehicle speed range, and sometimes, also from regulations (cf. [8]).

However, giving tight execution time bounds for complex applications such as perception applications is difficult on architectural and application-level. For the former case, this is due to (micro)architectural execution time variations and side-effects found in modern automotive multicore processing platforms (cf. Sect. 12.2.1.2). For example, for the target platform of our case study (see Sect. 12.2.2), microarchitectural timing ambiguity in low-level static analysis is additionally increased on architecture level by shared Level 2 caches and DMA transfers from accelerators. On application-level, specifying bounds for iterative and looped computations is mandatory when using static analysis. However, in perception problems, for example, at the early segmentation pipeline stages, execution time dynamics directly relate to the content of the input video data. This results in extreme overestimations in static analysis when assuming the theoretical worst-case, even if by far not observable during real drives. On the other hand, assuming worst-cases based on statistical data or from observations within test drives are inherently unsafe, as they give no formal guarantees. For later pipeline stages limitations might be easier (see [8]).

As confidence on such timing analysis has to be considered low, dynamic analysis and monitoring is additionally required to guarantee safe behavior on the road.

We propose a runtime monitoring approach coupled with a fail-operational mechanism: For each application to be concurrently executed we define a set of deadlines of different criticality where the criticality levels reflect different vehicle parameterizations (e.g., comfortable braking versus emergency braking). Within such a deadline, an application's pipeline is executed multiple times (iterations). Dividing it by the number of needed iterations that are required to decide on a reaction leads to an average per-iteration deadline that we use as an indicator of the current conformity of the real observed execution time versus the static per-application deadlines. In the case of our traffic light and TSR we require 15 iterations for deciding on a reaction and define two deadlines of different criticality, the first deadline allows for a comfortable vehicle reaction (here soft braking), while a shorter second deadline still allows for a safe reaction but at much less comfort (in our two-state case, this means parameterizing the vehicle for emergency braking).

Figure 12.6 shows the approach: A slight violation of the per-iteration average deadline which would allow for a comfortable vehicle reaction is interpreted as a signal, that an application might not meet its multi-iteration execution time requirement that would allow for a comfortable reaction. This leads to a state transition where the vehicle is prepared for a stronger, less comfortable, reaction that can still be executed within the remaining time period. Additionally, while such situation should never happen on the road, an overload indication period should also be implemented, detecting system failure and potentially triggering a limp-home/-aside mechanism. While we just implement two states. In a real deployment, more fine-granular vehicle parameterization with more states would be used. This allows for trading-off guaranteed performance against resource sharing potential, to provide safe behavior even when observing high system load (cf. Sect. 12.4.2.1).

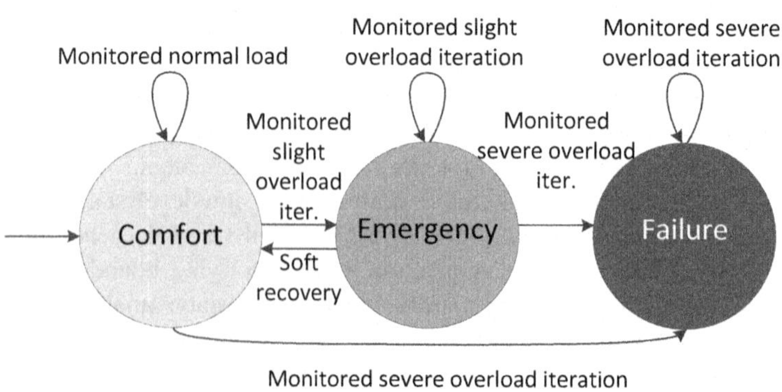

Fig. 12.6 Per-application fail-operational state machine: the indication of a potential overload causes a state change, still allowing for an emergency reaction

Additionally, between deadlines of different criticality, a recovery mechanism can be implemented, setting vehicle parameterization to a less critical level when observing deadline conformity for a number n of subsequent iterations.

12.3.3 Bridging the Development Gap Using Virtual Prototypes

Virtual prototypes promise to bridge the gap between the low efficiency of classic design processes and the complexity of modern systems. In this section, we present our simulation framework, which has already been applied to automotive ADAS production code and can be applied during most development stages.

An overview on the framework is shown in Fig. 12.7. Software execution on a processor, including a highly accurate timing approximation, is simulated by specialized software simulations. They are integrated with our hardware simulations. We discuss both in the following.

12.3.3.1 Simulation of Software Execution

We support binary code emulation using QEMU [5], which utilizes just-in-time compilation to achieve a high performance and supports a wide range of instruction sets. Furthermore we integrated a custom source level instrumentation tool into our framework. Both are augmented with a custom timing simulation library.

As in most prevailing timing simulations, our library advances time by consuming a certain number of processor cycles when a block of instructions, similar to a basic block, is executed. A novel concept of our approach is utilizing multiple

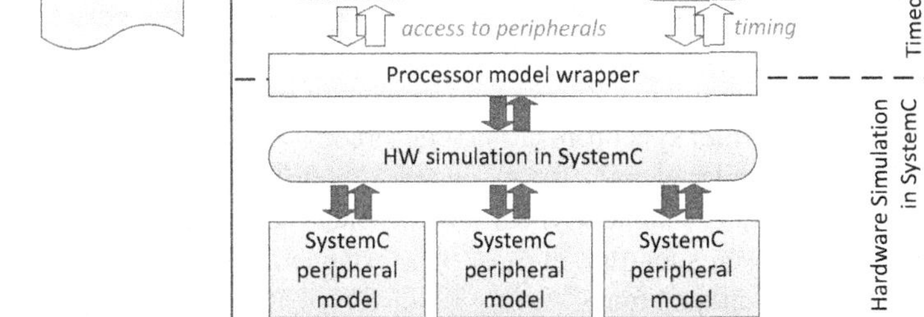

Fig. 12.7 Overview of our virtual prototyping framework

possible cycle counts for a single block. During simulation, values are selected based on the current context, an abstraction of preceding control flow. For example, for a function called from a loop, our simulation can differentiate the timing of blocks inside the function based on the loop iteration the function was called from.

For an ARM Cortex-M3 processor we have compared our approach to a context-insensitive simulation (i.e., the same cycle count is used on every block execution). Our results [27] demonstrate that even for this relatively simple processor, context-insensitive simulations can induce large errors in the simulated timing, whereas our context-sensitive simulation is highly accurate. More recent [28] results for an ARM Cortex-A9 suggest that our context abstraction is powerful enough to implicitly model caches. This suggests another advantage over context-insensitive simulation, which requires a dynamic cache model. These cache models can be a significant simulation performance bottleneck, slowing down BLS by 200 % [42] and SLS by nearly a factor of ten [18]. The context-sensitive simulation causes an overhead of only 70 % in comparison to a simulation without timing, which is only 20 % more than the 50 % [42] for a (extremely inaccurate) context-insensitive simulation that does not include a cache model.

Timing information is stored in the so-called timing database (TDB), which simplifies running multiple simulations (e.g., for different program inputs) with one set of timing data. The timing data itself can be obtained by static analysis or by tracing a single program execution on target hardware.

12.3.3.2 Simulation of Hardware

We build on the SystemC [15] standard for our hardware simulation. Therefore, many academic approaches as well as commercial models can be integrated with our simulations. For our in-house application we rely on the reference implementation of the SystemC kernel.

For software-focused simulations we often implement simplified models of the necessary peripherals, that exclude internal details of peripherals as well as functionality that is not utilized in a project. For hardware-focused simulation, we usually rely on more detailed models and in some cases even integrate RTL simulations when higher accuracy is required.

Transaction level modeling (TLM) [10] is applied to model inter-module communication in order to provide a flexible and reasonable trade-off between simulation accuracy and performance. We usually employ loosely timed synchronization, where modules may simulate without synchronization for a certain amount of time, called a quantum. Further synchronization is enforced on particular interactions with peripherals, for example, when software changes the interrupt mask in an interrupt controller. To accurately handle asynchronous events such as interrupts, a quantum must usually be chosen such that at most one instance of such an event can occur within each simulation quantum.

The software simulations are wrapped by specialized SystemC models to integrate them with the hardware simulation. They expose signals for asynchronous events (e.g., interrupts) and TLM interfaces for bus accesses. Multiple processors can simply be simulated by instantiating multiple processor modules in SystemC, which are each backed by an independent instance of our software simulation.

12.3.4 Bridging the Scalability Gap for Future Platforms

As discussed in Sect. 12.2.3.4, the lack of scalability of current automotive embedded architectures with regards to the growing number and nature of future applications may pose a serious threat for bringing highly or fully automated driving to a mass market soon. For general-purpose computing, the answer to scalability problems has been introducing more parallelism which eventually leads to large scalable multicore or even many-core solutions that contain some to several hundreds of CPU cores. As interconnect, the so-called NoCs are commonly used, connecting a regular structure of CPU tiles over a packet-based on-chip interconnect. Using such a regular structure together with sophisticated routing algorithms allows for large many-cores, where available chips easily contain 48 [19] or 64 cores [44].

Homogeneous many-cores, i.e., only containing one type of processing elements, have been evaluated for automotive perception and recognition problems in research work [6, 48] or [34]. While they all showed that in general complex single automotive functions can be run on such platforms, their absolute results lacked efficiency wrt. performance/watts and performance/#cores (performance per chip area), making them unfavorable for future A.D. functions.

Success of heterogeneous automotive SoCs suggests that instead, heterogeneous multi- to many-core solutions might be better suitable for bridging the scalability gap. An example for such a research architecture is the *HeMan* (*He*terogeneous *Man*y-core) architecture developed at FZI (see Fig. 12.8). It is a tiled multi- to many-core research platform that incorporates heterogeneous tiles each containing classic embedded cores with individual RAM and ROM memories but also allows for domain-specific (potentially programmable) hardware accelerators. Each tile has a special local memory used for message passing and providing data to the hardware accelerators. External memory is accessed by special tile types.

Specific measures in per-router message queues enforce message routing that can give latency guarantees for end-to-end message transfers between tiles based on providing virtual communication channels on top of the physical links.

HeMan employs privileged communication channels for system control messages such as heartbeats (see Sect. 12.3.2) where specific hardware measures prevent communication from non-privileged tiles on those channels. On top of these virtual communication channels, prioritization of messages is used. Such measures also provide the means for securing system-relevant communication, such as the boot process where the program code for application tiles is deployed by a system controller whereas for regular tiles, (accidentally) overwriting the program code of other tiles is forbidden. As the NoC is a shared resource, a guaranteed throughput and latency for individual communication can only be provided when access to this

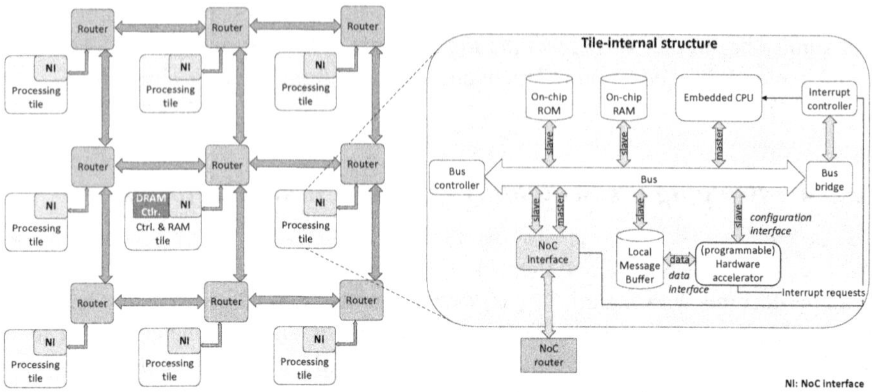

Fig. 12.8 Scalable heterogeneous automotive processing architecture: Processing tiles are connected via an NOC. Hardware accelerators provide required performance—while the interconnect allows for scaling towards future demands. Off-chip I/O and DRAM access is provided by special tile types

shared resource remains restricted: Thus the conformity of a tile to a previously assigned data plan (communication schedule) is controlled in hardware by specific traffic shaping components within each tile's network interface.

Hardware accelerators within tiles share a common interface as proposed in Sect. 12.3.1. The runtime approach from Sect. 12.3.2 can also be applied to such a tiled architecture: Then, tiles are grouped into a cluster that is augmented by one strategy controller. Also, redundancy can be applied such as adding lockstep cores for the CPU cores within each tile that execute the same instruction stream as the original core with a online self-tested comparator unit detecting mismatches that might occur because of hardware faults (dual-mode redundancy). In such a case, NoC messages to a safety management tile can be triggered automatically. This could especially be useful regarding the demanded further integration of more functionality per chip [22], where safety-relevant tiles, e.g., data fusion, planning or decision making, are executed on specifically hardened processing tiles, while functions which are not safety relevant are executed on generic processing tiles. On top, safety concepts on higher levels such as soft lockstep can be implemented where multiple tiles perform redundant calculations on the same input data whereas other tiles take the role of comparators. Also, hybrid approaches are possible: Here, the results from hardware accelerators are regularly checked by a software thread that runs on the tile's CPU during idle periods (self-test). For all these mechanisms, mixed system topologies are possible where only safety-relevant tiles provide redundancy, while processing-optimized application tiles are use the soft approaches.

12.4 Modern Platform-Based Automotive System Design

We now present our tooling framework, that shows how our individual solutions for realizing embedded A.D. functions from Sect. 12.3 can be efficiently combined. Additionally, we evaluate the discussed runtime concepts of Sect. 12.3.2 and demonstrate the accuracy of our timing simulation of Sect. 12.3.3 on an automotive function.

12.4.1 Tooling Framework

Creating and developing a complex heterogeneous target platform which integrates the concepts of Sect. 12.3 requires an efficient workflow, which can only be achieved with powerful tooling. Our tooling and workflow to achieve this goal is depicted in Fig. 12.9 and essentially consists of three steps: generating intermediate sources from a platform description (Platform Generation), creation of the actual simulation and FPGA platforms (Platform Creation), and finally, execution of the platforms (Platform Execution). These steps are performed iteratively during development.

Generating both, a virtual and an FPGA-based prototype, automatically from the very same description, enable a highly efficient development. We use the virtual

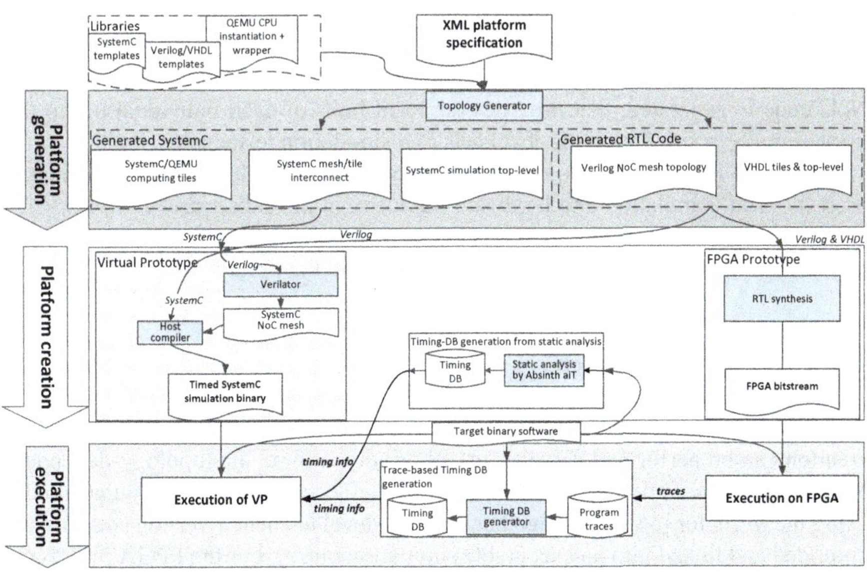

Fig. 12.9 Our tooling framework: simulation binaries and RTL code are generated from the same specification. The approach of Sect. 12.3.3 is used to simulate timing within tiles

prototype for early design space exploration beyond the chip area limitations of a physical FPGA and the time-consuming design of hardware and its debugging.

Furthermore, the virtual prototype is an ideal platform for efficient development of application software and, due to the integrated timing simulation, allows a continuous verification of timing requirements.

The hardware prototype allows extraction of non-functional properties, such as timing and power information which can be back-annotated to the virtual prototype. Additionally, we use the physical prototype for low-level hardware development and optimization, which is not possible on the virtual prototype due to the high level of abstraction, for example, to develop low-level interfaces to external components, such as CMOS sensors. Those results are then integrated into our RTL code template library making them available for automated platform generation.

12.4.1.1 Platform Generation

Platform Generation starts with a common platform description in XML. This description covers the topology of the mesh, together with parameters such as the desired number of virtual channels and the number, positions, and tile types for each individual tile. From a templated library of SystemC, Verilog, and VHDL source files, a custom tool called *Topology Generator* creates intermediate inputs for both, the virtual prototype and the hardware prototype. This includes the creation of the specified interconnect between tiles and the mesh. Two sets of source files are produced, SystemC sources and the RTL hardware descriptions: For SystemC, wrapper files integrate the simulations of individual CPU cores within tiles and links to our custom library that provides timed SystemC/QEMU co-simulation. Also, NoC code is generated, that provides the local links of each individual tile to the mesh. Finally, a simulation top-level acts as a simulation testbench.

RTL code includes the requested mesh topology in Verilog RTL code which is used for both, simulation and hardware platform, in the subsequent *Platform Creation* step. The RTL code for individual tiles, including the CPU core instantiations, is generated in VHDL, supported by templates from the library.

12.4.1.2 Platform Creation

Based on the intermediate results generated in the previous step, the actual platform creation can be performed. For the virtual prototype, first, the topology-dependent Verilog NoC description is translated into a cycle-accurate, yet fast SystemC model using the Verilator [45] tool. Together with top-level testbench Verilog code, this is compiled and linked into an executable simulation binary. For the FPGA prototype, the individual tiles, given in VHDL, and the mesh, given in Verilog, is synthesized, placed and routed using the vendor-specific synthesis toolchain.

12.4.1.3 Platform Execution

With the simulation and the hardware platform available, one can now utilize both: The FPGA prototype is executed by uploading first the bitstream, and then a software binary that includes the software of each individual tile. Our hardware implementation can fit up to nine tiles on our FPGA development board.

Providing target software binaries, the virtual prototype can be executed. Stimuli such as input data can be provided via dedicated testbench models. The SystemC simulation does not only provide a dramatic speedup compared to RTL (RTL running at approx. 100 simulated clock cycles per wall-clock seconds), but at the same time it also increases observability, as we can use target debugger instances to connect to each simulated processor and arbitrarily trace signals of the mesh. The simulation allows for systematic and fast software development as well as design space exploration, where topologies not fitting the FPGA can be explored.

If timing simulation is required, the co-simulation library reads context-sensitive timing information from a previously generated Timing Database that can either be extracted from recorded instruction traces of an execution on the physical prototype or from static analysis (cf. Sect. 12.3.3).

12.4.2 Evaluation

We now evaluate our dynamic resource management from Sect. 12.3.2 including our fail-operational strategy from Sect. 12.3.2.2 and our timing simulation from Sect. 12.3.3.

12.4.2.1 Runtime Resource Management

Our approach on runtime resource management has been evaluated on a Xilinx Zynq SoC, using it's two ARM Cortex-A9 cores and a hardware accelerator cluster of four SVM RBF kernels. However, it is retargetable to other architectures. The two applications from Sect. 12.3.2 each get assigned to one Cortex-A9 core for the software parts of their pipeline. As strategy controller, we use an embedded Xilinx MicroBlaze soft core. Signaling of heartbeats, requests for shared resources, and completion of accelerator executions to the corresponding strategy controller is done using interrupts to the strategy controller. We implemented different scheduling strategies (cf. [8]). For evaluation we use a combination of *static allocation* together with *earliest deadline first* where each application is statically assigned one accelerator exclusively while sharing the remaining ones using an *EDF* schedule. This provides a good trade-off between throughput and sharing potential. *EDF* always schedules the runnable tasks of an application that has least slack towards its deadline. For doing so, the already consumed time budget of an *application* is calculated from the heartbeat signals issued on each completion of a pipeline stage.

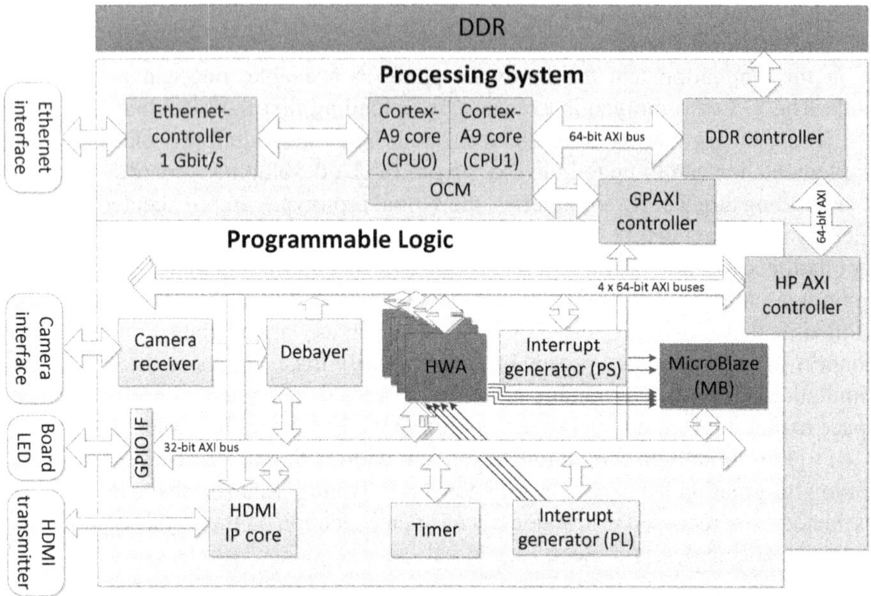

Fig. 12.10 Exemplary implementation of our runtime resource management: two ARM Cortex-A9 CPUs request hardware acceleration from the strategy controller (Xilinx MicroBlaze) which controls access to shared resources

Our FPGA-based evaluation system either can be used with a real CMOS sensor, where image preprocessing such as demosaicing is done in hardware, or as in-lab prototype where videos from test drives are provided using UDP. In the latter case, evaluation was performed subtracting the extra overhead introduced. Figure 12.10 shows the evaluation platform.

Our decoupling of processing from input data acquisition shows vast improvements compared to a static schedule where the time-slot period is set to the maximum observable execution time of a pipeline iteration during test drives. For the TSR, relative improvements range from a min./avg./max. of 37, 39, and 42 %, respectively. The traffic light application still shows minimum 8 % improvement with 36 % avg. and 49 % max. improvements. Total system power is below 6 Watts, including DRAM, HDMI, and GigE.

Runtime Fail-Operational Mechanism We evaluate our runtime fail-operational mechanism from Sect. 12.3.2 using the platform of Fig. 12.10. Heartbeat signals are used to notify completion of a pipeline stage in order to monitor the per-iteration execution time of each individual application, steering the per-application state machine. Dynamics in the TSR is much less as in the TLR. Driving from an outer area into the city, we could successfully demonstrate our approach using a ten-iteration soft-recovery, once switching to an emergency parameterization and, as expected, never reaching failure state (Fig. 12.11).

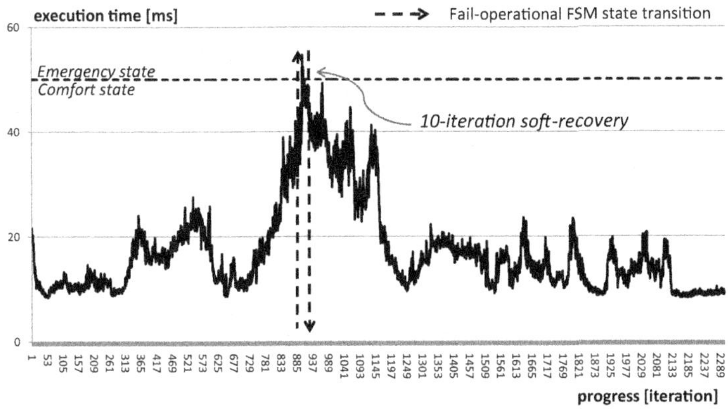

Fig. 12.11 Driving from an outer area to the inner city, we could successfully show the feasibility of our fail-safe approach including the soft-recovery from potential overload prediction

12.4.2.2 Platform Timing Simulation

As a preliminary evaluation of the timing simulation, we conducted a case study for the TSR application [28]. We simulate the execution of the TSR on an ARM Cortex-A9 using the framework presented in Sect. 12.3.3. The necessary timing database was generated by tracing the execution on real hardware. As simulation input we utilize several real pictures of traffic situations and 630 distinct combinations of the segmentation parameters.

The input used during hardware tracing influences the timing of the traced execution, which in turn influences the results of a simulation based on a TDB generated from that traced execution. To investigate this factor we generated several TDBs which were generated using default segmentation parameters and all but one image. One image was excluded, as it does not include any circle detected by the segmentation and a resulting TDB would thus not cover the classification code.

We use each TDB with all simulation inputs and evaluate the results to represent the following practical scenarios: (1) *Arb*: An execution for an arbitrary image is traced and the resulting TDB is used for all simulations. (2) *Del*: An execution for an image with all supported traffic signs is traced and the resulting TDB is used for all simulations. (3) *Perl*: Executions for all images are traced and each TDB is only used in simulations for the same image that was used in tracing. For an image with no signs, a visually similar image is used. Figure 12.12 shows our results. Even the Arb strategy produces results that allow for a practical application. Therefore in practice only straightforward factors need to be considered when applying our trace-based approach. For example, the TSR should be traced using an input image that includes a traffic sign. Putting more effort into tracing input selection can lead to further accuracy improvements.

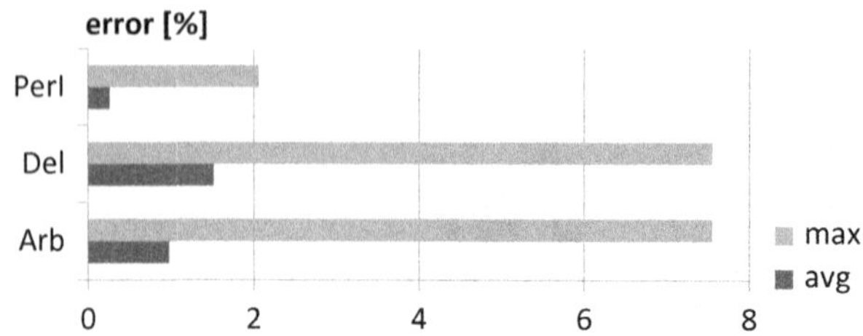

Fig. 12.12 Error in timing accuracy of our trace-based timing simulation against real hardware measurements on a traffic sign recognition perception/cognition function

12.5 Conclusions

In this chapter, we presented solutions that address the deficiencies in state-of-the-art development and realization of embedded automotive driving functions that pose a barrier towards automated driving. We believe that our contributions will bring significant improvements not only to individual challenges but especially when combined, where they might eventually lead to more efficient architectures for highly or fully automated driving, accelerating their introduction to a mass market.

Acknowledgements This work was partially funded by the State of Baden-Württemberg, Germany, Ministry of Science, Research and Arts within the scope of Cooperative Research Training Group EAES, and by the ITEA2/BMBF project MACH under grant 01IS13016B.

References

1. AbsInt Angewandte Informatik GmbH: aiT Worst-Case Execution Time Analyzers (2015), http://www.absint.com/ait/. Accessed 18 March 2015
2. ADAS Applications Processor TDA2x System-on-Chip Technical Brief (2013), http://www.ti.com/lit/ds/symlink/tda2.pdf. Accessed 18 March 2015
3. S. Altmeyer, B. Lisper, C. Maiza, J. Reineke, C. Rochange, WCET and mixed-criticality: what does confidence in WCET estimations depend upon? in OASIcs-OpenAccess Series in Informatics, vol. 47. Schloss Dagstuhl-Leibniz-Zentrum fuer Informatik (2015)
4. AUTOSAR, Specification of Timing Extensions, Release 4.2.2, Document 411 (2015). doi:411. http://www.autosar.org. Accessed 20 Oct 2015
5. F. Bellard, QEMU, a fast and portable dynamic translator, in *USENIX Annual Technical Conference, FREENIX Track* (2005)
6. J. Borrmann, A. Viehl, O. Bringmann, W. Rosenstiel, Parallel video-based traffic sign recognition on the intel SCC many-core platform, in *Proceedings of the 2012 Conference on Design and Architectures for Signal and Image Processing (DASIP)* (2012), pp. 1–2
7. J. Borrmann, F. Haxel, D. Nienhüser, A. Viehl, J. Zöllner, O. Bringmann, W. Rosenstiel, STELLaR - a case-study on SysTEmaticaLLy embedding a traffic light recognition, in *Proceedings of the 17th IEEE International Conference on Intelligent Transportation Systems (ITSC)* (2014), pp. 1258–1265. doi:10.1109/ITSC.2014.6957860

8. J. Borrmann, F. Haxel, A. Viehl, O. Bringmann, W. Rosenstiel, Safe and efficient runtime resource management in heterogeneous systems for automated driving, in *Proceedings of the 18th IEEE International Conference on Intelligent Transportation Systems (ITSC)* (2015)
9. O. Bringmann, W. Ecker, A. Gerstlauer, A. Goyal, D. Mueller-Gritschneder, P. Sasidharan, S. Singh, The next generation of virtual prototyping: ultra-fast yet accurate simulation of HW/SW systems, in *Proceedings of the 2015 Design, Automation & Test in Europe Conference & Exhibition* (EDA Consortium, Grenoble, 2015), pp. 1698–1707
10. L. Cai, D. Gajski, Transaction level modeling: an overview, in *Proceedings of the 1st IEEE/ACM/IFIP International Conference on Hardware/Software Codesign and System Synthesis* (ACM, New York, 2003), pp. 19–24
11. W. Ecker, V. Esen, R. Schwencker, T. Steininger, M. Velten, TLM+ modeling of embedded hw/sw systems, in *Proceedings of the Conference on Design, Automation and Test in Europe* (European Design and Automation Association, Dresden, 2010), pp. 75–80
12. ETAS GmbH, ASCET-SymTA/S End-to-end Timing Analysis for Electronic Control Systems (2015), http://www.etas.com/download-center-files/products_ASCET_Software_Products/ASCET_SYMTA_S_flyer_en.pdf. Accessed 01 June 2016
13. M. Geilen, S. Stuijk, Worst-case performance analysis of synchronous dataflow scenarios. in *Proceedings of the Eighth IEEE/ACM/IFIP International Conference on Hardware/Software Codesign and System Synthesis* (ACM, New York, 2010), pp. 125–134
14. R. Henia, L. Rioux, N. Sordon, G.E. Garcia, M. Panunzio, Integrating model-based formal timing analysis in the industrial development process of satellite on-board software, in *Proceedings of the 2nd International Conference on Model-Driven Engineering and Software Development (MODELSWARD)* (2014), pp. 619–625
15. IEEE Standard for Standard SystemC Language Reference Manual (2012). IEEE Std 1666-2011 (Revision of IEEE Std 1666-2005)
16. R. Kumar, V. Zyuban, D.M. Tullsen, Interconnections in multi-core architectures: understanding mechanisms, overheads and scaling, in *Proceedings of the 32nd International Symposium on Computer Architecture, 2005. ISCA'05* (IEEE, New York, 2005), pp. 408–419
17. K. Lu, D. Müller-Gritschneder, U. Schlichtmann, Hierarchical control flow matching for source-level simulation of embedded software, in *Proceedings of the 2012 International Symposium on System on Chip (SoC)* (IEEE, New York, 2012), pp. 1–5
18. K. Lu, D. Muller-Gritschneder, U. Schlichtmann, Fast cache simulation for host-compiled simulation of embedded software, in *Design, Automation & Test in Europe Conference & Exhibition (DATE), 2013* (IEEE, New York, 2013), pp. 637–642
19. T.G. Mattson, M. Riepen, T. Lehnig, P. Brett, W. Haas, P. Kennedy, J. Howard, S. Vangal, N. Borkar, G. Ruhl, S. Dighe, The 48-core SCC processor: the programmer's view, in *2010 ACM/IEEE International Conference for High Performance Computing, Networking, Storage and Analysis*, November 2010, pp. 1–11. http://ieeexplore.ieee.org/xpl/articleDetails.jsp?arnumber=5644880
20. MIRA Ltd, MISRA-C:2004 Guidelines for the use of the C language in critical systems (2004). www.misra.org.uk
21. Mobileye (2015), mobileye.com. Accessed 18 March 2015
22. A. Monot, N. Navet, B. Bavoux, F. Simonot-Lion, Multisource software on multicore automotive ECUs - combining runnable sequencing with task scheduling. IEEE Trans. Ind. Electron. **59**(10), 3934–3942 (2012)
23. Moving Closer to Automated Driving, Mobileye Unveils EyeQ4 System-on-Chip with its First Design Win for 2018 (2015), http://www.mobileye.com/blog/press-room/moving-closer-automated-driving-mobileye-unveils-eyeq4-system-chip-first-design-win-2018/. Accessed 18 March 2015
24. D. Nienhüser, Kontextsensitive Erkennung und Interpretation fahrrelevanter statischer Verkehrselemente. Ph.D. thesis, KIT Karlsruhe (2014)
25. nVidia, Tegra K1 Technical reference manual (2014), https://developer.nvidia.com/tegra-k1-technical-reference-manual. Accessed 18 March 2015

26. nVidia, Tegra X1 Super Chip (2015), http://www.nvidia.com/object/tegra-x1-processor.html. Accessed 18 March 2015
27. S. Ottlik, S. Stattelmann, A. Viehl, W. Rosenstiel, O. Bringmann, Context-sensitive timing simulation of binary embedded software, in *Proceedings of the 2014 International Conference on Compilers, Architecture and Synthesis for Embedded Systems (CASES)* (2014)
28. S. Ottlik, J.M. Borrmann, S. Asbach, A. Viehl, W. Rosenstiel, O. Bringmann, Trace-based context-sensitive timing simulation considering execution path variations, in *21st Asia and South Pacific Design Automation Conference (ASP-DAC)* (2016)
29. M. Pressler, A. Viehl, O. Bringmann, W. Rosenstiel, Execution cost estimation for software deployment in component-based embedded systems, in *Proceedings of the 17th international ACM Sigsoft Symposium on Component-Based Software Engineering* (ACM, New York, 2014), pp. 123–128
30. S.J. Russell, P. Norvig, *Artificial Intelligence: A Modern Approach*, 2. edn. Prentice Hall Series in Artificial Intelligence (Prentice Hall, Upper Saddle River, NJ, 2003)
31. Safer SoCs for safer driving. SemiWiki.com, the Open Forum for Semiconductor Professionals (2014), https://www.semiwiki.com/forum/content/3844-safer-socs-safer-driving.html. Accessed 18 March 2015
32. D. Sanchez, G. Michelogiannakis, C. Kozyrakis, An analysis of on-chip interconnection networks for large-scale chip multiprocessors. ACM Trans. Archit. Code Optim. (TACO) **7**(1), 4 (2010)
33. K. Schmidt, J. Harnisch, D. Marx, A. Mayer, A. Kohn, R. Deml, Timing analysis and tracing concepts for ECU development. Technical Report, SAE Technical Paper, 2014
34. T. Schonwald, A. Koch, B. Ranft, A. Viehl, O. Bringmann, W. Rosenstiel, Stereo depth map computation on a Tilera TILEPro64 embedded multicore processor, in *Proceedings of the 2012 Conference on Design and Architectures for Signal and Image Processing (DASIP)* (2012)
35. S. Stattelmann, *Source-Level Performance Estimation of Compiler-Optimized Embedded Software Considering Complex Program Transformations* (Verlag Dr. Hut, Munich, 2013)
36. S. Stattelmann, O. Bringmann, W. Rosenstiel, Fast and accurate source-level simulation of software timing considering complex code optimizations, in *Proceedings of the 48th Design Automation Conference (DAC)* (2011)
37. G. Stein, E. Rushinek, G. Hayun, A. Shashua, A computer vision system on a chip: a case study from the automotive domain, in *IEEE Computer Society Conference on Computer Vision and Pattern Recognition (CVPR'05) - Workshops*, vol. 3 (2005), p. 130. doi:10.1109/CVPR.2005.387. http://ieeexplore.ieee.org/lpdocs/epic03/wrapper.htm?arnumber=1565445
38. STMicro and Mobileye are Developing Third-Generation SoCs for Driver Assistance. John Day's Automotive Electronics - Insight for Engineers (2011), http://johndayautomotivelectronics.com/stmicro-and-mobileye-are-developing-third-generation-socs-for-driver-assistance/. Accessed 18 March 2015
39. Symtavision GmbH: Leading in real time - overview flyer (2015), https://www.symtavision.com/downloads/Flyer/Symtavision_Overview_Flyer_2015_SEPT.pdf. Accessed 20 Oct 2015
40. TDA2x SoC for Advanced Driver Assistance Systems (ADAS) (2015), http://www.ti.com/tda2x. Accessed 18 March 2015
41. TDA3 SoC Processor Advanced Driver Assistance Systems (ADAS) (2015), http://www.ti.com/product/tda3. Accessed 18 March 2015
42. D. Thach, Y. Tamiya, S. Kuwamura, A. Ike, Fast cycle estimation methodology for instruction-level emulator, in *Proceedings of the 2012 Design, Automation & Test in Europe Conference & Exhibition (DATE)* (2012)
43. The MathWorks, Inc. (2015), http://www.mathworks.com. Accessed 18 March 2015
44. Tilera goes pro with TilePro64 (2008), http://www.tgdaily.com/business-and-law-features/39408-tilera-goes-pro-with-tilepro64. Accessed 18 March 2015
45. Veripool Verilator (2015), http://www.veripool.org/. Accessed 18 March 2015
46. N. Weste, D. Harris, *CMOS VLSI Design: A Circuits and Systems Perspective*, 4th edn. (Addison-Wesley Publishing Company, New York, 2010)

47. XILINX: UG585 - Zynq-7000 all programmable SoC technical reference manual (2014), http://www.xilinx.com/support/documentation/user_guides/ug585-Zynq-7000-TRM.pdf. Accessed 18 March 2015
48. J. Zimmermann, Applikationsspezifische Analyse und Optimierung der Energieeffizienz eingebetteter Hardware/Software-Systeme. Ph.D. thesis, Universität Tübingen, Germany (2013). Website http://nbn-resolving.de/urn:nbn:de:bsz:21-opus-73952

Chapter 13
Systems Engineering and Architecting for Intelligent Autonomous Systems

Sagar Behere and Martin Törngren

13.1 Introduction

This chapter provides practical insights into specific systems engineering and architecture considerations for building autonomous driving systems. It is aimed at the ambitious practitioner with a solid engineering background. We envision such a practitioner to be interested not just in concrete system implementations, but also in borrowing ideas from the general theory of intelligent systems to advance the state of autonomous driving.

The practical development of autonomous driving systems involves domain specific algorithms, architecture, systems engineering, and technical implementation, as shown in Fig. 13.1. Of these, this chapter focuses on the latter three. Architecture and its development may possibly be considered as a part of systems engineering, but for the purpose of this chapter, we treat it as a distinct area. This is because of the extensive coverage of architecture in the chapter. The arrows in Fig. 13.1 represent an "impact" relationship. Thus, the arrow from architecture to systems engineering implies that architecture has an impact on systems engineering. The impact can be of various types, but the key point is that the practical development of autonomous driving systems must *holistically* consider all the areas and their impacting interrelations. Such a holistic view is not always within the scope of researchers "deep diving" into the specifics of a particular area. Nevertheless, for practicing engineers and concrete projects, it cannot (should not) be ignored. This chapter presents the three areas within a holistic context, with the aim of providing the practicing engineer with a grasp of key concepts in each area, and how they all come together for autonomous driving. Within each area, references to more detailed topics are also provided. This simultaneous consideration of

S. Behere (✉) • M. Törngren
KTH Royal Institute of Technology, Stockholm, Sweden
e-mail: behere@kth.se; martint@kth.se

D. Watzenig, M. Horn (eds.), *Automated Driving*,
DOI 10.1007/978-3-319-31895-0_13

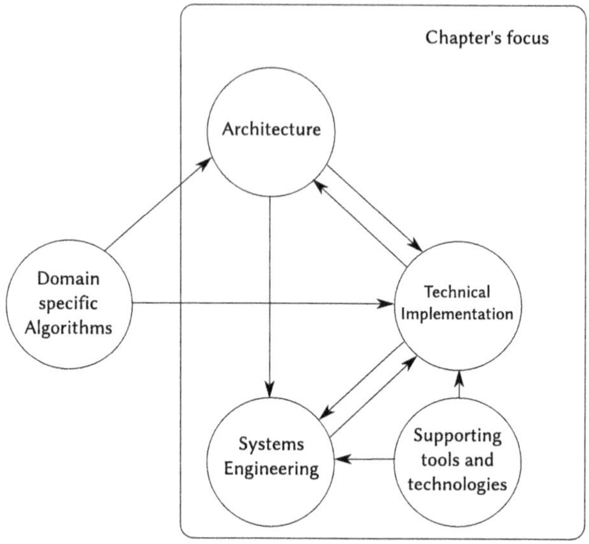

Fig. 13.1 The areas in focus of this chapter

three, typically disparate topics, is one of the innovative aspects of this book chapter. Very often in research, the technical aspects of the implementation are considered less important. In our experience with autonomous driving, the technical implementation aspects provide opportunities and strong constraints which *must* be considered during the processes of architecting and systems engineering. This prevents gaps imposed by "reality" when transitioning between different concerns and a subsequent weakness in the application of theoretical results. Therefore, this chapter devotes a complete section to the core tools and technologies supporting architecting, systems engineering, and implementation. This is a second innovative aspect of this chapter. The third innovation is an examination of key results in the areas of machine consciousness and the theory of mind, which are usually omitted from "hardcore" engineering discussions, because the gap between these areas and pragmatic, safety critical engineering is very large. But by covering these areas, we show how it can influence and provide guidelines for future developments in autonomous driving. The chapter deliberately emphasizes breadth of coverage rather than depth in one selected topic, because we have noticed a conspicuous lack of literature in the field that collects together the important considerations and topics relevant to the practicing engineer. It must be noted that some of the presented topics are not exclusive to the development of autonomous systems, but gain significant importance in that context.

The chapter thus contributes with a series of takeaways in the areas of architecture, systems engineering, technical implementation, and longer term evolution of autonomous driving. For architecture, the reader will find explicit descriptions of key functional components needed for autonomous driving and a three layered reference architecture showing the distribution of these components and their interconnections. For model based systems engineering (MBSE), the chapter outlines four classes of models to aid MBSE methodologies for concrete projects. The model classes may be viewed as a lightweight complement to methodologies advocated by the "V-model" and functional safety standards like ISO26262. For technical implementation, a development setup and key technologies are described. For longer term evolution, the chapter provides a brief overview of stronger autonomy concepts like machine consciousness and self-awareness, and relates them to current engineering practices. This is used to point out directions for evolution of current autonomous driving architectures. The impact of autonomy on architecting, systems engineering, and technical implementations in the automotive domain is also briefly discussed.

The scope is delimited to a broad treatment of functional architectures and systems engineering concerns relevant for autonomous driving. The emphasis is on early stages of development and prototyping. Concerns exclusively related to the engineering of safety critical systems, as well as human machine interaction, metrics for architecture and systems engineering, and topics related to engineering ethics are not covered.

The chapter is structured as follows: following this Introduction, Sect. 13.2 provides a description of the research method, followed by a summary of important terms in the text. Section 13.3 provides a quick description of important terms in the text. This is followed by Sect. 13.4 which provides a longer term perspective of the research area, exploring more abstract concepts for enabling machine autonomy. The paper then becomes progressively more concrete via Sect. 13.5 on Architecture, Sect. 13.6 on Systems Engineering, and Sect. 13.7 on Technical Implementation. These chapters cover the immediate and short term perspective in the field. The progression from abstract to concrete is deliberately selected to guide the reader's thinking from the more esoteric and principled notions of machine autonomy to a practical immersion in the engineering state-of-practice. Finally, Sect. 13.8 presents a synthetic discussion that reflects on each of the covered areas, and how they are affected by the characteristics of autonomy.

13.2 Research Method

Research methods of Engineering Design have been used to generate this chapter's content. Engineering design is one of the research methods in systems engineering [40] wherein researchers address a problem which is important and novel through the activity of designing a solution [41]. The knowledge developed is primarily for practical application. An additional outcome is some theoretical development based on generalization of design experiences. A potential weakness of engineering design, as a qualitative research method, is that of *external validity*. This is addressed here by a multitude of different case studies and engagement with experts in different domains. Since 2010, we have designed solutions for, and engaged in

Table 13.1 Projects contributing to this chapter's content

Year(s)	Projects	Vehicle	Partners	Outcome
2010–2011	GCDC 2011	Heavy duty commercial truck	Scania CV AB (OEM)	1. Autonomous longitudinal motion in platooning scenario [54]. 2. A reference architecture for cooperative driving [21]
2011–2012	CoAct 2012	Heavy duty commercial truck	Scania CV AB (OEM)	Second, different instantiation of above-mentioned reference architecture for cooperative driving
2013–2014	DFEA2020 + FUSE + ARCHER	Passenger cars	Volvo Car Corporation (OEM) + Scania CV AB (OEM)	Problem analysis, methods, and a reference architecture for autonomous driving [19, 20]
2014–	RCV	Novel research vehicle prototype	Departments within KTH (Academia)	Novel electric vehicle prototype with -by-wire control of steering and propulsion [77]
2015–	RCV-2.0	Novel research vehicle prototype	KTH + A private company	Novel vehicle prototype with full perception stack and urban autonomous driving capabilities (under development)

a variety of self-driving vehicle projects. The projects have involved a variety of commercial and research vehicles, in academic and industrial contexts. A concise summary is provided in Table 13.1. In 2010, an architecture for autonomous longitudinal motion control was designed and implemented on an R730 commercial truck from Scania CV AB. The truck participated in the Grand Cooperative Driving Challenge (GCDC) 2011, wherein vehicles operated autonomously on a public highway in a platooning scenario, with constant wireless communication between participating vehicles and the environment. The communication contained operating parameters of each vehicle, like its speed, acceleration, location, etc., as well as the states of infrastructural elements like traffic lights and prevailing speed limits. The architecture was refined and re-applied an year later, on a different truck (an R430 model), for the CoAct 2012 project. This project also involved a platooning scenarios similar to the GCDC 2011 event, but it included more demanding operational situations like splitting and merging lanes, and overtaking. The accumulated architecture underwent further evaluation and analysis in the course of three different projects with various industrial partners, including a potential application to passenger cars. One of these projects was DFEA2020—a large Swedish consortium project aimed at development of green, safe, and connected vehicles. Another project is FUSE—also a Swedish project, with a tighter focus on functional safety and architectures for autonomous driving. The third project is ARCHER—which investigates safety, reference architectures, and testing and verification techniques applicable to commercial trucks. The FUSE and ARCHER projects are still in progress. Starting from 2014, the architecture was then applied to a novel research concept vehicle (RCV) at KTH, with a view to endow it with autonomous driving capabilities. The RCV has an all-electric, drive-by-wire powertrain with a propulsion motor embedded inside each wheel, and active steering and camber control of all four wheels. The architecture was then adapted to a second variant of the RCV (RCV-2.0), where it serves as the foundation for autonomous urban driving capabilities in situations where a human driver is not expected to be available (or capable) of taking over vehicle control.

13.3 Essential Terminology and Concepts

In this section, we briefly describe our specific usage of some important terms that occur repeatedly in the text.

Autonomy is used in the practical sense as *a machine's ability to effectively (with respect to its goals) operate in an uncertain environment, without constant human supervision or intervention*. Effective operation implies that the desired goals are met in a safe manner with a desired level of performance. **Safety** implies absence of unacceptable risk.

Architecture is defined by ISO 42010:2011 as "..fundamental concepts or properties of a system in its environment, embodied in its elements, relationships, and principles of its design and evolution." The architecture is thus the "blueprint" of the system and one practical way to think of it is to decompose it into conceptual and technical design aspects [28, 29] as shown in Fig. 13.2. These aspects may alternatively be referred to as "views" of the architecture, a term recommended by ISO 42010 and pertaining to an architecture description from a specific "viewpoint." A detailed explanation of Fig. 13.2 can be found in [19]. Briefly, the process of architecting a system can be initiated by describing the functions which the

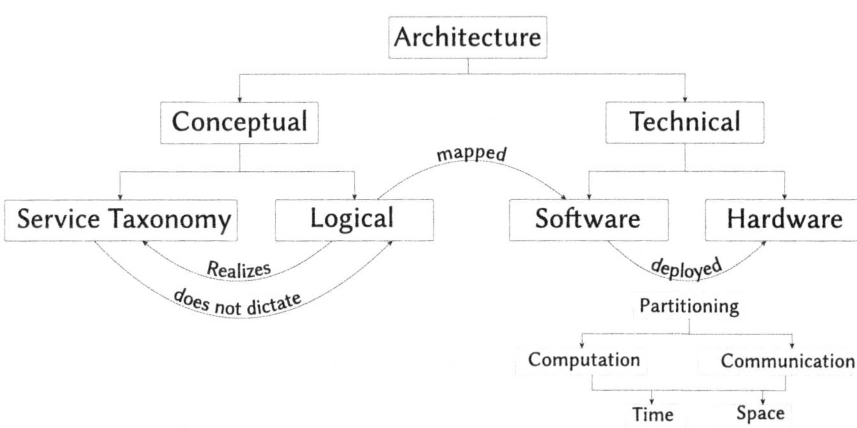

Fig. 13.2 An overview of system architecture

system shall offer to its user (service taxonomy), without stating how the functions are internally realized by the system. The service taxonomy is realized by the logical architecture, which shows the logical decomposition of the system into its constituent components, without specifying how those components are actually realized in hardware and/or software. The logical architecture components are subsequently mapped onto software elements, which are deployed on hardware computation units. The computation and communication systems may further be partitioned in time and space, depending on a variety of requirements related to performance, availability, safety, prototyping tools, etc.

Systems engineering is defined by INCOSE as *an interdisciplinary approach and means to enable the realization of successful systems*. It integrates all the disciplines and domain experts into a team effort forming a structured development process that proceeds from concept to production to operation. Functionality is defined early in the development lifecycle, requirements are documented followed by design synthesis, testing, verification, and validation all while considering the complete problem which the system solves. Systems engineering is especially relevant to the construction of large, complex, and safety critical systems. The formalized application of modeling to support various systems engineering tasks is termed as **model based systems engineering**.

13.4 The Context of Machine Consciousness

When thinking about autonomous driving, it is very easy to get lost in practical minutiae of sensors, hardware, programming, modeling, etc. However, staying entirely at this level of thinking can lead to "not seeing the forest due to all the trees." As the complexity of systems rises, it is worth taking a step back and asking, "What is it that we are really trying to do, and is this the right direction?"

In this section, we take a step back from the immediate engineering concerns, and look at autonomous driving within the larger and somewhat "philosophical" context of generally autonomous and intelligent machines. The engineering of autonomous driving systems rarely considers investigations into domains like machine consciousness [45], self-awareness, and theories of mind [34]. This is primarily due to two reasons: the technological/experiential background of the engineers involved, and lack of clear mappings from the mentioned domains to engineering concepts for certifiable, safety critical embedded systems. Nevertheless, we believe that keeping abreast of key results in these domains is valuable for practitioners of autonomous driving because it provides strategic guidance for future architecture concepts and identifies the gaps that prevents utilization of results from these domains.

The ultimate goal for autonomous driving is not just human-like driving but to go beyond human-like driving, in order to overcome human limitations and appreciably increase road safety, traffic efficiency, and environmental benefits. To achieve this goal, and to interact and coexist with human environments, machines would need a level of consciousness that approaches human (admittedly under tightly constrained notions). This is because consciousness is instrumental to reasoning, decision making, and problem-solving capabilities in the face of uncertainty and disturbances. There is a variety of literature [55] which suggests that robots' problem solving capacities would be enhanced by the ability to introspect. This is also a recurring theme across disciplines like computer systems-on-chip [68], programming languages [52], robotics and even explorations for fault-tolerant on-board computing for robotic space missions [81].

What would consciousness mean in the context of autonomous driving? A dictionary definition [8] of **consciousness** is "..the fact of awareness by the mind of itself and the world" where **awareness** is further defined as *"Knowledge or perception of a situation or fact."* Within the field of philosophy and cognitive sciences, however, consciousness is recognized as an umbrella term covering a wide variety of heterogeneous mental phenomena, often grouped under the categories of Creature Consciousness, State Consciousness, and Consciousness as an entity [76]. While comprehensive and precise definitions remain a topic of research and debate, from an engineering perspective we are interested in definitions only insofar as they help us to identify and characterize specific machine behavior. Indeed, as far back as in the 1950s, researchers like Alan Turing believed that questions about *actual* intelligence (and presumably consciousness) were too vague or mysterious to answer. Turing instead proposed a behaviorist alternative [75] wherein if a savvy human judge could not distinguish a computer's conversational abilities from those of a real person[1] at a rate better than chance, then we would have some measure of the computer's intelligence. This measure of perceived intelligence could be substituted for the computer's *real* intelligence or *actual* consciousness. A loose application to autonomous driving could be: If a savvy human judge can consistently accept a computer's driving abilities as equivalent to those of a competent human driver, then we would have some measure of the computer's driving capabilities. Indeed, on 4th May 2012, one of Google's self-driving Prius vehicles was granted a "driving license" by the Nevada state Department of Motor Vehicles, after the vehicle successfully passed driving tests similar to those administered to human drivers [6] .

At this point, it is useful to decompose consciousness into awareness of the external and the internal, from the machine's perspective. Awareness of the external world involves elements of sensing, data fusion, and perception, all of which demonstrate progressive and on-going improvements within the domain of autonomous driving (compare the sensors available for autonomous driving today, with those from just a decade ago). This awareness of the external world is usually explicitly represented in the internal world of the machine, by maintenance of data structures reflecting perception of the external world. From a philosophical view then, awareness of the external world is "absorbed" by the machine's internal states, to which engineering attention must be devoted, in order to achieve progressive

[1]This is the popular version. Turing actually framed a somewhat different test, as discussed in [37].

results in machine consciousness. However, for meaningful exploitation of any internal awareness, the machine needs to be aware of the awareness, i.e., it needs to be **self-aware**. This is supported by explicit conclusions in the domain of machine consciousness, for example, McCarthy states [55] "..some consciousness of their own mental processes will be required for robots to reach a level of intelligence needed to do many of the tasks humans will want to give them.. consciousness of self, i.e., introspection is ***essential*** for human level intelligence, not a mere epiphenomenon."

Which system characteristics and structures are instrumental for consciousness and self-awareness? One approach to answering this is to understand how the human brain functions, and mimic the biological structures found therein. The strongest approach to understanding human consciousness (and the only one relevant to autonomous driving) is that of *computationalism* [56] which is the theory that many relevant aspects of the human brain can be modeled as having a computational structure. This approach is the basis of the field of *Artificial Intelligence* (AI) which explores computational models of problem solving, although it reserves the possibility that digital computers and brains-as-computers may compute things in different ways. Research on computational models of consciousness has been driven by researchers like Hofstadter, Minsky, McCarthy, Dennet, Perlis, Sloman, and Cantwell Smith. A synthetic summary of their principal propositions is presented in [56], where the dominant proposition is that "..consciousness is the property a computational system X has if X models itself as experiencing things." Thus, central to the theory of computational consciousness is that introspection is mediated by models.

As far back as in 1968, Minsky [58] introduced the concept of a "self-model": "To an observer B, an object A* is a model of an object A to the extent that B can use A* to answer questions that interest him about A. A* is a good model of A, in B's view, to the extent that A*'s answers agree with those of A, on the whole, with respect to the questions important to B." This concept is semantically the same as the engineering definition of *model* defined by IEEE 610.12-1990, given in Sect. 13.3 above. Most engineers know that a model of a system is an abstraction of the system which provides answers about desired properties and behavior of the system, with desired accuracy. So we see that a specific category of models (the self-model) can be the theoretical basis for a meeting point between the more abstract reasoning of consciousness and the concerns of "everyday engineering."

The computing models based on results from cognitive theory and speculations on the structure of the human mind are explored in the field of *cognitive architectures*. There are a variety of cognitive architectures created to mimic specific aspects of human reasoning and decision making in machines. These include RCS, ACT-R, SOAR, CLARION, NARS, YMIR, and others. A review of the prominent architectures (with further references) is presented in [67] which is based on actively maintained online resources [3]. The review mentions 54 architectures, out of which 26 are described, and observes that virtually all take their origin from biological inspiration, and different approaches are remarkably similar in their basic foundations. A comparative review of these architectures with specific relevance to machine autonomy is provided in [72], which draws some important conclusions of relevance to practicing engineers working on autonomous driving and other safety critical systems. The comparison is made along four main themes of "Realtime," "Resource management," "Learning," and "Meta-learning." The conclusions highlight a prevalent gap between cognitive architecture design and concrete operation in real-world settings. This gap is fueled by the observation that cognitive architectures generally tend to ignore realtime operation and resource management aspects. Ignoring such practical matters not only delays useful technological application, but likely leads to flawed theoretical foundations. In turn, this limits the usefulness of the architectures to toy problems, "devoid of the complexity of the real world that human beings live and operate in." [72]. A similar conclusion is drawn by other researchers exploring challenges in the domain of embedded systems [46]. Fortunately, these limitations are not shared uniformly by the cognitive architectures, promoting the possibility of designing architectures with a more complete set of cognitive functions and usable operational capabilities.

The current approach to autonomous driving capabilities is bottom-up: based on refinement of features starting from the automatic transmission and cruise control, to adaptive cruise control, lane departure indications, and autonomous emergency braking, to traffic jam assist and advanced driving assistance systems, eventually all converging on autonomous driving capabilities. This approach has yielded excellent results so far; results which may not have been reachable (in this timeframe) with a top-down approach that started inquiring into the nature and structure of (human) consciousness. The current approach has heavily adopted results from the robotics and AI domains. The adoption has exclusively involved careful, manual construction of systems, where learning and decision making take place on the data/content or module levels. Such a hand-crafted approach is referred to as the *Constructionist Design Methodology*(CDM) in the AI domain [72, 73]. However, to create systems that approach human level intelligence, significantly larger and more complex architectures are necessary. There is a very real danger that methodologies based on constructionist AI will prove inadequate (Thórisson and Helgason [72] state more strongly "..are doomed") because of practical restrictions on complexity and size of software based systems designed and implemented by humans. These restrictions are captured in the term *cognitive complexity* [50], in the domain of embedded systems architectures. The cognitive complexity attribute of an architecture limits the ability of humans to hold the entire architecture in their head and reason about it. As cognitive complexity rises, it becomes increasingly more difficult (and costly) to verify and validate the systems, and assure properties like safety. Indeed, phenomena like *feature interaction* [48, 57] have already started occurring in automotive architectures (and generally, within cyber-physical systems) and their study and control is an emerging area of research.

Within the AI domain, a relatively new *constructivist* approach [71] has emerged, that advocates self-directed, introspective, learning, and dynamically adapting conscious architectures instead of the "carefully handcrafted" approach prevalent in the automotive domain. This is considered a paradigm shift, whereunder the machine may proactively invoke stimulus-response cycles to continuously form and maintain self-models, and reason about their characteristics. The evolution of existing engineering approaches towards constructivist concepts requires developments in at least four relevant areas [69]: (1) temporal grounding (2) feedback loops (3) pan-architectural pattern matching, and (4) small white-box components. Temporal grounding comprises of an awareness of time in the real world, as well as the time needed for the execution of software instructions, and how the two are correlated. Such an awareness is already important in the design of distributed realtime embedded systems, but as pointed out in [51], timeliness is a *semantic* property that is not well captured in popular programming paradigms. At a behavior level, temporal grounding involves predicting temporal result based on internal models, and updating the internal models if/when the predictions do not match the results. Traditional closed loop feedback is already prevalent in autonomous driving. These loops typically control short-time horizon actuation in response to the immediate sensing and perception of the environment. When moving towards constructivist AI, there need to be additional feedback loops that modify the control structures themselves, based on their perceived efficacy. This is closely related to the concept of self-modeling based on observations of stimulus-reaction experiments in a given context. Such modifications enable an expansion of existing skills and capabilities, which is an important characteristic of general purpose intelligence. Pan-architectural pattern matching is useful for the self-induced comparison of temporal versions of the architecture, as the system grows/evolves over time in response to some specification. Pattern matching is also useful for identification of contexts and operational scenarios, which in turn is useful for controlling mechanisms of attention and recall. These mechanisms are important because they enable the machine to filter out the large number of stimuli present in complex operational situations, and focus only on those that have been learned as being relevant. Existing autonomous driving architectures are composed out of integration of large components like localization, trajectory planning, propulsion, etc. These components are typically developed by different suppliers and are often "black-box" components with well-defined input/output interfaces. However their large size and black-box nature make it difficult for a machine to reason about their internals based purely on an observation of the input/output signals. This is a critical issue for self-organizing and self-aware systems, because it is difficult to reach self-awareness in the presence of large, opaque internal components.

Constructivist AI certainly presents new challenges, some of which lie in the determinism and predictability of state evolution, internal beliefs, and actions of truly autonomous systems. These challenges currently run counter to requirements on provable safety, determinism, and assurance of other critical properties, and the constructivist approach is unlikely to be adopted for autonomous driving until it evolves sufficient tools, methods, reference architecture patterns, etc., to create practical and demonstrably safe and predictable systems.

13.5 An Architecture for Autonomous Driving

In its current state, the tasks expected from an autonomous driving system are fairly well constrained and specific. Therefore, it is possible to create domain specific reference architectures for autonomous driving. Such architectures would include definitions of the various architectural elements needed by an autonomous driving system, the hierarchy and data-flows between these elements, and instantiation guidelines for specific use cases. In this section, we present a brief introduction to the required architectural components and some reasoning on their hierarchy and distribution. As with most domain specific reference architectures, we restrict the scope to the functional/logical views although some relevant technologies for instantiation are later described in Sect. 13.7.

13.5.1 *Main Architectural Components*

The main functional components needed for autonomous driving are summarized in Fig. 13.3. We have chosen to categorize these components into three categories

1. Perception of the external environment/context in which the vehicle operates
2. Decisions and control of the vehicle motion, with respect to the external environment/context that is perceived
3. Vehicle platform manipulation which deals mostly with sensing and actuation of the ego vehicle, with the intention of achieving desired motion

Each category has several components, whose functionality (from a strictly architectural perspective) will now be briefly described. A detailed description can be found in [20].

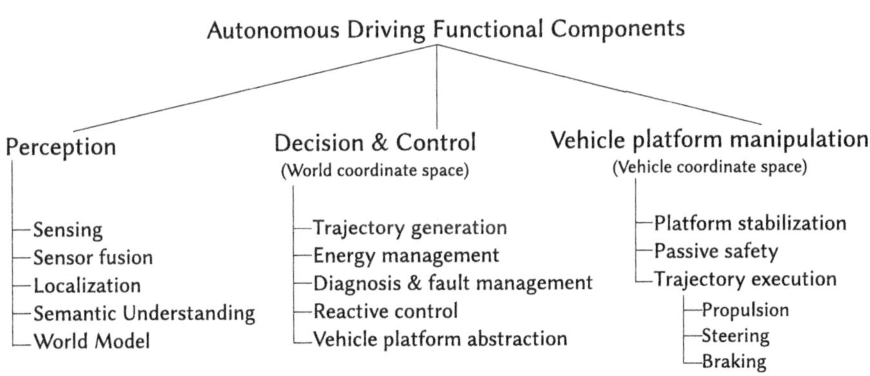

Fig. 13.3 Main components of an autonomous driving system

The **sensing** components are those that sense the states of the ego vehicle and the environment in which it operates. The **sensor fusion** component considers multiple sources of information to construct a hypothesis about the state of the environment. In addition to establishing confidence values for state variables, the sensor fusion component may also perform object association and tracking. The **localization** component is responsible for determining the location of the vehicle with respect to a global coordinate system, with needed accuracy. It may also aid the sensor fusion component to perform a task known as *map matching*, wherein physical locations of detected objects are referenced to the map's coordinate system.

The **semantic understanding** component is the one in which the balance shifts from sensing to perception. More concretely, the semantic understanding component can include classifiers for detected objects, and it may annotate the objects with references to physical models that predict likely future behavior. Detection of ground planes, road geometries, and representation of drivable areas may also happen in the semantic understanding component. In specific cases, the semantic understanding component may also use the ego vehicle data to continuously parameterize a model of the ego vehicle for purposes of motion control, error detection, and potential degradation of functionality. This component comes closest to incorporating the "self-model" needed for generating machine consciousness.

The **world model** component holds the state of the external (and possibly, internal) environment, as perceived by the ego vehicle. It can be characterized as either passive or active. A passive world model is more like a data store and may lack semantic understanding of the stored data. It cannot, by itself, perform physics related computations on the data it contains, to actively predict the state of the world given specific inputs. The active world model, on the other hand, may incorporate kinematic and dynamic models of the objects it contains, and be able to evolve beliefs of the world states when given a sequence of inputs.

The **trajectory generation** component repeatedly generates obstacle free trajectories in the world coordinate system and picks an optimal trajectory from the set. The **energy management** component is usually split into closely knit subcomponents for battery management and regenerative braking. Since energy is a system-wide concern, it is not uncommon for the energy management component to have interfaces with other vehicular systems like HVAC, lights, chassis, and brakes. The **diagnosis and fault management** components identify the state of the overall system with respect to available capabilities, in order to influence redundancy management, systematic degradation of capabilities, etc. **Reactive control** components are used for immediate (or "reflex") responses to unanticipated stimuli from the environment. An example is automatic emergency braking (AEB). These components typically execute in parallel with the nominal system, and if a threat is identified, their output overrides the nominal behavior requests.

The **vehicle platform abstraction** provides a minimal model of the vehicle, whose data is used to ensure that the trajectories being generated are compatible, optimal, and safe for the physics and capabilities of the actual vehicle.

The **platform stabilization** components are usually related to traction control, electronic stability programs, and anti-lock braking features. Their task is to keep the vehicle platform in a controllable state during operation. Unreasonable motion requests may be rejected or adapted to stay within the physical capabilities and safety envelope of the vehicle. The **trajectory execution** components are responsible for actually executing the trajectory generated by decision and control. This is achieved by a combination of longitudinal acceleration (propulsion), lateral acceleration (steering), and deceleration (braking). Most recent vehicles already incorporate such components and they may be considered "traditional" from the perspective of autonomous driving development.

13.5.2 A Reference Architecture

Having introduced the necessary functional components in the previous section, we now combine them into a suggested reference architecture, as shown in Fig. 13.4, where the arrows show directed data-flows between the architectural elements.

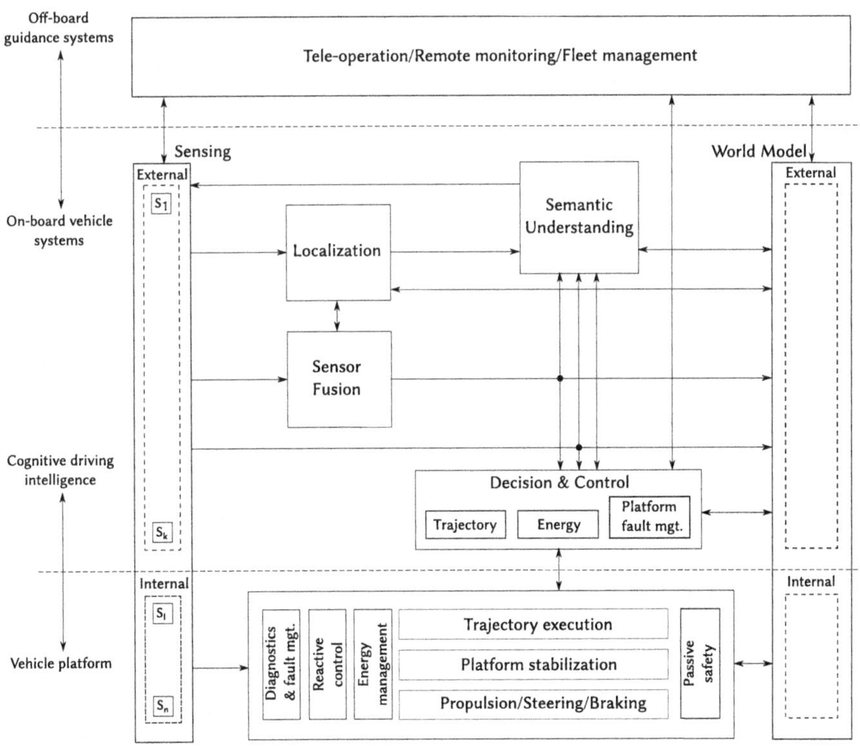

Fig. 13.4 A functional architecture for autonomous driving architecture

The architecture is organized into three layers: the vehicle platform and cognitive driving intelligence which are on-board the vehicle, and an off-board or "cloud" based layer for potential tele-operation, remote monitoring, and/or vehicle management. The off-board layer may be optional for many use cases, nevertheless it is almost always useful at least during the early phases of vehicle prototyping and testing. For heavy commercial trucks, some form of fleet management systems are usually provided to the fleet operators. Moreover, there are compelling drivers for including vehicle-to-infrastructure (V2I) communication to improve traffic efficiency and safety by sharing information acquired from multiple vehicles and other sources.

One of the key data flows in Fig. 13.4 is between the cognitive driving intelligence and the vehicle platform layers. Functionally, this contains motion requests in the form of desired, instantaneous vehicle velocities, accelerations (longitudinal and lateral), and deceleration. These are typically in some absolute, global coordinate space, rather than being relative to the vehicle's motion. In practice, it may contain a short-time series of these values (trajectory fragments) rather than individual requests, because knowing the anticipated future setpoints is helpful for achieving more optimal control of the actuators. It is also feasible to include two different sets of trajectory fragments: one which takes the vehicle to the desired destination, and the other which takes it to a safe(er) state in case of system errors. The safe(er) state trajectory is computed periodically and should ideally be executable by the vehicle platform in an open loop fashion. The output of the localization function contains at least the 2.5 dimensional vehicle pose consisting of the location in two dimensional space, as well as the heading. In practice, a lot more data and meta-data is provided which includes altitude, the latitude/longitude coordinates, the same information in a variety of coordinate spaces, detected known landmarks, number of satellites in the GPS fix, estimated accuracy, etc. The output of the sensor fusion function depends heavily on the actual sensors and algorithms being used. Since lidars are used very often it is not uncommon to see point clouds with associated meta-information. The addition of a camera leads to the inclusion of extracted image features, and possibly colored point clouds. The latter are especially useful for classification in the semantic understanding component. The usage of automotive grade radar is more interesting. Typically, automotive radar sensors come with an associated ECU which directly outputs (possibly classified) objects and their properties like relative velocity and distance. The radar outputs can be used for cross-correlation and plausibility testing of the data from the lidar and camera sensors. However, in projects with a strong emphasis on sensor fusion, the raw radar information may also be requested. It can be seen in Fig. 13.4 that the output of a functional component goes to other components as well as the world model. This is usually for performance reasons. In an ideal implementation, it would be possible to have a "star" data topology where all the components exchange data only with the world model. This is a simplifying abstraction, leading to a single source of information and other benefits. However, it raises concerns on performance, usually latency, as well as increasing the extra-functional requirements on the world model, such as robustness, reliability, and availability. The reference architecture permits both topologies and each instantiation may tune the amount of information each component receives from the world model, or directly from other components. The component interfaces are further refined in the technical architecture, considering the constraints imposed by the allocation of components to ECUs and the bandwidth of inter-component communications.

The on-board autonomous driving architecture admits a variety of distribution possibilities for the functional components. We choose to encourage a strong isolation between the vehicle platform and the rest of the driving intelligence. This isolation is beneficial from a number of perspectives. Firstly, from a legacy viewpoint, most automotive OEMs already have fairly sophisticated functionality in their existing vehicles for management of the vehicle motion. This includes features like (adaptive) cruise control, traction management, and brake management including possible regeneration (for hybrid and electric vehicle platforms), and in the case of heavy commercial trucks, additional features like control of multiple axles, external brake requests, and overall powertrain control. One of the most convenient ways to introduce autonomous driving functionality is to introduce an additional system (the cognitive driving intelligence) which generates the kind of operational setpoints that the various vehicle controllers are already setup to receive. These setpoints are typically in the form of commanded acceleration, velocity, and vehicle deceleration. Secondly, even if legacy is not a concern, isolating the vehicle platform enables a clean separation of concerns and system partitioning. The cognitive driving intelligence needs to generate desired vehicle motion in some world coordinate system. Thus, its concern is to specify the global motion parameters which the vehicle platform should fulfill. To do this, the cognitive driving intelligence requires only a minimum model (abstraction) of the vehicle dynamics and the vehicle platform configuration parameters. Consequently, the entity responsible for answering the question, "Where and how should the vehicle move in the next N units of time?" need not have an intimate knowledge of the various vehicle propulsion mechanisms and their continuous control. In turn, the vehicle platform is not required to have knowledge of how and why the motion requests are generated. Its responsibility is to fulfill the commanded motion requests while assuring the safety of the vehicle platform in terms of basic vehicle dynamics (limiting longitudinal and lateral accelerations, anti-lock braking, electronic stability control, etc.). This sort of encapsulation of functionality and separation of concerns reduce the cognitive complexity of the architecture and are recommended best practices in the field of systems architecting. Finally, the isolation also facilitates product and platform variability management. Especially in the domain of heavy commercial vehicles, it is quite common to find extreme variability in each manufactured vehicle, since it is specifically configured for each customer's needs. By separating the cognitive driving intelligence from the vehicle platform, it can be reused on different vehicle platforms.

The off-board layer, depending on its functionality, needs to tap into differing parts of the on-board architecture. In our experience, the maximum amount of data exchange occurs with the world model, since it holds practically all useful information needed by the off-board layer. This includes information regarding the current state of the on-board systems, the perceived external environment, as well as any upcoming motion decisions that may be in the execution pipeline. At the remote end, all received information is typically accumulated in a database, which in turn feeds application specific views of the gathered data. Active tele-operation [42] is foreseen in use cases where a fleet of autonomous vehicles is overseen by a command-and-control center. In such use cases, the vehicle may be able to "call home" when it gets stuck, or the remote center may actively claim control in potentially hazardous situations. In these cases, the tele-operation part of the system architecture needs to communicate with the decision and control part of the on-board systems. The commands sent are usually brief motion requests relative to the current location of the vehicle (in case the vehicle is stuck), or reprogrammed destinations. In our experience, the remote commands are directed at the cognitive intelligence and have relatively low bandwidth requirements. Direct control of the components in the vehicle platform requires significantly higher bandwidth and stricter timing constraints, which is rarely possible over large distances with existing wireless communication technologies.

13.5.2.1 Comparison with Similar Architectures

Comparisons of the reference architecture can be made with the architectures of Junior—Stanford's entry in the 2007 DARPA Urban Driving Challenge, the HAVE-IT project, and a Mercedes Benz autonomous car. These architectures are relevant because they represent a steady improvement of functionality and implementation, over the past decade. Junior is a successful example of a self-driving vehicle from the early days of the technology, and a largely academic proof-of-concept. The HAVE-IT project consortium had strong representations from OEMs and Tier 1 suppliers from the automotive domain, as well as independent research institutes and universities—the project focused on highly automated driving and advanced driver assistance systems. The Mercedes Benz autonomous car development had the automotive OEM Daimler AG as the majority stakeholder. The intent of the comparison is to highlight similarities and differences, rather than make claims of which architecture is "better." An architecture needs to be evaluated in its context, because the context imposes unique constraints with associated implications on the design. Thus, we choose to believe that every architecture that works has merits in its own context, and that there is rarely a definitively best solution to any given architectural problem.

Stanford University's DARPA Urban Challenge entry, Junior [59], provides an early example of an autonomous driving architecture. The interface to the VW Passat vehicle is via steering/throttle/brake controls, rather than direct longitudinal and lateral acceleration demands. This can probably be explained by the assumption that the autonomous driving architecture was designed exclusively for tight integration with one particular vehicle, which lacked general vehicle dynamics interfaces for setting acceleration and deceleration setpoints. The architecture is divided into five distinct parts for sensor interface, perception, navigation, user interface, and the vehicle interface. The localization is integrated into the perception part, and there seems to be no effort to classify detected obstacles. This architecture also explicitly includes a component/layer for "Global Services" dealing with functionality like file systems, and inter-process communication. We do not describe these services because they do not strictly fit into an architecture's functional view. The architecture is not strictly divided into layers, nor is there an explicit component to abstract the view of vehicle platform.

The layered approach to architectures and their description is also found in the European HAVE-IT project [12], which had its final demonstrations in June 2011. This project architecture consists of four layers: "Driver interface," "Perception," "Command," and "Execution." The Perception layer consists of environmental and vehicle sensors, and sensor data fusion. There is no mention of localization, perhaps because the system operates in close conjunction with a human driver. The Command layer contains a component named "Co-Pilot," which receives the sensor fusion data and generates a candidate trajectory. A "mode selection" component in the Command layer then switches between the human driver and the "Co-Pilot" as a source of the trajectory to be executed. The selected trajectory is then handed to the Execution layer in the form of a motion control vector. The Execution layer consists of the drivetrain control, which in turn controls the steering, brakes, engine, and gearbox. This execution layer corresponds closely to our vehicle platform layer in that it pertains to drivetrain control and "..to perform the safe motion control vector" [12]. Also similar is the usage of a motion control vector as an interface to the vehicle platform/execution layer. Our architecture additionally incorporates energy management as an explicit part of the decision and control component, which is especially valuable for electric and hybrid drivetrains, since then considerations of estimated range can be incorporated in the long term trajectory planning. The HAVE-IT architecture evolved in the context of Advanced Driver Assistance Systems (ADAS) with a strong reliance on the human driver and emphasis on driver state assessment components in the command layer; it remains unclear how well it can be adapted to L4 autonomous systems, where a human driver may not be present.

Close comparisons can be made with the architectural components of Bertha, the Mercedes Benz S-class vehicle, that recently (2014) completed a 103 mile autonomous drive from Mannheim to Pforzheim [80]. In the system overview presented in [80], components like perception, localization, motion planning, and trajectory control are clearly identified. These agree well with the components we have described in this paper, however, this is hardly surprising. Every autonomous driving system requires these functional components and they are likely to show up in practically every architecture for autonomous driving. The system overview in [80] does not explicitly acknowledge the existence of components for semantic understanding, world modeling, energy management, diagnostics and fault management, and platform stabilization. It is possible that some or all of these were present, but not mentioned. This is especially true for diagnostics and fault management. A part of semantic understanding related to classification of detected objects can (and often is) put in the Perception component, as in [80], but we see benefits in the explicit separation of sensor fusion and semantic understanding advocated by our architecture. The isolation of semantic understanding from raw sensor fusion enables faster and more independent iterations and testing of newer algorithms, without affecting the rest of the system. This is directly relevant to the engineering stakeholder concerns of independent development of individual subsystems, as well as their virtualized simulation and testing. Further, the raw object data from sensor fusion is still of value to the decision and control components, despite a lack of accompanying semantic understanding. This is because, although knowledge of whether a detected object is a pedestrian or motorcycle is useful for optimized path planning, collision with the object still needs to be avoided regardless of its classification. In a similar vein, incorporating a distinct component for world modeling enables incremental sophistication in internal representations while retaining (backwards compatible) interfaces. The existence of a distinct world model components makes it easier to answer questions like, "How will the world evolve if I perform action X instead of action Y?" Although [80] mentions the existence of a "Reactive Layer," it makes no mention of any other layers in the architecture and how the components are distributed across them. In our paper, we make a clear distinction between the vehicle platform and cognitive driving intelligence layers and provide a rationale for our proposed component distribution.

Comparison of these and a few other architectures with our proposed architecture leads us to believe that the explicit recognition of semantic understanding, world model, and vehicle platform abstraction components is unique to our architecture. This is not entirely coincidental, since our incorporation of these components is, to some extent, a deliberate action to resolve the short-comings we perceived during our early state of the art surveys. Furthermore, our architecture has been applied to a larger variety of vehicles (commercial trucks, passenger cars, as well as novel, legacy free designs) and therefore necessarily incorporates features related to greater isolation of functionality into distinct components and abstraction of vehicle interfaces. The aggressive partitioning of architectural components provides significant freedom to the component developers to modify and test new algorithms, without affecting the rest of the system. It also reduces the cognitive complexity of the system, and makes it relatively easy to foresee potential pitfalls and debug causes of objectionable behavior.

13.6 Systems Engineering

Systems engineering concerns cover the methodology and artifacts related to the engineering processes used throughout the development of the system. Ideally, these concerns commence with an investigation of the needs which the system intends to address, and then cover the entire development lifecycle until the system is deployed. Beyond development, systems engineering also looks at aspects related to the system's operation, maintenance, potential upgrades, and eventual decommissioning. In this chapter we restrict the scope to the development activities only. Within this restricted scope, we discuss the modeling steps and associated classes of models that are important for the development of autonomous driving systems. They are important not because the nature of autonomy directly necessitates an increased focus on systems engineering, but because autonomy requirements increase the complexity of the system being designed. The complexity, together with the safety critical nature of the system, requires careful attention to all aspects of the development processes, which is facilitated by systems engineering and its associated methodologies, tools, and artifacts. Thus, the topics covered in this section are *not exclusive* to autonomy, but they have a significantly increased importance within the context of autonomy.

In MBSE, the engineering processes are supported by an increasing set of models, some of which become increasingly detailed. To guide the systems engineering process, there exist a number of lifecycle development models and engineering methodologies. Most lifecycle models are grounded in one of three seminal models [38]: The Waterfall model [65], the Spiral model [23], and the V-model [43, 44]. Of these, the V-model and its variations have been extensively applied to systems engineering and development. These lifecycle models are leveraged by various MBSE methodologies, such as OOSEM, RUP-SE, Harmony-SE, and JPL State Analysis. An overview of these prominent methodologies is given in [38], which includes further information, including additional references for each methodology. More methodologies have since been identified and they are gathered and described online at the MBSE Wiki Page [13]. Beyond these relatively general methodologies, there are also approaches to specific parts of systems engineering that are more focused on embedded and automotive systems. These deal with topics like requirements engineering [27], formal semantics [31], multi-view modeling [30], etc. Among other things, the methodologies define a number of development activities. Example activities are the definitions of stakeholder needs, system requirements, logical architecture, and system validation and verification. In some methodologies, the execution of these activities may be referred to as "phases." Each methodology involves a slightly different grouping and sequencing of activities and phases. The MBSE methodologies may or may not recommend a specific language or tool framework, however, they all involve the creation of models or model views supporting each engineering phase. In this section, we propose four classes of models, the contents of which are independent of methodology, but which can be mapped to different activities within a selected MBSE methodology. Similarly, a project specific lightweight MBSE methodology may be created by selecting specific models from each class and specifying the sequence(s) of their creation using particular tools. Thinking in terms of model classes is valuable because classes easily encapsulate a diversity of modeling formalisms and tools. The diversity is necessitated, at the very least, due to the state of practice in modeling technologies—at the moment, there exists no comprehensive modeling formalism that can completely capture all relevant systems engineering stakeholder concerns. Therefore, the various concerns need to be captured using heterogeneous sets of possibly domain specific models.

Establishing and maintaining links between all the assorted models is an active area of research [74], but in the meantime, project specific inter-model links need to be selected manually by the practicing systems engineer, and these also need to be manually maintained by the engineering team. The links represent relations like refinement, allocations, correspondence, etc., between the model elements. An active risk to be guarded against is that in fast moving projects systems engineering models are assigned secondary importance. Therefore, either models are not created, or the models and links are not updated as the implementation changes. Over time, there is significant divergence between the implementations and the models. This leads to accumulation of systems engineering "debt." The debt accumulation can be partially mitigated by via two interrelated choices: the methodology and the tools used for the systems engineering process. Flexibility in the methodology and automation support from the tooling reduce the "cost" or effort of model creation and maintenance, making it more likely that the engineering team views systems engineering as less of a "burden" and more of a benefit.

Below, we discuss some key steps in systems modeling, sequenced as they would occur in an ideal development processes. Realistically, it is very important that the methodology and associated tools permit the engineering team to begin with modeling multiple, arbitrary concerns, which can later be extended and linked to other models in all directions. Thus it should be possible to start modeling the logical architecture of the system, or a software component, without having explicitly modeled the preceding requirements, actors, and their interactions. Such knowledge may be implicitly known/assumed by an engineer who is itching to sketch out an architectural solution—forcing her/him to explicitly model requirements and behavior first merely results in annoyance and rebellion.

The main systems engineering model classes that support the steps we describe are shown in Fig. 13.5. It should be noted that there may exist a variety of different models within each model class. The first step is to analyze and represent the system's users, their operational needs, and the usage scenarios. Who are the actors involved in interacting with the system? Actors can be human users as well as other entities involved in autonomous driving like other road users, road infrastructure like traffic lights, etc. Once the actors are defined, the next task is to determine what they want the system to do. What behavior and services do they expect from the system? What are the contexts within which the actors will interact with the system? What will be the modes of interaction, and how are they related to the operational contexts? Depending on the modeling languages and tools used, there

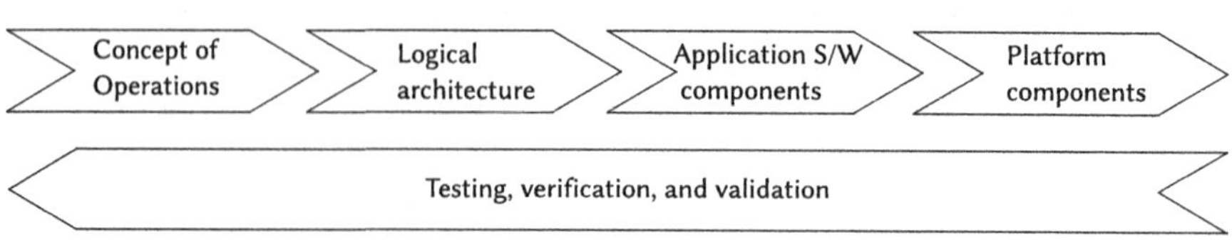

Fig. 13.5 Systems engineering model classes

can be a variety of diagrams, process definitions, allocation of actors to process artifacts, etc., for capturing and representing the ideas at this phase. This phase helps to identify the principal behavioral requirements expected from the system and its usage by the actors involved.

The second step is to identify the main system components, their contents, relationships, hierarchies, properties, and behavior. This constitutes the logical or functional architecture of the system and it excludes concerns related to technological implementation of the identified components. However, it does take into consideration all major extra-functional constraints (like safety, security, performance, reliability, etc.) so as to find a suitable compromise between them (to the extent this is possible without getting into implementation specific detail). The logical architecture defines the components and their interfaces, including formalization of all logical views and how these views are accounted for in the component designs. The behavior requirements from the previous step are refined and allocated to the identified architectural components. Links between requirements, operational scenarios, actors, and components are also established. The entire process of logical architecture can be applied repeatedly within the boundaries of a single logical component or its sub-components. Thus, the logical architecture may have multiple and differing "levels of zoom" for each of the components and their connections. Depending on the tools used, the logical architecture and its properties may be represented using a multitude of models, modeling tools and languages, or with a unified, all encompassing model, within which various related concerns are represented using specific views. Given a logical architecture including behavioral views, one could use tools for static analysis (e.g., checking interface compatibility) and perform behavioral analyses. These may include simulation and model checking to assure specific system properties like absence of deadlocks, reachability of specific states, etc. The specific analysis techniques of interest may very well dictate the choice of architecture representation languages. For example, it is not uncommon that a modeled architecture or its sub-component needs to be remodeled using a different language, just so that it can be verified by using a particular model checker.

The third step is the modeling and grouping of application software components within the system. The grouping may be in terms of individual software applications, or for the purpose of simulation in tools like Simulink. These models include representations of the behavior of the software components, their resource requirements, runtime characteristics, interfaces, properties and attributes, and communication specifications. At this stage it is increasingly common to see the usage of the so-called executable software models. Executable software models are those which can directly be transformed into compilable source code, which guarantees that when the source code is executed under assumed conditions, it should have the same behavior as the modeled software component. Dependencies on expected platform capabilities are usually explicitly mentioned for each software component model.

The final step is the modeling of the platforms on which the application software components execute. The platforms consist of a "stack" incorporating the computing silicon (micro-processors/controllers), an optional operating system which may or may not provide realtime guarantees or alternatively, a native language runtime for the chosen silicon, an optional middleware that abstracts the operating system and its services, and any libraries, daemons, and other services provided by the operating system and hardware. Multiple application software components may execute on the same or similar platforms, and in cases of advanced experimental architectures, application software components may migrate between similar platforms. In all cases, it is more convenient and useful to model a platform and its services as a whole, rather than embed this information into each application software component. However, we emphasize that this is largely a matter of preference.

Associated with all the four modeling steps is a continuous refinement of requirements, test cases, safety viewpoints, and documentation artifacts. A comprehensive taxonomy of models associated with each step is beyond the scope of this chapter, but each step must introduce additional models representing requirements and test cases relevant to that step. Requirements must be allocated to models that assure them, and both requirements and test cases need to be assigned to unique members of the engineering team. Safety considerations are usually generated by following processes established by safety standards like ISO26262. Additional models/views like functional safety architecture will be introduced, along with their refinements to technical safety architectures and associated redundancies and switching modes for application software and platform components. The number of models may grow and shrink at each step as the systems engineering process is iteratively applied during system development. Ultimately, the artifacts ideally exist as a web of interconnected models at each step and across the steps. Preserving the links between the models, and keeping the models up to date with the implementation is a significant challenge. One way to do this is via increasing toolchain automation, but given the relative lack of production ready tools, it becomes the responsibility of the architecting and systems engineering team to select and minimize the number of models used to represent the system. This in turn depends on a variety of technical and non-technical factors like the nature of the project and its maturity level, available tools, the importance attached to systems engineering by project management, the skills and qualifications of the people involved, etc. One recommendation based on our experience is to always synchronize the software and platform models with the actual technologies being utilized in the project. These will change as the project moves through stages of proofs-of-concept, prototyping, to certifiable implementations. The models need to be updated correspondingly. For example, during the early prototyping phase, if the entire application software is modeled in Simulink and the execution platform is a rapid prototyping system like the dSpace MicroAutoBox, it makes little sense to model the system in terms of software threads, tasks, operating system runtimes, etc., because the modeling concepts and technical implementation concepts just do not synchronize in terms of relevance. Here, it is better to restrict the models to Simulink and coarse dSpace specifications. When an eventual move is made to AUTOSAR or similar infrastructure, the associated models dealing with finer platform details may be created.

One of the conflicts that MBSE can help resolve is the need for fast development cycles, while guaranteeing consistency of the various process and product artifacts. In our opinion, this necessarily requires the support of advanced tooling and as such, the details are highly tool specific. But the general underlying patterns involve inter-model links, co-simulations, and model checkers. The inter-model links typically represent relationships between the linked models, for example, B "realizes" A or Q "is derived from" P, etc. An example of this is a facility offered by several commercial toolsuits wherein requirements can be linked to specific blocks in a Simulink diagram. When either of the models changes (the requirement or the Simulink block), a flag is raised to signify that the status of the link requires investigation. Depending on the tool support, additional data on what has changed may also be presented. Even a simple facility such as this helps to prevent undetected changes from propagating through the design. Prior to a release, the MBSE process may require that no inter-model links have unaddressed "change flags." Similarly, with the advent of technologies like Open Services for Lifecycle Collaboration (OSLC) [7], the concept of "round-trip" flows is gaining ground. In a round-trip, a model created using some particular modeling technology can be "exported" to a different tool, where it is enriched or analyzed, and the results can be sent back to the original tool. Techniques like OSLC or Functional Mockup Interface (FMI) [5] enable model exchange and co-simulation of heterogeneous models. Such co-simulations can be used to simulate an entire vehicle, comprised of different types of linked models. Thus, it becomes possible to easily examine the impact of changes deep within one particular model, on the overall vehicle behavior. For individual models, the usage of model checkers can help assure that desired properties are retained (for example, absence of deadlocks and unreachable states) after model modification.

Based on the modeling steps described, and the classes shown in Fig. 13.5, a partial view of modeling artifacts and their allocation links is shown in Fig. 13.6, where the rightmost column describes some of the functionality provided by the models. The dotted curved arrows crossing the vertical layers represent allocation links. Thus arrows between the architecture representation and application software components show the mapping between logical architecture elements and particular application software components. For the sake of clarity, the continuous refinement and accumulation of requirements and tests are not shown in the diagram. Examples for specification, structuring, and refinement of requirements can be found in [78, 79]. The tests description must include the tools and methods of conducting the tests, templates for data logging during testing, component behaviors, and data values/ranges that constitute an acceptable test result. Figure 13.6 represents a concise takeaway of the contents of this section on systems engineering.

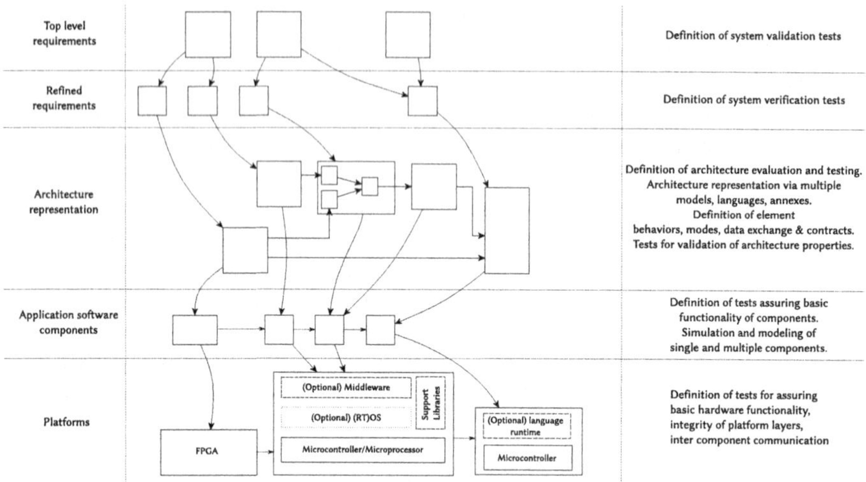

Fig. 13.6 Partial view of modeling artifacts and allocation links

13.7 Technical Implementation

In this section, we briefly discuss some selected technologies that have proven useful in our experiments for implementing autonomous driving systems. The emphasis is on the early prototyping phase, since that is usually when it is possible to experiment with novel technologies in a low risk manner. Similar to Sect. 13.6, the topics and technologies covered in this section are *not exclusive* to autonomy, but they have a magnified importance in the context of the development of autonomous driving systems, which makes them worthy for consideration.

Since the last 15 years, there has been a proliferation of various Architecture Description Languages (ADLs) in academia and industry. A systematic overview is provided in [47], which discusses 102 ADLs with 33 from Industry and 69 in Academia. Each has its own specific meta-model, notations, tools, and domain applicability, with little chance of interoperability between them. Moreover, the ADLs themselves have often undergone complete changes or remodeling between versions. Capturing the vast variety of stakeholder concerns within a single notation is exceedingly impractical, as is the aim of creating a "universal notation." Consequently, domain specific ADLs have emerged, which focus on the properties of a particular domain, and specific types of analyses and modeling environments [53]. When different concerns need to be modeled in differing languages, the individual models need to be synchronized, such that changes in one model are propagated and reflected in the others. There are efforts for tool based and automated synchronization but practically, this is still a manual process. Thus, domain specific ADLs need to strike a sweet spot where they are expressive enough to model all relevant stakeholder concerns in the domain, while minimizing syntax and remaining usable. This needs to be complemented with excellent tools, preferably those which also enable bi-directional synchronization not just between different models, but also between models and executable implementations. This is a steep challenge. We have identified three candidates that are broad and deep enough to address a non-trivial number (but not all!) of stakeholder concerns, in the field of automated driving: EAST-ADL2 [36], AADL [39], and ARCADIA [9]. Of these three, AADL and ARCADIA are generally applicable within the domain of embedded systems architecture, while EAST-ADL2 is specifically intended for automotive systems. All three enable the representation of requirements, behavior, structure, mapping to technical implementations as well as traceability links among them. Both EAST-ADL2 and AADL have reference tooling to import Simulink models to form architecture representations. This is important, because Simulink is the de facto simulation and modeling tool in the automotive industry. EAST-ADL2 is supported by a language specific core methodology [1] which can be complemented by additional extensions related to requirements traceability [18], formal analysis and verification [49], etc. Similarly, the ARCADIA method is embedded strongly in the recently open-sourced tool Capella [11]. In our opinion, Capella is the most mature, comprehensive, and user-friendly open-source tool available for MBSE at the time of this writing. It is built on the Eclipse ecosystem, and it been in use internally at the Thales company for over 5 years. The tool guides the user through a series of top-down modeling activities defined by the ARCADIA method. It also offers a series of default viewpoints for modeling, and within each viewpoint modeling can be done by means of a number of diagram types. Definition of custom viewpoints is also possible within Capella. Note that although all these tools have the word "architecture" in their names, their actual functionality goes beyond traditional architecture and into the various systems engineering modeling activities we have covered.

An important consideration in the selection of tools and languages used for systems engineering and other modeling is the support they have for the platforms and programming languages used in the system implementation. Automotive domain specific languages like EAST-ADL2 have built-in support for AUTOSAR concepts, enabling an easier mapping from the models to their implementation platforms.

For technical implementation of autonomous driving systems, we propose a setup represented in Fig. 13.7. The setup consists of aggregations of components, connected by a publish-subscribe bus with Quality of Service (QoS) filters. The aggregations are represented by the boxes in Fig. 13.7 and agree well with the architectural layers presented in Sect. 13.5. The overall intent is to support development of functionality in simulations and on target implementations, with the ability for seamless transitions. Even when certain functionality has been implemented on target hardware, it is still valuable to switch back-and-forth between simulations and propagate changes in both directions. The setup consists of a physical platform (a.k.a drivetrain) along with a simulation of it running on a general purpose computer (the "soft" simulation) and a version running on a hardware-in-the-loop (HIL) rig. Similarly, the various drivetrain controllers exist as soft simulations and implemented on target hardware. The cognitive driving intelligence layer is aggregated in an implementation, in addition to a synthetic 3D environment setup,

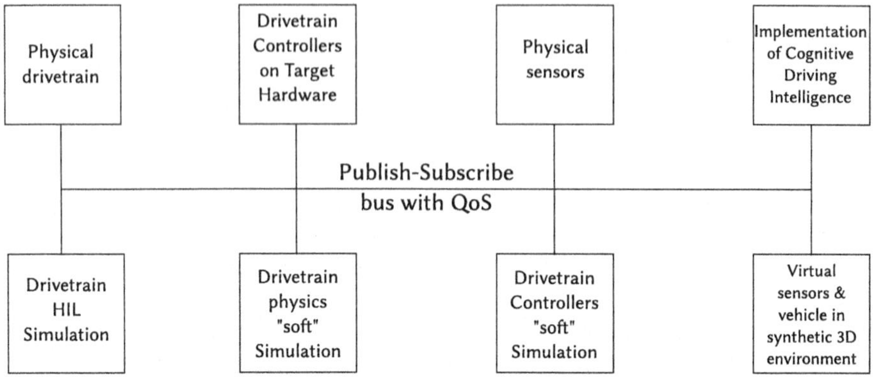

Fig. 13.7 Proposed implementation setup for autonomous driving

for example, within a 3D gaming engine like Unreal [16]. Within the synthetic environment, techniques like ray-tracing are used to emulate radar, lidar, and camera sensors. The environment incorporates physics engines for calculations of solid body dynamics, collisions, etc. The setup allows for combinations like driving the real, physical vehicle in a real environment, driving a virtual vehicle in virtual environment using the real, implemented driving intelligence, testing the implemented drivetrain controllers on a simulated drivetrain, etc. Of course, depending on the constraints and resources of a particular project, not all aggregations need to be present.

The technical implementation platforms for the cognitive driving intelligence and the virtual environments are typically (during prototyping) commercial off-the-shelf (COTS), general purpose, multi-core computers with best-of-breed COTS graphics processing units (GPUs). GPU vendors also provide a variety of graphics and parallel processing libraries to leverage the capabilities of their platforms. The computers usually execute the Linux operating system, with middleware like ROS [64] or OROCOS [32]. Of these, ROS is more popular, with a large, open-source community. It provides facilities for computation and communication across processes running on the same and different computers. ROS does not support hard realtime operation, but for prototyping this is usually not a limitation, provided the hardware has sufficiently fast processors and enough memory. Hard realtime requirements are usually more critical for the drivetrain control, compared to the cognitive driving intelligence. However, Linux does provide some hardened timing guarantees with the help of the `PREEMPT_RT` kernel patch [15] or dual kernel approaches like Xenomai [17]. The OROCOS middleware provides a component framework that can leverage the hardened realtime properties and it also supports inter-component communication on the same and different computers. The programming language of choice for the cognitive driving intelligence is usually C++, a choice often mandated by the used middleware and useful libraries[2] like OpenCV [26] and the Point Cloud Library (PCL) [66]. The usage of Java for high level tasks is not uncommon either, although there is a lingering perception that it suffers from performance problems in comparison with C++. A modern alternative to Java is the Scala programming language [61], which in our opinion, deserves greater interest. Scala allows for strict functional programming, object-oriented procedural programming, as well as a free combination of the two. It runs on top of the Java Virtual Machine (JVM) enabling it to leverage existing Java libraries. The functional programming paradigm emphasizes stateless, side-effect free functions as building blocks, which leads to greater ease of side-effect free composition of functionality.

The "soft simulations" of the drivetrain and its controllers are usually made in Simulink, typically running on the Windows operating system on general purpose COTS computers. The reason for choosing Windows here is that these simulations are often executed on rapid prototyping systems like the Simulink RealTime and dSpace autobox, the tooling for which is often Windows only.

There is a greater variety of platforms for the drivetrain controllers, where constraints of realtime, scheduling, input–output, and computing resources are more prominent. Relevant operating system standards here are OSEK/VDX [14], parts of which are standardized in ISO 17356, and AUTOSAR [10], which reuses large parts of OSEK. There are a large and diverse number of vendors supplying customized realtime operating systems, and middleware stacks with varying capabilities. The programming languages of choice are usually C or C++, where large portions of the code are autogenerated from tools like Simulink. For both languages, safety critical subsets are defined by the Motor Industry Software Reliability Association (MISRA), in the form of approved language usage rules and there are tools to generate and verify C/C++ code against the MISRA rules. In our recent projects, the usage of the Ada 2012 programming language is receiving greater attention. This is partially because of its reputation as the programming language of choice in other safety critical domains, as well as its syntactic support for notions of contract based programming in the 2012 version. The latest incarnation of the safety critical subset of Ada, SPARK 2014, also deserves rising interest, in our opinion, not least because of its accompanying tooling for code analysis, coverage and testing. Ada provides a number of runtime profiles, including a multi-tasking profile named Ravenscar [33]. The language then provides built-in support for tasking and inter-task communication, often negating the need for an explicit operating system on embedded microcontrollers. The dominant embedded microcontrollers in this domain are based on the ARM platform, which provides both uni- and multi-core processors in 32 and 64 bit configurations. For drivetrain controllers, the matching of the hardware platform with the operating system (OS), and the OS support for low level hardware and peripheral drivers is a more important consideration than

[2]Although some libraries do provide bindings for other languages, in our experience C++ still dominates the scene.

for the platforms used for the cognitive driving intelligence, since there is less standardization in this area.

Communication middleware has not made significant inroads in the automotive domain, beyond the functionality provided by AUTOSAR with its Virtual Function Bus (VFB) concept. At a lower level, the CAN bus remains the most common technology for connecting distributed vehicle controllers. Autonomous driving systems impose greater bandwidth requirements on the in-vehicle communication links, which is especially true when high resolution sensors like cameras and multi-beam lidars are utilized. The usage of gigabit Ethernet is exceedingly common in autonomous driving projects, especially for connecting the computers executing the cognitive driving intelligence. Considering that these computers use general purpose operating systems and programming libraries, they are able to utilize advanced communication middleware, beyond bare TCP/IP or UDP/IP socket communication in client–server scenarios. Three facilities provided by advanced communication middleware are: automatic de/serialization of data structures into wire representation, high level functions for sending and receiving the data using fine-grained QoS guarantees, and automatic routing and service discovery within the network. The middleware uses either the brokered or broker-less architectures and often uses uniform APIs regardless of whether the communication is between threads in a single process, between multiple processes on the same computer or across computers. It also makes available smarter communication patterns like publish-subscribe, push-pull, N-to-M, fan-in, fan-out, etc. A relatively recent communication middleware standard becoming increasingly common for critical, distributed realtime systems is the Object Management Group's Data Distribution Service (OMG-DDS) [62]. DDS supports message structure definition using an interface definition language (IDL) similar to CORBA, and thenceforth manages all aspects of data transmission and wire representation with a large number of QoS settings, using the publish-subscribe pattern. Another efficient communication middleware is ZeroMQ [4], which provides no built-in support for data de/serialization (it transmits given "binary blobs"), but supports substantially more patterns than just publish-subscribe. ZeroMQ also provides bindings for over 40 programming languages.

13.8 Discussion

In this chapter, we have touched upon some relevant topics on the nature of autonomy, architectures, and systems engineering aspects for autonomous driving, as well as implementation technologies for prototype systems. This section presents a synthetic discussion highlighting how these areas affect each other, as well as the how autonomy influences each of them.

13.8.1 A Holistic View

A holistic view, relevant for both long-term perspectives and contemporary engineering, is shown in Fig. 13.8. The figure shows that each of Concepts, Architecture, Systems Engineering, and Technical Implementation can be divided into two parts. One part is for the relatively low level concepts of contemporary engineering, and the other is for the higher level concepts arising from theories of general autonomous and intelligent systems. The former part is shown in the inner circle of Fig. 13.8, while the latter part is shown in the outer circle. Thus, the basic functional components for the architecture, as described in Sect. 13.5.1, and their interconnections form part of the inner architecture circle. The higher level concepts like runtime system level reasoning, which would be useful to have, but are currently not included in most contemporary architectures, are shown in the outer architecture circle. Figure 13.8 also shows that the concepts and systems engineering areas drive developments of architectures and technical implementations, for both high and low level concepts. The state of the art in autonomous driving still lies mostly within the inner circle, with very brief and occasional excursions to the higher levels.

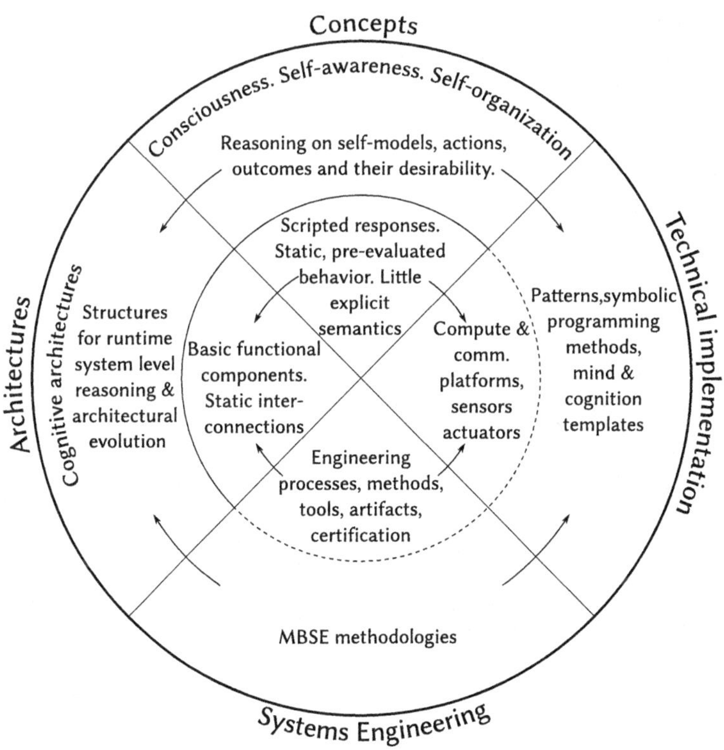

Fig. 13.8 A holistic view of the concepts covered in this chapter

The construction of autonomous driving systems today very much follows the Constructionist Design Methodology (CDM) which involves integration of a large number of functionalities that must be carefully coordinated to achieve coherent system behavior. For autonomous driving, the coordination is manual, and notions of coherency are mostly implicit in the heads of the designers. This approach limits how well systems scale with rising system complexity. These limits are already acknowledged in research on AI systems integration [70], which is exploring a fundamental shift from manually designed to self-organizing architectures that can learn and grow [69]—the so-called constructivist approach. However, given the safety, reliability, determinism, and realtime requirements of autonomous driving, it is not feasible to abandon the constructionist approach entirely. Therefore, it is necessary to reason on the gap between the two approaches to determine the direction in which autonomous driving architectures need to evolve. One way to explore the gap is to think in terms of "missing components" and how they can be injected into the existing architecture patters without breaking them. A key missing component is an explicit representation of a "Self" or "Ego" within the vehicle, along with a "Goals" module. Today, the automobile architecture is made up of a variety of basic subsystems like the engine, the transmission, the brakes, steering, etc., and some hierarchically higher subsystems related to traction control, traffic jam assist, and so on. These systems have limited knowledge of the existence, purpose, and functioning of the other subsystems (following the excellent architectural principal of "separation of concerns") and consequently, in the presence of severe uncertainty or disturbances within the operational environment or due to internal failures, there is no particular subsystem that can perform system-wide reasoning, revise previously held beliefs about the system capabilities and its environment, and act/adapt accordingly. When the operational situation exceeds the limits envisioned by the system's human designers, or when the carefully scripted responses fail, the system as a whole also fails. The presence of explicit "Ego" subsystems could be a first step in addressing this problem. The "Ego" would be aware of the primary behaviors the vehicle is expected to fulfill, as well as the various subsystems present and how they work together to generate the desired behavior. The architecture of the "Ego" is an interesting topic in itself: it could range from a "thick central" ego component, to a hierarchical network of ego components in each subsystems. The self-awareness would be continuously represented and maintained in the form of a self-model. The set of self-models can be "seeded" at the time of the architecture deployment and the "Ego" would be free to evolve them and synthesize other models throughout vehicle operation. Preliminary methods for automatic synthesis of multiple internal models already exist (see, for example, [24]) and have been proven to successfully increase machine resilience [25]. This directly contributes to characteristics like fail-operational behavior and survivability, which are desirable in autonomous driving. Initially, the operation of the Ego subsystem would be limited to (de)activation of particular subsystems, dynamic reconfigurations of the connections between the systems, transfer of functional responsibility between the subsystems, and triggering specific modes in individual subsystems. With today's technological means, all permissible system configurations need to be proven for coherence, consistency, and safety in advance. The eventual transition to constructivist architectures creates the possibility of runtime reasoning in the Ego components, leading to runtime verification of the desired properties. The AI domain is already reporting some progress towards agents, design methodologies, and programming paradigms following constructivist approaches [35, 60, 63]. We feel that this is a good time to engage with researchers in the domain of AI systems integration, in order to highlight the requirements relevant to practical embedded systems. This is intended to prevent a repetition of the problem seen in the area of cognitive architectures, wherein real world concerns are rarely accounted for in theoretical development [72].

As noted in the outer circle of Fig. 13.8, there have been efforts to create higher level concepts like consciousness and cognitive architectures, that mimic some aspects of human decision making. However, corresponding developments in systems engineering are conspicuously missing. We anticipate that an expansion to the scope of existing systems engineering to constructivist architectures will be adequate, without needing a revolution of the underlying methodologies. However, particularly strong challenges still exist within the area of MBSE. These challenges are in two general directions: maintaining consistency between various models/views, as well as between the models and implemented artifacts. The usage of constructivist AI designs in upcoming architectures will also impact the testing, verification, and validation aspects of the systems engineering processes. This is because, if the system behavior is partly synthesized at runtime, and could take forms not anticipated during design, it would be challenging to find testing and verification methods that demonstrate correctness of behavior prior to deployment.

In our experience, the availability of suitable implementation platforms and technologies is not a strong limitation in autonomous driving, with a possible exception being in the area of sensors for perception and localization. The capabilities of state of the art silicon, programming tools, communication technologies are adequate to implement the elements found in existing and proposed autonomous driving architectures. This is not to say that better tools for analyzing code, model checking, translation of models to executables, etc., are not desired. All of this holds true within the constructionist approach. The shift to a constructivist approach, however, introduces paradigm shifts in implementation, notably for capabilities of programming languages. The programming of autonomy requires language constructs supporting autonomous knowledge acquisition, realtime and any-time control, reflectivity, learning, and massive parallelization [60]. Such constructs are missing from the most common programming languages currently employed in the development of safety critical embedded systems.

The four steps for systems engineering proposed in Sect. 13.6 are complementary to the systems engineering processes based on the V-model, or those recommended by ISO26262. The former chiefly recommends activities, the latter introduces views and requirements to analyze and assure functional safety. Models, views, and other artifacts underlie both. In our experience, systems engineering and associated modeling are usually given secondary importance during the prototyping phase. Once a functional prototype is available, the focus shifts to writing extensive requirements and acceptance criteria, which are then off-loaded to vendors and suppliers for process adherent development. The usage of early lightweight modeling with subsequent refinement and additional models/views has the potential to keep development consistent not only during the prototyping phase, but also for making a smoother transition to high maturity implementations.

13.8.2 The Influence of Autonomy

For autonomous driving, the **impact of autonomy on architecture** is mostly related to the need for specific functional components. These components are responsible for perception of the external world and its internal representation, localization of the vehicle with respect to some coordinate system, finding collision free trajectories between detected obstacles while staying on a drivable surface, and reacting to unexpected external and internal events with the intention of reaching a safer state. Also present are more "traditional" components governing basic motion of the vehicle viz. lateral and longitudinal accelerations and deceleration. Finally, overall energy management of the entire vehicle is also a concern. Beyond functional components, the architecture is driven by needs of safety, survivability, fault-tolerant operation, and cost. This includes aspects of redundancy management, systematic degradation of available functionality in the presence of faults, movement of software functionality between similar computational platforms, and dynamic changes to inter-component data-flows depending on context and failure conditions. Among these, while redundancy has been a well-established pattern in the architecture of safety critical systems, the rest are not yet established. For example, it is common in AUTOSAR to statically specify inter-component data-flows and their contents. In case an architecture needs to modify the data-flow routing and content at runtime, all the possible combinations need to be determined and statically specified, along with the conditions under which each combination will occur. Then, all the possible combinations and their transitions would need be proven safe, to meet certifiability and standards requirements. This is an excellent example of the "careful, manually crafted" characteristic of the constructionist approach. It is also the approach taken by the next generation Integrated Modular Avionics (IMA-2G) architectures, where an architecture reconfiguration is intended to be triggered by a component referred to as the "reconfiguration supervisor" [22]. The approach of prior static definition of all possible combinations is easier to certify, compared to the case where the reconfiguration supervisor contains algorithms to determine a configuration's safety at runtime, and uses this to develop potentially new configurations at runtime, which were not statically evaluated during the system design phase. This latter approach of evaluating circumstances and generating viable new solutions "on-the-fly" is a characteristic of human level reasoning, emulated by experimental constructivist AI, but is still a far cry from being implemented in domains like autonomous driving.

Given the rapid development of autonomous driving technologies and ever shortening time-to-market needs, a key challenge for automotive architects is to incorporate newer learning and keep the architecture relevant for the expected lifetime of the vehicle. Continuous deployment has hitherto not been a great concern in the automotive domain, but autonomy is pushing the envelope in this regard. A big obstacle to continuous deployment is the verification of any proposed changes, to assure system level safety properties are retained. To some extent, accelerated testing via virtualization techniques is one mitigating solution, but techniques for correctness-by-construction via contract based design, modularization, and composability are needed to first reduce the amount of testing and verification required for any given change.

For autonomous driving, we see the **impact of autonomy on systems engineering** to be mostly in the areas of testing, verification, and validation, and corresponding requirements management. This is driven by three characteristics of autonomous driving architectures: Newer types of perception sensors, growing system state space, and complexity of expected operational scenarios. Perception sensors like cameras, lidars, radars, etc., each have multiple failure modes under various operating conditions. Complex, probabilistic sensor fusion guards against full perception failure to some extent, but in turn leads to a more complex assurance process and surprising common points of failure. For example, in one of our projects, similarity in filter time constants between pre-processing algorithms of two different types of sensors was questioned as a potential common failure mode. Such information is sometimes not even available to the systems integrator. The growth of system state space and complexity of operational scenarios implies that capturing requirements comprehensively is an additional challenge. Traditional testing and verification methods cannot yield sufficient test coverage within reasonable timeframes. One solution is accelerated testing by means of 3D virtual worlds and scripted unit tests. For example, the evaluation of a new trajectory planning algorithm was conducted in one of our industrial projects by letting the algorithm drive a virtual vehicle in a synthetic 3D world. An indication of usefulness of this technique is that approximately a thousand tests were executed in parallel on a GPU based computing cluster, within a few hours. Each test executed a different scenario from a scenario library (for example, traffic intersections, unexpected pedestrians running in front of the vehicle, etc.) and metrics were gathered for each test. The metrics utilize parameters like resulting minimum distance to obstacles, accelerations within the cabin, etc., and help in rapid evaluation of the planning algorithm. To the best of our knowledge, such testing infrastructure is not common in traditional automotive OEMs. Beyond testing and verification, we see little immediate influence of autonomy on systems engineering processes, although autonomy may demand stricter discipline in their execution. For example, we see comparatively lower influence to the process of gathering, analyzing, structuring, and documenting requirements, beyond a rigorous enforcement of the process itself. Autonomous driving may, however, give an added push to development of Intelligent Transport Systems (ITS) and connected systems-of-systems, increasing the number of stakeholders in the systems engineering process.

The **impact of autonomy on technical implementation** tools for autonomous driving is principally the introduction of technologies and platforms from other domains to the automotive industry. This introduction would, in turn, drive various ways to make the technologies and platforms more robust, in line with demands of the automotive domain (e.g., greater emphasis on verification, diagnosis, error handling, etc.). The introduction of Linux in the form of Android is already occurring in the area of automotive infotainment, but its use for propulsion guidance, navigation, and control purposes is a relatively new phenomenon. The Automotive Grade Linux (AGL) [2] is a Linux Foundation workgroup with over 50 members including OEMs, Tier 1s, and system integrators. Although Linux may not be the final implementation of choice for safety critical ADAS functions, it is the de facto platform for prototyping, and also making inroads into solutions for HMI and telematics, which are crucial support functions for automated driving. The usage of silicon (GPUs), programming libraries, and 3D engines from the computer gaming industry is instrumental for accelerated testing and verification tooling. Computation middleware like OROCOS and ROS from the robotics domains, and communication middleware like DDS and ZeroMQ used in distributed information systems are also useful implementation technologies for autonomous driving. These middleware enable concrete component based software engineering techniques, as well as smarter communication patterns like push-pull, router-dealer, N-to-M, fan-in, and fan-out. This in turn introduces more options for technical architecture implementations.

13.8.3 Concluding Remarks and Future Work

This chapter attempted to provide an overview of the architecture and systems engineering for autonomous driving, aimed towards the ambitious practitioner. We started by placing autonomous driving within the greater context of general purpose intelligent, autonomous systems, and highlighted some philosophical and practical gaps between two in terms of approaches and architectures. Despite the fact that the automotive industry practices a bottom-up approach to the engineering of autonomous driving systems, we believe it is valuable for practitioners to get ideas from the theories of Artificial Intelligence, Theory of Mind, Machine Consciousness, and Self-awareness.

We then presented a functional architecture for autonomous driving, by introducing the key functional components and a way of connecting them together in an example architecture. Supporting the architecting efforts are practices for MBSE. We introduced four modeling steps to aid the systems engineering process, each of which needs to be supported by requirements and test cases for verification and validation. Finally, we touched upon some key technologies used to prototype autonomous driving systems.

Future work lies along two dimensions: scientific and documentary. The scientific part needs to elaborate on the theories and algorithms for adapting concepts from different domains to autonomous driving, keeping in mind concerns of safety, certifiability, and engineering processes. The documentary part should add to our ambition for eventually creating a "Handbook for Autonomous Driving." This then needs to provide greater and in-depth treatment of model taxonomies, listings of architectural and technological options, as well as guidance for selection and application.

The engineering of Human Machine Interfaces was left unaddressed in this chapter. From an engineering viewpoint, the functionality offered by the system is somewhat distinct from the way it interacts with its users (HMI). But from the viewpoint of the system's users, the HMI **is** the functionality, making HMI issues of critical importance to autonomous driving. The topics of design space exploration and extra-functional metrics for architecture evaluation, like cost, reliability, performance, flexibility also deserve an in-depth treatment.

Finally, the importance of standards and certification of autonomous driving systems cannot be underestimated. Existing functional safety standards like ISO26262, as well as industry specific norms like MISRA C, AUTOSAR, etc., need to be re-evaluated and upgraded to meet the requirements of autonomous driving.

References

1. Methodology Guidelines When Using EAST-ADL2. Public deliverable 5.1.1 of the ATESST2 project (2010), http://www.atesst.org/home/liblocal/docs/ATESST2_Deliverable_D5.1.1_V1.1.pdf
2. Automotive Grade Linux (2015), https://www.automotivelinux.org/
3. Biologically Inspired Cognitive Architectures (BICA) Society. The MAPPED Repository (2015), http://bicasociety.org/mapped/
4. Distributed Messaging with Zero MQ (2015), http://zeromq.org/
5. FMI: Functional Mock-up Interface (2015), https://www.fmi-standard.org/
6. IEEE Spectrum. How Google's Autonomous Car Passed the First U.S. State Self-Driving Test (2015), http://spectrum.ieee.org/transportation/advanced-cars/how-googles-autonomous-car-passed-the-first-us-state-selfdriving-test
7. OSLC: Open Services for Lifecycle Collaboration (2015), http://open-services.net/
8. Oxford Dictionaries (2015), http://www.oxforddictionaries.com/definition/english/consciousness
9. The ARCADIA MBSE method for systems, hardware and software architectural design (2015), https://www.polarsys.org/capella/arcadia.html
10. The AUTOSAR Platform (2015), http://www.autosar.org/
11. The Capella Graphical Modeling Workbench (2015), https://www.polarsys.org/capella/index.html
12. The HAVE-it EU project. Deliverable D12.1 Architecture document (2015), http://haveit-eu.org/LH2Uploads/ItemsContent/24/HAVEit_212154_D12.1_Public.pdf
13. The OMG MBSE Wiki page on Methodology and Metrics (2015), http://www.omgwiki.org/MBSE/doku.php?id=mbse:methodology
14. The OSEK-VDX portal (2015), http://www.osek-vdx.org/
15. The PREEMPT_RT patch for the Linux kernel. Real-Time Linux Wiki (2015), https://rt.wiki.kernel.org
16. The Unreal gaming engine (2015), https://www.unrealengine.com/
17. The Xenomai solution for real-time Linux (2015), http://xenomai.org/
18. A. Albinet, J.L. Boulanger, H. Dubois, M.A. Peraldi-Frati, Y. Sorel, Q.D. Van, Model-based methodology for requirements traceability in embedded systems, in *Proceedings of*

3rd European Conference on Model Driven Architecture Foundations and Applications, ECMDA'07, Haifa (2007), https://hal.inria.fr/inria-00413488
19. S. Behere, M. Törngren, Architecture challenges for intelligent autonomous machines: an industrial perspective, in *Proceedings of the 13th International Conference on Intelligent Autonomous Machines, IAS-13* (Springer International Publishing, Berlin, 2014). http://link.springer.com/chapter/10.1007%2F978-3-319-08338-4_120
20. S. Behere, M. Törngren, A functional architecture for autonomous driving, in *Proceedings of the First International Workshop on Automotive Software Architecture, WASA '15* (ACM, New York, NY, 2015), pp. 3–10. doi:10.1145/2752489.2752491. http://doi.acm.org/10.1145/2752489.2752491
21. S. Behere, M. Törngren, D. Chen, A reference architecture for cooperative driving. J. Syst. Archit. **59**(10, Part C), 1095–1112 (2013). doi:http://dx.doi.org/10.1016/j.sysarc.2013.05.014. http://www.sciencedirect.com/science/article/pii/S1383762113000957. Embedded Systems Software Architecture
22. P. Bieber, F. Boniol, M. Boyer, E. Noulard, C. Pagetti, New challenges for future avionic architectures. Aerosp. Lab (4), 1–10 (2012)
23. B.W. Boehm, A spiral model of software development and enhancement. Computer **21**(5), 61–72 (1988)
24. J. Bongard, H. Lipson, Automatic synthesis of multiple internal models through active exploration, in *AAAI Fall Symposium: From Reactive to Anticipatory Cognitive Embodied Systems* (2005)
25. J. Bongard, V. Zykov, H. Lipson, Resilient machines through continuous self-modeling. Science **314**(5802), 1118–1121 (2006)
26. G. Bradski, The OpenCV library. Dr Dobb's J. Softw. Tools (2000). http://www.drdobbs.com/open-source/the-opencv-library/184404319
27. P. Braun, M. Broy, F. Houdek, M. Kirchmayr, M. Müller, B. Penzenstadler, K. Pohl, T. Weyer, Guiding requirements engineering for software-intensive embedded systems in the automotive industry. Comput. Sci. Res. Dev. **29**(1), 21–43 (2014)
28. M. Broy, Model-driven architecture-centric engineering of (embedded) software intensive systems: modeling theories and architectural milestones. Innov. Syst. Softw. Eng. **3**(1), 75–102 (2006). doi:10.1007/s11334-006-0011-y. http://link.springer.com/10.1007/s11334-006-0011-y
29. M. Broy, Two sides of structuring multi-functional software systems: function hierarchy and component architecture, in *5th ACIS International Conference on Software Engineering Research, Management & Applications (SERA 2007)* (2007), pp. 3–12. doi:10.1109/SERA.2007.129. http://ieeexplore.ieee.org/lpdocs/epic03/wrapper.htm?arnumber=4296910
30. M. Broy, Software and system modeling: structured multi-view modeling, specification, design and implementation, in *Conquering Complexity*, ed. by M. Hinchey, L. Coyle (Springer, London, 2012), pp. 309–372
31. M. Broy, F. Huber, B. Paech, B. Rumpe, K. Spies, Software and system modeling based on a unified formal semantics, in *Requirements Targeting Software and Systems Engineering*, ed. by M. Broy, B. Rumpe. Lecture Notes in Computer Science, vol. 1526 (Springer, Berlin, Heidelberg, 1998), pp. 43–68
32. H. Bruyninckx, Open robot control software: the OROCOS project, in *Proceedings 2001 ICRA IEEE International Conference on Robotics and Automation* (Cat No01CH37164), vol. 3 (2001), pp. 2523–2528. doi:10.1109/ROBOT.2001.933002. http://ieeexplore.ieee.org/lpdocs/epic03/wrapper.htm?arnumber=933002
33. A. Burns, The Ravenscar profile. Ada Lett. **XIX**(4), 49–52 (1999). doi:10.1145/340396.340450. http://doi.acm.org/10.1145/340396.340450
34. D. Chalmers, *The Conscious Mind: In Search of a Fundamental Theory*. Oxford paperbacks (OUP, New York, 1996)
35. A. Chella, M. Cossentino, V. Seidita, C. Tona, An approach for the design of self-conscious agent for robotics, in *Active Media Technology*, ed. by A. An, P. Lingras, S. Petty,

R. Huang. Lecture Notes in Computer Science, vol. 6335 (Springer, Berlin, Heidelberg, 2010), pp. 306–317
36. P. Cuenot, P. Frey, R. Johansson, The EAST-ADL architecture description language for automotive embedded software, in *Model-Based Engineering of Embedded Real-Time Systems* (springer, Berlin, Heidelberg, 2010), pp. 297–307. http://link.springer.com/chapter/10.1007 %2F978-3-642-16277-0_11
37. D. Davidson, Turing's test, in *Modelling the Mind*, ed. by K. Said (Oxford University Press, Oxford, 1990)
38. J.A. Estefan et al., Survey of model-based systems engineering (MBSE) methodologies. INCOSE MBSE Focus Group 25:8 (2007)
39. P.H. Feiler, B.A. Lewis, S. Vestal, The SAE Architecture Analysis and Design Language (AADL) a standard for engineering performance critical systems, in *2006 IEEE Conference on Computer Aided Control System Design, 2006 IEEE International Conference on Control Applications, 2006 IEEE International Symposium on Intelligent Control* (2006). doi:10.1109/CACSD-CCA-ISIC.2006.4776814
40. T. Ferris, On the methods of research for systems engineering, in *Annual Conference on Systems Engineering Research* (April 2009). http://cser.lboro.ac.uk/papers/S10-62.pdf
41. T. Ferris, Engineering design as research, in *Research Methodologies, Innovations and Philosophies in Software Systems Engineering and Information Systems, IGI Global*, ed. by M. Mora, O. Gelman, A.L. Steenkamp, M. Raisinghani (2012). doi:10.4018/978-1-4666-0179-6. http://services.igi-global.com/resolvedoi/resolve.aspx?doi=10.4018/978-1-4666-0179-6
42. T. Fong, C. Thorpe, Vehicle teleoperation interfaces. Auton. Robots **11**(1), 9–18 (2001)
43. K. Forsberg, H. Mooz, The relationship of systems engineering to the project cycle. Eng. Manag. J. **4**(3), 36–43 (1992)
44. K. Forsberg, H. Mooz, Application of the "vee" to incremental and evolutionary development, in *Systems Engineering in the Global Market Place* (1995), pp. 801–808
45. D. Gamez, Progress in machine consciousness. Conscious. Cogn. **17**(3), 887–910 (2008)
46. T. Henzinger, J. Sifakis, The embedded systems design challenge, in *FM 2006: Formal Methods*, ed. by J. Misra, T. Nipkow, E. Sekerinski. Lecture Notes in Computer Science, vol. 4085 (Springer, Berlin, Heidelberg, 2006), pp. 1–15
47. S. Hussain, Investigating architecture description languages (ADLs) a systematic literature review. Master's thesis, Linköpings universitet, Linköping, 2013
48. A.L. Juarez Dominguez, Feature interaction detection in the automotive domain, in *2008 23rd IEEE/ACM International Conference on Automated Software Engineering* (IEEE, 2008), pp. 521–524. doi:10.1109/ASE.2008.97. http://ieeexplore.ieee.org/lpdocs/epic03/wrapper.htm?arnumber=4639390
49. E.Y. Kang, E.P. Enoiu, R. Marinescu, C. Seceleanu, P.Y. Schobbens, P. Pettersson, A methodology for formal analysis and verification of east-ADL models. Reliab. Eng. Syst. Saf. **120**, 127–138 (2013). doi:http://dx.doi.org/10.1016/j.ress.2013.06.007
50. H. Kopetz, The complexity challenge in embedded system design, in *2008 11th IEEE International Symposium on Object and Component-Oriented Real-Time Distributed Computing (ISORC)* (IEEE, 2008), pp. 3–12. doi:10.1109/ISORC.2008.14. http://ieeexplore.ieee.org/xpls/abs_all.jsp?arnumber=4519555 http://ieeexplore.ieee.org/lpdocs/epic03/wrapper.htm?arnumber=4519555
51. E.A. Lee, Computing needs time. Commun. ACM **52**(5), 70–79 (2009). doi:10.1145/1506409.1506426. http://doi.acm.org/10.1145/1506409.1506426
52. P. Maes, Concepts and experiments in computational reflection, in *Conference Proceedings on Object-Oriented Programming Systems, Languages and Applications, OOPSLA '87* (ACM, New York, NY, 1987), pp. 147–155. doi:10.1145/38765.38821. http://doi.acm.org/10.1145/38765.38821
53. I. Malavolta, H. Muccini, P. Pelliccione, D. Tamburri, Providing architectural languages and tools interoperability through model transformation technologies. IEEE Trans. Softw. Eng. **36**(1), 119–140 (2010). doi:10.1109/TSE.2009.51

54. J. Mårtensson, A. Alam, S. Behere, The development of a cooperative heavy-duty vehicle for the GCDC 2011: team scoop. IEEE Trans. Intell. Transp. Syst. **13**(3), 1033–1049 (2012). doi:10.1109/TITS.2012.2204876. http://ieeexplore.ieee.org/lpdocs/epic03/wrapper.htm?arnumber=6236179 http://ieeexplore.ieee.org/xpls/abs_all.jsp?arnumber=6236179
55. J. McCarthy, Making robots conscious of their mental states, in *Working Notes of the AAAI Spring Symposium on Representing Mental States and Mechanisms*, Menlo Park, CA, 1995
56. D. McDermott, Artificial intelligence and consciousness, in *The Cambridge Handbook of Consciousness*, ed. by P.D. Zelazo, M. Moscovitch, E. Thompson (Cambridge University Press, Cambridge, 2007), pp. 117–150. http://dx.doi.org/10.1017/CBO9780511816789.007. Books Online
57. A. Metzger, Feature interactions in embedded control systems. Comput. Netw. **45**(5), 625–644 (2004). doi:10.1016/j.comnet.2004.03.002. http://linkinghub.elsevier.com/retrieve/pii/S138912860400043X
58. M. Minsky, Matter, mind, and models, in *Semantic Information Processing*, ed. by M.L. Minsky (1968)
59. M. Montemerlo et al., Junior: the Stanford entry in the urban challenge. J. Field Rob. **25**(9), 569–597 (2008). http://onlinelibrary.wiley.com/doi/10.1002/rob.20258/abstract
60. E. Nivel, K. Thórisson, Towards a programming paradigm for control systems with high levels of existential autonomy, in *Artificial General Intelligence*, ed. by K.U. Kühnberger, S. Rudolph, P. Wang. Lecture Notes in Computer Science, vol. 7999 (Springer, Berlin, Heidelberg, 2013), pp. 78–87
61. M. Odersky, L. Spoon, B. Venners, *Programming in Scala: A Comprehensive Step-by-Step Guide*, 2nd edn. (Artima Incorporation, Walnut Creek, 2011)
62. G. Pardo-Castellote, OMG data-distribution service: architectural overview, in *Proceedings of the 2003 IEEE Conference on Military Communications - Volume I, MILCOM'03* (IEEE Computer Society, Washington, DC, 2003), pp. 242–247
63. F. Perotto, R. Vicari, L. Alvares, An autonomous intelligent agent architecture based on constructivist ai, in *Artificial Intelligence Applications and Innovations, IFIP International Federation for Information Processing*, vol. 154, ed. by M. Bramer, V. Devedzic (Springer US, 2004), pp. 103–115. http://link.springer.com/chapter/10.1007%2F1-4020-8151-0_10
64. M. Quigley, K. Conley, B. Gerkey, J. Faust, T.B. Foote, J. Leibs, R. Wheeler, A.Y. Ng, ROS: an open-source robot operating system, in *ICRA Workshop on Open Source Software* (2009)
65. W.W. Royce, Managing the development of large software systems: concepts and techniques, in *Proceedings of the 9th International Conference on Software Engineering, ICSE '87* (IEEE Computer Society Press, Los Alamitos, CA, 1987), pp. 328–338
66. R.B. Rusu, S. Cousins, 3D is here: Point Cloud Library (PCL), in *2011 IEEE International Conference on Robotics and Automation (ICRA)* (IEEE, 2011), pp. 1–4. doi:10.1109/icra.2011.5980567
67. A.V. Samsonovich, Toward a unified catalog of implemented cognitive architectures, in *Proceedings of the 2010 Conference on Biologically Inspired Cognitive Architectures 2010: Proceedings of the First Annual Meeting of the BICA Society* (IOS Press, Amsterdam, 2010), pp. 195–244. http://dl.acm.org/citation.cfm?id=1893313.1893352
68. S. Sarma, N. Dutt, P. Gupta, A. Nicolau, N. Venkatasubramanian, On-chip self-awareness using cyberphysical-systems-on-chip (CPSOC), in *Proceedings of the 2014 International Conference on Hardware/Software Codesign and System Synthesis, CODES '14* (ACM, New York, NY, 2014), pp. 22:1–22:3. doi:10.1145/2656075.2661648. http://doi.acm.org/10.1145/2656075.2661648
69. K.R. Thórisson, From constructionist to constructivist ai, in *AAAI Fall Symposium: Biologically Inspired Cognitive Architectures* (2009)
70. K.R. Thórisson, *A New Constructivist AI: From Manual Methods to Self-constructive Systems* (2012). doi:10.2991/978-94-91216-62-6_9
71. K.R. Thórisson, A new constructivist ai: from manual methods to self-constructive systems, in *Theoretical Foundations of Artificial General Intelligence, Atlantis Thinking Machines*, vol. 4, ed. by P. Wang, B. Goertzel (Atlantis Press, Amsterdam, 2012), pp. 145–171

72. K.R. Thórisson, H.P. Helgason, Cognitive architectures and autonomy: a comparative review. J. Artif. Gen. Intell. **3**(2), 1–30 (2012). doi:10.2478/v10229-011-0015-3. http://dx.doi.org/10.2478/v10229-011-0015-3
73. K.R. Thórisson, H. Benko, D. Abramov, A. Arnold, S. Maskey, A. Vaseekaran, Constructionist design methodology for interactive intelligences. AI Mag. **25**(4), 77 (2004)
74. M.Törngren, A. Qamar, M. Biehl, F. Loiret, J. El-khoury, Integrating viewpoints in the development of mechatronic products. Mechatronics **24**(7), 745–762 (2014)
75. A.M. Turing, Computing machinery and intelligence. Mind **59**(236), 433–460 (1950). http://www.jstor.org/stable/2251299
76. R. Van Gulick, Consciousness, in *The Stanford Encyclopedia of Philosophy*, Spring 2014 edn., ed. by E.N. Zalta (2014). http://plato.stanford.edu/entries/consciousness/
77. O. Wallmark et al., Design and implementation of an experimental research and concept demonstration vehicle, in *Vehicle Power and Propulsion Conference (VPPC), 2014* (IEEE, 2014), pp. 1–6. doi:10.1109/VPPC.2014.7007042
78. J. Westman, M. Nyberg, A reference example on the specification of safety requirements using ISO 26262, in *SAFECOMP 2013 - Workshop DECS (ERCIM/EWICS Workshop on Dependable Embedded and Cyber-physical Systems) of the 32nd International Conference on Computer Safety, Reliability and Security*, France, ed. by M. Roy p NA (2013). https://hal.archives-ouvertes.fr/hal-00848610
79. J. Westman, M. Nyberg, M. Törngren, Structuring safety requirements in ISO 26262 using contract theory, in *Computer Safety, Reliability, and Security*, ed. by F. Bitsch, J. Guiochet, M. Kaâniche. Lecture Notes in Computer Science, vol. 8153 (Springer, Berlin, Heidelberg, 2013), pp. 166–177
80. J. Ziegler et al., Making bertha drive: an autonomous journey on a historic route. IEEE Intell. Transp. Syst. Mag. **6**(2), 8–20 (2014). doi:10.1109/MITS.2014.2306552
81. H.P. Zima, M.L. James, P.L. Springer, Fault-tolerant on-board computing for robotic space missions. Concurr. Comput. Pract. Exp. **23**(17), 2192–2204 (2011). doi:10.1002/cpe.1768. http://dx.doi.org/10.1002/cpe.1768

Chapter 14
Open Dependable Power Computing Platform for Automated Driving

Andrea Leitner, Tilman Ochs, Lukas Bulwahn, and Daniel Watzenig

14.1 Introduction

Future automated driving systems need to be able to handle any possible traffic situation safely without relying on human supervision. These systems have two seemingly contradictory characteristics: first, they must be verifiably highly reliable because malfunctions could endanger passengers and other traffic participants, and second, they consist of complex and computer-intensive application software components. Figure 14.1 shows that an automated vehicle will operate in an ecosystem of infrastructure, cloud, and back-end systems, such as traffic lights, authority services, software distribution services, or map services. The vehicle will connect to those services via wide-area wireless internet links or vehicle-to-vehicle communication media defined by ETSI.[1] Such an automated driving system is composed of a network of redundant electronic control units (ECUs)

[1] www.etsi.org/

A. Leitner (✉)
Virtual Vehicle Research Center, Graz, Austria
e-mail: andrea.leitner@v2c2.at

T. Ochs
BMW AG, Munich, Germany
e-mail: Tilmann.Ochs@bmw.de

L. Bulwahn
Lukas Bulwahn, BMW AG, Munich, Germany
e-mail: Lukas.Bulwahn@bmw.de

D. Watzenig
Virtual Vehicle Research Center and Graz University of Technology, Institute of Electrical Measurement and Measurement Signal Processing, Graz, Austria

D. Watzenig, M. Horn (eds.), *Automated Driving*,
DOI 10.1007/978-3-319-31895-0_14

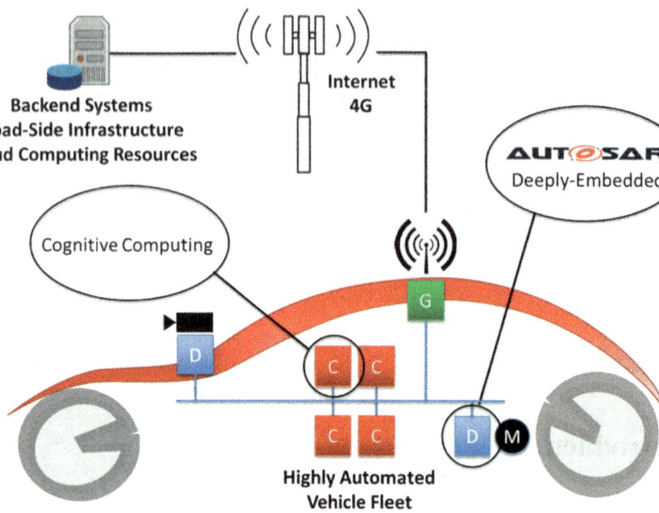

Fig. 14.1 Schematic architecture of a highly automated vehicle

that are connected to the vehicle's sensors and actuators through a highly reliable, adequately responsive vehicle network.

Cars are becoming cyber-physical systems that need to respond in real time to dynamic and complex environmental situations. Given the current and future road infrastructure, it is mandatory to reach a safe vehicle state, i.e., pulling over a car at a situation-dependent speed to a safe stop on every road in every condition in the event of a single safety-critical system malfunction. This is especially challenging because of the high complexity and the partially probabilistic nature of intermediate results. Especially for active driving functions, shutdown of the system will not be a safe state; instead, the system must tolerate certain failures and support a fail-operational mode.

Automated driving therefore requires the introduction of a new type of automotive systems called cognitive systems (marked as "C" in Fig. 14.1). Figure 14.2 illustrates the differences between this new type of system and well-known control systems (marked as "D" in Fig. 14.1). A control system processes information from inside the vehicle, while cognitive systems in addition receive and process environmental information (Table 14.1). The basic elements of a cognitive system are thus environment perception, decision making, and initiation of actions based on machine learning and information fusion. In contrast to traditional deeply embedded control systems, which are generally perceived as limited in scope, cognitive systems require much more computational resources, e.g., to solve complex optimization problems in real time. Furthermore, software architectures for such systems are significantly more sophisticated than for traditional control systems. This paradigm shift necessitates the development of new platforms that support these applications in an efficient and effective manner.

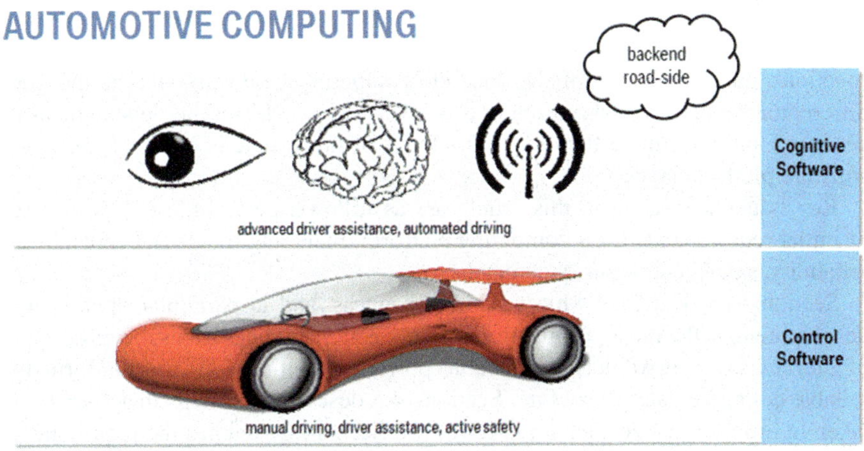

Fig. 14.2 Control software vs. cognitive software in automotive computing

Table 14.1 Characteristics of control and cognitive software

Control software	Cognitive software
State machine and controller	Dynamic models and artificial intelligence
Mature state of the art	Rapidly evolving technology
Static software structure and configuration	Dynamic software structure and configuration
Automotive microcontrollers	High-performance mainstream hardware

The current state-of-the-art software platform for deeply embedded control ECUs and early cognitive computing ECUs is AUTOSAR.[2] The main concern with the current AUTOSAR platform in this context is the static runtime environment [1] together with the signal-oriented communication infrastructure (i.e., restricted types and data formats). In particular, fixed scheduling and fixed communication relationships are not ideal in a multi-core environment because they restrict the efficiency of these systems. A future dependable power-computing software platform should therefore provide a more efficient solution. Due to its monolithic and highly optimized design, a partial update of the AUTOSAR specification is difficult. This means that the future of AUTOSAR will be an additional, new AUTOSAR Adaptive Platform [2] supporting the needs for automated driving. Activities in this direction have already been started but are of course currently only in the initial stages.

The software architecture for a cognitive computing ECU can be decomposed into two layers: the application software layer and the software platform layer. The higher-level application software layer contains the main cognitive software functions (e.g., high-level recognition of traffic situations, prediction of other traffic participants' behavior, planning of the vehicle's maneuvers) for automated driving.

[2]http://www.autosar.org/

This part is the differentiating part for car manufacturers and suppliers. The software platform layer provides basic services, e.g., communication between the software functions, and abstraction of the concrete computing hardware—this is the non-differentiating part. Collaboration at this level will accelerate the development of advanced and innovative functions. The focus of this article is on this lower-level software platform layer.

Key issues addressed in this article are as follows: Sect. 14.2 addresses a set of initial requirements for a computing platform, including functional capabilities, reliability, safety, and security demands.

Section 14.3 discusses why we think openness, and in particular open-source development, is the most promising solution. Section 14.4 describes the basic ideas of the envisaged in-vehicle dependable power-computing software platform for reliable cognitive data processing. Section 14.5 describes the steps that need to be taken in order to realize such a platform, and Sect. 14.6 sketches the requirements for an open-source development process in a safety-critical context. Section 14.7 concludes the paper.

14.2 Requirements for an Open Dependable Power Computing Platform

Cognitive systems and reliable cognitive data processing for advanced assisted and automated driving functions have significantly different requirements compared to those for systems using traditional automotive control algorithms. The most important requirements are (i) very large and highly dynamic resource demands with respect to communication bandwidth, memory and mass storage, and raw CPU power; (ii) in-vehicle update of parameter data, models, and algorithms; and (iii) the probabilistic nature of some basic algorithms.

Here we can only present a preliminary set of initial requirements for such a platform, since the specific requirements will only become available in the course of product development. We have extrapolated resource demands and use cases from existing systems and research projects [3].

Openness in the context of a dependable power-computing software platform can be understood in two ways. First, it must be open from a production point of view (pushing toward an open-source approach), and second, the system must be open to deal with the fast evolution of cognitive software, because continuous learning from field experience and short iteration cycles with frequent rollouts into vehicles are essential for usability, robustness, safety, and security.

We strongly believe that future market differentiation will come from up-to-date information and advanced functional software (application software) rather than vehicle E/E architectures and software platforms. The latter are becoming increasingly non-differentiating, and therefore joint development will help to share significant development efforts.

The benefits of an open-source approach furthermore include the possibility to create a community able to contribute to the evolution of the platform as well as the development of applications without depending too much on tool-chain providers. Using an open-source approach enables research institutions and SMEs to develop and test their applications on existing state-of-practice platforms. Currently, research projects seem to be slowed down by focusing too much on the software architecture [4, 5] instead of the more innovative part: the application. An open-source platform would thus reduce time to market for new innovative functions.

Complexity Software for automated driving functions has tens of thousands of lines of code and makes use of complex and dynamic data structures, such as graphs or sparse matrices. In order to support the development of this kind of software, the platform needs to support (i) the integration of loosely coupled application software components to minimize covert interference, (ii) a rich interface definition language to specify application software component interfaces unambiguously, (iii) high-level programming languages to reduce code complexity, and (iv) mainstream software engineering methods and technology. The latter fosters the use of mature and state-of-the-art technology in order to leverage experience and progress in the general IT industry (or other domains) while minimizing friction for application development due to immature or unfamiliar niche products.

Computing Performance Research projects for automated driving use modern personal computers to their full capacity. This means memory usage in the gigabyte range, CPU usage in the gigaflop range, and heavy use of hardware acceleration and parallel data processing. Hence, the platform should provide (i) a hardware-independent programming interface for acceleration hardware to enable easy porting of application software components, (ii) symmetric multiprocessing to leverage component level parallelism on mainstream processors, and (iii) performance-optimized processor architectures to maximize computational throughput.

Typical control software utilizes predefined fixed scheduling in order to ensure correct timing behavior. This is not meaningful for cognitive software with its high computational power requirements. These systems require multi-core systems, which are very inefficient with fixed scheduling. A possible solution could be the specification of dedicated timing properties (e.g., deadlines, criticality, etc.) in order to support dynamic scheduling and thus more efficient utilization of computing power.

Mixed Criticality Automated driving functions are highly safety critical because malfunctions can cause harm to passengers and other traffic participants. In order to assure the safe operation of an automated vehicle, a safety concept has to be defined at the vehicle level and decomposed to derive safety requirements for all relevant vehicle components. Following the ISO 26262 standard [6] for the development of safety-critical automotive E/E systems, vehicle components are classified according to four integrity levels from the lowest Automotive Safety Integrity Level (ASIL) A to the highest level ASIL D, depending on the consequences of failures. The platform should be able to host application software components with different integrity levels (mixed criticality) and must ensure that those with lower integrity level do not interfere with higher-integrity components. This requirement is called "freedom from interference" in ISO 26262, and it means that it must be ensured that shared memory is protected from access violation, tasks are executed in the correct order, and functions are able to finish within their real-time boundaries.

To execute mixed-critical software functions, the platform should support (i) the coexistence of application software components with different integrity levels (mixed criticality), (ii) misbehavior detection for application software components and restart or shutdown of affected components to increase robustness, (iii) control flow monitoring in application software components and misbehavior detection, (iv) data flow monitoring between application software components and misbehavior detection, and (v) standardized software and error propagation models to support automated safety analyses.

Decomposition Based on personal communication with various hardware vendors, we assume that a platform will at most provide integrity levels up to ASIL B due to a lack of high-integrity processing hardware with the required performance. Furthermore, development of high-integrity software is very time-consuming and limited to a certain level of complexity. Therefore, cognitive software should be of limited ASIL, while hard safety functions should be shifted to more traditional and static high-integrity computer systems.

If required, higher-integrity levels can be achieved by vehicle-level redundancy. This concept is called ASIL decomposition in ISO 26262 and means that ASIL D can be attained by using two redundant, independent ASIL B components.

Of course, it would also be possible to qualify the platform itself, or the applications implemented on it, at a higher level. Techniques such as safety kernels [7], both hardware and software, allow failures of the cognitive processing software to be tolerated and hence a higher ASIL to be achieved. Using architectural patterns recommended by IEC 61508 [8] and other standards, an overall level of ASIL C can be supported. Even ASIL D would be possible with formal proof of the protection mechanisms (e.g., the safety kernel). Wika et al. [9] have demonstrated the feasibility of such a proof.

In order to support decomposition, the platform should support (i) the execution of application software components with integrity levels up to ASIL B and (ii) failure detection for hardware and platform software components and silent shutdown of affected partitions/containers to support vehicle-level redundancy,

Time Sensitivity The timing properties of automotive systems are among the most important properties. It is not only important to demonstrate that the system meets its timing requirements; the developers also need to show that activities are performed in the correct order, the system does not deadlock or live-lock, and the system degrades in case of a failure.

Automated driving systems have additional real-time requirements because outdated environment data or lagging maneuver planning can lead to oscillating dynamic behavior and collisions. From our experience, we assume that a sporadically occurring maximum jitter below 100 μs in the application components will not lead to a hazardous event. Recent work by the Open Source Automation Development Lab (OSADL)[3] shows that interrupt latency below 100 μs can be achieved on mainstream processing hardware using Linux with the RT Preempt patch. Hence, the software platform must provide (i) deterministic timing behavior with a maximum jitter of 100 μs to ensure replicable behavior, (ii) real-time scheduling to meet the timing requirements of application software, and (iii) timing monitoring of application software components and detection of timing violations with 100 μs tolerance.

Volatility Research in computer vision and artificial intelligence will evolve rapidly in the near future, and several research projects are currently developing infrastructure for automated driving.[4,5] As already argued in the paragraph on openness, we assume that car manufacturers must supply software updates through a remote connection to provide the latest functionality and integration of locally available infrastructure services. Hence, the platform should support (i) agile development practices for software components to mitigate low concept maturity, (ii) safe and secure remote update of application software components to maintain software continuously, (iii) a modular safety concept to integrate independently developed application software components, and (iv) safe and secure remote addition and removal of application software components to provide extensibility.

Automotive Specifics Automated driving systems are integrated into vehicle networks, are managed by standardized diagnostic functions, and interact with vehicle state management and energy management. Hence, the platform should support (i) standardized diagnostic and vehicle management functions, (ii) integration of vendor-specific diagnostic and vehicle management functions, and (iii) automotive-specific communication protocols such as Automotive Ethernet [10], CAN, or FlexRay.

14.3 Why Qualifiable Open Source Is Appropriate

There are several possible approaches to provide software platforms. In the following, we discuss the advantages and disadvantages of the different approaches and explain why we think that open source is the best choice:

- *Custom*—The advantage of this approach is that the entire design and development approach can be tailored to the specific application, but most companies do not have the required specialist skills. Furthermore, it entails significant cost.

---

[3] http://www.osadl.org

[4] http://kofas.de/

[5] http://www.simtd.de/

- *COTS*—Many COTS-based systems have the advantage of more "formal" software engineering methods behind them than open-source development, but this comes with significant upfront and recurring costs. Additionally, the user is dependent on the vendor to implement required changes.
- *Qualifiable open source*—Another approach is to build the platform from existing open-source software and further develop and maintain it using an open-source approach. With the qualifiable open-source approach, not only the delivered software but also the software lifecycle, including specification, documentation, and safety case, is provided using free/libre/open-source software (FLOSS) licensing concepts [11]. This approach has several advantages compared to proprietary solutions:
  - Higher quality: widely used software matures more quickly, since it incorporates experiences from various use cases.
  - Higher confidence: everybody, including educational institutions, can assess and rate risk classifications and the effectiveness of safety measures.
  - Higher agility: innovative car manufacturers, application software developers and integrators who need additional platform capabilities can collaborate on new features, implement them a dedicated branch, and use them in their product development before the changes are integrated into the main branch.
  - Lower cost: shared costs for development and qualification of this complex platform.

The most important and famous open-source operating system is the Linux operating system (OS). A major advantage of the Linux real-time OS (RTOS) is that it has been demonstrated as adequately (in terms of flexibility and reliability) supporting a wide range of different platforms and applications. In particular, it has been proposed and evaluated for use in critical systems [12, 13] and has been used in cognitive systems and reliable cognitive data processing applications [14]. From an economic point of view, the Linux operating system is the most successful open-source project and has been developed in a joint manner. It is hard to estimate the cost of its development and consequently its worth. Nevertheless, there have been studies that investigated what it would have cost to develop Linux in a typical company in the USA. David A. Wheeler [15] determined in 2002 a cost of more than \$1.2 billion. An updated study in 2008 showed that it would cost \$10.8 billion to develop the Fedora 9 Linux distribution at this time by traditional proprietary means. This shows the tremendous value which can be created by the open-source community and which could be exploited by the automotive domain. For these reasons, Linux RTOS is an efficient and cost-effective option.

Of course, the Linux RTOS currently has one big disadvantage: there is presently no appropriate qualification approach for the Linux RTOS itself and no certification strategy for safety-related systems using this operating system.

Open-source projects, such as the Linux operating system, have the big advantage that they employ stringent development processes [16, 17] and deliver software of very high quality, but they do not fulfill the requirements of current safety standards, such as ISO 26262, for software in automotive systems. These standards impose strict demands on project management, developer qualification, risk management, requirements management, quality assurance, and documentation. Therefore, open-source software cannot be used in safety-critical systems without further activities. Nevertheless, a related project called SIL2LinuxMP [18] is attempting to show that open-source processes fulfill most of the requirements but by different means.

Still, the following issues need to be resolved: (i) a qualification strategy for Linux RTOS, which also supports the co-evolution of software and safety case, (ii) a configured and ASIL B qualified version of Linux RTOS suitable for use in safety-related cognitive systems, and (iii) guidance on how this configuration can be maintained.

The idea of using an open-source platform in a safety-critical context is not new. An initiative in the railway domain has resulted in a project called OpenETCS,[6] which aims to develop a software kernel for the European train control system based on open source and open proof. Another project called SIL2LinuxMP [18] plans to certify the base components, i.e., the boot loader, root file system, Linux kernel, and C library bindings, of an embedded GNU/Linux real-time operating system compliant with safety integrity level 2 (SIL2) according to safety standard IEC 61508 [8], which roughly corresponds to ISO 26262 integrity level ASIL B.

For the automotive domain, GENIVI[7] is another example for an open-source initiative, though not in the context of a safety-critical system. Many industrial partners have joined forces to develop a reusable, open-source in-vehicle infotainment (IVI) platform.

14.4 Considerations for the Platform Architecture

Figure 14.3 shows an architectural overview of the dependable power-computing software platform. The dependable power-computing software platform presumes high-performance processor hardware that extends current mainstream microcontrollers for mobile devices by additional integrity mechanisms and reliability figures that are able to fulfill the requirements of the ISO 26262 safety standard for integrity levels up to ASIL B. The combination of this hardware and the dependable power-computing software platform delivers the overall dependable power-computing platform. An API built on the software platform abstracts the underlying implementation from the application which uses it.

A dependable power-computing software platform consists of two layers: (i) the platform foundation layer and (ii) the platform services layer. The platform foundation layer consists of system software modules that expose APIs to application software components and implement basic platform capabilities, such as hardware abstraction, mass storage, network communication, power management,

[6]http://openetcs.org/

[7]http://www.genivi.org/

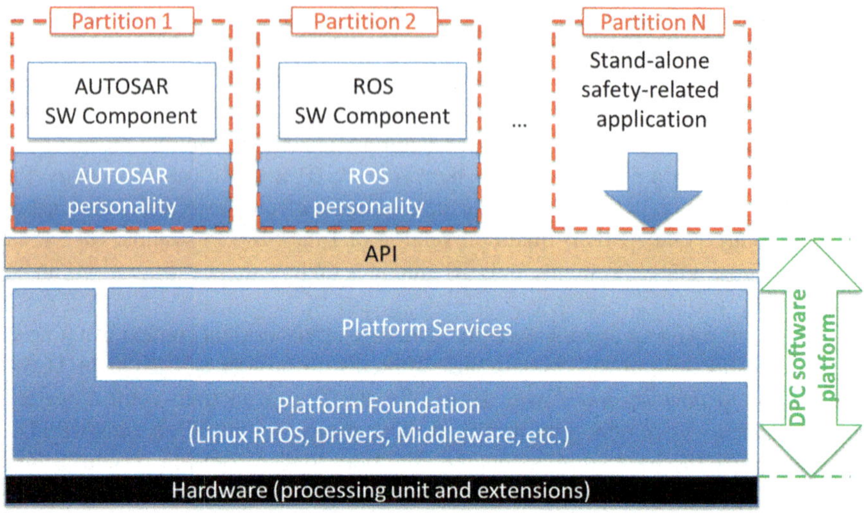

Fig. 14.3 Envisioned architecture

and process control. In addition, low-level safety and security mechanisms, such as temporal and spatial isolation, mandatory access control and runtime monitoring are provided here. A large part of this platform foundation layer can be implemented by a qualified open-source RTOS based on the Linux operating system. The platform service layer consists of software components that implement higher-level management and monitoring functions, such as state management, graceful degradation, update over the air, diagnostics, and real-time intrusion detection. This layer should also be implemented using an open-source approach and should reuse as much existing software as possible.

The differentiating part for OEMs will still be the application software. The dependable power-computing software platform also allows the partitioning of application software components and provides protection mechanisms for resilience against malicious attacks, design flaws, and hardware faults. Personalities allow same base platform to be adapted to different application domains (e.g., IMA for aerospace, ROS for robotics, etc.).

14.5 Steps Toward an Open Dependable Computing Platform

For a driverless vehicle, a core component such as an operating system along with its supportive environment (e.g., development tools) needs to be qualified. Any qualification is built around a well-specified set of traceable processes and procedures. The qualification of Linux as an operating system in a safety-critical context requires two main things: first, selection of the kernel and core components to be qualified, and second, definition of a qualification strategy for open source that is compliant with the automotive safety standard ISO 26262.

Linux is an existing component and has not only a broad range of functionality but also varying maturity of concepts and implementations. The starting points for OS selection are well-defined selection criteria and evaluation of available operational data sources. While many of the criteria will need to remain qualitative criteria, quantification is important for managing the dynamics of open-source projects, which often exhibit change rates of multiple changes sets per hour. Definition of quantitative selection criteria and implementation of their automation can significantly ease the utilization of open-source components in safety-related and dependable systems. Based on these criteria, the Linux kernel and core OS components to be used as the qualification target can be selected.

Relevant functional safety standards, notably IEC 61508 and ISO 26262, are built around well-defined processes to mitigate risks related to systematic faults in complex systems. Selection of methods from the set of accepted state-of-the-art methods is a first step, but not all of the traditional methods developed for custom systems are suitable for open-source components. Thus, there might be a need for new methods, processes, and procedures for the formal qualification of a use-case-driven subset of functional and nonfunctional capabilities of Linux, open-source tools, and support libraries.

A starting point for qualifying open-source is to develop a complete mapping/interpretation of standards. Because coverage of systematic faults relies on the development process, the main need is to develop documentation of the detailed procedures that are used and follow-up gap analysis of the current development life cycle. This should be based on the evaluation of available process data and assurance criteria for data and procedures.

As mentioned before, the platform foundation and the platform service layer should be built from existing and operationally proven open-source software wherever possible. This seems to be reasonable, because open-source is a vast resource pool of technology. Nevertheless, a key obstacle to utilizing it is the lack of an accepted generic procedure to integrate these components into qualified systems. Although key standards do give some high-level guidance, there is a lot of room for interpretation. Narrowing this range of interpretation is a key risk mitigation strategy, in order to simplify the use of open-source components in industrial projects in general. The open-source community has well-established methods and policies, some of which need minor adjustments to satisfy the qualification needs set out in relevant standards. One key aspect therefore is active feedback of modifications and procedural changes to the open-source community in order to achieve the long-term objective of a maintainable qualification methodology.

14.6 Open-Source Software Development Process

As mentioned before, open-source projects employ stringent development processes, which might need to be adapted for the use in a safety-critical context. Figure 14.4 describes the roles and interactions of an open-source development process and the actions necessary to transform a common open-source process into a qualifiable open-source process.

In any case, we distinguish between collaborative development, which is done by the open-source community, and product development, which is subsequently done within a company based on the jointly developed deliverables.

A central aspect of any open-source development is the management organization (e.g., the Linux foundation), which provides the infrastructure (e.g., repositories, etc.) and defines and monitors the processes.

The contributors of course play an important role because they do the actual work on the deliverables according to the defined process. Before they are able to contribute, they have to accept a Contributor License Agreement which governs intellectual property rights. The user (OEM or Tier 1) has to accept the open-source license and usually drives the development. The user therefore engages service providers (software experts) to contribute to the open-source project. This potentially leads to a huge business opportunity for service providers, who can offer

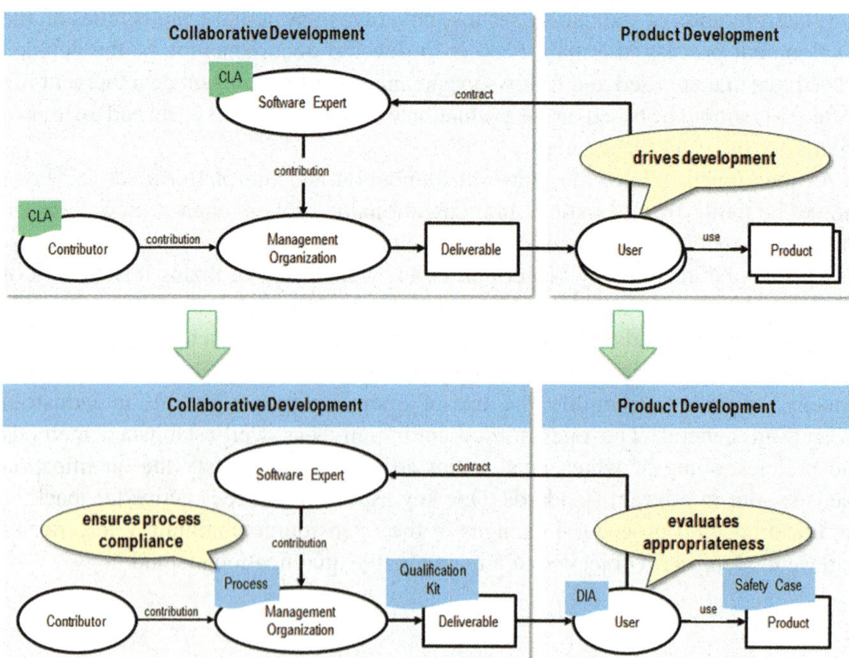

Fig. 14.4 From open-source software to qualifiable open-source software

their services to the users. The big advantage for the user is that required changes or adaptations can be implemented efficiently without depending on a single supplier.

A qualifiable open-source process needs to be enhanced by a more stringent safety development process. This needs to be driven by the management organization, which also needs to ensure compliance with the processes. The management organization is furthermore responsible for establishing a safety culture. This means that it must be clearly described and communicated to the contributors how functional safety will be part of the product and how it should be handled within the development process.

One important document in a qualifiable process is the Developer Interface Agreement (DIA) as defined in ISO 26262. The DIA is a multilateral contract that determines each party's responsibilities. Especially in a safety-critical context, there is a need for additional deliverables which document the respective safety activities. This is especially important because the final responsibility lies with the user (in this case the OEM or Tier 1) who integrates the open-source development in the final product. At the end, there needs to be a safety case, which documents that all relevant risks have been identified and that the corresponding mitigation measures have been taken.

14.7 Conclusion

This article discusses the needs of an automated driving computing platform and highlights the advantages of open-source development. Nevertheless, we are fully aware that these goals are not easy to achieve without major efforts by various parties. There are several open research questions which need to be targeted for the successful implementation of complex cognitive systems. One of these questions is related to functional safety and how it can be addressed in this complex and highly dynamical context. The high degree of automation requires a transition from fail safe to fail operational, because it is nearly impossible to determine a safe state for an automated vehicle. This requires more advanced redundancy patterns for automotive software architectures, as for example shown in Fig. 14.5.

Another important issue is the agility of cognitive systems, which requires a modular safety model. Currently, a system has to be considered as a monolithic entity. In order to support system properties required for future highly innovative vehicle functions, for example, remote updates (or regular functional updates in general) modularization is a key feature.

Cognitive systems are essential for automated driving but are still a challenging topic. This new type of system requires completely new and more advanced software platforms, which are able to cope with newly emerging requirements.

In this article, we describe the main requirements for such a qualifiable platform and why we think that joint development using an open-source approach is the way forward.

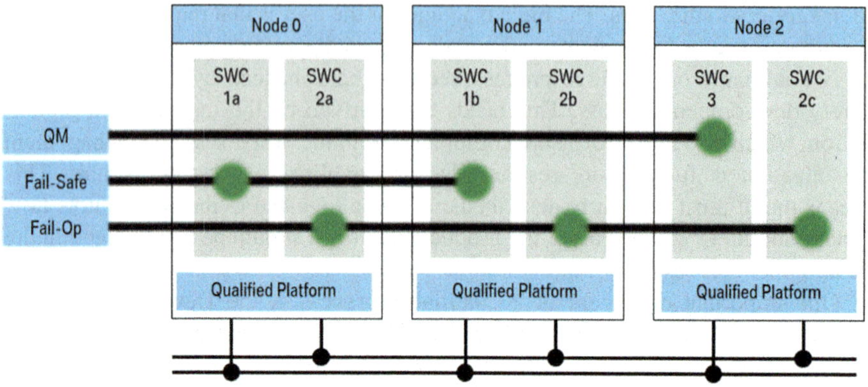

Fig. 14.5 Sample architecture for fail-safe and fail-operational tasks

References

1. H. Martorell, J.-C. Fabre, M. Roy, R. Valentin, Improving Adaptiveness of AUTOSAR Embedded Applications, in *Proceedings of the 29th Annual ACM Symposium on Applied Computing* (SAC '14) (ACM, New York, NY, 2014), pp. 384–390
2. M. Bechter, Softwareplattform für zukünftige automotive Anwendungen, http://www.informatik.tu-cottbus.de/ase2015/Abstract01.pdf. Accessed 12.10.2015
3. S. Holder, M. Hörwick, H. Gentner, Funktionsübergreifende Szeneninterpretation zur Vernetzung von Fahrerassistenzsystemen. Automatisierungssysteme, Assistenzsysteme und eingebettete Systeme für Transportmittel (AAET 2012) (2012)
4. S. Sommer, A. Camek, K. Becker, C. Buckl, A. Zirkler, L. Fiege, M. Armbruster, G. Spiegelberg, A. Knoll, Race: A Centralized Platform Computer Based Architecture for Automotive Applications (2013)
5. M. Goebl, M. Althoff, M. Buss, G. Faerber, F. Hecker, B. Heissing, S. Kraus, R. Nagel, F.P. Leon, F. Rattei, M. Russ, M. Schweitzer, M. Thuy, C. Wang, H.-J. Wuensche, Design and Capabilities of the Munich Cognitive Automobile, in *Proceedings of IEEE Intelligent Vehicles Symposium*, pp. 1101–1107, Eindhoven, The Netherlands, Jun 2008
6. ISO 26262:2011, Road vehicles—Functional safety International Standard (Parts 1–10). International Organization for Standardization, 1st edn (2011)
7. J. Rushby, Kernels for Safety? in T. Anderson (ed.), *Safe and Secure Computing Systems* (Blackwell Scientific Publications, Malden, 1989), Chapter 13, pp. 210–220
8. CENELEC, IEC 61508 Functional Safety of Electrical/Electronic/Programmable Electronic Safety-Related Systems. International Electrotechnical Commission (IEC), 2nd edn (2010)
9. K. Wika, J. Knight, On the Enforcement of Software Safety Policies, in *Proceedings of the 10th Annual IEEE Conference on Computer Assurance*, Jun 1995
10. Broadcom Corporation, BroadR-Reach® Physical Layer Transceiver Specification for Automotive Applications, http://www.ieee802.org/3/1TPCESG/public/BroadR_Reach_Automotive_Spec_V3.0.pdf. Accessed 12.10.2015
11. K.-R. Hase, Open Proof' for Railway Safety Software—A Potential Way-Out of Vendor Lock-In Advancing to Standardization, Transparency, and Software Security, in *FORMS/FORMAT 2010*, ed. by E. Schnieder, G. Tarnai (Springer, Berlin, 2011), pp. 5–38
12. CSE International Limited, Preliminary Assessment of Linux for Safety Related Systems, HSE (2002)

13. R. Kammerer, *Linux in Safety-Critical Applications*. OSADL Academic Works (OSADL, Heidelberg, 2011)
14. M. Goebl, G. Farber, A Real-Time-Capable Hard-and Software Architecture for Joint Image and Knowledge Processing in Cognitive Automobiles, in *Intelligent Vehicles Symposium*, 2007 IEEE, pp. 734–740, Jun 2007
15. D.A. Wheeler, More Than a Gigabuck: Estimating GNU/Linux's Size (2001), http://www.dwheeler.com/sloc/redhat71-v1/redhat71sloc.html
16. J. Corbet, *How the Development Process Works* (The Linux Foundation, San Francisco, 2011)
17. A. Mockus, R.T. Fielding, J.D. Herbsleb, Two Case Studies of Open Source Software Development: Apache and Mozilla. ACM Trans Softw Eng Methodol **11**(3), 309–346 (2002)
18. OSADL Project: SIL2LinuxMP, http://www.osadl.org/SIL2LinuxMP.sil2-linux-project.0.html. Accessed 12.10.2015

Part V
Active and Functional Safety in Automated Driving

Chapter 15
Active Safety Towards Highly Automated Driving

Klaus Kompass, Markus Schratter, and Thomas Schaller

15.1 Introduction

15.1.1 Motivation for Automated Driving

The increase of comfort, safety and efficiency is the motivation for introducing Highly Automated Driving from the point of view of BMW. Highly Automated Driving [1] allows the driver to withdraw from the driving task to a certain degree, depending on the required takeover time needed by the driver [2].

Comfort
Maximum comfort is achieved when the driver can exploit his or her optimal performance in any driving situation. The performance is strongly dependent upon the situation and can be divided into three states: under-demanded, balanced and over-demanded. The transition between these states is smooth and variable, dependent on the current state of the driver. The relationship between the driver's workload and its performance is derived from the Yerkes–Dodson curve [3] and visualised from the point of view of the driver assistance in Fig. 15.1.

There is no need for automated driving functions when the driver is in the optimum range (competence). The driver can take advantage of his full capability and perceives driving as pleasant. But, the need for and the opportunity to assist the

K. Kompass • T. Schaller
BMW Group, Munich, Germany
e-mail: thomas.sa.schaller@bmw.de

M. Schratter (✉)
Virtual Vehicle Research Center, Graz, Austria
e-mail: markus.schratter@v2c2.at

D. Watzenig, M. Horn (eds.), *Automated Driving*,
DOI 10.1007/978-3-319-31895-0_15

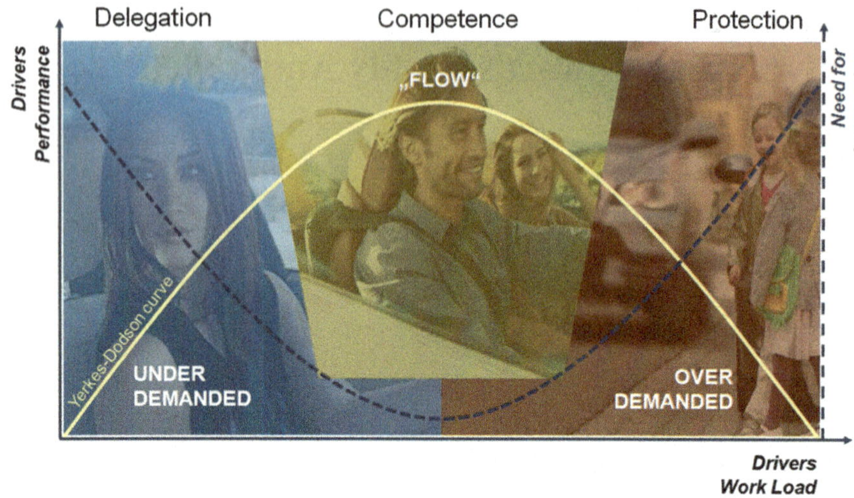

Fig. 15.1 Yerkes–Dodson curve—driver performance/driver workload

driver with HAD (Highly Automated Driving) functions and Active Safety Systems[1] exist when the driver is over- or under-demanded. Driving in a traffic jam or a monotonous traffic on the motorway can result in under-demand. In these situations, the driver does not use his potential performance. In this case, potential exists for Highly Automated Driving, if the driver delegates the driving task to the vehicle and uses his additional available mental capacity for other tasks.

Safety
A goal of HAD functions is to increase traffic safety. For this reason, a vehicle for Highly Automated Driving must be at least as safe as current traffic is; there must be no worsening in the traffic safety compared to today's global safety level. To fulfil this requirement and to bring HAD functions on the road, new and advanced technologies are necessary (for further details, see Sect. 15.2). These technologies can be used to improve Active Safety Systems for non-Highly Automated Driving and thus address the "protection" workload (over-demanded) of the Yerkes–Dodson curve (see Fig. 15.1).

Efficiency
A further motivation for Highly Automated Driving is to increase the efficiency by reducing the energy consumption. Automated driving functions, for example, can regulate more accurately in regard to upcoming speed limits or vehicles in front. Route planning can be improved by vehicle connectivity; information provided about the traffic flow of road sections collected by other vehicles can be considered

[1]Active Safety Systems in this chapter are systems which provide information to the driver or intervene in the vehicle control.

to calculate an optimal route. Thus, a microscopic efficiency enhancement can be achieved in the short term. In the long term, optimised route planning with traffic flow management is feasible, where vehicles receive a proposed route from a global instance to achieve an optimal total traffic volume. Thus, efficiency enhancements are also possible on a macroscopic level.

15.1.2 Development of Automated Driving Functions

The first steps have already been taken to bring Highly Automated Driving to the road. In 2013, the partially automated driving function, traffic jam assistant [4], was introduced by BMW. More automated driving functions will follow in the next few years to finally achieve—step by step—the preliminary goal: Highly Automated Driving.

The expectations regarding the timeline should not be too high. It must not be forgotten, with each increase in the degree of automation, the complexity of the overall system increases considerably [5] and legal issues are still not solved.

15.1.3 Introduction of Highly Automated Driving: Motorway

As the first use case for Highly Automated Driving, BMW sees HAD functions exclusively on the motorway (Fig. 15.2). This is explained by the following two reasons:

1. Motorways are in many cases monotonous and do not make high demands of the driver. In such situations on the motorway, where the driving task is not enjoyable, the driver can be assisted by HAD functions. The driver dedicates his mental capacity for other tasks.

FUNCTION OVERVIEW

- Remains in the lane.
- Overtakes.
- Cooperative characteristics on entrances to the motorway.
- Observes all traffic rules.
- Speed 0 – 130 km/h.
- In appropriate situations the driver can delegate the entire driving task to the vehicle.
- Highly automated changes between motorways.

Fig. 15.2 Function overview—Highly Automated Driving on the motorway

Function Overview

- Remains in the lane
- Overtakes
- Cooperative characteristics on entrances to the motorway
- Observes all traffic rules
- Speed 0–130 km/h
- In appropriate situations, the driver can delegate the entire driving task to the vehicle
- Highly automated changes between motorways

2. The "low" complexity of the use case on the motorway is likely to be manageable in the near future. Demonstrations with prototypes of different vehicle manufacturers [6–8] and suppliers have already shown on public roads.

   The number of possible driving manoeuvres and interactions with other road users is limited, compared with manoeuvres on urban and rural roads. Complex crossing manoeuvres or situations with pedestrians/cyclists can be excluded as far as possible. In addition, motorways have less complex road geometries and also a more stationary environment with no permanent changes. Thus, a highly accurate localisation of landmarks (e.g. guardrails, bridge piles, lane markings, etc.) is easier to implement. Despite of the relatively simple use case on the motorway, the vehicle validation is much more complex in comparison to today's available driver assistance systems. The average driver drives safely on the motorway; a fatal traffic accident occurs statistically approximately every 12 million km on German motorways [9]. These circumstances make it very difficult to validate the safety of HAD functions with traditional methods. Multitudes of test kilometres on motorway would have to be covered to obtain a statistically validated statement of evidence of safety [9].

15.1.4 Direct Safety Benefit

HAD functions must not have negative effect on the traffic; there must be no worsening in the traffic safety, compared to today's global safety level. As a result, a direct increase in safety on the motorway is expected, but this will not lead to a large reduction of traffic accidents and fatalities. The following three points explain why HAD will not lead to a considerable reduction in road fatalities on the motorway:

1. **The motorway is not an accident black spot**

   Accident and traffic statistics from Germany show that 11 % of traffic fatalities are caused by accidents on the motorway [10], although 30 % of all journeys are covered on motorways [9]. This shows clearly that on motorways, fewer accidents occur per kilometre as on urban or rural roads (Fig. 15.3).
2. **Accidents addressed with ADAS on market on motorway**

   In the beginning, no great benefit is to be expected for road safety by means of Highly Automated Driving. Relevant accidents can be already avoided

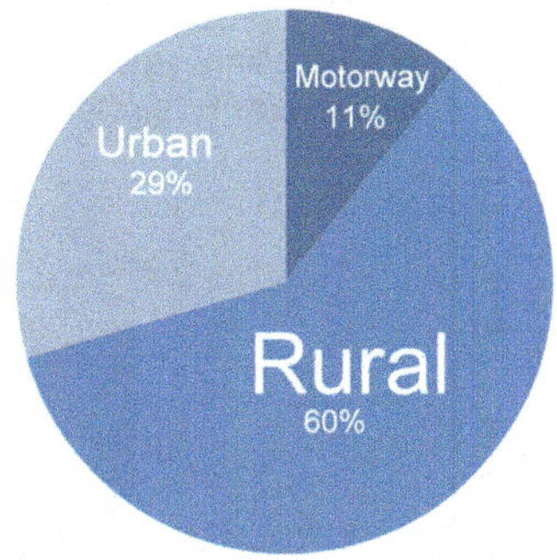

Fig. 15.3 Traffic fatalities on German roads

or mitigated on motorway by ADAS offered today [11]. The distribution of accidents on the motorway shows that 60 % of the accidents happen between vehicles moving along in carriageway and 27 % of accidents are caused by driving accidents, in which no other road user contributed to the cause of the accident [12]. Most of the accidents on motorways are already addressable by current ADAS, as they are available in the current BMW 7 series [13]:

- Adaptive cruise control with emergency brake
- Lane keeping assistant with active side collision protection
- Steering and lane control assistant
- Driver drowsiness detection
- Traffic jam assistant
- Advanced eCall
- Active protection

3. **Low market penetration**
   In the beginning, the market penetration of Highly Automated Driving functions will be limited to high-priced fully equipped vehicles. The market for this class of vehicle is small and, therefore, a low market penetration of HAD functions is initially expected in the entire vehicle fleet.

15.1.5 Indirect Safety Benefit

In addition to a direct increase in safety, there will be an indirect increase in safety with the usage of technology from Highly Automated Driving. From the beginning, HAD technology can be used for improving Active Safety Systems while the vehicle is not in Highly Automated Driving mode. In the long term, a cost-down effect will additionally result in a greater market penetration of HAD technology and derived Active Safety Systems.

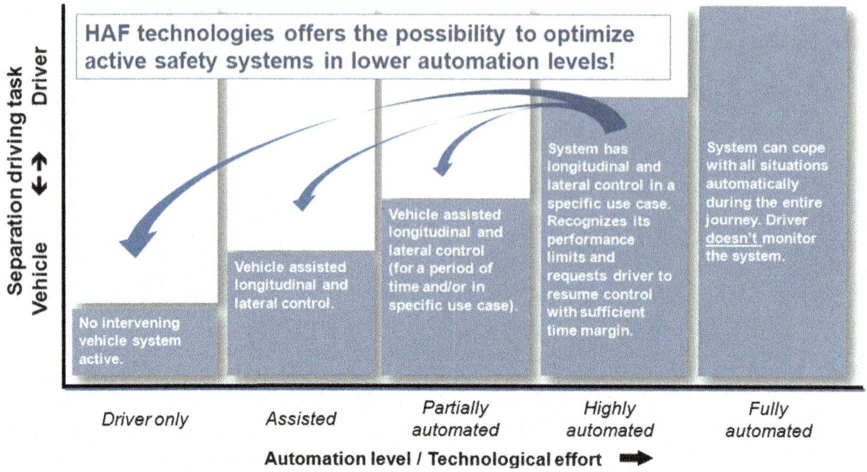

Fig. 15.4 Possibility to use HAD technologies during non-automated driving mode [14]. Automation Levels (BASt classification [1])

1. **Highly Automated Driving Technologies for non-Automated Driving**
   The first use case for Highly Automated Driving on the motorway does not address the majority of accidents, because most traffic accidents occur in situations on urban and rural roads. Precisely, in these situations there is the possibility to assist the driver by means of lower levels of automation (see Fig. 15.4). Vehicles with HAD functions will be equipped with extensive technology for environment recognition and automated vehicle control. These technologies can be used to improve Active Safety Systems. A more accurate detection of the environment around the vehicle and a more detailed interpretation of the traffic situation are possible, which can be used for the design of advanced functions (e.g. higher speed range, stronger deceleration) and for new Active Safety Systems (e.g. evasion assistant). The potential of optimisation of Active Safety Systems is discussed in Sect. 15.2.
2. **"Democratisation" of HAD Technology**
   In the long term, a higher market penetration of Highly Automated Driving functions can be expected in all vehicle classes. The past has shown how new safety systems were introduced in upper class vehicles first, then were successively optionally available in all models and finally became part of the standard equipment for all passenger vehicles. Examples with a high impact on vehicle safety are the airbag, ABS, DSC (ESP) and, most recently, the Autonomous Emergency Braking (AEB) Systems.

   The comfort function Adaptive Cruise Control (ACC) was available for the first time in the BMW 7 Series in 2000. Meanwhile, ACC is available in each vehicle class offered by BMW in different variants. These variants, depending on the vehicle class, use different sensor layouts from a mono camera to a fusion system with stereo camera and radar for object detection. The safety function

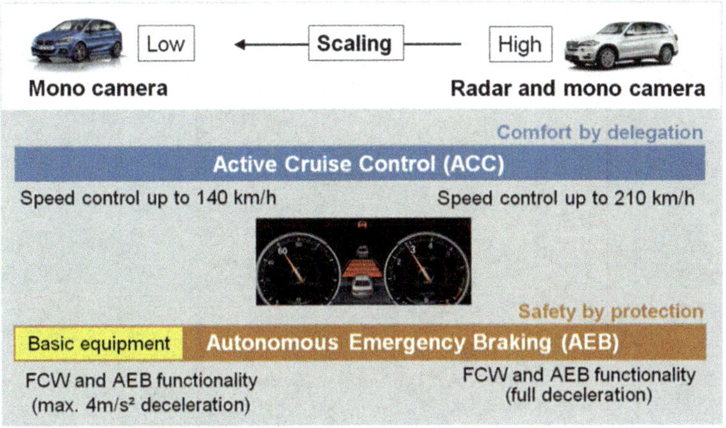

Fig. 15.5 Scaling system characteristics for ACC and AEB

Forward Collision Warning (FCW) including Autonomous Emergency Braking assistant (AEB) has been derived from the comfort function ACC. In case of a potential rear-end collision, the system detects the critical situation, warns the driver and initiates autonomous braking. This safety function has high demands on the overall system: it must be reliable; the strength of deceleration can vary depending on the identification and validation of the object in question up to full deceleration levels and must be controllable for the rear traffic at any time. Despite these high safety requirements, this function has become part of the basic equipment (e.g. in the BMW 2 Active Tourer) and will have an even higher market penetration in future.

Different scalable variants of comfort and safety systems can be offered across all vehicle classes; the variants differ in the offered speed range and the maximum possible deceleration of the emergency brake assistant. Figure 15.5 shows exemplary differences between the basic variant in the 2 Series Active Tourer and the high-end variant in the X5.

15.2 Future Development of Active Safety Systems

15.2.1 Required Technologies: Highly Automated Driving

In order to offer HAD functions, it is not sufficient to combine existing driver assistance systems. Necessary requirements pertaining to environment recognition, trajectory planning and vehicle management cannot sufficiently be fulfilled by available driver assistance systems today. Existing systems must be improved and new technologies need to be integrated into the vehicle. If these premises are fulfilled, HAD functions can bring safety on the road. Figure 15.6 shows the necessary captured environmental parameters and the required simplified technologies.

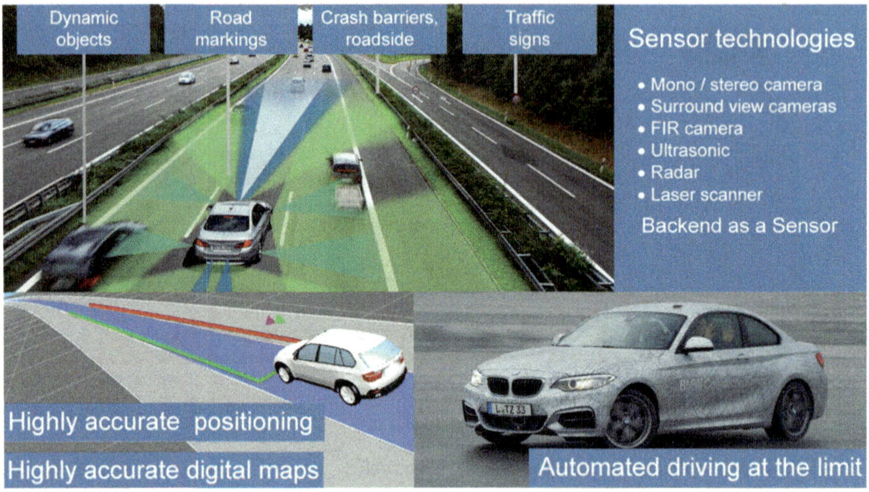

Fig. 15.6 Detected environment and required technologies

15.2.2 Difference Between Highly Automated Driving and Assisted Driving

From a vehicle's safety point of view, the main difference between Highly Automated Driving and assisted driving is that the driver must not monitor the system all the time in every driving situation. As a result, critical situations must be solved by the system and not by the driver. To meet this requirement, all variables which are necessary to describe the traffic situation adequately within the limits of the system must be determined and interpreted correctly.

During manual driving, a critical situation arises by means of a chain of interacting parameters, which are ignored or not recognised by the driver. Driving too fast for a specific situation creates a risk, but does not necessarily lead to an accident. The accident happens only if additional critical parameters occur, such as, e.g. an obstacle after a blind curve. This chain leads to an accident, which is generally due, in approximately 90 % of cases, to misbehaviour of the driver [12]. However, it must be noted that the driver solves a critical situation with the right response in most of the cases.

Even in Highly Automated Driving mode, it is not possible to drive entirely risk free because road transport is per se risky, since unexpected hazards cannot be excluded. Therefore, an action is already initiated when a defined risk level exceeds a limit, in order to reach a safe state, for example, speed reduction or to abort a lane change manoeuvre.

Perhaps, Active Safety can or must learn from the described strategy in the above paragraph, because a critical situation cannot be addressed by automated intervention in the vehicle dynamics at any time. In an early stage, intervention in the vehicle dynamics cannot be undertaken due to the uncertainties in the interpretation of the traffic situation. The uncertainties would result in a too high false positive rate. An excessive frequency of false reactions would lead to reduced acceptance and would cause a negative impact on road safety [15]. Therefore, early intervention can only occur if the critical situation is unavoidable and the driver has not addressed the danger as such.

15.2.3 Benefits for Active Safety Systems

Detailed recognition and correct interpretation of the environment are essential for Highly Automated Driving functions to perform a safe trajectory planning and to respond to oncoming threats [16]. To fulfil these requirements, a **360° environment recognition** is needed, which provides much more details of the traffic situation than systems available today are able to provide. Sub-areas are covered by different sensors (such as camera, radar, laser scanner); this leads to intended overlapping areas in the sensor view (redundancy). Thus, a plausibility check of the provided data is possible and an advanced function design can be implemented, such as a higher deceleration or to initiate steering torque to evade an obstacle.

HAD vehicles will be connected with a **backend**. Necessary information is exchanged in both directions via data connection to safely perform HAD functions. The backend is a server infrastructure and provides data such as updated map data, hazard points or track clearance for HAD functions. As an extended sensor, the backend can indicate threats which are not collectable from the vehicle or are out of the detection range of the onboard sensors. The backend collects data from different sources and evaluates and processes this data to provide useful information for other vehicles. A typical application could be the recognition and warning of a traffic jam. The backend analyses the traffic jam from real time data and forwards this information to relevant vehicles. If this information is available, a safe and efficient speed reduction can be realised, as shown in [17]. Another way to provide useful data is by learning from fleet data. From recorded vehicle data, driver assistance-related parameters are learned by the backend, such as speed limits, cornering speeds or hold lines at intersections. These useful parameters can be used for a better assessment of the traffic situation [17]. This requires strict compliance to the legal framework for data protection.

Highly accurate digital maps contain information such as, e.g. lane-accurate track models, traffic signs, lane markings and landmarks. This information offers the opportunity to gain a better understanding of the traffic situation. Other road users can be associated on the road with the help of the environment recognition and the highly accurate digital map [16]. Available spaces for movement and possible points of conflict can be determined. The map is continually updated by the backend.

By use of **highly accurate positioning** [18] and highly available actuators, **automated vehicle control** at the physical limits [19] can be used to showcase Active Safety Systems and were in fact already demonstrated by BMW [20]. Emergency braking, evasion, automated lateral and longitudinal vehicle control at higher speeds and a safe hold on the roadside in case of an emergency [21] are possible.

Driver monitoring can be used for an advanced function of Active Safety Systems. By analysing gaze behaviour, it is possible to determine whether a critical traffic situation was theoretically recognised or not by the driver. In case of an existing information deficit, the driver can be specifically warned without increasing unjustified warnings. As a result, greater effectiveness at an acceptable false positive rate can be achieved [22]. Upcoming manoeuvres can be recognised with the help of a driver's gaze behaviour. The possible movement of the vehicle can be restricted; with this information, a better interpretation of the traffic situation can be realised [23].

15.2.4 Development Process for Active Safety

Available Active Safety Systems already address a high proportion of the most common accidents. However, not all accidents can be addressed with the current system design, because today a complete interpretation of the traffic situation is not possible. This limitation is based on a too large variation in traffic situations and too many parameters (road topology, road users, driver condition, etc.) which define a traffic situation. Therefore, the system design of Active Safety Systems is focused on common types of accidents (e.g. rear-end collisions) with few possible causes of accidents (e.g. insufficient stopping distance). This approach works very well and a high coverage in the field is reached, e.g. for rear-end collisions, as described in [11].

A detailed analysis of accident data has shown [24] that there are 5313 meaningful combinations of types and causes of accidents, which covers all possible road accidents. Interestingly, 50 % of all accidents are covered by the 26 most common combinations of types and causes of accidents (see Fig. 15.7).

An additional linear increase of the coverage of types of accidents and causes of accidents leads to an exponential increase in the effort that is necessary for developing Active Safety Systems. To gain a greater coverage of accidents, this approach is no longer practicable. Too great effort in the development of individual Active Safety Systems is required to cover only a small number of accidents. The number of different system designs of Active Safety Systems would lead to a no longer manageable complexity of the overall system. For a manageable system, new approaches are needed, which are able to address a larger variation of critical situations.

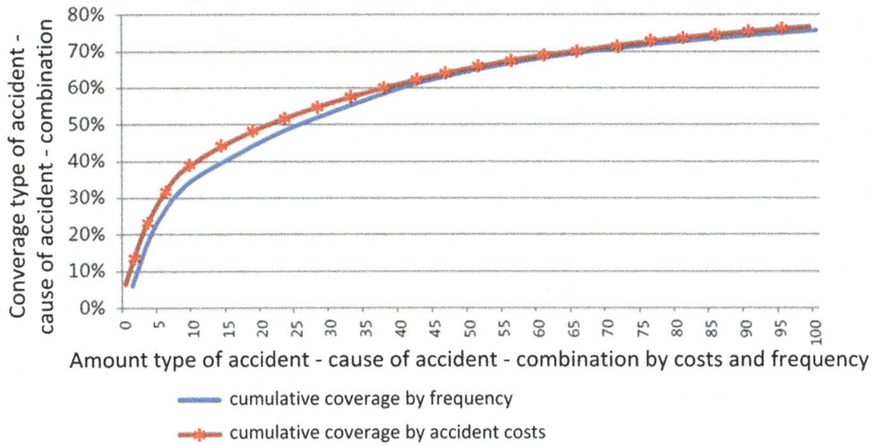

Fig. 15.7 Cumulative coverage of accidents depending on the type of accident and cause of the accident combination [24]

15.2.5 Future Requirements and Perspectives

For a greater coverage of critical situations, the variation of potential traffic situations is too great and can only be solved if all relevant properties are included in the calculation of the criticality of the traffic situation. New approaches are required for the functional development of Active Safety Systems, especially when more and more properties of the traffic situations under consideration are recognised, due to the increasing degree of automation. Particularly in the interpretation and evaluation of traffic situations, Rodemerk et al. [25] and Wachenfeld and Winner [26] demand a more generic approach. Future approaches should not be restricted to a few types of accident with a limited variation. They should interpret the traffic situation as a whole.

To address more accidents, an earlier resolution of the cause of the accident is needed and not only a reaction at the latest possible time, when a collision is unavoidable. To achieve this, it is necessary to interpret the traffic situation correctly and to derive the potential cause of accident. In the first step, all relevant properties of the traffic situation must be determined and assessed; see Fig. 15.8. The first challenge is to determine which properties are relevant (such as road topology) and how these properties can be determined (e.g. with highly accurate digital maps). Only if all relevant properties are captured, the cause of the critical situation can be identified and addressed in the next step.

Not all properties can be captured without additional new technologies. A pedestrian who is hit by a vehicle, accident black spots on a road section or the road conditions are not fully identifiable with currently available systems in the vehicle.

In the next step, the captured properties of the traffic situation need to be evaluated so that the critical variables of this situation can be derived. An example

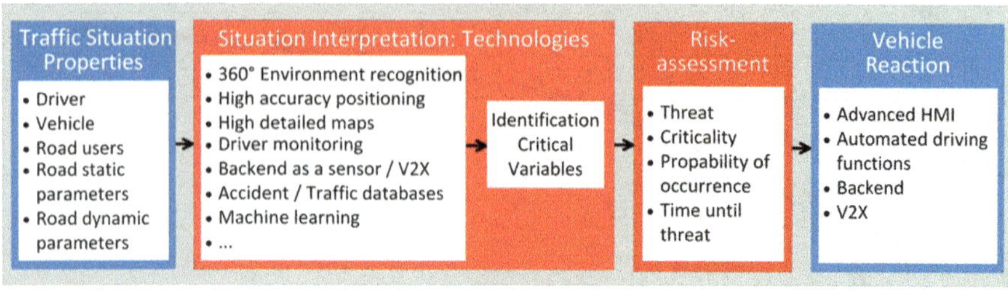

Fig. 15.8 Systematic information flow of ADAS

would be: pedestrian is partially hit, crossing trajectory, wet road, driver distracted, school on the left side, 07:09, etc.

After identifying the critical variables, an assessment of risk is performed in the next step. The risk assessment must not only cover selected types of accidents but must be carried out using a generic approach. As a result, factors such as the type of threat, criticality, probability of occurrence and time until threat are provided by the risk assessment.

In the final step, the vehicle reacts on the basis of the risk assessment. Depending on the escalation level, the reaction can range from warning information to an automated driving manoeuvre. The challenge here is not to disturb or to overstrain the driver with too much information. To solve this problem, only dedicated information can be distributed to the driver when there is an information deficit.

New technologies from Highly Automated Driving move the function development of Active Safety Systems in three directions:

1. An improvement of current Active Safety Systems (e.g. evasion assistant [27]), with an interpretation of more properties of the traffic situation and an advanced actuation strategy.
2. New Active Safety Systems (e.g. emergency stop assistant), which resolve critical situations when the driver is no longer able to control the driving task.
3. The development of a continuous risk assessment of the entire traffic situation to enable an earlier resolution of a critical situation by means of a better interpretation of the traffic situation. Thus, the driver can mitigate the critical situation by himself with information provided specifically for this purpose, if an information deficit exists [23].

15.3 Prospective Evaluation of the Effectiveness of Active Safety Systems and HAD Systems

15.3.1 Challenges

Highly Automated Driving functions and Active Safety Systems are safety relevant; Daimler [28] also discussed the challenges and the safety-relevant validation associated with HAD functions. It is clear that the development and the usage of Active Safety Systems require prospective and quantitative statements regarding the impact of Active Safety Systems on traffic safety, requested in [15]. For homologation of HAD functions, it has to be proven that there is no negative effect on traffic safety. Available driver assistance systems must provide evidence that the driving functions are safe and manageable. This validation of available driver assistance systems is still manageable; the traffic situations in which the systems operate are complex, but the systems evaluate and operate the driving function only based on a limited number of properties of the traffic situation.

It is not feasible to perform the validation of the safety aspects of Highly Automated Driving functions on the road: according to Winner [9], HAD functions for the highway would have to cover around 100 million km road tests to generate a statistically valid statement that the system can be classified as safe. Any changes to the system require a renewed assessment. Both technically and economically, it is not possible to provide the necessary evidence using road tests alone.

Another problem is the almost infinite number of possible traffic situations. This necessitates a validation of perception and cognitive capability of the system. With currently available methods, the validation of Highly Automated Driving functions or complex Active Safety Systems is not possible.

15.3.2 Variability at the Model Design

HAD functions and Active Safety Systems are designed to perform automated vehicle guidance and to avoid and minimise the consequences of critical situations. To achieve this, all relevant parameters pertaining to the driver, the traffic situation and the system itself need to be captured and necessary reactions need to be performed. This leads to a widespread of possible variations, shown in Fig. 15.9. Each variation can entail thereby in principle positive or negative effects on traffic safety and must, therefore, be considered. For example, a stochastic modelling is recommended to cover the variation of the relevant influencing factors [15].

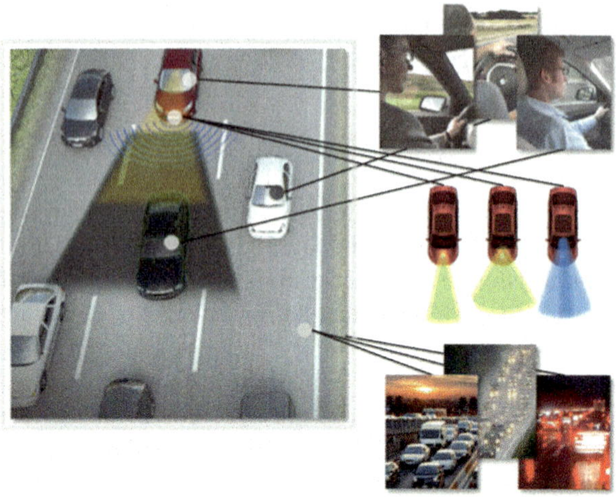

Fig. 15.9 Necessary variation in modelling, for example, emergency brake assistant [14]

15.3.3 Evaluation of the Effectiveness

An objective metric, as required by Kompass et al. [15, 29], is necessary to determine a quantifiable prognosis regarding traffic safety benefits. Different interests of vehicle manufacturers, suppliers, official decision makers, insurance companies and consumer protection organisations need to be considered. This requires a careful definition of a metric, which must deal with all necessary interests of stakeholders as stated above to achieve the necessary acceptance.

In addition to the safety benefit of new driving functions, also possible risks need to be considered to enable an assessment of traffic safety. Helmer shows in [30] a way to perform an assessment of safety benefits and negative effects with the help of an objective metric. By use of this, a driving function can be assessed in terms of traffic safety and optimised in the development process.

The overall consideration of a system must take into account the varying boundary conditions of the situations contemplated. A complete assessment of HAD functions and future safety functions can only be achieved with the help of virtual testing and validation, where realistic traffic scenarios must be taken into account. A possible process for effectiveness assessment of HAD functions and Active Safety Systems is described in detail in Chap. 20.

To come closer to a harmonised methodology, the "Harmonisation Group" was founded with representatives from different domains in 2012 [31]. A pilot usage of the methodology for the effectiveness analysis is applied in the project AdaptIVe [32] to analyse the iterative process in the development and validation phase of HAD functions.

15.4 Conclusion

Highly Automated Driving will increase comfort and safety on the roads. In the first HAD use case on motorways, the number of traffic accidents will not decrease enormously, because of the low number of accidents on the motorway compared to urban or rural roads, the addressing of today's available Active Safety Systems and the low market penetration of Highly Automated Driving functions.

Highly Automated Driving can be the enabler for safety functions, which could additionally increase the safety in the field of assisted driving in any traffic situation. Current Active Safety Systems can be improved and new Active Safety Systems can be implemented, which are able to address a large number of different accidents.

The effectiveness analysis of Active Safety Systems and the validation of HAD functions require new and until now unresolved challenges. Currently available methods do not sufficiently demonstrate evidence of the safety benefits. Prospective effectiveness analyses are needed to solve this problem so that HAD function and improved Active Safety Systems can be brought on the road.

References

1. T.M. Gasser, D. Westhoff, BASt-Study: Definitions of Automation and Legal Issues in Germany, German Federal Highway Research Institute, 2012
2. D. Damböck, M. Farid, L. Tönert, K. Bengler, Übernahmezeiten beim hochautomatisierten Fahren, 5. Tagung Fahrerassistenz, Munich, 2012
3. R. Yerkes, J. Dodson, The relation of strength of stimulus to rapidity of habit-formation. J. Comp. Neurol. Psychol. **18**(5), 459–482 (1908)
4. S. Lüke, O. Fochler, T. Schaller, U. Regensburger, Stauassistenz und -automation, in *Handbuch Fahrerassistenzsysteme*, 3. Auflage, ed. by H. Winner, S. Hakuli, F. Lotz, C. Singer (Springer Vieweg, Wiesbaden, 2015)
5. B. Filzek, Schlüsseltechnologien zum Automatisierten Fahren, AAET 2014, Braunschweig, 2014
6. BMW Group PressClub Global, Ready for Takeover! (2011), [Online] https://www.press.bmwgroup.com
7. AUDI AG, Long-Distance Test Drive Successfully Completed: Audi A7 Sportback Piloted Driving Concept Arrives in Las Vegas Following 560 Mile Drive (2015), [Online] http://www.audi-mediacenter.com
8. Daimler AG, Pioneering Achievement: Autonomous Long-Distance Drive in Rural and: Mercedes-Benz S-Class INTELLIGENT DRIVE Drives Autonomously in the Tracks of Bertha Benz (2013), [Online] http://media.daimler.com/
9. H. Winner, Absicherung automatischen Fahrens, 6. FAS-Tagung München, Munich, 2013
10. DESTATIS, Road Accidents 2012, Statistisches Bundesamt, Wiesbaden, 2013
11. T. Hummel, M. Kühn, J. Bende, A. Lang, An Investigation of Their Potential Safety Benefits Based on an Analysis of Insurance Claims in Germany, UDV (German Insurers Accident Research), Berlin, 2011
12. DESTATIS, Road Accidents 2013, Statistisches Bundesamt, Wiesbaden, 2014.
13. BMW Group, BMW 7 Series (2015), [Online] http://www.bmw.com/7series
14. T. Schaller, Active Safety on the Road Towards Highly Automated Driving, in *8th Graz Symposium Virtuelles Fahrzeug*, Graz, 2015

15. K. Kompass, T. Helmer, L. Wang, R. Kates, Gesamthafte Bewertung der Sicherheitsveränderung durch FAS/HAF im Verkehrssystem: Der Beitrag von Simulation, in *Haus der Technik—Tagung Fahrerassistenz und Aktive Sicherheit*, Essen, 2015
16. N. Aeberhard, S. Rauch, M. Bahram, G. Tanzmeister, J. Thomas, Y. Pilat, F. Homm, W. Huber, N. Kaempchen, Experience, results and lessons learned from automated driving on Germany's highway. IEEE Intell. Transp. Syst. Mag. **7**(1), 42–57 (2015)
17. F. Klanner, C. Ruhhammer, Backendsysteme zur Erweiterung der Wahrnehmungsreichweite von Faherassistenzsystemen, in *Handbuch Fahrerassistenzsysteme*, 3. Auflage, ed. by H. Winner, S. Hakuli, F. Lotz, C. Singer (Springer, Wiesbaden, 2015)
18. R. Krzikalla, A. Schindler, M. Wankerl, R. Wertheimer, *Vernetztes Automobil: Mehr Sicherheit durch Positionsbestimmung mit Satelliten und Landmarken* (Springer, Wiesbaden, 2014)
19. P. Waldmann, D. Niehues, Der BMW Track Trainer—Automatisiertes fahren im Grenzbereich auf der Nürburgring Nordschleife, Garching, 4.Tagung Sicherheit durch Fahrerassistenz, Germany, 2010
20. BMW Group PressClub Global, BMW at the Consumer Electronics Show: Highly Automated Driving at the Limit (2014), [Online] https://www.press.bmwgroup.com/
21. N. Kaempchen, M. Aeberhard, P. Waldmann, M. Ardelt, S. Rauch, Der BMW Nothalteassistent: Hochautomatisiertes Fahren für mehr Sicherheit im Straßenverkehr, Elektronik Automotive, 2011
22. M. Liebner, C. Ruhhammer, F. Klanner, C. Stiller, Generic Driver Intent Inference Based on Parametric Models, in *IEEE Conference on Intelligent Transportation Systems, Proceedings*, Netherlands, 2013
22. M. Liebner, F. Klanner, Driver intent inference and risk assessment, in *Handbook of Driver Assistance Systems*, ed. by H. Winner, S. Hakuli, F. Lotz, C. Singer (Springer, Switzerland, 2015)
24. T. Heinrich, J. Ortlepp, J. Schmiele und H. Voß, Infrastrukturgestützte Fahrerassistenz, UDV (German Insurers Accident Research), Berlin, 2011
25. C. Rodemerk, S. Habenicht, A. Weitzel, H. Winner, T. Schmitt, Development of a General Criticality Criterion for the Risk Estimation of Driving Situations and Its Application to A Maneuver-Based Lane Change Assistance System, Alcala de Henares, in *IEEE Intelligent Vehicles Symposium*, Spain, 2012
25. W. Wachenfeld, H. Winner, Lernen autonome Fahrzeuge? in *Autonomes Fahren—Technische, rechtliche und gesellschaftliche Aspekte*, ed. by M. Maurer, J. Christian Gerdes, B. Lenz, H. Winner (Springer Vieweg, Berlin, 2015)
27. A. Seewald, C. Haß, M. Keller, T. Bertram, Emergency steering assist for collision avoidance. ATZ Worldwide **117**, 14–19 (2015)
28. R. Herrtwich, Autonomous Automobile Future, in *8th Graz Symposium Virtuelles Fahrzeug*, Graz, 2015
29. K. Kompass, L. Wang, T. Helmer, Prospektive Wirksamkeitsanalyse von aktiven Sicherheitssystemen und HAF, VDA Technischer Kongress, 2015
30. T. Helmer, Development of a Methodology for the Evaluation of Active Safety Using the Example of Preventive Pedestrian Protection, Springer Thesis, 2015
31. Y. Page, F. Fahrenkrog, A. Fiorentino, J. Gwehenberger, T. Helmer, M. Lindman, O. Lex van Rooij, S. Puch, M. Fränzle, U. Sander, P. Wimmer, A Comprehensive and Harmonized Method for Assessing the Effectiveness of Advanced Driver Assistance Systems by Virtual Simulation: The P.E.A.R.S. Initiative, in *International Technical Conference on the Enhanced Safety of the Vehicles*, Gothenburg, Sweden, 2015
32. AdaptIVe, European Research Project: AdaptIVe, Automated Driving (2015), [Online] https://www.adaptive-ip.eu/

Chapter 16
Functional Safety of Automated Driving Systems: Does ISO 26262 Meet the Challenges?

Helmut Martin, Kurt Tschabuschnig, Olof Bridal, and Daniel Watzenig

16.1 Introduction

Science fiction stories about autonomous cars have inspired the imagination for many years. In the early 1980s, the television series *Knight Rider* presented the self-driving and artificial intelligent car named KITT,[1] and the slogan went, 'Knight Rider—A shadowy flight into the dangerous world of a man who does not exist'. Techies of the time were fascinated by the possibility of a technology and imagined that it would be possible to drive or simply travel in cars of the kind in the near future. Today, some decades later, that vision is starting to be made a reality, which will change and further influence the common understanding of the existing human road mobility system. For the last 30 years, the main innovations of vehicle technologies have been achieved by E/E systems in the automotive industry [1], e.g. anti-lock braking system (ABS) in 1978, electronic stability program (ESP) in 1995 and up to collision avoidance systems in 2010 (see Fig. 16.1).

New generations of the advanced driving assistance systems (ADAS) are more complex than ever before in two aspects: firstly from a technical point of view in

[1] Knight Industries, 2000.

H. Martin (✉)
Virtual Vehicle Research Center, Graz, Austria
e-mail: helmut.martin@v2c2.at

K. Tschabuschnig
Magna Steyr Engineering AG & Co KG, Graz, Austria

O. Bridal
VOLVO Group Trucks Technology, Gothenburg, Sweden

D. Watzenig
Virtual Vehicle Research Center and Graz University of Technology, Institute of Electrical Measurement and Measurement Signal Processing, Graz, Austria

D. Watzenig, M. Horn (eds.), *Automated Driving*,
DOI 10.1007/978-3-319-31895-0_16

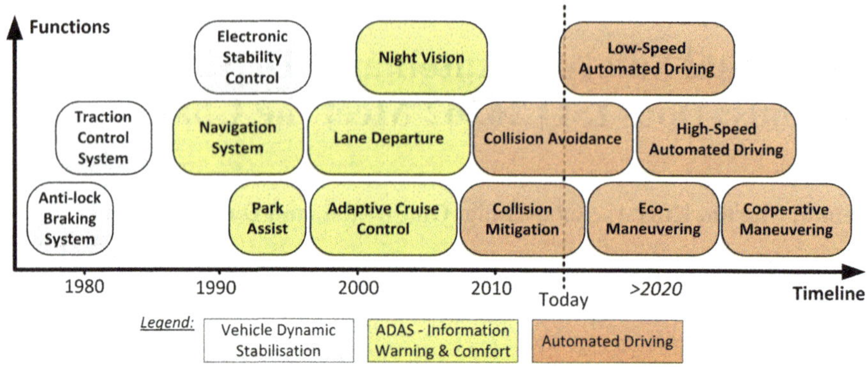

Fig. 16.1 Evolution of advanced vehicle functions

the context of the introduction of new technologies for implementing the functions required and secondly from an organisational point of view concerning the whole supply chain including the suppliers involved for a different kinds of services and products during the lifecycle of an automotive vehicle. In this chapter, we will focus on the technical aspect as well as on the discussion about the challenges of automated driving functions and of how to apply the existing version of the ISO 26262 [2] standard concerning automotive functional safety.

16.1.1 From Driver Assistance to Highly Automated Driving Systems

Today, almost every car in the market provides driver assistance systems (e.g. electronic stability control—ESC). For safety reasons, high-class vehicles are equipped with various additional ADAS functions (e.g. adaptive cruise control—ACC). The introduction of such systems has helped to reduce the number of fatal accidents [3, 4]. However, more than 90 % of accidents still occur as a result of human misbehaviour or mistakes. Thus, it is an important topic for the European Union to reduce the number of human-caused accidents by introducing the next generation of ADAS for our cars, which are referred to as automated driving systems (ADS).

The different definitions of driving automation for on-road vehicles by SAE in the standard J3016 [5] and recommendations provided by BASt[2] and NHTSA[3] are shown and compared with each other in Fig 16.2. The comparison between the

[2]Germany Federal Highway Research Institute (BASt)—http://www.bast.de.

[3]US National Highway Traffic Safety Administration (NHTSA)—http://www.nhtsa.gov/.

Level	Name	Narrative definition	Execution of steering and acceleration/ deceleration	Monitoring of driving environment	Fallback performance of *dynamic driving task*	System capability (*driving modes*)	BASt level	NHTSA level
***Human driver* monitors the driving environment**								
0	No Automation	the full-time performance by the *human driver* of all aspects of the *dynamic driving task*, even when enhanced by warning or intervention systems	Human driver	Human driver	Human driver	n/a	Driver only	0
1	Driver Assistance	the *driving mode*-specific execution by a driver assistance system of either steering or acceleration/deceleration using information about the driving environment and with the expectation that the *human driver* perform all remaining aspects of the *dynamic driving task*	Human driver and system	Human driver	Human driver	Some driving modes	Assisted	1
2	Partial Automation	the *driving mode*-specific execution by one or more driver assistance systems of both steering and acceleration/deceleration using information about the driving environment and with the expectation that the *human driver* perform all remaining aspects of the *dynamic driving task*	**System**	Human driver	Human driver	Some driving modes	Partially automated	2
***Automated driving system* ("system") monitors the driving environment**								
3	Conditional Automation	the *driving mode*-specific performance by an *automated driving system* of all aspects of the *dynamic driving task* with the expectation that the *human driver* will respond appropriately to a *request to intervene*	System	**System**	Human driver	Some driving modes	Highly automated	3
4	High Automation	the *driving mode*-specific performance by an *automated driving system* of all aspects of the *dynamic driving task*, even if a *human driver* does not respond appropriately to a *request to intervene*	System	System	**System**	Some driving modes	Fully automated	3/4
5	Full Automation	the full-time performance by an *automated driving system* of all aspects of the *dynamic driving task* under all roadway and environmental conditions that can be managed by a *human driver*	System	System	System	**All driving modes**		

Fig. 16.2 Definition of SAE driving automation levels for on-road vehicles and comparison with BASt and NHTSA [36]

levels proposed by the various standards/recommendations is possible up to the BASt Level 4 'fully automated' (see blue line in Fig 16.2).

Evolution of driving systems (based on the definition by BASt/Lx . . . Level x):

L0. *Driver only*—Driver assistance comfort system (e.g. speed limiter).

Responsibility: Driver.
Safe State: Driver always has the control of the vehicle.

L1. *Assisted*—Advanced driver assistance provides safety improvement; ADAS supports the driver (e.g. EBA,[4] ACC, LKA[5]).

Responsibility: Driver.
Safe State: Driver takes over full control of the vehicle.

L2. *Partly automated*—Driving system controls laterally and longitudinally for a certain time in few situations (e.g. motorway assistant).

Responsibility: Driver.
Safe State: Driver takes over full control of the vehicle.

L3. *Highly automated*—Driving system controls lateral and longitudinal movement for a certain time in specific situations (e.g. motorway chauffeur).

Responsibility: Driver *or* system.

[4]Emergency Brake Assist.

[5]Lane Keeping Assist.

Safe State: Driver takes over full control of the vehicle within a specific timeframe *or* system has to control the vehicle in defined driving situations, if the driver did not take over full control.

L4. *Fully automated*—Driving system has complete control of lateral and longitudinal movement within a specified situation of the application (e.g. motorway pilot).

Responsibility: System.
Safe State: System controls the vehicle in some driving situations.

In SAE J3016, the highest level is 'Full Automation', which means from our perspective an autonomous vehicle that is able to drive without a driver. This level is not reached in this chapter because this scenario is too far away from today's technical practice.

The role of the driver will continue to be important for the introduction of automation functions in vehicles over the next few years. For high levels of automation, the driver should not be required to cope with any critical driving situation. In such cases, the ADS should be able to handle any kind of driving situation autonomously—but this is still a future perspective expected that is expected to become reality around the years 2025–2035.

In the past, vehicle manufacturers realised their particular ADAS functions independently on a do it alone basis and using different OEM[6]-specific trade names (e.g. Adaptive Cruise Control (ACC), Active Cruise Control (ACC), Cooperative Adaptive Cruise (CACC), Distronic Plus). The function itself as well as the handling and the user interaction typically slightly differed from each other to guarantee OEM-specific originality. The levels of automation have to be harmonised for the introduction of ADS functions; otherwise, the driver will not be able to operate different systems in the required way without training or a special extended driving licence for automated vehicles as recommended by NHTSA [6]. One important aspect for handling the challenges is the standardisation and harmonisation of ADS functions of all OEMs on the market. The standardisation must include not only the vehicle itself but also the overall aspects concerning the ecosystem that are required to realise ADS functions like infrastructure (e.g. map data) or environment (e.g. secure C2X[7] communication). In aviation, the rulemaking advisory committee ARAC[8] harmonises all the aviation-specific standards (e.g. for system failures, underdetermined air traffic situation and human factor faults). The awareness of the need for such a rulemaking advisory committee for road vehicles is also given in the automotive industry as an automated vehicle will not be a closed system as was the situation in the past.

[6]Original equipment manufacturer.

[7]Car-to-x means a communication between the car and any other external system, e.g. other cars C2C or the infrastructure C2I.

[8]Aviation Rulemaking Advisory Committee—http://avstop.com/legal/2.htm.

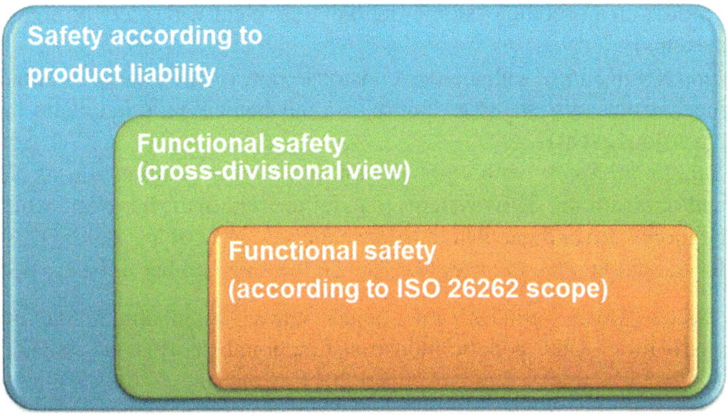

Fig. 16.3 Overview of different safety levels

If we compare the situation of aviation with the road mobility standards concerning safety, ISO 26262 today covers only a subset of those system safety regulations. As an example, we wish to mention the interaction of ADS with the driver in aspects such as warning of the driver, supporting the driver so that an appropriate reaction can occur and feedback to the driver concerning his/her reaction. Only if the reaction of the vehicle is clearly defined and the driver knows which actions are carried out by the vehicle on its own, the right decision or reaction can be expected from the driver within a specific driving situation when needed.

16.1.2 Functional Safety According to ISO 26262

Safety is one of the key issues of road vehicle development. New innovative vehicle functionalities are not only introduced as driver assistance functions. Concerning propulsion, vehicle dynamics control and active and passive safety systems increasingly enter the domain of system safety engineering. Development and integration of these functionalities will enforce the need for a serious consideration of safety within the system development and the need to provide evidence that all reasonable system safety objectives are reached [5].

There are different levels of safety (LoS) (see Fig. 16.3):

LoS1. *Safety with Respect to Product Liability*[9] where safety aspects of any kind must be covered in order to achieve the permission for the launch of a

[9]For example, Austrian Federal Act—Governing the Liability for Defective Product/Product Liability [7]: §5. (1) A product §5. (1) A product shall be deemed defective if it does not provide the safety which, taking all circumstances into account, may be reasonably expected, in particular with respect to: (1) the presentation of the product, (2) the use to which it can reasonably be expected that the product would be put and (3) the time when the product was put into circulation.

product on a specific customer market (e.g. electrical safety of high-voltage systems)

LoS2. *Functional Safety* with a cross-divisional view of any type of malfunction in mechatronic systems (e.g. failure of a mechanical part that could lead to a hazardous event)

LoS3. *Functional Safety* with emphasis on any kind of malfunction of electrical and/or electronic (E/E) systems (e.g. failure within the hardware which must be monitored and handled to achieve the safe state of a system). This means for the automotive industry, the ISO 26262 standard has to be applied.

ISO 26262 'Road Vehicles—Functional Safety' is an automotive industry-specific derivation of the generic industrial functional safety standard IEC 61508 [8]. ISO 26262 was released in November 2011 as the state of the art international standard for E/E systems in passenger cars. It provides a structured and generic approach for the complete safety lifecycle of an automotive E/E system, including design, development, production, service processes and decommissioning. ISO 26262 defines the Automotive Safety Integrity Level (ASIL) as a risk classification parameter for the safety-critical hazardous situation of an item.[10] This is an important parameter for all subsequent safety activities in the safety lifecycle. The ASIL can be seen as a parameter that indicates a reduction of risk requirement in order to achieve a tolerable risk level.

The overall systems engineering must cover all kinds of system properties such as reliability, availability, maintainability, security and (functional) safety. Reliability engineering is closely related to safety engineering and to system safety. Both use common methods for their analyses and may require inputs from each other. Reliability engineering typically focuses on costs through failure caused by system downtime, cost of spares, repair equipment, personnel and the cost of warranty claims. Safety engineering normally does not emphasise costs but rather the preservation of life and nature. Therefore, it deals only with particular safety-critical and dangerous system failure modes [9]. Safety and reliability are different properties. A system can be reliable and unsafe while it can also be unsafe and reliable (see Fig. 16.4). Furthermore, in some cases, these properties even come into conflict with each other. Leveson discusses this problem with very interesting examples from the military as well as the avionic and chemical industries [10].

The ISO 26262 standard states 'ISO 26262 does not address the nominal performance of E/E systems, even if dedicated functional performance standards exist for these systems (e.g. active and passive safety systems, brake systems, Adaptive Cruise Control)'. ASIL is not a nominal performance metric for other system properties (e.g. maintainability, reliability, availability) of ADS functions. Specific metrics for other concerns need to be examined in certain analyses of the particular scope (e.g. mean time to repair (MTTR) for maintainable systems).

[10] An item is a system or array of systems for implementing a function at vehicle level, to which ISO 26262 is applied.

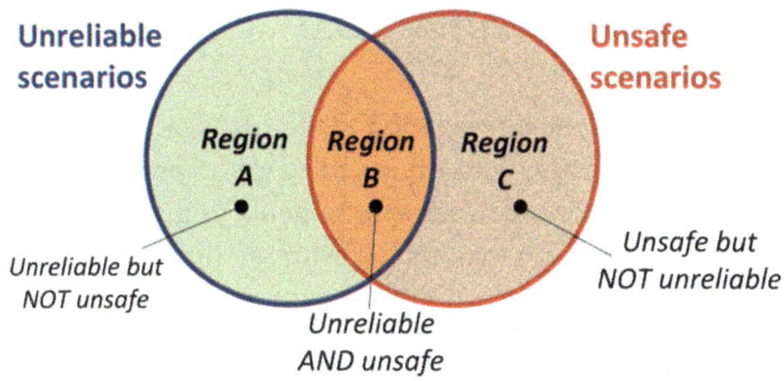

Fig. 16.4 Relation of unreliable and unsafe scenarios

The ISO 26262 standard provides guidance by introducing requirements and recommendations to reduce the risk of systematic development failures and to handle the complexity of E/E systems. Nevertheless, compliance with the standard presents a significant challenge for companies, because ISO 26262 sets requirements and recommendations but does not explicitly define how they should be implemented in an efficient way in the context of a particular application. To implement the requirements and recommendations of the ISO 26262 in a particular application, expert knowledge in functional safety must create a thoughtfully argued and documented interpretation of the ISO 26262 for the particular application.

ISO 26262 provides a systematic top-down engineering approach based on the V-model.[11] A specification starts from the system-of-systems (SoS) level down to the subsystem and component level and subsequently to the implementation level of hardware (HW) and software (SW) modules. After the implementation and verification of HW and SW, the integration a bottom-up approach follows on at the right side of the V-model: integration of HW and SW modules in components (e.g. HW+SW in ECU), components in subsystems (e.g. ECU in HV battery), subsystems to system (e.g. HV battery in powertrain) and system in SoS (e.g. powertrain in vehicle).

16.2 General Challenges of ADS

Some challenges are particularly relevant for automated systems in general terms (compared to 'classic' automotive electronic systems) and are related to complexity, availability and reliability. This section provides an overview of different kind of challenges that must be investigated for the development of safety-critical aspects of ADS.

[11] See definition at http://v-modell.iabg.de/v-modell-xt-html-english/index.html.

16.2.1 Increasing Complexity of ADS

A system can be described as an aggregation of elements or components concerning their cooperation and interaction with others to function properly. Interactions in a system are exchange processes between components realised by flow of material, energy and information (component relationships). In the event of failure, the system should be able to react in a fault-tolerant manner, which means that the system is able to trap a fault—'the system and its intended functions are able to survive' [11].

Safety is a system property intended to avoid system faults or malfunctions from causing any substantial damage (e.g. injuries to people or damage to the environment), which requires precise error detection. If an error is detected, the system must switch into a passive safe state with the consequence that the system is no longer available or reliable, but it is safe (failure integrity). The influence of system attributes such as availability, reliability, safety and security[12] must be harmonised, and a kind of trade-off is required, because the ADS can be safe but that does not mean that the system is available or secure.

If a system is required to guarantee high availability and fail-operational characteristics, the system architecture is expected to have higher complexity of implemented functions. This means that the system grows in terms of the number of components and the interactions between them. The effort involved for the additional system safety causes increasing complexity. In addition unexpected effects arise when repetitive interactions are effected by increasing non-linear functions between the components. The most important attributes [12] of complex systems are:

- *Nontransparency*—state, interconnection and behaviour of a system and its components are only partly known.
- *Sensitivity*—interference of results in the case of unexpected input changes.
- *Instability*—smallest disturbances cause unknown, unwanted behaviour of the system.
- *Internal dynamics*—continuous change of the system's state by the system itself without any external influence.

The mentioned attributes promote the appearance of additional faults and complicates their identification. Despite simplest components and interactions, the whole system generates forms, patterns and behaviour dynamics that could not be derived from particular components. This property is referred to as emergence,[13] which arises from various signal feedbacks of the system components.

One popular development method is to abstract the reality, which means building a model to simplify or reduce the reality and capture the interesting major behaviour

[12]See also 'dependability'—umbrella term to describe different quality attributes of a system.

[13]Emergent entities (properties or substances) 'arise' out of more fundamental entities and yet are 'novel' or 'irreducible' with respect to them [13].

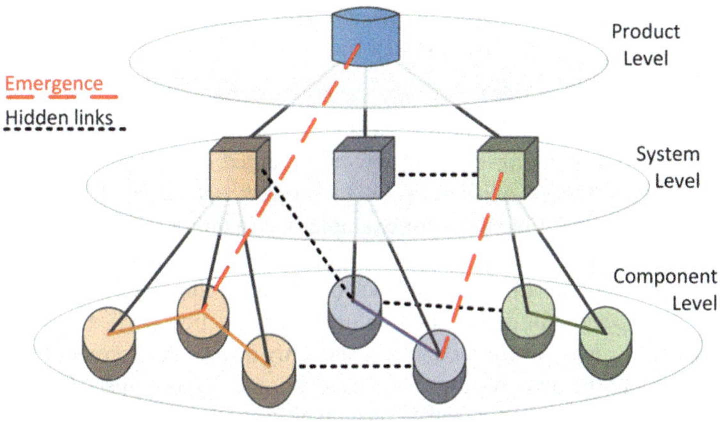

Fig. 16.5 Latent linkages between system components and integration levels

of the system. The state space of a model is always smaller than the state space of the real world because not all parameters such as temperature and friction that affect the components are considered. The synthesis of the component models does not show all operating states or all linking conditions. In particular, undesirable effects and hidden links could occur (see Fig. 16.5).

Unidentified coupling of components and over different integration levels may lead to systematic faults during the modelling of the systems. These nontransparency links and reactions to signals are the cause of unpleasant effects such as emergence (spontaneous system behaviour caused by smallest state changes on lowest level without direct derivation), common cause effects (single fault, cause simultaneously multiple components failure), powered run away (activation of a not provided function and is not designated in the conception or signal flow) and hidden links (unwanted operation states in the system, not identified as failure). In these cases, the system works incorrectly, while a faulty state is not visible. That could cause the loss of all safety reserves in the system. The implemented safety mechanisms are ineffective and cannot be activated because the functional chain is unknown. These nontransparency links must be discovered during the system design.

The mission of mastering complex systems is to control the above-mentioned impacts in time and to prevent injurious effects. This could be done by a safe system design and increasing system transparency. The quality, robustness and fault tolerance of the design depends on prediction potential of the applied development procedure.

16.2.2 Strict Requirements Concerning Availability and Reliability of ADS

A high degree of automation means that many—and potentially all—of the tasks usually carried out by the driver will now be executed by control systems in such a manner that the driver relies fully on the correct operation of these systems. The unavailability of a function—for example, the inability to perform automated braking or automated steering—is more critical when the driver is 'not in the loop' than it would be if the driver is 'in the loop'. Regarding safety, it is generally considered as acceptable that a semi-autonomous function such as conventional cruise control or adaptive cruise control is suddenly deactivated, provided the driver is informed about the deactivation. The deactivation could be caused by a detected error in the system, by the activation of a stability function such as ESP or ASR or by some other triggering condition. The sudden loss of the vehicle's ability to drive autonomously, perhaps after several hours of fully autonomous driving, would typically be considered highly critical concerning safety, even if the driver is forced to take over control of the vehicle. In an extreme case, the vehicle continues to operate fully autonomously and to the extent that the driver does not even have any possibility to take over control.

Thus, the closer we approach towards fully autonomous vehicles, the more important it becomes to ensure that automated functions are fully available. Classical 'fail-safe' design solutions that rely on deactivating a function and informing the driver are no longer sufficient. Instead of fail-safe designs, fault-tolerant designs will be needed so that functions remain operational even when a fault is encountered in the system.

In context of criticality of potential failures of functions for highly automated driving, it is clear that systems providing the functions are able to significantly affect the vehicle behaviour. Potential failures can cause very bad effects, and highly autonomous functions are, therefore, typically associated with strict requirements on safety integrity. However, it should be noted that many conventional systems also require high levels of safety integrity, for example, brake systems and steering systems. So, this aspect is not a *fundamental* difference between ADS functions and other vehicle functions. In general, automated functions tend to be associated with stricter safety requirements.

16.3 Challenges to ADS Concerning Functional Safety

For relatively high levels of automation (i.e. closer to 'autonomous driving' than 'driver warning functions'), a complexity issue must be faced that makes the safety analysis more difficult than that of conventional systems. In a 'classic' vehicle, the driver is responsible for coordinating all the vehicle functions (propulsion, deceleration, steering, headlamps, direction indicators, etc.). In principle, this means that each independent system function can be investigated separately with respect to functional safety and taking into account the possibilities that exist for the driver to handle a particular malfunction of that vehicle function. But with higher degrees of automation, the driver is no longer the overall coordinator, which means that any malfunction need to be handled by another function. In fact, the limits between these functions become blurred and difficult to define since the interaction between the different functions grows which is now more complex. The ISO 26262 approach of looking at one function (or 'item', which is the real or imagined system that provides the function) at a time is less appropriate when the functions are heavily dependent on each other. In the following section, more safety-related topics will be discussed that must be taken into account for the engineering of ADS.

The innovations of today's vehicles follow a continuing evolutionary approach. The development of future technologies is based on existing automotive engineering best practices and does not only reuse the existing ones. Some of these evolutionary aspects will be discussed in the following.

16.3.1 Vehicle Platform for Basic Driving Functions

Many of the current discussions on ADS are concerned with the functional level to replace the single driver tasks by additional ADS functions. Further important issues that need to be covered are the basic actuation functions, such as accelerating, braking and steering, to implement the required vehicle movement. For these functions, today's vehicles provide function-specific assistance for the human driver through means such as force support in braking systems by a hydraulic or an electromechanic brake. Systems for automated driving functions need to be improved to support the fully required brake force without a human driver. Furthermore, the safety concepts of existing systems must be updated because the ECU (e.g. of the steering system) needs to detect any kind of malfunction and their effects have to be mitigated, because without a driver the system has to monitor, decide and react on its own. The steering system's safety goal can be formulated like, 'Avoid the reversible and irreversible steering request from the steering system affected by any of the involved E/E systems' (e.g. steering angle sensor or ECU) [14]. The 3-level monitoring concept (EGAS concept) provides a possible technical solution, which is a standardised principle for safety designs for vehicle engine controls published by German OEMs [15].

Future vehicle architectures will introduce new safety concepts in the automotive industry (e.g. steer-by-wire systems will change safety concepts in contrast to the systems nowadays). In the event of any fault, a deactivation in a fail-silent mode as a safe state will not be possible (e.g. a fail-operational mode can be realised by redundant system architecture). As a conclusion, it is obvious that the implementation of ADS functions in existing vehicle platforms cannot be seen as only add-ons to existing functions. The overall safety concept of vehicles has to be updated for upcoming requirements concerning fault-tolerant and fault-operational behaviour of highly automated vehicles.

Issue: *Are existing vehicle platforms ready for ADS*?

16.3.2 From ADAS to ADS Functions

Today, ADAS functions are used as a basis for the realisation of ADS functions. However, these ADAS functions concern specific aspects of specific automotive use regarding:

- *Scenarios*: from simple to complex scenarios (e.g. from keeping a driving distance by ACC on the motorway to city chauffeur at traffic crossing)
- *Vehicle speed*: from low to high speed (e.g. from park steering assist to high-speed motorway chauffeur)
- *Vehicle Safety Risk*: from 'normal' to 'low' risk (e.g. from emergency braking assist to automated driving on the motorway)

The challenge is the combination and interaction of these basic functions. All kinds of interactions between these basic functions need to be analysed and handled in such manner that no unintended interactions concerning timing and value could occur. Any kind of functional and technical interaction must be dealt with during the system design phase.

Issue: *Is reusing of existing ADAS possible*?

16.3.3 Share of Sensors and Actuators

Different vehicle functions share the same sensors and actuators, and all functional and technical condition has to be met. Sensor signals and actuator command signals may not be faulty in the case of feature interaction and synchronisation. In many applications an adequate fusion of sensor data and a voter mechanism for actuator command signals are required. In particular, any kind of unwanted interactions has to be handled so that no hidden links could affect any malfunction behaviour.

Issue: *Is the available technology sufficient and adequate for the required functions*?

16.3.4 From Many ECUs to Host ECUs

Today, more and more functions of vehicles are implemented on existing single-core ECUs. These existing technologies slowly reach their limits (e.g. clock frequency, heat dissipation, size of gates). The following challenge approach is a shift from single-core to multicore ECUs, which means a shift from distributed functions with many ECUs to a few multicore host ECUs. The latter offer many different functions, but this rather new technology also requires new safety features. For safety-critical applications according to ISO 26262, these multicore ECUs with shared resources have to support specific safety measures in hardware (e.g. use of lockstep core or memory protection). Furthermore, safety measures have to be supported by the software and software engineering constrains. Real-time (e.g. loads of cores), functional (e.g. sequences) and safety (e.g. spatial redundancy) aspects have to be considered by the operating system and the application software. Many new algorithms from different vendors have to be integrated in these platforms, and coordination, configuration and documentation pose a further challenge. All these aspects have to be compliant to ISO 26262 and require safety evidence for the assessment of those applications.

Issue: *Is new technology ready for safety-critical applications*?

16.4 Importance of the Concept Phase

The concept phase defined in ISO 26262 focus on the functional abstraction of a specific item by (1) definition of the item, (2) assessment of the potential risks of that item by performing the hazard analysis and risk assessment (HARA), (3) determination of the ASIL for each hazardous event, (4) definition of high-level functional safety requirements as safety goals and (5) derivation of a functional safety concept (FSC), which covers all relevant safety measures to achieve functional safety for the defined item. In the following, each of these activities is described and relevant steps will be discussed in more detail.

16.4.1 Item Definition

This activity covers the definition of the item, the required functionalities, the intended behaviour, the interaction with other items/systems of the vehicle and the interaction with the external environment of the vehicle. ISO 26262 is intended as an automotive-specific functional safety standard, and it should be usable for any kind of E/E system in a vehicle. This can be slightly different when considered beyond the scope of specific items. For example, if we compare a hybrid powertrain system component such as a high-voltage battery system with an automated driving systems for a motorway assistant (MWA): The MWA contains much more complex and networked functionalities that must to be coordinated with external items (e.g. other vehicles) and environmental systems (e.g. traffic signs) and furthermore with vehicle internal functions related to fundamental vehicle platform functions.

16.4.2 Hazard Analysis and Risk Assessment

In the concept phase, the functional abstraction allows to have an abstract view of the system. Functional safety concerns unintended behaviour of the item. Safety analyses should be carried out in that phase to identify potential hazards of the item (e.g. HAZOP[14] or Concept FMEA[15]) followed by risk assessment.

The following steps describe activities that need to be done during the HARA including some proposed further extensions concerning ADS functions; these are written in bold letters and described in more detail:

Step 1: Elaboration of Hazardous Events

- Step1.1: Driving scenarios by situation analysis
  - Driving situation (e.g. manoeuvre at crossroads)
  - Infrastructure (e.g. communication between car and environment)
  - Environmental condition (e.g. weather)
  - Operating mode of the vehicle (e.g. acceleration)
  - Traffic participants involved (e.g. pedestrian)
  - **Driver presence** (e.g. driver in the loop/or not)
- Step 1.2: Hazard identification (e.g. by HAZOP)
  - From malfunctions
  - To malfunction behaviour
  - To hazard
- Step 1.3: Derivation of hazardous events
  - Combine driving situation with hazards
  - Potential source of harm to specific group of traffic participants at risk

Step 2: Classification of Hazardous Events

- Step 2.1: Severity classification
- Step 2.2: Exposure classification
- Step 2.3: **Controllability classification**

Driver Presence and Controllability Classification Each hazardous event is classified by the risk parameters severity (S), probability of exposure (E) and controllability (C) during the HARA. Parameter C denotes the estimation of controllability of a hazardous event by the driver or other persons potentially at risk. Controllability classes are C0 to C3, where C0 meaning 'controllable in general' and C3 meaning 'difficult to control or uncontrollable'. In the specific context of risk assessment for automated driving functions, the parameters depend on the role of the driver within a specific driving situation, which is why an ASIL should be

[14] Hazard and operability study.

[15] Failure mode and effects analysis.

determined for any potential hazardous event. For ADAS and partially automated functions, the driver must always be able to take over control of the vehicle within a defined reaction time. Concerning functionality, for highly or fully automated functions, it is not required that the driver monitors the driving situation. Thus, it might not be possible for the driver to consider any kind of controllability of the vehicle. This may lead to a classification of C3, which would result in ASIL C/D[16] worst case.

16.4.3 Determination of ASIL and Safety Goals

The next steps concern the rating of ASIL and the definition of safety goals:

Step 3: ASIL Derived from Risk Parameters
- ASIL = f (S, E, C) based on ISO 26262, part 3, Table 4

Step 4: Elaboration of Safety Goals
- Formulation of Safety Goals
- **Definition of Safety Goal attributes** (e.g. safe state)

Definition of Safety Goal Attributes A safety goal must be specified as a top-level safety requirement. We want to avoid any unreasonable risk of a possible hazardous event (e.g. 'unwanted acceleration shall not occur'). Safety goals are not expressed in terms of technological solutions but in terms of functional objectives. If a safety goal can be attained by transitioning to, or by maintaining of one or more safe states, then the corresponding safe state(s) shall be specified. Further relevant parameters regarding a safety goal are safe state, fault-tolerant time interval (FTTI),[17] diagnostic test interval (DTI),[18] fault reaction time (FRT)[19] and safe tolerance time (STT)[20] to maintain safe state before a possible hazard may occur (see Fig. 16.6).

$$\mathrm{FTTI} = \mathrm{DTI} + \mathrm{FRT} + \mathrm{STT}$$

The definition of these parameters is very important in the case of FRT being required to have critical driving situations handled by the system or by the driver to maintain the defined safe state (e.g. ADS function level 2 defines safe state as 'driver takes over control').

Further Influences to Define a Safe State The complexity of the driving situation must be considered for the definition of safe states. Another important requirement

[16]Depending on the classification as S and/or E.

[17]Time span in which fault(s) can occur in a system before a hazardous event ([2], Part 3, 1.45).

[18]Amount of time in which a safety mechanism takes online diagnostic tests ([2], Part 3, 1.26).

[19]Time span between detecting a fault and reaching the safe state ([2], Part 3, 1.44).

[20]Amount of time between achieving the safe state before a hazard could occur.

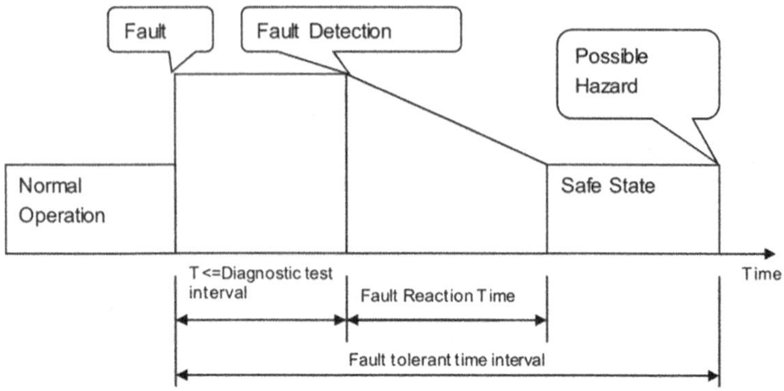

Fig. 16.6 Fault reaction time and fault-tolerant time interval [2]

Table 16.1 Overview of exemplary influences on the safe state

Item definition	Low ADS	Mid ADS		High ADS
Driver presence	YES	YES	NO	NO
System availability	Deactivation not available	Not available	Available	Available
Safe place	–	–	Stop vehicle on the same lane	Stop at the rightmost lane
Safe state scenario	Driver must take over	Driver must take over	Vehicle must stop at safe place	Vehicle must stop at safe place

in ISO 26262 concerning the safe state is '**8.4.2.4.** If a safe state cannot be reached by a transition within an acceptable time interval, an emergency operation shall be specified'.

Based on this requirement, further constraints have to be taken into account:

- *Item Definition*—provided functionality of ADS to maintain safe state (e.g. low ADS level, only comfort functions vs. high ADS level, self-driving).
- *Driver Presence*—difference between driver in the loop or not (e.g. driver's hands on the steering wheel vs. checking e-mails at the touchscreen).
- *System Availability*—possible or required degradation function depends on the level of ADS and the driver reaction in the case of malfunction.
- *Safe Place*—reachable safe place depends on the current driving situation and environmental conditions (e.g. safe state required during overtaking on the third lane of the motorway).
- *Safe State Scenario*—accessible safe state in specific driving situations including all constraints.

An overview of different influences is given in Table 16.1.

16.4.4 Functional Safety Concept

The objective of the functional safety concept is to derive functional safety requirements from the safety goals and to allocate them to preliminary architectural elements of the item or to external measures.

The following aspects have to be addressed in FSC:

- Error detection and failure mitigation
- Transition to a safe state
- Warning and degradation concept
- **Fault tolerance mechanisms**
- **Error detection and driver warning**
- **Arbitration logic**

The last three aspects will be discussed in the following in more detail:

Fault tolerance mechanisms means that a fault does not directly lead to the violation of the safety goal(s). The mechanism maintains the item in a safe state with or without any kind of degradation.

Error detection and driver warning are important to reduce the risk exposure time to an acceptable interval (e.g. engine malfunction indicator lamp, ABS fault warning lamp).

Arbitration logic is required to select the most appropriate control request from multiple requests generated simultaneously by different functions and is particularly important for the interacting functionalities of ADS.

However, not all of these aspects are always relevant for every system. Some systems do not offer any fault tolerance and some systems do not need any arbitration logic. The relevant safety measures concerning error detection, driver warning and transition to the safe state are important topics that must be considered in that phase.

16.4.4.1 Examples of FSC for Different ADS Levels

Depending on the type and degree of automation, there are several different strategies for ensuring safe operation despite faults in associated systems. This is illustrated in Fig. 16.7, which shows three potential event sequences unfolding after the occurrence of an error. From top to bottom, these can be described as follows:

An **assisted or partially automated function** can no longer be trusted to fully function and as a consequence the driver is alerted to (re)take control of the vehicle. During and after the handover, the partially automated function is prevented from working unsafely, perhaps by deactivating that function completely.

Example: Cruise control is deactivated due to a detected error. The driver is informed and takes control of the longitudinal motion of the vehicle.

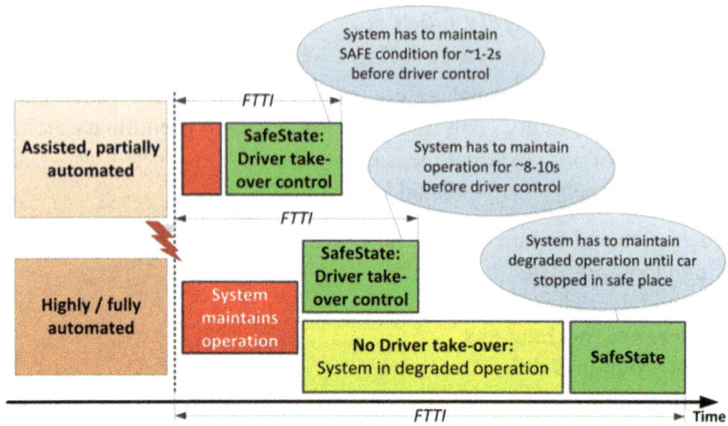

Fig. 16.7 Different concepts for transition to safe state

A **highly or fully automated function** determines that the driver needs to take over due to a detected error. The driver is informed about the need for handover of control. Due to the expected relatively long time for the handover, the automated function needs to continue to operate fully or almost fully for some time. Note: This means that the handover is initiated when the automated function is still either fully or almost fully operational.

Example: An autonomous driving system detects an error that indicates that an additional (subsequent) fault may lead to unsafe system behaviour. The driver is informed and takes control of the vehicle.

A **fully automated function** without any possibility for the driver to take over control determines that the vehicle shall be stopped in a defined time interval to avoid any hazardous event. As in the previous case, this means that the handover is initiated when the automated function is still fully or almost fully operational.

Example: An autonomous driving system detects an error that indicates that an additional (subsequent) fault may lead to unsafe system behaviour, so the automated function brings the vehicle to a safe stop within a few minutes or possibly seconds.

For the second and the third case described above, i.e. in the lower part of Fig. 16.7, it is shown that the automated function needs to be fully—or almost fully—operational for several seconds after an error occurs. If there is no driver to take over, the function has to remain operational, albeit potentially degraded, for several minutes. Thus, the implementation of such highly or fully automated functions needs to be fault tolerant in the sense that full or degraded functionality is possible even when a fault occurs in the system.

16.4.4.2 Vital Role of the Driver in the FSC

ISO 26262 sets requirements concerning error detection, driver warning and reaction of the driver. For today's automotive E/E systems, the role of the driver can be regarded as almost being covered in a cooperative manner. The driver must be able to control the vehicle on every trip (in Europe see also: Vienna Convention). By contrast, *how* the automated vehicles operate in a standardised way and *how* safety-critical aspects should be handled in a standardised way is *not* defined.

Thus, the driver needs to be familiar with different specific automated driving systems because the behaviour vehicles may differ. The training of the driver is required for specific ADS functions to ensure the driver's correct reaction within the required reaction time.

An additional aspect that must be taken into account here, and this is the 'habituation effect', i.e. the introduction of ADAS and ADS functions, will change the driving experience and require different skills of the driver. In HARA, the parameter C for controllability might change to 'uncontrollable'. In the near future, a driver may be unable to handle a critical vehicle situation without assistance systems within the required reaction time because of the lack of experience. Special driving licences for automated driving systems could be a possible scenario. However, they may not be accepted by customers who may hinder the introduction of such systems.

At present we do not train drivers to be able to deal with either a total brake failure or total loss of steering capabilities. Both braking and steering systems are extremely safe and reliable as a result so that the drivers do not need to worry about such problems at all. An alternative solution is simply to make the future ADS so safe and reliable that the drivers can fully rely on them at all times.

16.5 Supporting Methods to Handle Complexity of ADS

The complexity of these safety-critical systems must be considered, and negative effects need to be detected and mitigated by fault identification and fault mitigation techniques. Today, in the development of automotive electronic systems, there are established methods and technologies for safety activities available (e.g. safety analyses such as HARA for ASIL determination [2], failure mode and effect analysis (FMEA) [16], fault tree analysis (FTA) [17]).

The available technologies that need to be improved and developed further for their practical application in systems engineering:

- Formal/semiformal specifications by model-based systems engineering
- Formal verification by contract-based design
- Simulation and co-simulation

16.5.1 Model-Based Systems Engineering

The following definition of MBSE can be found in Friedentahl [18]:

'Model-based systems engineering (MBSE) applies systems modelling as part of the systems engineering process ... to support analysis, specification, design, and verification of the system being developed'.

The MBSE approach is a semiformal methodology to support engineers in the specification phase with analysis of the system and reduction of reality to an abstract model representation. The requirements for a specific level are defined and a virtual solution for the system is elaborated and hierarchically divided into representative components from system of systems, systems, subsystems and components. Models at a lower hierarchy level provide more specific details concerning the realisation. During the modelling phase, a separation of intended and unintended functions (= fault behaviour) is required, which is represented by specific functional properties and safety-related properties of the system. The model-based engineering approach is highly recommended by ISO 26262, part 6, for software development at ASIL C and D. This approach should be enhanced for the system level of such software-intensive systems. One of the major standardisation working groups concerning MBSE is the Object Management Group (OMG), which is an international, open membership, not-for-profit technology standards consortium. OMG task forces develop enterprise integration standards for a wide range of technologies and industries. Various standardised general purpose modelling languages are available for the system level (e.g. SysML,[21] MARTE[22] or EAST ADL[23]). These modelling languages have been elaborated, improved, applied and evaluated by many EU research initiatives by academia, research and industry partners. MBSE presents many possibilities for how to model a system through the use of different modelling elements, but for practical application, the reduction of the number of elements to a subset and provision of guidance and modelling constraints for engineers are requirements. A model-based systems engineering method[24] is a method that implements all or part of the systems engineering process and produces a system model as one of its primary artefacts. A system model provides the basis for specification of the intended behaviour of the system and is further used for identification and derivation of error models. An error model handles fault propagation over different hierarchy levels from singular components up to hazards at vehicle level. Different safety analysis methods (e.g. FTA or FMEA) can be supported by applying the error model. The output of the safety analysis defines safety measures by safety requirements for mitigation of any potential fault by detection, prevention,

[21] Systems Modelling Language—http://www.omg.org/spec/SysML/.

[22] Modelling and Analysis of Real Time and Embedded systems—http://www.omgmarte.org/.

[23] Electronics Architecture and Software Technology—Architecture Description Language—http://www.east-adl.info/.

[24] A method is a set of related activities, techniques, conventions, representations, and artefacts that implement one or more processes and is generally supported by a set of tools.

degradation or warning actions in the safety concept. A possible approach for the automotive domain by using SysML is described by Martin et al. in the SAE technical paper [19].

Biggs et al. [20] present a profile for a conceptional meta-model to cover relevant aspects of system safety and describes safety stereotype based on SysML (e.g. hazard, harm, harm context, etc.). The profile models common safety concepts from safety standards and safety analysis techniques. As a profile of SysML, it can be used to directly model the safety-related information of a system in the same model as that system's design. Furthermore, the profile supports communication between safety engineers and system developers; in order to improve the understanding on both sides of the risks, a system is vulnerable to and the features the system uses to mitigate those risks.

The MBSE approach by using SysML covers the following concerns [18]:

- *Provide a common and standardised description language* to improve the communication between system engineers and engineers from other disciplines.
- *Support of the performance of different kinds of checks* of the system model for the verification of specification rules (e.g. for the system design, to achieve correctness and completeness).
- *Improve the processing of the system modelling artefacts* by using transformation of the system model to another description model and extension with other relevant aspects (e.g. error modelling).
- *Traceability of relevant safety artefacts* is provided, and so the change management and impact analysis of particular safety concerns are possible. A further benefit of MBSE is the possible reuse of existing best practices by different kinds of patterns for requirements definition, safety design and safety argumentation.

16.5.2 Formal Verification by Contract-Based Design

Contract-based design (CBD) is a formal method for specifying what a component/system is able to offer (e.g. service, data, information, energy) for its environment by means of so-called guarantees and what a component/system requires (e.g. service, data, information, energy) from its environment by 'assumptions' [21]. Guarantees may be the performance and restrictions of output interface/channels which are only valid if all assumptions are confirmed. Assumptions define the environmental constrains for the input interface or channel of a system or component. The coupling of software-intensive systems and their components is hard to handle. It is difficult to handle all potential hidden links that could affect the safety of a system. CBD is able to guarantee that the system model only engages defined system states. By applying CBD, only specified system states are allowed and the coupling and communication of systems is only permitted via defined and well-known channels.

It is possible to provide patterns to assume and guarantee contracts which are defined for different characteristic such as timing, safety, security, etc. or patterns that are formalised to be checked automatically. The sum of all the system patterns defines all possible contracts.

CBD describes system components to be black boxes and defines their behaviour via interfaces with other system components. All kinds of dependability aspects are formulated as contracts, for example, timing (e.g. real-time contracts), safety (e.g. ASIL x or reaction time) and security (e.g. authentication certificates) and are manageable by this means.

Different hierarchical levels of contracts are defined as follows:

- Contracts between different SW modules
- Contracts between SW modules and HW components
- Contracts between different HW components
- Contracts between HW components and subsystems

CBD is able to coordinate interoperability and boundary limits of components and services they provide and also data over different hierarchical organisations. By modularization, it is possible to reduce the complexity of the components during system design. Every component is described by a limited catalogue of properties and constraints which establish safety. Conflicts between contracts are found very easily by means of a consistency test, if all contracts are free of any contradictions. Satisfactory tests check whether the implementation of a component is consistent with the contract. Adequate tooling support is now finally available today (e.g. for model checking). Several publications discuss the use of contracts in context of the requirements of the engineering and safety standards such as ISO 26262 [22].

A new methodology to support the development process of safety-critical systems with contracts is presented by Baumgart et al. [23]. They compared existing meta-models also stating their shortcomings in relation to their approach, and they introduced the semantic foundation of our meta-model. They described their concepts of abstraction levels, perspectives and viewpoints and provided a proof of concept with exemplary use cases.

Westman et al. [24] shows that safety requirements can be characterised by contracts for an item and its elements with guarantees that constitute the safety requirements, by providing explicit requirements on their environments as assumptions. A contract therefore enriches a safety specification for an item/element by explicitly declaring what each element/item expects from the environment to ensure that the safety requirements are satisfied. Furthermore, they showed that consistency and completeness of safety requirements can be ensured through verifying the dominance property of contracts.

Past and recent results as well as novel advances in the area of contracts theory are presented by Benveniste et al. [25]. They show that contracts offer support to certification by providing formal arguments that can assess and guarantee the quality of a design throughout all design phases. Furthermore, they showed that contracts can be used in any design process: Contracts provides an 'orthogonal' support for all methodologies and can be used in any flow as a supporting technology in composing and refining designs.

16.5.3 Simulation and Co-simulation

Simulation methods are commonly used in the automotive industry where complex embedded systems from different cooperative disciplines are referenced to realise highly interdependent functions. In this context, simulation methods allow engineers to predict the behaviour of complex embedded systems without an available prototype of the entire system. Complex systems like ADS require a data structure that considers the behavioural interactions within the system because of their multidisciplinary nature. A combination of simulation and MBSE methodology supports modelling activities and improves the integration of simulation activities in the design process. This combination supports a system presentation for addressing the overall behavioural aspects of the product (multi-physics, local and global behaviours) and thus considers several system levels.

The ISO 26262 standard recommends the use of simulation methods for verification on different system integration levels (e.g. ISO 26262 part 3 for verification of the controllability parameter of HARA [26]). For system design verification, ISO 26262, part 4, Table 3 suggests simulation as a highly recommended method and a technique, e.g. fault injection and back-to-back test for ASIL C and D.

A model-based workflow for safety-critical embedded system is shown by Karner et al. [27]. Their approach covers three main aspects during the development of safety-critical systems, namely, system modelling, system simulation and system verification based on simulation. By using the Software Process Engineering Metamodel (SPEM), the workflow is defined in a consistent and seamless way, allowing continuity from preliminary concepts up to the final system verification report. Aligned with requirements given by ISO 26262, the demonstrated workflow enables safety verification at system level during an early stage of development by using modelling and simulation.

A system modelling-based approach for the integration and test of automotive embedded systems is proposed by Krammer et al. [28]. A V-model is introduced, targeting process-oriented needs for safety, and indicates whether modelling languages in favour can be applied best. To establish a link between safety goals and the structure of simulation models, the initial model is enriched with necessary information and transformed to a language suitable for advanced simulation tasks. SystemC has the capabilities to support this approach for hardware and software evenhandedly. The integration of SystemC into a co-simulation environment also enables the usage of external simulation models within the proposed architecture. The proposed system modelling-based approach enables safety verification and validation at an early stage of development.

Graignic et al. [29] propose a software framework based on a data model that manages complex system structures. This data model structures behavioural information that considers three major interactions: interactions between components simulation models, interactions considering multilevel behaviours (e.g. use of component simulation for a module simulation) and interactions between domain behaviours (e.g. thermal impact on mechanical components) in a so-called co-simulation environment. Such methods can be used to perform early validation of the specifications by the MBSE approach to provide early validation feedback of adequate safety measures.

In the context of automated driving, different aspects beyond embedded systems behaviour are simulated such as the interaction of a vehicle with its environment, other vehicles or systems (e.g. Simulation of Urban MObility—SUMO [30, 31]), the interaction of a vehicle with a driver or the interaction of vehicle subsystems for dynamic proof of a specified behaviour of systems and components [32].

16.6 Further Safety-Related Topics

In the following section, more safety-related topics will be discussed that must be taken into account for the engineering of ADS.

16.6.1 Influence of Security on Safety Functions

One objective of system development is to ensure 'freedom of unreasonable risks' in any operational condition. This objective has different meanings depending on whether safety or security aspects are considered. From the safety point of view, the risk to the environment arising from inside of the system must be minimised (and this apart from a system including humans). This can result in a technical failure in the system, for example, fire hazard due to a high-voltage battery system of an electric vehicle or an accident because of an unintended acceleration of the ADS. Regarding security, potential threats to the system through the environment, which could result from intentional manipulations, e.g. a hacker attack, must be minimised. While the term safety represents the system view on any potential hazards of the system to the world outside, security concerns by contrast the aspects from the outside world to the inside of the vehicle and the influence on the vehicle internal systems. The goal of security measures is to protect the system from unauthorised use and manipulation (hacker, low-cost spare parts, etc.). The discipline of security in the automotive industry concerns the growth in vehicle functions and the innovation potentials in the networking of vehicles with the environment (e.g. other cars) or Internet of things (e.g. cloud services). The particular challenge on the one hand is the linking of the two disciplines' safety and security for utilising synergies and on the other hand the prevention of conflicting effects. Different motivations for unauthorised access scenarios in vehicles are possible [33]:

- Manipulation of the vehicle and its components as well as the corruption or deactivation of vehicle functions—attacking of 'availability of a service' (e.g. change of torque limits of the electric machine that could damage the powertrain)

- Vehicle tuning by changing functional properties—attacking of 'functional integrity' (e.g. chip tuning, manipulation of the speedometer or deactivation of warning messages)
- Illegal attempts to obtain personal data—attacks on 'personal integrity' (e.g. the driving behaviour of the user, preferences for shops, restaurants or hotels)

ISO 26262 provides guidance for automotive development process issues concerning functional safety lifecycles. However, a process for security concerns is not state of practice for automotive engineering. Many similarities exist between safety and security on a common abstraction level, and it would appear to be useful to interweave ISO 26262 development processes with security concerns. After defining a security item, the result of these considerations could be the consideration of security risks and the preparation of hazard analyses. Security goals with corresponding security measures can, hence, be derived from the analyses. After system design, verification and validation, a joint assessment should take place to rate the functional safety level reached according to ISO 26262 and any safety threat on the security side. Based on the similarities of these two disciplines, it would appear to be wise and necessary to expand the ISO 26262 framework by aspects of security topics. The extent to which these suggestions or other methods are expedient will be established in the course of an ongoing discussion in different standardisation communities [33].

16.6.2 Liability of ADS

Liability is a crucial topic in the context of future automated vehicles because legal authorities need an answer to the question, 'who was responsible?' in case of an accident.

Different responsibilities can be found under the law [34], e.g.:

- *Liability of the vehicle keeper*: Any operational risk in connection with an automobile is born by the vehicle keeper—ADS will not change the liability for the operation of automatic systems in motor vehicles.
- *Liability of the driver*: In damage event a fault of the driver is legally assumed (under civil law) until proof of the contrary is provided. In case of a fault of the ADS, the driver still has the option to insist on proof of exoneration.
- *Motor vehicle liability insurance*: If a harmed third party raises claims against vehicle keepers or drivers, they will be covered by the insurance—ADS will not cause any relevant change of the liability principles of the motor vehicle liability insurance.
- *Product liability of the manufacturer*: The OEM is liable if a defective product was brought to the market being subject to product liability. The OEM must provide evidence that the product was not defective and did not cause damage. The drivers must be instructed carefully in order to reasonably influence their expectations about the system's capabilities and to encourage drivers to perform

any necessary overriding functions. The safety of the system design is closely linked to the instructions given to the driver.

- *Liability of the infrastructure*: Future highly and fully automated vehicle functions will require precise data. These data will refer to local conditions too and will require a time stamp. The vehicle infrastructure should be able to provide all necessary information and is also liable for safe and secure functionality.

Ethical aspects will also play a role. In complex driving situations, events may occur that are difficult to handle by human drivers and that could lead to so-called dilemma situations. Sometimes, it is not possible to manage critical situations without harming any people. Thus, a decision has to be made to determine the minimum of harm. A decision between 'plague or cholera?' is a difficult one for humans to make, but it is even more difficult for machines. Future highly and fully automated systems will need certain risk determination algorithms that can rise to situations of this kind.

For this reason an 'event data recorder' in the vehicles will be a requirement for recording relevant information about crashes or accidents. Information from these devices is collected and analysed after a crash to help in determining exactly what happened. This will be similar to the 'black box' found in airplanes, which records all critical data in the course of a flight. Further research is needed for the assessment and classification according to the level of abstraction and degree of automation for a standardised definition and understanding.

16.6.3 Validation of ADS Functions

Systematic testing methods are very important for the validation of ADS functions (e.g. concerning safety aspects) [37]. For such complex systems, test methods must comprise a combination of simulation and real-world testing for different levels of integration like xiL (x in the loop) and model/software/processor/hardware/vehicle in the loop approaches. The most widely used approaches for the validation of driving functions are based on the V-model, endurance testing, xiL testing, open-loop offline perceptions tests, 'Trojan horse' tests, stepped implementation tests, complex tests and so on. All these testing methods have different potentials and disadvantages, for example, 'Trojan horse' tests are functional tests without hazardous effects in serial cars [35].

A further issue of ADS functions is that a strategy for safety confirmation cannot be implemented because a malfunction mechanism cannot be caused by the function but by decisions of the system. Although a test is able to characterise safety-relevant system states, there could be system reactions during automated driving situation where the decisions cannot be affected by the ego-vehicle alone. The actions and reactions of other road users must be anticipated, but a 100 % expectation cannot be ensured. Adequacy here cannot yet be reached on basis of road user reaction models. The system reaction is going to be probabilistic and the decision on accuracy will become time dependent and ascertainable only in simple situations. The first development of automated function was concentrated on technology goals. But without appropriate validation steps for safety-critical automated functions, the vehicles cannot hope to be established on the consumer market.

16.7 Conclusion

The ISO 26262 standard is intended to be an automotive functional safety standard for handling hazards caused by malfunctioning behaviour of E/E safety-related systems including interaction of these systems. It does not address the nominal performance of E/E systems such as powertrain control or any kind of ADAS. For this reason the ISO standard is also applicable to any level of automated driving. But the complexity of such systems is much higher than today's engineers are used to deal with, because of the high degree of networking functionalities that must be handled. Different kinds of challenges must be considered to realise ADS functions in an adequate manner. Following challenges were discussed in this chapter: Increasing complexity of highly interconnected functions and influence of system attributes, such as availability, reliability, safety and security, must be harmonised.

The concept phase of ISO 26262 becomes more important for ADS functions, because the development of ADS requires the engineering approaches and technologies beyond state of the art.

In particular, influence of the driver in the HARA, definition of safety goals and corresponding attributes for specific levels of ADS (e.g. safe state) as well as the changes of the functional safety concept from fail-safe to fail-operational strategies.

Today, several methods are available to support complex systems but they must be improved for the development of ADS. Possible technologies were discussed to handle the increasing complexity: model-based systems engineering, formal verification by contract-based development, as well as simulation and co-simulation. Which of those methods are adequate and applicable to meet a specific safety-critical demand still has to be defined and argued in the individual safety cases with respect to the specific context.

An enhancement of ISO 26262 that provides guidance for handling such highly complex systems would be useful. In the near future, that kind of application-specific guidance has to be discussed within the working group of ISO 26262 for the upcoming enhancement of the standard. This enhancement should be included in the upcoming revision of the standard which is scheduled to be released by the beginning of 2018. In particular part 3 of the standard needs additional guidance to classify hazardous events during the hazard analysis and risk assessment to determine the ASIL and the system level activities to handle highly networked systems.

16.8 Acknowledgements

The research work has been funded by the European Commission within the EMC^2 under the ARTEMIS JU. The authors acknowledge the financial support of the COMET K2—Competence Centres for Excellent Technologies Programme of the Austrian Federal Ministry for Transport, Innovation and Technology (BMVIT), the Austrian Federal Ministry of Science, Research and Economy (BMWFW), the Austrian Research Promotion Agency (FFG), the Province of Styria and the Styrian Business Promotion Agency (SFG).

References

1. K. Bengler et al., Three Decades of Driver Assistance Systems: Review and Future Perspectives, in *Intelligent Transportation Systems Magazine*, IEEE 6.4, 2014, pp. 6–22
2. International Organization for Standardization, ISO 26262—Road Vehicles—Functional Safety, Part 1–10. ISO/TC 22/SC 32—Electrical and Electronic Components and General System Aspects, 15 Nov 2011
3. European Commission, *CARE Project: Road Safety Evolution in the EU*, Mar 2015, [On-line] http://ec.europa.eu/transport/road_safety/pdf/observatory/historical_evol.pdf. Accessed 12 Oct 2015
4. O. Carstena et al., Vehicle-based studies of driving in the real world: the hard truth? Accid. Anal. Prev. **58**, 162–174 (2013)
5. SAE International, SAE J3016—Taxonomy and Definitions for Terms Related to On-Road Motor Vehicle Automated Driving Systems. J3016-201401, 1 Jan 2014
6. National Highway Traffic Safety Administration (NHTSA), Preliminary Statement of Policy Concerning Automated Vehicles, 30 May 2013, [On-line] http://www.nhtsa.gov/staticfiles/rulemaking/pdf/Automated_Vehicles_Policy.pdf. Accessed 12 Oct 2015
7. Austrian Federal Act, Governing the Liability for a Defective Product (Product Liability Act). 21 Jan 1988, [On-line] www.ris.bka.gv.at/Dokumente/BgblPdf/1988_99_0/1988_99_0.pdf. Accessed 12 Oct 2015
8. International Electrotechnical Commission, *IEC 61508—Functional Safety of Electrical/Electronic/Programmable Electronic Safety-Related Systems*, 2nd edn. TC 65/SC 65A—System aspects, 4 Apr 2010
9. R.W.A. Barnard, What is wrong with Reliability Engineering? INCOSE Int. Symp. **18**, 357–365 (2008). doi:10.1002/j.2334-5837.2008.tb00811.x
10. N. Leveson, *Engineering a Safer World: Systems Thinking Applied to Safety*. MIT Press, Jan 2012, [On-line] https://mitpress.mit.edu/books/engineering-safer-world. Accessed 12 Oct 2015
11. H. Butz, Safety and Fault Tolerance in a Complex Human Centred Automation Environment. Innovation Forum Embedded Systems, Munich, 24 Apr 2009, [On-line] http://bicc-net.de/events/innovation-forum-embedded-systems. Accessed 12 Oct 2015
12. H. Butz, Systemkomplexität methodisch erkennen und vermeiden, in *Anforderungsmanagement in der Produktentwicklung*, R. Jochem, K. Landgraf (Hrsg) (Symposion Publishing GmbH, Düsseldorf, 2011), pp. 183–217
13. Stanford Encyclopedia of Philosophy, Emergent Properties, 28 Feb 2012, [On-Line] http://plato.stanford.edu/archives/spr2012/entries/properties-emergent. Accessed 12 Oct 2015
14. D. Campos et al., Egas–collaborative biomedical annotation as a service. Proc. Fourth BioCreative Challenge Evaluation Workshop **1**, 254–259 (2013)

15. IAV GmbH—Ingenieurgesellschaft Auto und Verkehr, Standardized E-Gas Monitoring Concept for Gasoline and Diesel Engine Control Units, Version 6, 22 Sept 2015, [On-Line] https://www.iav.com/en/publications/technical-publications/etc-monitoring-concepts. Accessed 12 Oct 2015
16. International Electrotechnical Commission, IEC 60812—Analysis techniques for system reliability—Procedure for failure mode and effects analysis (FMEA), TC 56—Dependability, 26 Jan 2006
17. International Electrotechnical Commission, IEC 61025—Fault tree analysis (FTA). TC 56—Dependability, 13 Dec 2006
18. S. Friedenthal, A. Moore, S. Rick, *A Practical Guide to SysML: The Systems Modeling Language*, 3rd edn. (Morgan Kaufmann, Amsterdam, 2014)
19. H. Martin et al., Model-based Engineering Workflow for Automotive Safety Concepts. No. 2015-01-0273, SAE Technical Paper, 2015
20. G. Biggs et al., A profile for modelling safety information with design information in SysML. Softw. Syst. Model **15**(1), 147–178 (2014). Springer
21. B. Meyer, Applying 'design by contract'. Comput. IEEE **25**(10), 40–51 (1992). 2015
22. J.-P. Blanquart et al., Towards cross-domains model-based safety process, methods and tools for critical embedded systems: The CESAR approach, in *Computer Safety, Reliability, and Security*, ed. by F. Flammini, S. Bologna, V. Vittorini. Lecture Notes in Computer Science, vol. 6894 (Springer, Berlin, 2011), pp. 57–70
23. A. Baumgart et al., A model-based design methodology with contracts to enhance the development process of safety-critical systems, in *Software Technologies for Embedded and Ubiquitous Systems*, ed. by S.L. Min, R. Pettit, P. Puschner, T. Ungerer. Lecture Notes in Computer Science, vol. 6399 (Springer, Berlin, 2011), pp. 59–70
24. J. Westman et al., Structuring safety requirements in ISO 26262 using contract theory, in *Computer Safety, Reliability, and Security*, ed. by F. Bitsch, J. Guiochet, M. Kaâniche (Springer, Berlin, 2013), pp. 166–177
25. A. Benveniste et al., Contracts for System Design. INRIA, Rapport de recherche RR-8147, Nov 2012, [Online] http://hal.inria.fr/hal-00757488. Accessed 12 Oct 2015
26. M. Fischer et al., Modular and scalable driving simulator hardware and software for the development of future driver assistance and automation systems, in *New Developments in Driving Simulation Design and Experiments*, 2014, pp. 223–229
27. M. Karner, et al., System Level Modeling, Simulation and Verification Workflow for Safety-Critical Automotive Embedded Systems. No. 2014-01-0210, SAE Technical Paper, 2014
28. M. Krammer, H. Martin et al., System Modeling for Integration and Test of Safety-Critical Automotive Embedded Systems. No. 2013-01-0189, SAE Technical Paper, 2013
29. P. Graignic et al., Complex system simulation: Proposition of a MBSE framework for design-analysis integration. Proc. Comput. Sci. **16**, 59–68 (2013)
30. D. Krajzewicz, Traffic simulation with SUMO—Simulation of urban mobility, in *Fundamentals of Traffic Simulation, Series: International Series in Operations Research and Management Science*, ed. by J. Barceló, vol. 145 (Springer, Berlin, 2010)
31. J. Erdmann, Lane-Changing Model in SUMO. German Aerospace Center (2014), [On-Line] http://elib.dlr.de/89233/1/SUMO_Lane_change_model_Template_SUMO2014.pdf. Accessed 12 Oct 2015
32. A. Rousseau et al., Electric Drive Vehicle Development and Evaluation Using System Simulation, in *Proceedings of the 19th IFAC World Congress*, 2014, pp. 7886–7891
33. M. Klauda et al., Automotive Safety und Security aus Sicht eines Zulieferers, 4 Oct 2013, [On-line] http://subs.emis.de/LNI/ Proceedings/Proceedings210/13.pdf. Accessed 12 Oct 2015
34. T. M. Gasser, Legal consequences of an increase in vehicle automation. Bundesanstalt für Straßenwesen, 2013, [On-Line] http://bast.opus.hbznrw.de/volltexte/2013/723/pdf/Legal_consequences_of_an_increase_in_vehicle_automation.pdf. Accessed 12 Oct 2015
35. H. Winner, W. Wachenfeld, Absicherung automatischen Fahrens, in *6.FAS-Tagung München*, 29 Nov 2013, [On-Line] http://tubiblio.ulb.tu-darmstadt.de/63810/. Accessed 12 Oct 2015

36. B. Walker Smith, *SAE Levels of Driving Automation*. The Center for Internet and Society at Stanford Law School, 18 Dec 2013, [On-line] http://cyberlaw.stanford.edu/loda. Accessed 12 Oct 2015
37. H. Winner et al., Handbuch Fahrerassistenzsysteme, 3. Auflage. ATZ/MTZ-Fachbuch, (Springer Fachmedien, Berlin, 2015)

Part VI
Validation and Testing of Automated Driving Functions

Chapter 17
The New Role of Road Testing for the Safety Validation of Automated Vehicles

Walther Wachenfeld and Hermann Winner

17.1 Introduction

There is a difference between *developing* a vehicle that is driving itself as safely as today's cars are driven and *assessing* this vehicle in terms of safety. But, in order to introduce highly automated vehicles to public traffic, both must be done and remaining open questions for both must be answered. When trying to introduce these vehicles not just for a few users with limited use cases but rather as a mass product for a whole society, the safety has to be shown for many kilometers over many years resulting in a vast number and combination of different situations.

For the introduction of highly automated vehicles that cover the SAE levels 4 and 5 ("High Automation" and "Full Automation") [1] like an "Interstate Pilot Using Driver for Extended Availability," a "Full Automation Using Driver for Extended Availability," or a far advanced mobility concept like the "Vehicle on Demand" [2], special challenges arise for assessing the vehicle in terms of safety as the controller, being the human driver, isn't available anymore to correct the automation's behavior [3]. In each and every situation, the automation needs to choose an adequate behavior out of manifold possibilities.

In order to discuss the question: "How can road testing contribute to the safety assessment of highly automated vehicles?" in the first part of this chapter, we will specify the goals for assessment before explaining the challenges for road testing. Other test tools besides road testing are briefly introduced, and their demand for road tests is derived. The second part of this chapter will lay down a new train of thought on how to proceed with the "approval-trap" for highly automated vehicles. The

W. Wachenfeld (✉) • H. Winner
Institute of Automotive Engineering at TU Darmstadt, Otto-Berndt-Straße 2, 64287 Darmstadt, Germany
e-mail: wachenfeld@fzd.tu-darmstadt.de; winner@fzd.tu-darmstadt.de

D. Watzenig, M. Horn (eds.), *Automated Driving*,
DOI 10.1007/978-3-319-31895-0_17

statistical approach will be developed further to get a first idea how an introduction strategy for a highway pilot could look like.

17.2 The Goal of the Validation in Terms of Safety

Wachenfeld and Winner [3] describe the necessary relation V_{acc} between additional risk R_{add} and avoided risk R_{avo} to human beings that emerges from the introduction of automated vehicles.

$$V_{\mathrm{acc},p} = \frac{R_{\mathrm{add}}}{R_{\mathrm{avo}}} \tag{17.1}$$

The goal of the validation is to prove this relation. The exact level of this relation that needs to be met is unclear and depends on the benefit that comes with the introduction. As long as the validation can prove that the advantages offset the disadvantages for every stakeholder p, the vehicle can be released for production and can be introduced into traffic. The benefits caused by automated vehicles separate the people into two groups of stakeholders. On the one hand, the users or passengers directly benefit from the technology and can choose whether they want to use the technology or not. On the other hand, there are those affected by the technology outside the vehicle without being able to choose being affected or not. When assuming that the ones affected are exposed uniformly to the technology, we define these as the society. As both groups benefit and are affected in different ways, the validation has to take this into account. Dividing up the goal on users and society considers the freedom of individuals to also take a higher risk for themselves with the restriction not to endanger others above the currently accepted threshold. Examples are the usage of motorcycles and cars compared to trains for this freedom of choice.

At this point, we propose two different goals for the validation of automated vehicles in terms of safety:

1. A user or passenger who benefits from the use of the automated vehicle can be exposed to a higher risk than the one taken when driving by himself (reference risk) as long as he is informed about this additional risk. The user should get the option to weigh conservatively defined advantages and disadvantages.
2. As long as no major additional benefit for society is agreed on, we demand that the society is not exposed to a higher risk than to being part of today's traffic.

17.3 Challenges for Safety Validation of Automated Vehicles Based on Road Testing

Road testing for safety validation of highly automated vehicles means bringing a nearly final version of the object under test (OUT) into real traffic. In focus are systems that could endanger humans if not functioning properly. It is a safety critical unsupervised/non-correctable system [4] that will be released after the validation step. As it is unknown if the required safety is reached, a trained test driver is necessary to supervise the OUT although it is not originally made for being supervised (like all demonstrators of self-driving vehicles presented up to now).

The idea of road testing is that the OUT is confronted with *enough* relevant situations in real traffic. If the OUT passes these situations without the intervention of test drivers, the assumption is made that it will be able to do this in everyday use as well. The question is: How many situations in real traffic are enough? To answer this question, the properties of road testing, meaning the way situations are formed in public traffic, need to be studied. The most important property of road testing is that the reality is not manipulated, as the validity of the test would suffer. For example, there should not be some kind of actor (traffic participant) that imitates behavior to test the OUT. As he or she is conscious about what happens, by definition the behavior and especially his or her reactions will be artificial. If no manipulation on real traffic occurs, parameters of different kind form situations. For example, the authors in [5] distinguish dynamic objects, scenery, and actions that are described by parameters like position in time, width of a lane, or a deceleration value.

These parameters can be classified by the possibility to consciously select or cover situations that are formed by these parameters:

- By choosing the route (spatial), static objects and correlated effects can be gathered consciously. For example, if a certain form of tunnel is relevant for testing, planning a route along this tunnel easily leads to the OUT confrontation with this situation.
- By choosing a certain time (temporal), the parameters like lighting, weather conditions, season of nature, etc. can be selected consciously.

To a certain extent, both categories are predictable and as a result easy to aim for. On the other hand, constellations and situations that can be influenced by a living being, such as the two that follow, are more difficult to aim for:

- The exposure to high traffic density can, for example, be increased by choosing the route and time in combination. This is possible as the large number of living beings evens out different behavior. Some of many human beings, e.g., need to commute back home at a certain time, thus making it predictable.
- What is difficult to experience is behavior that is not correlated to time and space, such as a human overlooking an obstacle or being distracted from another passenger or his mobile phone.

Different driving behavior and thus different trajectories are caused by different combinations of these parameters. As not all parameters can be chosen actively without manipulating the road test, trying to cover dangerous situations with low exposure is challenging. It is more like a random selection that a certain combination of parameters is experienced during a test drive. The higher the number of kilometers or time driven, the more probable it gets to come upon a certain combination of parameters. The question of how many kilometers need to be driven on the road to assure sufficient coverage of combinations now arises.

It is relevant to highlight that the number of parameter combinations between the different levels of automation (SAE 0–2 vs. SAE 3–5) differs significantly. Today's vehicles (SAE 0–2) need to be assessed [3] for being a controllable tool supervised by a driving human. The possibility for the driver to brake, steer, and accelerate don't depend on the surrounding vehicles, whereas for more highly automated vehicles (SAE 3–5), the whole vehicle behavior also depends on the other vehicles' behavior as well as the environmental conditions. Accordingly, the challenge explained in the following is also valid for today's vehicles but the human capabilities to perceive and interpret the surrounding vehicles simplify the assessment and reduce the relevant parameter combinations strongly. This human driver is removed for highly automated vehicles, and therefore this simplification is missing.

The challenge for assessment consequently follows from the high level of safety that is defined by the benchmark and by the random occurrence of situations for road testing. Transferring the high level of safety into the theory of parameters forming situations in real traffic means that parameter combinations that end in challenging situations for the OUT appear rarely.

One benchmark is accident statistics that provide numbers (see Table 17.1) on average distances between two accidents of different severity levels [7].

Accidents can consequently be seen as parameter combinations that are relevant for safety assessment. To answer the question of how many kilometers need to be driven, the probability distribution function of a random number is necessary in addition to the average distance between two events. When the number of road accidents for a given amount of kilometers driven can be seen as a Poisson distributed number [8, 9], a statistical argumentation can be laid out how many kilometers are necessary to scientifically prove safety. This leads to the so-called "approval-trap" [3]. The approval-trap explains that although a safe vehicle is developed, there is no economical way to prove its safety. For the example of the

Table 17.1 Average distance between two accidents for different severities on German highways ([6] according to [7])

Severity level	Average distance between two accidents of this level (km)
S3	$660.0 \cdot 10^6$
S2	$53.2 \cdot 10^6$
S1	$12.5 \cdot 10^6$
S0	$7.5 \cdot 10^6$

Highway-Pilot, this means that although a vehicle is developed that is twice as good as today's traffic, approximately 10 times the distance between two events is necessary to have a chance of 50 % to prove (level of significance 5 %) that the developed vehicle is safer than the comparison group. For severity S3 (fatality), this leads to 10 times $660 \cdot 10^6$ km $= 6.6 \cdot 10^9$ km, which is economically not feasible. The safety can't be shown by just driving on German highways, and thus the lack of feasible safety assessment hinders the vehicle release. Up to this point, we haven't even discussed changes that would require retesting affected kilometers of the road test.

A first conclusion can be drawn: Road testing before start of production (SOP) will not be suitable to statistically prove the same safety level of higher automated vehicles compared to today's traffic. Thus, other ways out of the approval-trap need to be found. The next two sections will focus on alternative ways. Firstly, alternative approaches replacing the road test will be discussed before a new interpretation of the statistical approach is motivated.

17.4 Challenges for New Approaches on Safety Validation

Due to rising costs that are a consequence of increasing complexity of additional functionality that is implemented in new vehicles, the automotive industry is looking for a way to optimize verification and validation (V&V) processes. The question is whether these approaches can solve the approval-trap for highly automated vehicles, explained above.

17.4.1 New Approaches on Safety Validation

From the road test perspective, three approaches to optimize V&V processes are described in literature:

First of all, one could improve road testing itself. Above, different parameters of road testing were explained. Some of these can be actively aimed for. The goal of this approach is to reduce irrelevant kilometers and to aim for potentially relevant kilometers/events in real traffic. In [10], this idea is laid out in more detail for advanced driver assistant systems (ADAS). Based on a road test, events are monitored and interpreted as relevant or irrelevant, and future test routes are planned based on this monitoring.

Besides improving road testing, approaches are described to substitute road testing by artificial or virtual tests [11]. For higher automated driving, Fig. 17.1 structures on a generic level the different tools used. A more detailed classification can be found in [12]. The distinction between real, artificial, and virtual is illustrated by the example of a human as part of the environment: The human can either behave

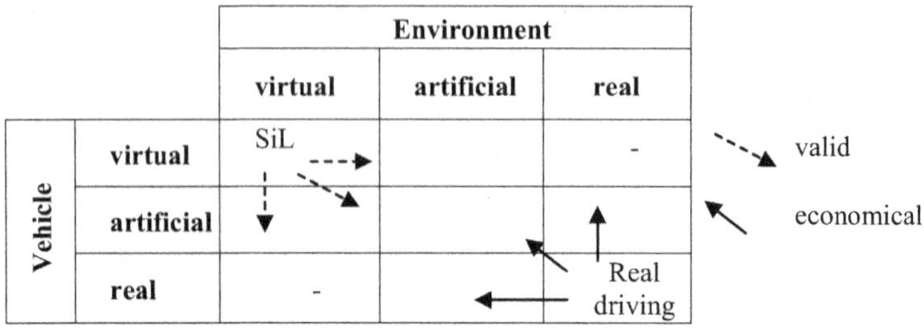

Fig. 17.1 Classification of tools for testing higher automated driving [3]

like he would in reality, or he could show artificial behavior like a person being aware that he is monitored. In both cases the human is a living being. On the other hand, the human can also be replaced by a technical system like an artificial dummy. The third approach would be a virtual representation of the human in software. In this case, the behavior of the human is modeled by software in a virtual world.

The different tools like test tracks, Vehicle-in-the-Loop, Hardware-in-the-Loop, and Software-in-the-Loop (SiL) replace some of the environmental or vehicle parts with artificial or virtual parts. Virtual and artificial parts can be manipulated and observed more easily with the motivation to set up relevant test cases and neglect irrelevant ones. Additionally, these tools try to reduce additional risk that is introduced when testing technical systems. A special motivation for SiL is the independence of time and hardware. Simulations can be parallelized and accelerated, limited only by computational power.

The third approach tries to combine road testing with other tools like SiL. The approach "Virtual Assessment of Automation in Field Operation" (VAAFO [13]) as well as other methods described in literature [14–16] follow a similar method: Human- or ADAS-controlled vehicles are equipped with hardware to perceive and interpret the situation as the OUT would. The virtual behavior of the OUT is assessed in all situations covered by the real human driver. As the OUT cannot act on the actuators, this method can be executed without introducing additional risks to public traffic. Not only test drivers, but in principle every driver with a driver license can execute this combined road and SiL testing.

All three approaches are based on simplifications and assumptions to either replace real parts of a test with artificial/virtual ones or neglect irrelevant situations/parameter combinations. These simplifications and assumptions can be invalid when applied on OUT assessment.

17.4.2 Validation of Alternative Approaches by Road Testing

To avoid using simplifications and assumptions that are not proper for OUT assessment, real driving such as road testing is necessary. This time road driving is used for safety validation of test tools and for safety validation of assumptions. The validation of tools for a defined number of test cases seems possible. But again, who can tell whether the selection met the necessary situations? Therefore, we come back to the challenge raised by the statistical train of thoughts. How can we show that the tools and assumptions are valid for OUT safety assessment?

An advantage for tool and assumption validation is that the number of kilometers doesn't need to be driven with the OUT. This simplifies the collection of kilometers. Another factor that would reduce the huge amount of possible situations that need to be covered for tool or assumption validation could be the independence of different parameters of a situation. For example, properties of traffic models are independent of properties of radar sensor models and therefore don't need to be modeled and validated in combination. This independence doesn't exist for the validation of the OUT as errors in real sensors lead to different behavior depending on the surrounding traffic. A disadvantage is that even more things need to be validated. For example, the behavior of other road participants needs to be reflected by the tool, at least to a certain extent.

Until now we have not seen any proof that the advantages outweigh the disadvantages resulting in less kilometers necessary to be driven (no matter who collects the stated amount). On the one hand, the more components are replaced and the more cases are neglected, the more validation effort for tools and assumptions has to be made. On the other hand, the more cases are left for road testing, the more validation of the OUT has to be done on the road. This seems to be a trade-off between OUT validation on the one hand and tool validation for OUT validation on the other hand. Additionally, the long-term perspective has to be considered as well. It can be that the first validation of tools needs a higher effort as the road testing itself, but when validating another version, vehicle type, or new generation, the overall effort could be reduced by orders of magnitude. An example for that effort reduction is described in [11] for ESC testing.

At this point, a second conclusion can be drawn: When pursuing approaches to replace or reduce road testing, road tests will still be of interest as these approaches need to be validated. At least until now, it's unclear whether or not other approaches reduce the validation effort for the first vehicle.

Of course, if a tool or an assumption is validated, its advantages and potential to increase efficiency can be utilized. But up to that point, validation activities based on real driving are and will be necessary.

17.5 A Confession About the First Introduction of Automated Vehicles

The proof of safety of the OUT by simply road testing before SOP is economically infeasible with statistical significance. For alternative approaches, it is at least uncertain if the required validation effort is reduced. Tool and assumption validation could equal out the reduction of OUT validation.

This leads to the conclusion that, from a statistical perspective, the first vehicles that will be introduced will not satisfy a scientific proof of comparable safety.

This seems to be an obstacle on the way to everyday automated driving. But, if safety cannot be proved, the scientific question to ask is if the hypothesis "the automated vehicle achieves the requirements in terms of safety" can be falsified. If using a conservative approach to test for being worse without a result, one could take the chance and the risk to introduce the vehicle.

For this introduction, the goals defined in the beginning of this chapter must still be achieved.

- The risk for society shall not be increased.
- The individual as the passenger should be able to weigh imminent risks and, depending on that weighting, be able to choose whether or not to use the vehicle.

In the following, we will explain an argumentation that has the potential to enable introducing automated vehicles based on the same statistics that before have led to the approval-trap.

17.6 Argumentation for Introduction of Automated Systems Motivated by Statistics

The following argumentation is based on the assumption that accidents are Poisson distributed. This Poisson distribution is explained before the generic theory is laid out and, based on that, the introduction of automated driving on German autobahn is illustrated with examples.

To represent the distribution of accident events, we use the Poisson distribution [8, 9]:

$$P_\lambda(k) = \frac{\lambda^k}{k!}e^{-\lambda} \tag{17.2}$$

This distribution assumes that the occurrence of an accident is an independent and non-exhaustive random process $P_\lambda(k)$. In the equation, k corresponds to the number of accident events and λ to the expected value with which this event occurs. The expected value λ is defined by the quotient

$$\lambda = \frac{d_{\text{test}}}{SP}, \tag{17.3}$$

Whereby d_{test} stands for the observed test kilometers and SP for the performance of the system. The performance represents the expected number of kilometers between the accidents. Two distributions of a probability distribution function with different expected values are depicted in Fig. 17.2. For this example, it is assumed that a certain number of kilometers d_{test} were driven and one event occurred. Can we now

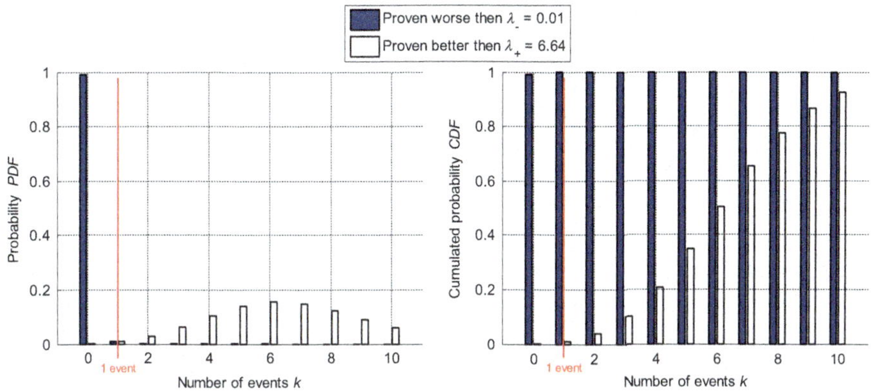

Fig. 17.2 Poisson probability distribution function (*left*) and cumulative distribution function (*right*) for two different expected values: one for the proof of being worse and one for the proof of being better when one event occurred

define a worse and a better performance level *SP* from this test? Equation (17.3) connects this with the search of an expected value λ. Based on a probability of error value[1] $e = 1\ \%$, these questions can be mathematically formulated with two equations:

$$P_{\lambda_-}(k > 0) \leq 1\ \% \tag{17.4}$$

For which expected number λ_- is the probability to have at least[2] one accident less than or equal to 1 %? A numerical search provides the value $\lambda_- = 0.01$. This tells us that when one accident occurs after d_{test} kilometers, we statistically prove with $e = 1\ \%$ error probability that the vehicle is *worse* in terms of safety compared to a performance level of $SP = \frac{d_{\text{test}}}{\lambda_-} = \frac{d_{\text{test}}}{0.01}$. In other words, Eq. (17.4) tells that with a probability of 99 %, no event occurs assuming $\lambda_- = 0.01$.

$$P_{\lambda_+}(k \leq 1) \leq 1\ \% \tag{17.5}$$

The second equation (17.5) asks for which λ_+ the probability that one or no accidents happened is at most 1 %. In this case, the numerical search provides the

[1]The value (5 %, 1 %, 0.1 % etc.) that is taken needs further considerations but is just one variable in that theory. For this chapter, we intentionally use 1 % and thus a different value as used in [3]. We don't want to be misunderstood that one or the other is the "right" value for that.

[2]Please be aware that $P(k > a) = \sum_{a+1}^{\infty} \frac{\lambda^k}{k!} e^{-\lambda}$ is the cumulative distribution function. The same counts for $P(k < a) = \sum_{0}^{a-1} \frac{\lambda^k}{k!} e^{-\lambda}$.

value $\lambda_+ = 6.64$. This tells us that when one accident occurs after d_{test} kilometers, we statistically prove with $e = 1\ \%$ that the vehicle is *better* in terms of safety compared to a performance level of $SP = \frac{d_{\text{test}}}{\lambda_+} = \frac{d_{\text{test}}}{6.64}$. Both probability functions (PDF and CDF) are plotted in Fig. 17.2. With a chance of 99 %, the expected value of the OUT is

$$\lambda_- \leq \lambda_{\text{OUT}} \leq \lambda_+. \tag{17.6}$$

As the smaller the expected value, the better the OUT, the value for λ_- defines a kind of best case and the value for λ_+ defines similar a worst case in that sense.

Based on that theory of accidents being Poisson distributed, we explain in [3] why using road tests for the proof being better than today's human drivers are economically not feasible.

In the following, we will use the test to the "other hand," meaning how and when one can be sure and state that the vehicle that is introduced into traffic contravenes the goals stated above.

17.6.1 Universal Theory on a "Brave Introduction" of Automated Systems

Based on the explanation of the Poisson distribution of accidents, we will now lay out a universal theory on the introduction of automated systems where certain events (accidents) exist which have to be considered that happen after a certain time or distance of travel. Additionally, it is assumed that a comparison level (benchmark) is known.

Let's assume that the performance level for the benchmark is $SP = SP_{\text{bench}}$. So, in statistical average, after SP_{bench} kilometers, one of the relevant events should happen. Now we want to introduce the OUT after it was tested for

$$d_{\text{test}} = \frac{SP_{\text{bench}}}{\tau} \tag{17.7}$$

kilometers and one event occurred. τ describes the ratio between the benchmark and the test kilometers. We know from Eq. (17.4) that the performance level of the OUT is equal or worse:

$$SP_{\text{OUT}} \leq \frac{d_{\text{test}}}{\lambda_-}. \tag{17.8}$$

Combining both Eqs. (17.7) and (17.8), the performance level of the OUT is

$$SP_{\text{OUT}} \leq \frac{SP_{\text{bench}}}{\tau \cdot \lambda_-}. \tag{17.9}$$

Equation (17.9) tells us that the OUT is worse than $\frac{1}{\tau \cdot \lambda_-}$ times the benchmark. On the other hand, we can only tell that the vehicle is better than

$$SP_{\text{OUT}} \geq \frac{SP_{\text{bench}}}{\tau \cdot \lambda_+}. \tag{17.10}$$

With this test, we can't prove that we are less safe than $\frac{1}{\tau \cdot \lambda_-}$ times the benchmark, and on the other hand we only prove that we are safer than a $\tau \cdot \lambda_+$ times worse system.

Why should this system not be introduced into traffic when we can't prove it is less safe than $\frac{1}{\tau \cdot \lambda_-}$ the benchmark? Why should we take the risk to introduce the system when we just know that it is safer than a $\tau \cdot \lambda_+$ times worse system? Let's compare the result of the test with the goals stated above separately for the user and the society.

17.6.1.1 User's Perspective

From the user's perspective, the system has a certain benefit. This benefit comes not from safety but from mobility, freedom, money, etc. The user knows that he or she probably needs to take a higher risk to access the benefits, but as long as the counted events don't lead to a λ_- below the line of Fig. 17.3, less safety can't be proven. This actually leads to an introduction that matches the goal for the user.

For the operating company of the system, there should be an option to stop and block the automated mode when from a monitoring of relevant events a non-acceptable safety can be proven. This limit is not necessary the whole area below the line as the users' acceptance defines this limit. How this monitoring should look and how the option to stop should be implemented is intentionally left open.

Fig. 17.3 Area of proven less safety than benchmark system for $e = 1\ \%$

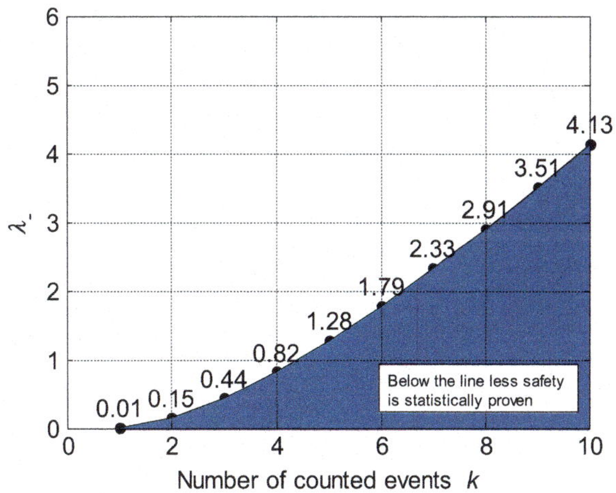

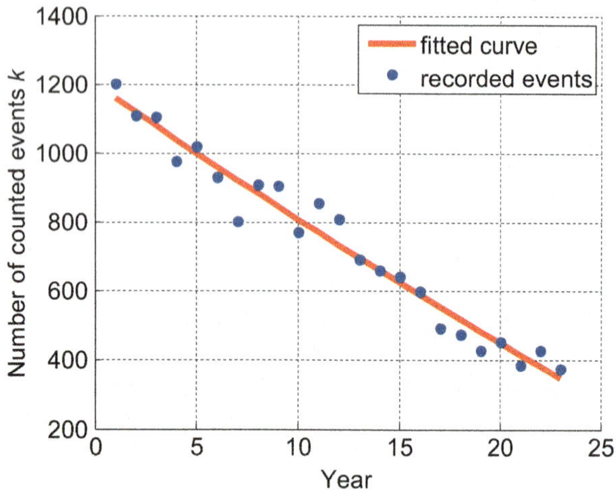

Fig. 17.4 Number of events recorded each 23 years. These numbers were fitted by a quadratic equation

17.6.1.2 Society's Perspective

On the other hand, for the society that in the first place has no direct benefit from the individual using the automated system, it is justified to ask if the society is exposed to a higher risk. The question is how to prove that society is exposed to that higher risk. One way to study the risk this society faces when using this technology would be to study the number of events that happened before the technology was introduced. This number of events, recorded, for example, each year, should not be affected negatively by this introduction. Figure 17.4 displays such an example. Each year, a discrete number of events were recorded. Over the year, the number decreased following a certain monotonic trend. This trend is not given but can be fitted by, for example, a least square approximation of a suitable mathematical function like a quadratic function[3] f_{fit}. This is done for the example by the (red/solid) line (see Fig. 17.4). All points differ from the trend line. These deviations are still independent from the technology we want to introduce. This fact leads to the question: How does the next number of recorded events have to differ from the values in the past, fitted by the trend line, to be sure that it was affected negatively?

[3] $k = 0.18 \cdot i^2 - 41.16 \cdot i + 1202.$

To answer this, we derive the standard deviation of these events compared to the trend line.

$$\sigma = \sqrt{\frac{1}{N}\sum_{i=1}^{N}(k_i - f_{\text{fit}}(i))^2} \tag{17.11}$$

In this equation, N is the number of years and k_i is the number of events recorded in year i. We now define that

$$k_{N+1} \leq f_{\text{fit}}(N+1) + \frac{\sigma}{\beta} \tag{17.12}$$

is indistinguishable for society, where $\frac{\sigma}{\beta}$ expresses the number of events caused by the introduction of the new technology. We assume this being indistinguishable for society as it is β times smaller than the standard deviation and therefore disappears in the noise of numbers each year. There is actually no way to prove or detect that the trend is affected negatively, as the number is too small and lies below the limit of detection.[4]

To match the goal for society, we now want to prove that the risk will not increase above the stated limit of $\frac{\sigma}{\beta}$. This can be done by limiting the numbers of kilometers driven on public roads. The slower the technology is introduced, the fewer events will happen in a single year. The number of systems or correspondingly the time of usage or number of kilometers will be limited by this approach. But how many systems I_{OUT} can be introduced?

If we assume that one system covers an average of $\bar{d}$ each year by introducing I_{OUT} systems, we get $d_\Sigma = I_{\text{OUT}} \cdot \bar{d}$ driven during 1 year. With the requirement

$$P_{\lambda_\Sigma}\left(k \leq \frac{\sigma}{\beta}\right) \geq 99\ \% = 1 - e \tag{17.13}$$

where $\lambda_\Sigma = \frac{d_\Sigma}{SP\_OUT}$. Equation (17.13) requires that the number of events caused by the OUT should be smaller than $\frac{1}{\beta}$ of the standard deviation with a probability of error of 1 %. The value for λ_Σ can again be derived numerically. The performance level of the OUT can be estimated from the test drive with Eq. (17.10) which, together with the result from Eq. (17.13), results in

$$d_\Sigma \leq \lambda_\Sigma \cdot SP_{\text{OUT}} = \lambda_\Sigma \cdot \frac{SP_{\text{bench}}}{\tau \cdot \lambda_+}. \tag{17.14}$$

Equation (17.14) shows that if less than d_Σ kilometers with the new technology are driven during 1 year, the danger to contravene the requirements from society is smaller than 1 %.

[4]There exists more theory to define this limit but at this point we use a variable to keep it simple.

17.6.2 EXAMPLE: Introducing Highly Automated Driving on German Autobahn

Let's try to apply this introduction approach on automated driving. Events in this case can be accidents with different levels of severity. The most challenging and best recorded number of accidents are the ones that involve fatalities. Focusing on Germany, we can discuss this example for the highway pilot, as numbers are given for this use case (see Table 17.1).

$$SP_{\text{bench}} = 660 \cdot 10^6 \text{ km} \tag{17.15}$$

Equation (17.15) expresses that on average on German autobahn, a fatality occurred every 660 million km. This defines the benchmark for the introduction. When assuming that the OUT is tested for τ times less kilometers and maximum one accident with fatality was generated, the test delivers two results. First, the vehicle in best case has a certain maximum level of performance in terms of safety [conclusion drawn from Eq. (17.9)]. Second, Eq. (17.10) shows the performance level is reliably higher than

$$SP_{\text{OUT}} \geq \frac{SP_{\text{bench}}}{\tau \cdot \lambda_+} = \frac{660 \cdot 10^6 \text{ km}}{\tau \cdot \lambda_+}. \tag{17.16}$$

At this point, a conservative approach is chosen as the worst case estimation is selected. Now we know the users' perspective for the introduction of the highway pilot.

For the society's perspective, we refer to [17] where the accident statistic (Destatis) for the years 1992–2014 is reported. These numbers and the fitted curve are depicted in Fig. 17.4. From Eq. (17.11)

$$\sigma \approx 48 \tag{17.17}$$

can be calculated. Combining the performance level from Eq. (17.16) with the standard deviation from Eq. (17.17) and defining a value for β, the numerical search delivers a certain λ_Σ (see Table 17.2). Together with Eq. (17.14), one gets

$$d_\Sigma \leq \lambda_\Sigma \cdot \frac{SP_{\text{bench}}}{\tau \cdot \lambda_+}. \tag{17.18}$$

All vehicles together shouldn't activate the highway pilot for more than this amount of kilometers d_Σ. Based on the numbers of 2013 where 52.4 million vehicles were registered and 224.2 billion km where driven on autobahn

$$\bar{d} = \frac{224.2 \cdot 10^9 \text{ km}}{52.4 \cdot 10^6 \text{ km}} \approx 4278 \text{ km}. \tag{17.19}$$

Table 17.2 Four chases for the introduction of automated driving on German autobahn. Assumptions are used to calculate examples

Parameter	Case 1	Case 2	Case 3	Case 4
β	1	1	10	1
τ	100	100	100	1000
$e \rightarrow \lambda_+$	1 % → 6.64	5 % → 4.75	1 % → 6.64	1 % → 6.64
$e \rightarrow \lambda_\Sigma$	1 % → 34.2	5 % → 38.08	1 % → 1.79	1 % → 34.2
Results				
d_Σ	$34 \cdot 10^6$ km	$52 \cdot 10^6$ km	$1.7 \cdot 10^6$ km	$3.4 \cdot 10^6$
I_{OUT}	7947	12384	414	794

This number can be converted into a number of vehicles:

$$I_{\text{OUT}} = \frac{d_\Sigma}{\bar{d}} \tag{17.20}$$

assuming the usage doesn't change.

With these numbers for the German autobahn, we get different cases for introduction depending on the variables we have defined in the theoretical part (see Table 17.2).

17.7 Conclusion

Besides the efforts spend on test tracks, today, road testing is the most important tool to validate the safety defined for a new vehicle. A huge effort is made to manage new functionalities resulting in higher complexity of new vehicles. Since the statistical proof of safety for highly automated systems (SAE-level 4 and 5) by classical road testing is economically infeasible, other approaches are requested. Three ways can be distinguished: improving road test efficiency; applying simplifications as well as virtualizations to reduce the effort with tools like test tracks, test benches, or XiL; and combining both with tools like the VAAFO concept. Unfortunately, up to now the evidence that these approaches can reduce the effort for the validation of highly automated systems is missing due to the reason that the approaches also need to be validated. Today, this seems to be a trap for the approval of the vehicles.

But, when looking back to the safety requirements, these can be separated for a user that benefits and a society that is ideally not burdened with additional risks. With these two goals, the statistical approach can be used for another second way of argumentation: (1) Whether we only introduce a new technology when we have unambiguously proven that the new technology does not introduce additional risks independently of every additional advantages or (2) we focus on the additional advantages and introduce the technology at a conservative estimate to profit from the advantages until one can reliable prove we are really introducing inappropriate risks. Both ways of argumentation are valid interpretations from today's perspective, but their consequences differ significantly. Argumentation two could pave the way for highly automated vehicles should it withstand the open debate of relevant stakeholders.[5]

We have seen that road testing was and will be relevant for the introduction of new vehicles. However, road testing will be used in a different way. One way will be the validation of alternative approaches, their tools, and simplifications. This will also lead to a big effort before one can profit the new approach. A second way will be the described partial validation for a cautious introduction strategy. Whether or not these kilometers need to be run on a real road or with other tools remains open.

Acknowledgement The research work has been funded by the Daimler and Benz Foundation.

References

1. SAE International Standard J3016: Taxonomy and Definitions for Terms related to On-Road Motor Vehicle Automated Driving Systems, SAE International (2014)
2. W. Wachenfeld et al., Use Cases for Autonomous Driving, in *Autonomous Driving: Technical, Legal and Social Aspects*, ed. by M. Maurer, J.C. Gerdes, B. Lenz, H. Winner (Springer, Berlin, 2016)
3. W. Wachenfeld, H. Winner, The Release of Autonomous Vehicles, in *Autonomous Driving: Technical, Legal and Social Aspects*, ed. by M. Maurer, J.C. Gerdes, B. Lenz, H. Winner (Springer, Berlin, 2016), pp. 425–449
4. W. Wachenfeld, H. Winner, Do Autonomous Vehicles Learn? in *Autonomous Driving: Technical, Legal and Social Aspects*, ed. by M. Maurer, J.C. Gerdes, B. Lenz, H. Winner (Springer, Berlin, 2016), pp. 451–471
5. S. Geyer, M. Baltzer, B. Franz, S. Hakuli, M. Kauer, M. Kienle, S. Meier, T. Weißgerber, K. Bengler, R. Bruder, Concept and development of a unified ontology for generating test and use-case catalogues for assisted and automated vehicle guidance. IET Intell. Transp. Syst. **8**(3), 183–189 (2013)
6. German Federal Statistics Agency (DESTATIS): Verkehrsunfälle 2013
7. H.-P. Schöner, Challenges and Approaches for Testing of Highly Automated Vehicles, in *CESA 3.0—Congress on Automotive Electronic Systems*, Paris, 4 Dec 2014
8. A. Nicholson, Y.-D. Wong, Are accidents poisson distributed? A statistical test. Accid. Anal. Prev. **25**(1), 91–97 (1993)
9. M. Gründl, Fehler und Fehlverhalten als Ursache von Verkehrsunfällen und Konsequenzen für das Unfallvermeidungspotenzial und die Gestaltung von Fahrerassistenzsystemen (2005)
10. P. Glauner, A. Blumenstock, M. Haueis, Effiziente Felderprobung von Fahrerassistenzsystemen, in *8. Workshop Fahrerassistenzsysteme*, ed. by UNI DAS e.V., pp. 5–14, Walting (2012)
11. U. Baake, K. Wüst, M. Maurer, A. Lutz, Testing and simulation-based validation of ESP systems for vans. ATZ Worldw. **116**(2), 30–35 (2014). doi:10.1007/s38311-014-0021-6

[5]This is what we would like to encourage at that point. Please give positive and negative feedback on this introduction strategy to overcome the hindrances for introduction of higher automated systems.

12. F. Schuldt, T. Menzel, M. Maurer, Eine Methode für die Zuordnung von Testfällen für automatisierte Fahrfunktionen auf X-in-the-Loop Simulationen im modularen virtuellen Testbaukasten, in 10. *Workshop Fahrerassistenzsysteme*, ed. by UNI DAS e.V., Walting (2015)
13. W. Wachenfeld, H. Winner, Virtual Assessment of Automation in Field Operation. A New Runtime Validation Method, in *10. Workshop Fahrerassistenzsysteme*, ed. by UNI DAS e.V., Walting (2015)
14. H. Winner, Einrichtung zum Bereitstellen von Signalen in einem Kraftfahrzeug, Patent DE 101(02), 771 (2001)
15. B. Hoye, D. Lambert, G. Sutton, Autonomous Driving Comparison and Evaluation, US Patent 20,150,175,168 (2015)
16. http://www.heise.de/newsticker/meldung/Autonome-Autos-Danke-dass-Sie-das-Auto-von-morgen-testen-2760591.html. Accessed 23 Jul 2015
17. http://www.dvr.de/betriebe_bg/daten/unfallstatistik/de_autobahn.htm. Accessed 21 Aug 2015

Chapter 18
Validation of Highly Automated Safe and Secure Systems

Michael Paulweber

18.1 Introduction

The increasing importance of advanced driver assistance systems is visible in the intensive development in this area of automotive OEMs as well as in research activities from nonautomotive companies. A well-known example of later activities is Google's self-driving car project, which works successfully towards a fully automated vehicle.

The potential of advanced driver assistance systems to actively increase the safety of vehicles as well as the comfort of their passengers is currently discussed in many newspapers and reports in various TV programs. Therefore, it is likely that new passenger cars as well as heavy duty vehicles will be legally obliged to have certain Advanced Driver Assistance Systems (ADAS) installed in the coming next years. As a step in this direction, the EU has required new cars to be equipped with an Electronic Stability Program (ESP) from 2011 onwards. Five years later, in 2016, emergency braking assistant systems will be mandatory for new heavy duty vehicles in the European Union.

The OEMs have actively taken on this trend and equip new vehicle models with more and more advanced driver assistant systems. This development is driven by two main societal challenges: The increasing age of the population, especially in the western countries, and the continuing growth of the population on earth.

Mobility is one of the biggest values of the western countries. This can be seen in the fact that many families, after buying a house or an apartment for living, spend the second largest amount of money on vehicles. As people get older, they want to keep the personal mobility they have been used to all their life. In order to keep the trend of fatalities decreasing, it is necessary to support elderly people when driving

M. Paulweber (✉)
AVL List GmbH, Graz, Germany
e-mail: michael.paulweber@avl.com

D. Watzenig, M. Horn (eds.), *Automated Driving*,
DOI 10.1007/978-3-319-31895-0_18

cars even at age of 80 and older. Without advanced driver assistance systems, elderly people have to stop driving or have an increased risk of fatal accidents. A statistics from [1] shows this trend, which can be reversed by the use of advanced driver assistant systems.

The growing global population makes space for one of the most precious goods especially in megacities such as London, Beijing, or New York. Traffic jams are normal in those cities. They do not only annoy the drivers but also increase the CO_2 and exhaust emissions, which contribute to global warming. As it is nearly impossible to build more or bigger streets in those cities, the only alternatives are public transportation or taking better advantage of the available space of existing streets for the personal mobility. Semi-automated vehicles which know where to park a car as well as jam pilots are first steps to keep personal mobility attractive also under the difficult environmental conditions of megacities. Vehicle-to-vehicle and/or infrastructure communication as well as the usage of map data allow to find energy optimal control schemes for vehicles, which can be implemented in (partially) automated vehicles. This can reduce the CO_2 emissions in the transport sector (especially in the transportation of goods using heavy duty trucks) significantly.

This indicates that the introduction of new and more sophisticated advanced driver assistant system up to fully automated vehicles will continue.

Depending on the degree of automation provided by ADAS, it partially or completely takes over the control of the transversal and/or longitudinal vehicle movement in many different situations, which may occur in the traffic environment. This requires the correct and reproducible reaction of those ADAS in all possible traffic scenarios, which might occur. Therefore, the validation of ADAS is already very complex and will get even more complicated when the degree of automation of the vehicles will continue to increase. Today, there are no test systems and validation methods available, which can guarantee the complete functional safety under all possibly traffic and environmental conditions. Current validation methods use catalogues of test scenarios, which define the traffic scenario as well as the expected correct reaction of the ADAS in this situation [2]. OEMs test vehicles with ADAS functions use scenarios of the catalogues either in proving grounds, in real traffic situations, or more and more in virtual (simulated) environments.

Testing on proving grounds or in real traffic does not require sophisticated simulation environments and is therefore often used at early functional prototypes of new ADAS or in partially or highly automated vehicles (in the remaining part of this chapter, ADAS shall also include the control systems of partially, highly, or fully automated vehicles). Unfortunately, those testing methods have severe drawbacks: Tests are difficult to reproduce. Therefore, it takes long testing times to check, if corrections in ADAS functions really solved identified problems of previous tests. Many traffic scenarios of the validation databases are difficult to reproduce or even dangerous for the test driver to execute. This also contributes to the high validation costs and long validation duration.

The development of simulation environments for validation in virtual traffic scenarios on the other hand is also very costly. A prominent company developing advanced driver assistant functions predicts the effort to develop validation environments for highly automated vehicles 10–20 times higher than the effort to develop the vehicle automation function to be validated. The ratio increases for fully automated vehicles to 20–50 times higher.

This chapter tries to summarize the most promising upcoming trends in the validation of advanced driver assistant systems [2].

18.2 Complexity of Automated Vehicles

ADAS controllers offer unprecedented values to the drivers of vehicles: ADAS were initially introduced to improve the safety of vehicles (active safety measures compared to previously passive safety measures). But they also increase the comfort in releasing the driver from stressful situations such as driving in traffic jam. Additionally, they are used to decrease the negative impact of vehicles to the environment (e.g., traffic light assistant minimizes the energy consumption and exhaust emissions in optimizing the vehicle speed depending on the current and future status of the upcoming traffic signals on the route of the vehicle).

Figure 18.1 depicts the basic structure of ADAS. Contrary to conventional control units in automotive vehicles such as engine control units (ECU) or battery control units (BCU), ADAS control units (ACU) communicate with a significantly more complex environment using sensors which are new in the automotive industry.

Many of these sensors provide information about the outside world in object lists updated periodically (video sensors, radar sensors, LiDAR sensors, ultrasonic sensors). Vehicle-to-vehicle communication as well as vehicle-to-infrastructure communication also sends information about the environment around the vehicle. A third source of information for ADAS controllers is the information of accurate

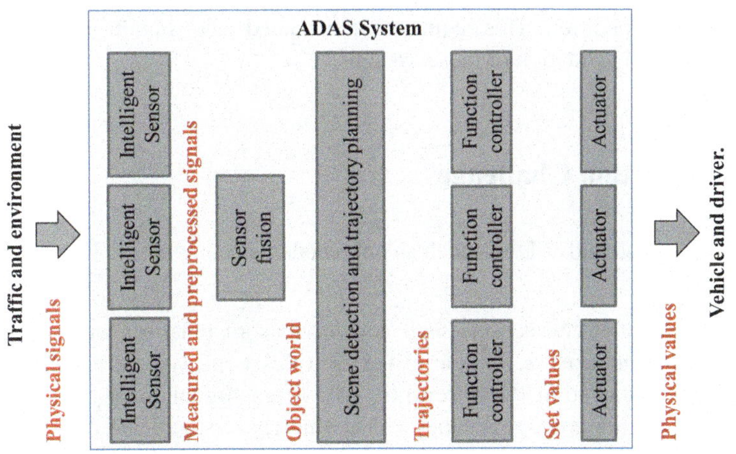

Fig. 18.1 Structure of ADAS

maps together with GPS sensors. All these information together with classical online data in vehicles, e.g., engine speed, velocity, gear number, etc., are used to get an image of the world surrounding the car as accurate as possible.

Unfortunately, all these information sources have their insufficiencies. Therefore, sensor fusion is used to combine the information of various sources and improve the image of the outside environment. This sensor fusion increases the complexity of these systems significantly, as it requires the synchronized acquisition of data from many sensors to test or to reproduce the result of sensor fusion and the calculations in ADAS controllers.

The physical signals are in most cases already preprocessed in the sensor data acquisition subsystems and periodically sent to the sensor fusion algorithms. The results of the sensor fusion are used to generate an image (position, velocity, type of object, etc.) of the outside objects at the current instance of time as well as a projection into the future. The next step is the calculation of an optimal trajectory for the vehicle in order to fulfill the requested mission (e.g., driving from point A to point B or parking the car in a parking space). The controllers for the lateral and longitudinal movements of the vehicle calculate the set value for the various actuators in the vehicle as steering actuator, throttle value, brake pressure, etc.

The control strategies of ADAS controllers have to interact with many conventional cars around the vehicle which is steered by the ADAS (ego-vehicle), as well as with human beings, animals, and other objects such as traffic signs, road borders, stones, or objects flying in the wind.

Additionally, the accuracy and trustworthiness of the different ADAS sensors are heavily influenced by the environmental conditions (like rain, snow, fog, night, glaring sun light, etc.). These effects need to be simulated in virtual environments for validation of ADAS too.

An additional difficulty with ADAS are the increasing threads of security breaches from vehicle-to-vehicle or vehicle-to-infrastructure communication. Vehicles with ADAS, for example highly automated cars, have to be safe under every environmental condition. This requires sophisticated ADAS control algorithms as well as extensive validation of these systems.

18.3 Validation Challenges

Vehicles equipped with ADAS reach an unprecedented complexity. This has various reasons:

- There are a lot of new sensors such as video sensors, radar sensors, LiDAR sensors, ultrasonic sensors, vehicle-to-vehicle (V2V) and vehicle-to-infrastructure (V2I) communication, GPS sensors, data from digital maps, etc. Many of these sensors are intelligent subsystems, which have significant data processing capabilities built in. Validation systems have to cope with streams of object lists generated by these sensors.

- As all of the new sensors have their weaknesses, sensor fusion is needed to create an image of the outside world. This image includes information about surrounding objects as well as their most likely future trajectories, which has two impacts on validation: it is necessary to either simulate all sensors or stimulate them simultaneously. Especially, if applying physical signals to the sensors of the ADAS via stimulators, which are directly connected to data from an online simulation of the surrounding environment, all time delays of these different actuators have to be compensated. Otherwise, the sensor fusion algorithms might lead to wrong results as the data from the different sensors are not exactly time aligned.
- ADAS equipped vehicles interact with the external world. Other vehicles driven by human drivers change their trajectory based on the reaction of the ADAS automated vehicle (e.g., whether it brakes rapidly at a road crossing or it smoothly decelerates and stops). Therefore, it is necessary to take also different behaviors of human actors into account when validating ADAS.
- Weather conditions heavily change the behavior of ADAS sensors. This requires tests under many different environmental conditions.
- Vehicle security has to be tested too, as security breaches may lead to wrong images of the outside world, which creates unwanted trajectories and might even result in accidents.

Vehicles equipped with ADAS functionalities (e.g., fully automated cars) have to be safe under uncountable environmental conditions and scenarios. Therefore, most of the validation of automated research vehicles is currently done on the road.

Prof. Winner from the Technical University in Darmstadt, for instance, estimated the distance that needs to be tested with validation vehicles on German roads in order to prove that an automated vehicle is as safe as a manual driven car [3]. The calculated distance of 100 million km takes into account the average number of fatal accidents on German more than roads and the average driving performance on these roads. Considering that OEMs develop up to 25 different variants of one vehicle model and the fact that complex software systems, such as ADAS, typically have a new software version update every few months, very large test vehicle fleets would be required. Even if not all variants have different ADAS software, some modifications will lead to additional validation effort. The test vehicles would have to perform more than 100 million km testing several times a year.

When trying to perform validation road testing in short periods of time, it is difficult to ensure that almost all possible environmental conditions (hot, cold, low or high altitude, snow, rain, fog, ice, etc.) occur within this time frame. Additionally, many validation sequences require dangerous maneuvers, which are safety critical to test drivers piloting vehicles with new ADAS versions. Also costs of the additional objects (other vehicles, test robots, etc.) needed to validate ADAS systems in different scenarios are significant.

The replacement of pure road testing by validation in a virtual environment can reduce the validation time significantly, if eliminating those portions of test time on the road, where no safety relevant events occur, in the tests executed in the simulated

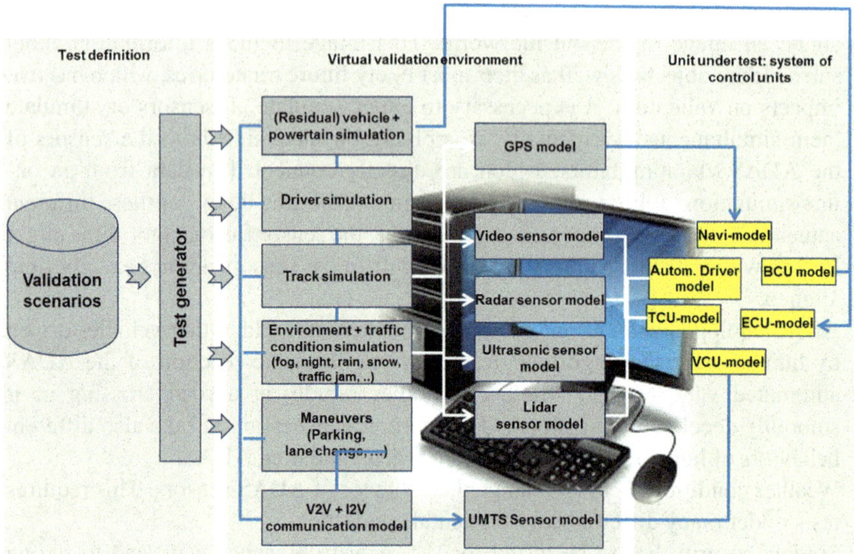

Fig. 18.2 MIL (Model in the loop) or SIL (Software in the loop) validation environment

environments. Currently, several projects deal with the collection of databases of safety relevant scenarios for the validation of ADAS-based vehicles with many control units as transmission control unit (TCU), vehicle control unit (VCU), engine control unit (ECU), and many more.

In order to cope with the unrealistic required test time on roads discussed so far, it is common agreement that validation of ADAS require testing in simulated environments (see Fig. 18.2), which allow to perform testing faster than wall clock time and provide the possibility to stimulate different environmental conditions during the validation sequences. Currently, the required complex simulation models, which can simulate realistically "all" safety critical traffic and environment scenarios for ADAS equipped vehicles, do not yet adequately exist.

In order to get a better understanding if validation in a virtual environment can solve the problem, the following rough calculation is done:

- Road testing: The duration to perform 100 million km road testing with 100 test vehicles assuming 18 hours per day driving lead to a test duration of 8 years
- Virtual environment testing: Assuming testing of 300 safety relevant scenarios on 25 test beds under all combinations of environmental conditions will lead to test duration of 2 years (see Fig. 18.3)

This indicates that validation in a simulated environment is clearly a way to investigate in more detail; nevertheless, 2 years of testing is too long to perform validations for several variants of vehicles and several new ADAS software versions per year. The validation of highly automated systems is therefore still an unsolved

Fig. 18.3 Rough estimation of duration of validation in virtual environment

Test duration per safety relevant scenario	5	min
Number of safety relevant validation scenarios	300	scenarios
Number of virtual validation testbeds used	25	testbeds
number of testing hours per day	18	h
Number of variations of environmental conditions		
# of vehicle variants	1	variants
Clear / Rain / Snow	3	variations
No fog / Fog	3	variations
Day / night	3	variations
warm / cold	3	variations
low altitude / high altitude	3	variations
Driver type	3	variations
heavy traffic / low traffic	3	variations
Low GPS reception	2	variations
Vehicle load	3	variations
Number of scenarios under different env.conditions	3,936,600	scenarios
time to test all scenarios in all conditions	**2.0**	**years**

problem; the "state-of-the-art" validation methods and tools are not sufficient anymore.

Safety means the absence of unacceptable or unreasonable risk. Validation is the activity to determine that the requirements are the right requirements and that they are complete. Verification checks ensure that a developed systems fulfills all defined requirements. Therefore, it is necessary to foresee all potential dangerous situations during the requirement phase of a new product. As ADAS heavily interact with the outside world, an uncountable number of safety critical situations can occur. It is very likely that many of them are not foreseen at specification time of a new system. Therefore, the most difficult validation challenge in highly automated vehicles is the functional insufficiency [4]. It results from unknown requirements, which are consequently neither implemented nor verified. As validation shall ensure that all relevant requirements are foreseen, it is necessary to think about concepts to handle functional insufficiencies.

18.4 Validation Concepts

In order to reduce the validation time even further, concepts from combinatorial testing together with validation in a virtual environment should be applied. This leads to the validation tool chain depicted in Fig. 18.4.

Validation scenarios derived from several sources:

- Test vehicles record data from ADAS sensors, environment and road data, and vehicle data (e.g., engine speed, vehicle velocity). These data are then analyzed and checked whether safety critical events occurred and whether these events are already part of a data base of safety critical ADAS scenarios. If it is not already existent, a new validation scenario is generated from the recorded data and added to the data base.
- Scenarios defined from official bodies (e.g., consumer organizations as EURO-NCAP) are also added to the data base of safety critical ADAS scenarios.
- Safety analysis of ADAS will also lead to additional validation scenarios.

Especially the first source, which derives validation scenarios from data recorded on the road, allows to create a learning cycle over time, which helps to overcome the problem of functional insufficiency. It is thinkable to extend this learning cycle also over data recorded during operation of ADAS equipped vehicles as long as the buyers of those vehicle agree to this procedure.

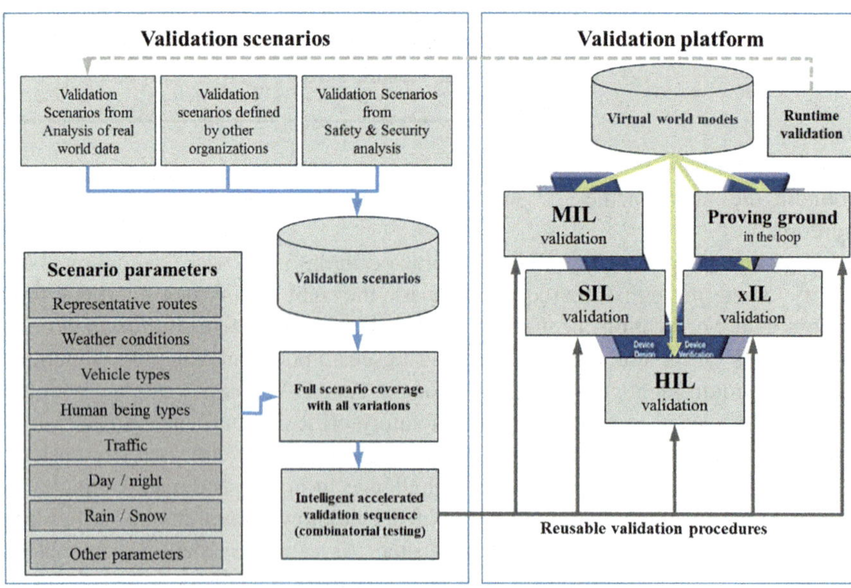

Fig. 18.4 Validation tool chain for ADAS equipped vehicles

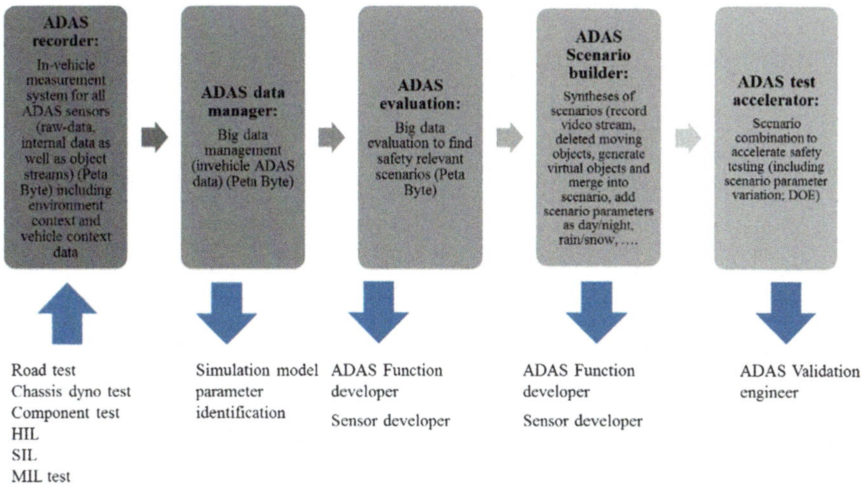

Fig. 18.5 ADAS Validation tool box

The test infrastructure components in Fig. 18.5 allow to set up ADAS validation environment to reduce the validation time to a reasonable length. In order to allow the use of combinatorial testing concepts, all ADAS scenarios need the possibility to execute them in different combinations of environmental conditions. They are described by different sets of values of the scenario parameters as amount of rain (0–100 %), fog (0–100 %), daylight (0–100 %), etc. All ADAS scenarios contain a test maneuver sequence with external inputs for the scenario parameters and a procedure to calculate, if an executed validation scenario has successfully passed.

When preparing the test sequence for the validation, those ADAS validation scenarios are taken from the ADAS scenario database, which are relevant for the ADAS function to be validated. An intelligent test generator based on methods of design of experiments (e.g., combinatorial testing methods as described in [5]) defines a set of values for the scenario parameters used during the execution of the scenario. Not all scenarios are tested with all possible combinations of scenario parameters. This reduces the number of necessary tests significantly.

In order to execute the test sequences, simulation models for the non-existing components of the ADAS equipped vehicle and the outside environment are needed. They are taken from the data base of environment models. The environment model components can also have the additional scenario parameter inputs (similar to the scenario sequences).

The last step to reduce the validation effort is the reuse of test sequences in different steps of the development V-process. This requires, on the one hand, a standardized scenario sequence description and, on the other hand, a simulation environment, which always provides the same functionality in combination with the unit under test components available in reality. The standardized scenario description language can be based on well-established standards as ASAM-ODX

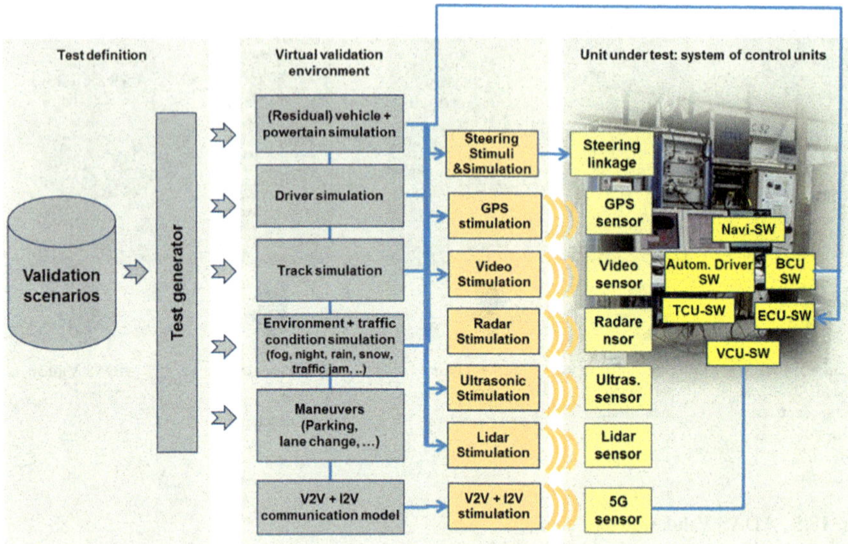

Fig. 18.6 HIL (HW in the loop) validation environment

[6] or Open Scenario [7], which will require extensions to cover the needs of ADAS validation.

The test sequences are reused during the development process:

- In MIL (model in the loop) environment during functional development in early phases of the V-model (see Fig. 18.2)
- In SIL (software in the loop) environment during ADAS software development in early phases of the V-model (see Fig. 18.2 MIL (Model in the loop) or SIL (Software in the loop) validation environment)
- In HIL (hardware in the loop) environment during ADAS development in early phases of the V-model (see Fig. 18.6)
- In power in the loop (xIL) environment during ADAS vehicle integration and validation in early phases of the V-model (see Fig. 18.7)
- Potentially also in vehicle in the loop (VIL) environment during vehicle validation when using remotely controllable platforms as described in [8].

18.5 Virtual Validation Environment

The accelerated validation procedure described in the previous chapter requires a complex simulation environment. It combines real components with simulated components in different variations. The real components shall be equivalent in the behavior of the simulated components functionally as well as in its timing behavior.

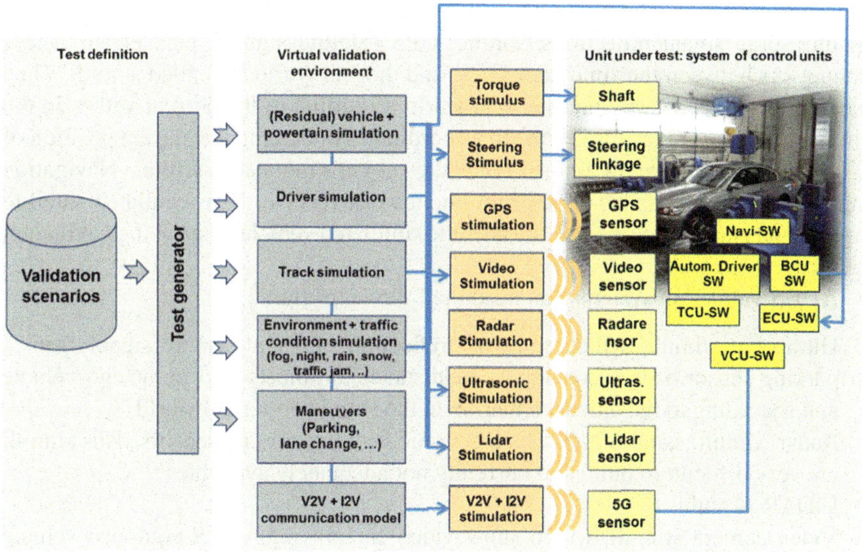

Fig. 18.7 PIL (Power in the loop) validation environment

Component \ Testbed type	MIL (Model in the loop)	SIL Software in the loop)	HIL HW in the loop)	xIL (Powertrain in the loop)	Proving ground	Road
Vehicle	Simulated	Simulated	Simulated	Simulated	Real	Real
Powertrain	Simulated	Simulated	Simulated	Real	Real	Real
Driver	Simulated	Simulated	Simulated	Simulated	Real	Real
Stimuli needed	no	no	yes / no	yes / no	no	no
Sensors	Simulated	Simulated	Real / Simulated	Real / Simulated	Real	Real
Sensor fusion	Simulated	Real	Real	Real	Real	Real
Trajectory building	Simulated	Real	Real	Real	Real	Real
Controllers	Simulated	Real	Real	Real	Real	Real
Actuators	Simulated	Simulated	Real / Simulated	Real / Simulated	Real	Real
Road	Simulated	Simulated	Simulated	Simulated	Real	Real
Traffic and environment objects (pedestrians etc.)	Simulated	Simulated	Simulated	Simulated	Real / Simulated	Real
Environmental conditions (rain, snow, fog, night, ice, etc.)	Simulated	Simulated	Simulated	Simulated	Simulated	Real

Fig. 18.8 Simulated/real components in ADAS validation environments

Figure 18.8 shows the combination of real, simulated, and emulated objects, which are used at the different test bed types described above. The table also indicates where a conversion from simulated values to real physical quantities such as ultrasonic signal reflections, torque, GPS satellite signals, etc., exists. These connectors between the simulated world and the real world are called stimuli. They have to introduce energy into the real world according to the current values in the simulated world. An example are GPS coordinates indicating the current position of a simulated world driving on a simulated road via simulated satellites. Navigation units in vehicles but also on mobile phones can receive these emulated satellite signals and show the correct position of the simulated vehicle when using navigation software packages.

ADAS validation systems can need the following stimuli:

- Ultrasonic stimuli, which send the reflection of an ultrasonic signal from a parking sensor back according to the distance to objects next to the ego-vehicle and according to the current situation in the simulated virtual world
- Radar stimuli: same function as ultrasonic stimuli for radar sensors. This stimuli are very difficult to build and currently not adequately available
- LiDAR stimulus
- Video camera stimuli, which show video cameras of ADAS equipped vehicle the image of the currently surrounding environment via replayed or artificially created video sequences on video screens
- GPS stimulus, which convert simulated GPS coordinates in a simulated environment to HF signal, which an antenna of a navigation system would receive from one or several GPS satellites.
- Vehicle-to-infrastructure and vehicle-to-vehicle communication stimuli, which simulate the communication between ego-vehicle and infrastructure control center and surrounding vehicles
- Torque stimulus as used in powertrain test beds
- Steering system stimulus, which emulates the mechanical feedback from the wheels on the road to the steering system sensor of a vehicle even if the wheel is not turning (either on a chassis dyno test bed or a powertrain test bed)
- Climatic stimulus (climatic chamber)

Stimuli need to ensure that the simulated output of a component matches the real measureable input of a physical component. Therefore, a lot of energy is needed in many cases to minimize the control error of a stimulus.

The connection between different simulated components requires a co-simulation framework, the combination of different simulated and real components needs a special co-simulation framework which is for example described in [9]. In order to mix and match the vehicle components, the environmental components, and the components to build ADAS controller and the driver models, it is necessary to define several interfaces:

- Inputs and outputs of real/simulation components. They connect the outputs of simulated components with the inputs of other simulated components. In case of a connection to a real component, a stimulus is needed in between the output of the simulated components and the physical input of the real component.

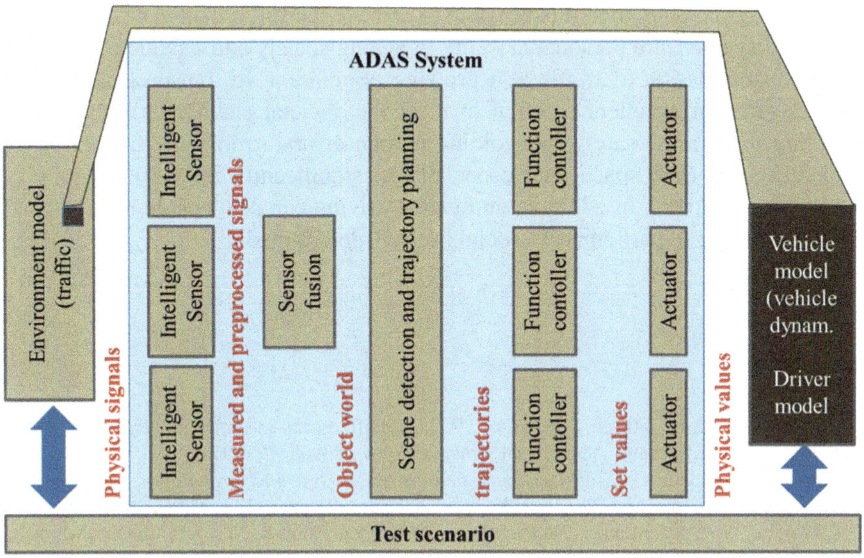

Fig. 18.9 Validation simulation environment

- Inputs of scenario parameters. Scenario parameters are changed at the beginning of a test scenario, but stay unchanged during the test sequence of the scenario. Also stimuli must react to scenario parameters (e.g., modification of a video image in case a scenario is tested in simulated rain).

In order to allow also ADAS function development on MIL systems, the main function blocks of ADAS are also needed as simulation components (e.g., sensor fusion block). Figure 18.9 shows these main building blocks in a simulated environment. It also indicates that the simulated ego-vehicle model interacts with the traffic and environment simulation (outside upper gray connection between left and right side of the figure).

The last validation building block is an executor of the test sequence. It is of advantage, if the test sequence is stored in a standardized format, which allows easily to reuse test sequences in different phases of the development process of ADAS. This saves significant costs as the creation of validation environments, and test sequences is 5–25 times more expensive than the development of the ADAS function itself.

18.6 Conclusion

This chapter presented a summary of the challenges to validate the functionality and especially the safety of ADAS for vehicles. It explained why new approaches are required to allow the validation of ADAS in reasonable time frames and at reasonable costs. Only then it is possible to achieve the full potential of ADAS and consequently to ensure personal mobility to ageing society and a reduction of CO_2 and exhaust emission of traffic at a growing population. In transport, automated vehicles offer an excellent solution to meet the societal challenges due to their capability to increase safety by avoiding human errors, to improve efficiency by better usage of road space, and especially to significantly reduce emissions in mobility applications. In addition, automated driving can also enable handicapped or elderly people to participate in social life self-determined.

References

1. J. Oxley, B. Fildes, E. Ihsen, J. Charlton, R. Day, Differences in traffic judgements between young and old pedestrians, Accid. Anal. Prev. 26–29(6), 839–847 (1997)
2. H. Winner, S. Hakuli, G. Wolf, *Handbuch Fahrerassistenzsysteme: Grundlagen, Komponenten und Systeme für aktive Sicherheit und Komfort* (Vieweg+Teubner, Wiesbaden, 2011)
3. H. Winner, W. Wachenfeld, Absicherung automatischen Fahrens, in *6.FAS Tagung*, Munich, Germany, 2013
4. Müller, Bernd; Robert Bosch AG, Challenges in Demonstrating Safety for Highly Automated Systems, in *Safetrans Industrial Day 2015*, Renningen, 2015
5. L. Yu, Y. Lei, N. Kacker, R. Kuhn, ACTS: A Combinatorial Test Generation Tool, in *IEEE 6th International Conference on Software Verification and Validation*, Neumunster Abbey, Luxembourg, 2013
6. A. eV, ASAM Connects—Standard Details, ASAM eV, May 2015, [Online] http://www.asam.net/nc/home/standards/standard-detail.html?tx_rbwbmasamstandards_pi1%5BshowUid%5D=525. Accessed 10 08 2015
7. V.S. GmbH, OPen Scenario Bringing Content to the Road, VIRES Simulationstechnologie GmbH (2015), [Online] http://www.openscenario.org/. Accessed 10 08 2015
8. M. Schmidl, H. Steffan, ADAS Testing with UFO Test System, in *Autonomous Vehicle Test & Development 2015*, Stuttgart, 2015
9. G. Stettinger, J. Zehetner, H. Kokal, M. Paulweber, M. Wierse, B. Toye, Control of an Engine Test-Bench via Hardware-Software-Co-Simulation, in *14.Internationales Stuttgarter Symposium Automobil- und Motorentechnik*, Stuttgart, 2014

Chapter 19
Testing and Validating Tactical Lane Change Behavior Planning for Automated Driving

Simon Ulbrich, Fabian Schuldt, Kai Homeier, Michaela Steinhoff, Till Menzel, Jens Krause, and Markus Maurer

19.1 Introduction

19.1.1 Motivation

During the last 25 years, the driving abilities of automated vehicles have progressed rapidly. This went along with a huge increase of complexity for automated vehicles, regarding the multiplicity of interacting components being required to implement the functionality of automated vehicles.

To keep this complexity manageable, many teams use modularity and hierarchical abstraction to break down the overall driving task into separate modules. One possible abstraction is described by Matthaei and Maurer [8].

For efficient and targeted testing, it is useful to test modules of an automated vehicle separately to ensure their correct functionality. Above this, the interoperability of modules needs to be tested on different integration levels, culminating in a test of the system as a whole (system test). A possible development and test process for those systems is the V-model, which is described, e.g., in the ISO 26262 standard development process [6, Part 3]. The V-model describes the development and test

S. Ulbrich (✉) • F. Schuldt • T. Menzel • M. Maurer
Institute of Control Engineering, Technische Universität Braunschweig, Hans-Sommer-Str. 66, 38106 Braunschweig, Germany
e-mail: ulbrich@ifr.ing.tu-bs.de; schuldt@ifr.ing.tu-bs.de; menzel@ifr.ing.tu-bs.de; maurer@ifr.ing.tu-bs.de

K. Homeier • J. Krause
Volkswagen Group Research, Berliner Ring 2, 38440 Wolfsburg, Germany
e-mail: kai.homeier@volkswagen.de; jens.krause1@volkswagen.de

M. Steinhoff
IAV GmbH, Carnotstraße 1, 10587 Berlin, Germany
e-mail: michaela.steinhoff@iav.de

D. Watzenig, M. Horn (eds.), *Automated Driving*,
DOI 10.1007/978-3-319-31895-0_19

Fig. 19.1 "Jack," the *Audi A7 piloted driving concept* vehicle

process of systems on the abstraction levels of unit tests, module tests, integration tests, system tests, and acceptance tests. These levels are also used in the test process for the tactical behavior planning, which will be described in this chapter.

The test levels are realized with different simulation-based tests and real world driving tests. Real world driving tests often come along with significant cost and time efforts. Additionally, real vehicles or prototypes are needed for testing. Often, such prototypes are not available at an early stage in a research project. Therefore, real world driving tests alone are often insufficient.

Accordingly, the authors propose to use situation-based open-loop tests and scenario-based closed-loop tests for module and integration tests. Simulation-based testing in a software-in-the-loop setup does not require prototypes and can be executed faster than in real time. Thus, the effects of little changes in the software can be tested with less efforts. Hence, simulation-based testing can be used during the development process and allows to tailor scenarios specifically to the needs of the developers.

The test procedure presented in this article has been used by our team to develop "Jack," the *Audi A7 piloted driving concept* vehicle shown in Fig. 19.1. "Jack" has been demonstrated at the Consumer Electronics Show 2015 in the USA[1] and in

[1] www.audi.com/content/com/brand/en/vorsprung_durch_technik/content/2014/10/piloted-driving.html

February[2] and April[3] on a German highway to the media. For this, the vehicle drove a stretch of 550 miles on a highway from Stanford to Las Vegas and around Braunschweig as well as around Ingolstadt.

19.1.2 Article Outline

This article is structured as follows: In Sect. 19.2 we define the terms scene, situation, and scenario. Section 19.3 gives an overview about the usage of scenario-based and situation-based testing. Section 19.4 presents testing on four suggested levels of abstraction. A case study with an exemplary implementation for lane change behavior testing is presented in Sect. 19.5. Last of all, Sect. 19.6 finalizes this article with conclusions and a research outlook.

19.2 Definition of Terms Scene, Situation, and Scenario

To differentiate separable test levels and to have unified interfaces between different components, it is necessary to define the terms *scene*, *situation*, and *scenario*. A more extensive discussion as well as a literature review can be found in Ulbrich et al. [19]. The authors define the term *scene* as follows:

> *A scene describes a snapshot of the environment including the scenery and dynamic elements as well as all actors' and observers' self-representations, and the relationships among those entities. Only a scene representation in a simulated world can be all-encompassing (objective scene, ground truth). In the real world it is incomplete, incorrect, uncertain, and from one or several observers' points of view (subjective scene).*

Similar to Geyer et al. [5], the term *scenery* subsumes all geo-spatially stationary elements of the scene. Dynamic elements are elements that are moving, or have the ability to move. The scene representation is completed by a *self-representation* containing the current skill levels and general system skills as well as the states and attributes of all actors and observers (cf. Maurer [9] and Bergmiller [3]).

Vice versa, we define a *situation* by:

> *A situation is the entirety of circumstances, which are to be considered for the selection of an appropriate behavior pattern at a particular point of time.*[4] *It entails all relevant conditions, options, and determinants for behavior.*[5] *A situation is derived from the scene by an information selection and augmentation process based on transient (e.g., mission-specific) as well as permanent goals & values. Hence, a situation is always subjective by representing an element's point of view.*

---

[2] www.stern.de/auto/news/jack-das-selbstfahrende-auto-von-audi-erstmals-auf-einer-deutschen-autobahn-2174446.html

[3] www.volkswagenag.com/content/vwcorp/info_center/en/news/2015/04/dobrindt.html

[4] Cf. Wershofen and Graefe [20].

[5] Cf. Meyer [4]. Determinants as in determining factors.

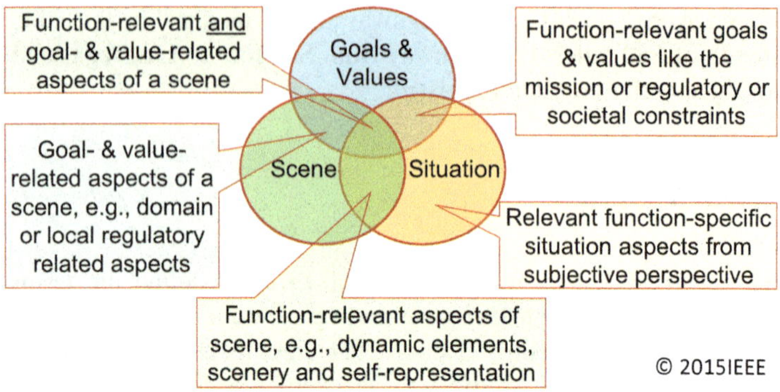

Fig. 19.2 Venn diagram of scene, situation, and an element's goals & values [19]

A situation consists of several situation aspects to be interpreted or comprehended by situation assessment modules. A situation can be an input and output of such modules at once. Within these situation assessment modules, the situation is augmented with additional information.

According to the authors' definition of a situation, it can be fully derived from a scene and the system's goals & values, as illustrated by the Venn diagram in Fig. 19.2. There is a wide overlap between a scene and a situation to include, e.g., all relevant parts of the scenery, all relevant dynamic elements, and all relevant aspects of the self-representation. This *information selection* helps to simplify the situation representation and by this the driving function development.

Moreover, the situation is implicitly or explicitly *augmented*, e.g., by goals and values. For instance, by explicitly labeling the usefulness of roads or lanes to reach the mission goal or implicitly by characterizing a playing child on the side to be more relevant than a flying around plastic bag. The remaining part of the situation, not overlapping with the scene or the goals & values, represents situation aspects evaluated and populated with information by situation assessment modules.

According to Fig. 19.3, a *scenario* contains scenes, actions, and events as well as goals & values. The authors suggest the following definition:

> *A scenario describes the temporal development between several scenes in a sequence of scenes. Every scenario starts with an initial scene. Actions & events as well as goals & values may be specified to characterize this temporal development in a scenario. Other than a scene, a scenario spans a certain amount of time.*

Depending on what a scenario is used for, it may also be sufficient to specify *only* situations instead of entire scenes plus goals & values. This may be true for a test setup solemnly designed to test, e.g., a situation assessment as in the situation-based open-loop test described in Sect. 19.4.

Furthermore, the functional description of the system (use-case) needs to be defined in the early phases of the system design according to the V-model, e.g., in the ISO 26262 standard development process [6, Part 3]. A use-case entails a

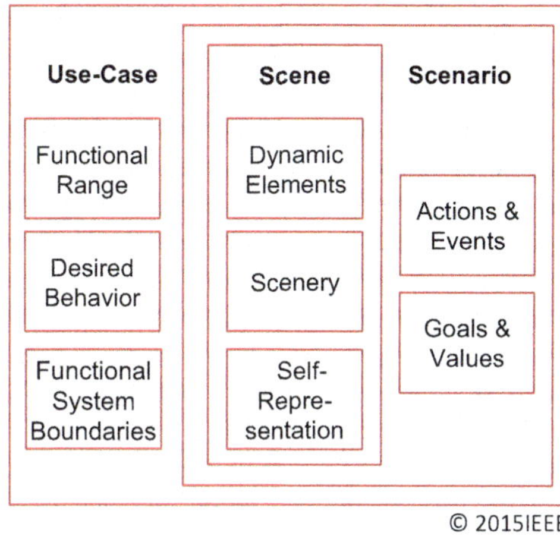

Fig. 19.3 Elements of a scene, scenario, and use-case [19]

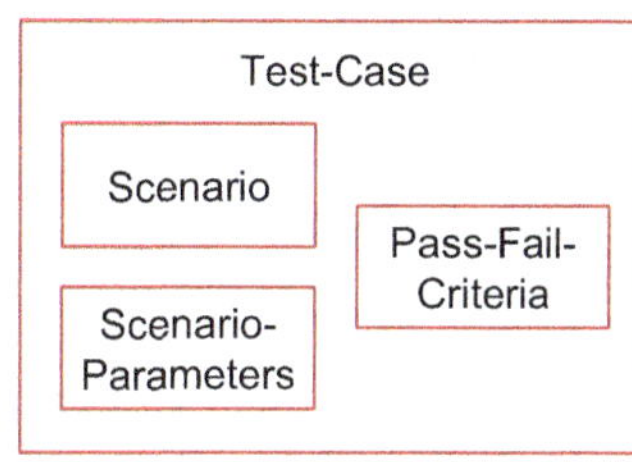

Fig. 19.4 Elements of a test-case

description of the functional range and the desired behavior, the specification of system boundaries, and the definition of one or several usage scenarios.

In software testing, scenarios need to be specified to generate specific test-cases. Thus, each test-case entails a scenario with a specific set of parameters and pass–fail criteria to evaluate it (see Fig. 19.4).

Therefore, it is possible to detail use-cases with scenarios, scenes, parameters to generate specific test-cases to test the requirements, and the desired behavior within and beyond the system boundaries.

19.3 Background

The introduction pinpoints the significance of testing during system development and for system validation. For this purpose, test methods can be differentiated by their open-loop or closed-loop nature.

Among often used closed-loop test techniques are x-in-the-loop techniques, like software-in-the-loop testing, driver-in-the-loop testing, vehicle-in-the-loop testing, or hardware-in-the-loop testing [7, 11]. The fraction of simulated versus not-simulated, physically real aspects may vary in any of them. The differentiation of x-in-the-loop techniques rather addresses the physical test setup than the method of testing itself: For instance, scenario-based closed-loop testing may be applied to any of the x-in-the-loop test techniques. In this article it will only be demonstrated in its application for software-in-the-loop testing.

A second dimension is the scope of testing, referring to the level of abstraction from units to the entire system as in the V-model in the ISO 26262 [6, Part 3]. The right half of the V-model subsumes fine granular unit and model tests up to higher level integration, system and finally acceptance testing. Thus, test methods may be differentiated by their level of abstraction, open/closed-loop nature, and fraction of simulated vs. real world aspects. The interrelation of the abstraction levels in the V-model and situation-based open-loop as well as scenario-based closed-loop testing methods is described in Sect. 19.4 of this article.

Winner [22] illustrates the infeasibility of pure real world driving tests (cf. 240 million test kilometers) for sufficient test coverage of a software release for an automated vehicle for highway usage. Bridging the gap between the infeasible amount of real world driving tests and open-loop testing, Winner [21] patented the idea of real world driving, open-loop testing by integrating a—to be tested—function into an existing vehicle sold to customers. The function obtains real world input data but the output is only recorded and not translated into real driving maneuvers. By running such a system in big fleets of vehicles, it is possible to achieve a high test coverage for a later released system. The scope of this concept is limited to, e.g., intervening driving functions; driving functions focused on a closed-loop interaction with the environment cannot sufficiently be tested by this approach.

By using recorded measurement data during the development process for testing algorithms, another variant of real world open-loop testing is implemented to the core of development platforms like the Robot Operating System (ROS)[6] or the Automotive Data and Time-Triggered Framework (ADTF).[7] A measurement-data-based open-loop testing and evaluation of the lane change situation assessment algorithms in "Jack" have been demonstrated by Ulbrich and Maurer [17, 18], already. Thus, it is not further pursued for this article.

Arnold et al. [1] survey the use of scenarios in different industrial projects. They show that scenarios can be used for generating test-cases in different domains, e.g., in medical systems, for telecommunication systems, financial systems, and also in software development. They analyze the kinds of the scenarios, the content of the scenarios, and the kind of representation of the scenarios. The broad usage of the scenario-concept indicates a transferability of the different test levels towards other domains.

Ryser and Glinz [10] describe a scenario-based approach for validating and testing software. They suggest the so-called scenario-based validation and test of software by systematically traversing paths in scenario state charts. Other than in this article, Ryser and Glinz primarily focus on the structured generation of scenarios.

---

[6] www.ros.org

[7] www.elektrobit.com/products/eb-assist/adtf

Bai et al. [2] and Tsai et al. [15, 16] elaborate the concept of scenario-based testing. Tsai et al. [16] use scenario-based testing as an integration test method. Complex scenarios are compiled of simpler sub-scenarios. According to Bai et al. [2], a sub-scenario is instantiated based on a template data structure. Bai et al. and Tsai et al. suggest to structure scenarios in multiple levels of scenario groups. Tsai et al. [15] focus on integrating scenario-based testing into an object-oriented test framework. They demonstrate the feasibility of their approach by a case study for a banking transaction system. Touseef and Qaisar demonstrate in [14] an approach for scenario-based tests on an overall system level. It is demonstrated in a case study of an inventory system.

The authors are only aware of few references for situation-based open-loop testing. Bai et al. [2] mention the so-called looping scenarios as a special case of scenario-based testing, where the same scenario is repeated over and over again. Given that a situation is based on a scene in a scenario, looping a scenario without considering the outputs of a device under test will result in something similar to a situation-based open-loop testing.

Tadjine et al. [13] mention to use a scene-based test catalog for the development of a pedestrian detection driver assistance system. Given the open-loop nature of the perception part and given that their "scenes" are tailored specifically to the goals & values of the driving function (which we consider as a "situation," as in Sect. 19.2), Tadjine et al.'s approach seems like an example of what the authors call situation-based open-loop testing.

Unit tests are a commonly used test approach in software engineering. A literature review is out of scope for this publication. Sommerville [12] provides details on different aspects of unit tests in software engineering.

19.4 Integrating Unit Tests, Situation-Based Open-Loop Testing, and Scenario-Based Closed-Loop Testing into the V-Model

This section provides an overview over the testing and simulation efforts to ensure a correct functionality of an item under test. The testing and validation efforts follow a four-step procedure as illustrated on the right side in Fig. 19.5. This four-step procedure can be integrated into the testing in the V-model [6, Part 6] according to the checkmarks in Fig. 19.5.

On a very basic level, unit tests are executed to test and ensure the correct functionality of aspects of atomic functions. As a next step, a situation can be generated as a stimulus to test a driving function as a whole in an open-loop test by a situation-based testing. Such testing is limited in its scope to the driving function itself; it does not test the situation extraction from a scene. If this situation extraction needs to be tested as well, the authors propose a scene-based open-loop

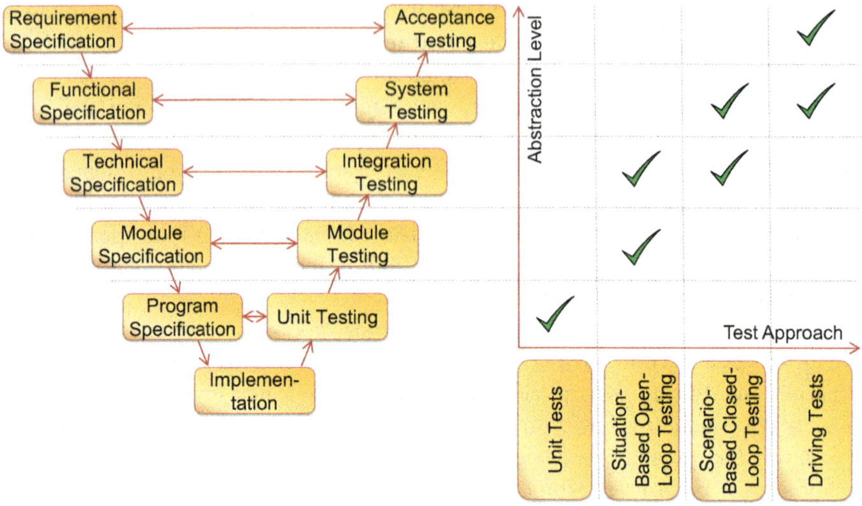

Fig. 19.5 Four-step test procedure for software validation and testing

testing. Anyway, in the test process for *Jack* it was not necessary to implement the intermediate step of scene-based open-loop testing.

When all situation-based test-cases are passed successfully, scenario-based closed-loop testing is used to test an item under test in its interaction with strategic level modules and stabilization level modules as a whole. As a last step, testing is completed by real world driving tests. The test steps will be explained in detail in the following subchapters. The situation-based open-loop testing and scenario-based closed-loop testing are presented in Sect. 19.5 by the case study of a lane change planner module.

Figure 19.6 illustrates the differences between miscellaneous levels of testing. Unit tests allow only to test particular code parts of the driving function. This is depicted by single parts of a jigsaw puzzle of the driving function in Fig. 19.6. Situation-based open-loop testing generates situations for one or several different timestamps from the test-case description and evaluates the behavior response without feeding this behavior response back into future situations. Scenario-based closed-loop testing specifies an entire scenario in a test-case. This includes scenes, events to alter the following scenes, and goals & values for situation extraction and as an input for the driving function. The control and behavior response from the driving function is used to influence future scenes and by this implicitly future situations as well. A test system generates the inputs in Fig. 19.6.

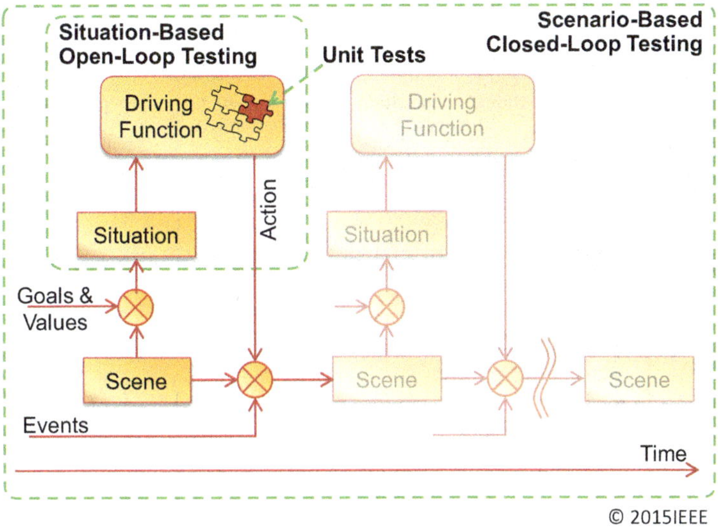

Fig. 19.6 Illustration of the differences between unit tests, situation-based open-loop testing, and scenario-based closed-loop testing [19]

19.4.1 Unit Tests

On a basic level, unit tests are executed to test basic software functionalities. These unit tests are of particular value to reduce the amount of errors in basic functions, e.g., to calculate distances, time gaps, time to collisions, or similar physical-law-based numbers that follow clear calculation rules and numeric properties. However, unit tests are not an eligible method for testing situation assessment functions as they typically require knowledge about the past development of a situation. Thus, it is hard to evaluate the results of a single processing cycle of situation assessment. To test these more abstract functions, it is far easier to broaden the scope and use a situation-based testing instead of a traditional unit testing scheme.

19.4.2 Situation-Based Open-Loop Testing

The situation-based open-loop testing uses a broader scope for testing. While unit tests focus on testing single functions and lines of code, the focus for the situation-based testing is wider: According to Fig. 19.7, a situation data structure is generated as a mock-up for a particular simplified real driving situation. The situation is fed unchanged into the tactical behavior planning module in the guidance block.

In the guidance block, the tactical behavior planning module evaluates the situation and derives tactical driving decisions accordingly. The driving decisions

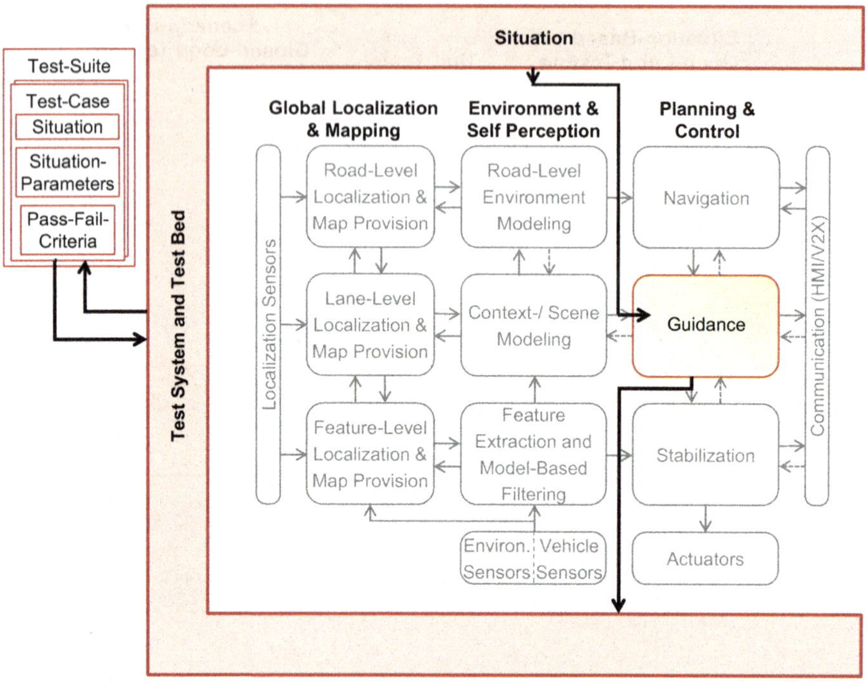

Fig. 19.7 Situation-based open-loop testing as a middle ground between unit tests and scenario-based testing in a generalized, not project-specific functional system architecture based on [8]

from the tactical behavior planning modules are compared with an a-priori-known ground truth of correct driving decisions. Deviations from that ground truth of expected behavior are evaluated and marked as a pass or fail of such a particular test. Other than in traditional unit tests, the same situation may be used repeatedly as a stimulus for a planning module. Thereby, steady states of dynamic, model-free filtering components (e.g., low pass filters) can be reached and tested. Moreover, modules can be tested as a whole, not only single classes of them.

Advantage of this test method is its applicability for fast testing of gradual software changes. Additionally, the test suite can easily be expanded with new test situations. Moreover, the tests can be executed faster than in real time. Thus, situation-based open-loop testing is a versatile tool for testing during the development process and a necessary step to be executed and passed for any releases.

A limitation of this test procedure is that the temporal development of a situation is currently not predicted by a situation prediction model. This renders it possibly insufficient for testing components containing model-based filters. Moreover, the approach is clearly limited by its open-loop nature: The situation is not modified and predicted based on the tactical behavior decisions of the module under test.

19.4.3 Scenario-Based Closed-Loop Testing

Scenario-based closed-loop testing lifts some of the above-described limitations of the situation-based open-loop testing. A test-case in scenario-based closed-loop testing specifies an entire scenario with its parameters and pass–fail criteria. This includes scenes, events to alter these scenes, and goals & values used for situation extraction and as an input for the driving function. The device under test is not only one module like the tactical behavior planning module but rather any subset of modules of the automated vehicle as a whole. Figure 19.8 illustrates a scenario-based closed-loop testing of the feature extraction, scene modeling, guidance, and stabilization.

By the scenario-based closed-loop testing, the interaction of ideally the whole chain can be tested. The control and behavior response from the driving function is used to influence future scenes and by this—implicitly—future situations.

As illustrated in Fig. 19.6, scenes will be modified over the course of the scenario according to prediction models in the test system. Thus, this mode of testing allows to test model-based filtering approaches. Other than in the situation-based open-loop testing, objects may move ahead and initiate maneuvers over the course of the simulation time. Such maneuvers may either be triggered by driver models in the test system or defined externally by events in the scenario.

The strengths of scenario-based closed-loop testing are short development cycles for driving function development. In fact, a certain complexity level of scenarios

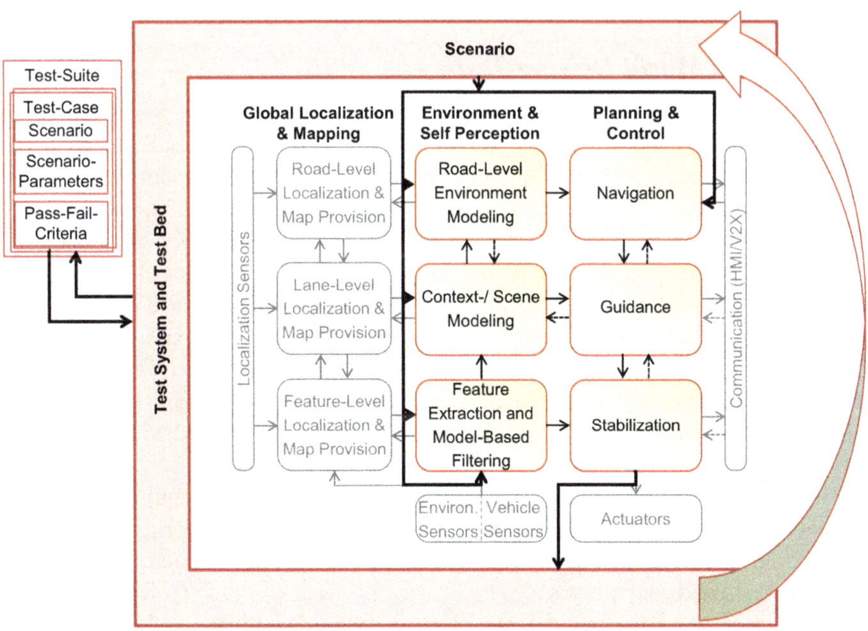

Fig. 19.8 Example for scenario-based closed-loop testing of perception and planning and control modules in a generalized, not project-specific functional system architecture based on [8]

can only be tested in a simulation framework in a resource efficient manner. This test technique allows to create scenarios to match the particular needs of a developer within minutes. The closed-loop test allows to test the interaction of several modules and helps to identify and analyze signal latencies or functional instabilities.

The authors see potential in using scenario-based closed-loop testing to obtain meaningful test results for system validation, but proofs for it are yet to be provided.

Among the limitations of scenario-based closed-loop testing is the representativeness of the results. In fact, the simulation is based on several models of the real world; starting with behavior models and vehicle dynamic models, via sensor models, up to the design patterns of the scenery. The big caveat of such a simulation-based design process is that a solution may get particularly well tailored to a simulation environment but not necessarily to the challenges in the real world. At the time of writing, the modeling of perception-induced measurement errors and uncertainties is not yet sufficiently close to the real issues. While there are certain sensor models to emulate certain artifacts of certain sensor systems, today's error models still lack the modeling of high-level errors like false classifications, false segmentations, model-incompliant movement behavior, or false semantic associations between perceived entities.

Moreover, a sophisticated simulation tool chain comes along with a significant overall complexity. Therefore, it often takes a lot of time to achieve the intended results. In particular, switching back and forth between a development based on recordings from driving in a real vehicle and a simulation-based development is currently often a time-consuming issue.

19.4.4 Real World Driving Tests

Real world driving tests are the last step in the presented test process. They are used for system and acceptance tests in the V-Model and simulation model verification. Necessitating a real vehicle or prototype, they should be executed after successfully passing all situation-based open-loop and scenario-based closed-loop tests.

A test of driving functions with real environment conditions and resulting uncertainties is only possible in real world driving tests. Real world driving tests require also test-cases with scenarios, scenario-parameters, and pass–fail criteria. The scenario parameters are given indirectly, e.g., by actually driven trajectories or a hereof resulting physical arrangement of other traffic participants. Figure 19.9 illustrates the modules in a system architecture, which can be tested with real world driving tests.

The advantage of real world driving tests is that they do not require simulation models and thus do not induce errors by incorrect models. Real world driving tests are to some extend random tests. Thus, many scenarios can be tested, without any particular test-case generation.

The aspect of—to some extent—random testing simultaneously imposes a severe limitation of the test method: Due to the random behavior of other traffic

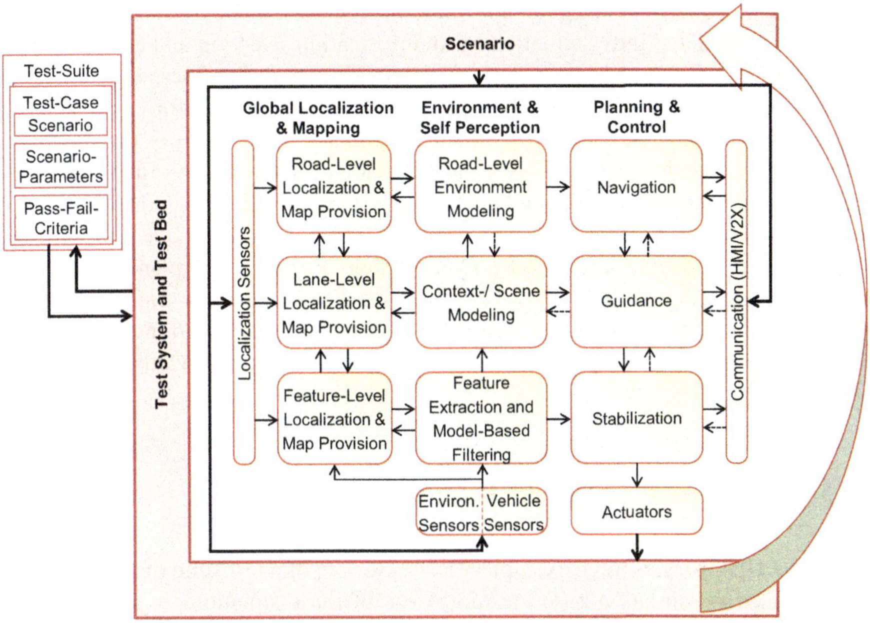

Fig. 19.9 Real world driving testing in a generalized, not project-specific functional system architecture based on [8]

participants, the test-cases are not fully reproducible. Moreover, the test-cases cannot be performed faster than in real time. Because of this, executing a large number of test-cases is directly linked to high costs and expenditure of time.

19.5 Case Study: Testing and Validating Tactical Lane Change Behavior Planning

This section presents a case study for testing and validating a module for tactical lane change behavior planning. Section 19.5.1 describes the module under test in brief. After this, the simulation-based testing for module and integration tests is demonstrated. Section 19.5.2 presents situation-based open-loop testing for this modules. Section 19.5.3 illustrates a scenario-based closed-loop testing for the lane change behavior planning. Real world driving tests are excluded from this paper. They can be found in Ulbrich and Maurer [17, 18].

19.5.1 Item Under Test: Behavior Planning for Lane Changes

To demonstrate testing on the earlier introduced levels, a module for tactical lane change behavior planning is used as an item under test. This module is part of the guidance block in a functional system architecture as in Figs. 19.7, 19.8, or Fig. 19.9.

Within the guidance block, a goal- and value-independent scene from the perception modules is translated into a situation by an information selection and augmentation process specifically based on the automated vehicle's goals & values. This situation is used as an input for the module for tactical behavior planning for lane changes. It entails a situation assessment to estimate non-measurable aspects of the situation. Moreover, the module encompasses the behavior decision making itself. This is based on a situation prediction and a cost and reward model. This model estimates how future behavior alternatives and resulting future situations have an affect on the possibility and benefit to execute a lane change. Based on this planning into the future, the immediate next behavior action is used to command a trajectory target point, e.g., in front of the automated vehicle in the ego lane, on a neighbor lane, or with a targeted velocity and position difference to adjust towards a gap on a neighbor lane.

19.5.2 Situation-Based Open-Loop Testing

Figure 19.10 illustrates three exemplary test-cases out of a test suite of 29 test-cases that are used for situation-based testing. Each of these situations is generated by a set of support functions. Each situation is fed into the tactical behavior planning module for planning lane changes for several cycles until a steady state of any low pass filtering component can be assumed. For the moment, each situation is repeated 400 times resulting in an evaluation speed five times faster than real time (3.2 s per test-case). After such a cycle the tactical behavior of the lane change planning module is compared to the expected behavior noted in the test-case. If the behavior is identical, a test-case is passed. In this case study, the situation contains value-continuous and value-discrete elements. To evaluate the behavior response of the lane change planning module, value-discrete behavior choices were evaluated.

It is possible to execute a human-designed set of the 29 most essential test-cases on a standard computer in less than 2 min. This fact renders this test procedure very efficient for iterative testing even after minimal source code changes. Apart from the test-cases itself, Fig. 19.10 illustrates the expected resulting behavior for each of these test-cases and if it matches the obtained resulting behavior from the lane change planning module. As indicated in the figure, all three test-cases have been passed successfully.

19.5.3 Scenario-Based Closed-Loop Testing

For the scenario-based closed-loop tests, Vires' Virtual Test Drive (VTD)[8] is used as a test system. It is a simulation tool chain for road traffic, railroad, and flight

[8]Vires Virtual Test Drive, www.vires.com

Nr.	Test Case	Expected Result	Passed
TC 22)	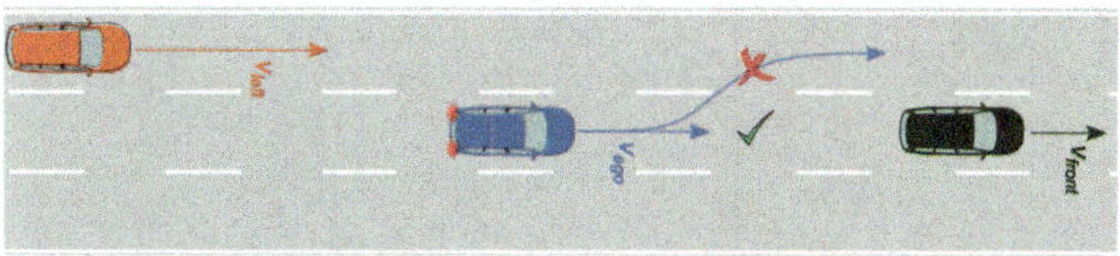 The ego vehicle drives with $v_{ego} = 33.3\,m/s$ on the center lane of a three lane highway with no speed-limit. An object drives with a longitudinal offset of $s_{front} = 100\,m$ and a velocity $v_{front} = 16.6\,m/s$ ahead. Another object drives on the left lane with a velocity of $v_{left} = 50\,m/s$ and a longitudinal distance of $s_{left} = -80\,m$ behind the ego vehicle.	No lane change	✓
TC 23)	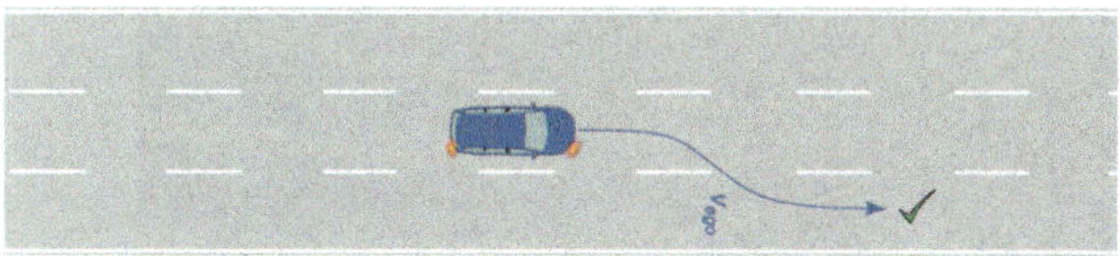 The ego vehicle drives with $v_{ego} = 33.3\,m/s$ on the center lane of a three lane highway with a speed limit of $v_{speedlimit} = 33.3\,m/s$ in a country with a right lane driving order.	Lane change right	✓
TC 27)	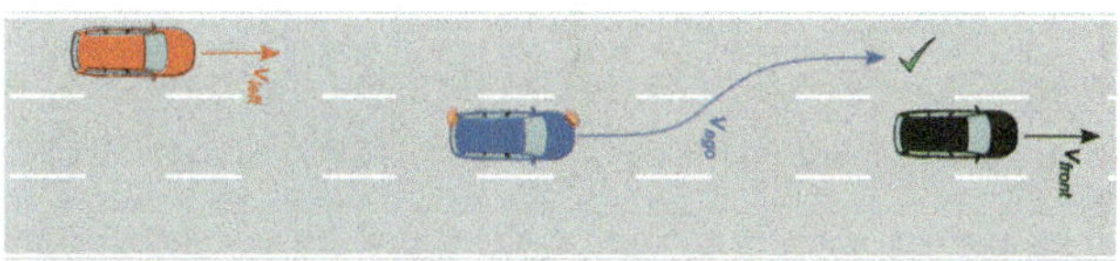 The ego vehicle drives with $v_{ego} = 25\,m/s$ on the center lane of a three lane highway with a speed limit of $v_{speedlimit} = 33.3\,m/s$. A slower object with a velocity of $v_{front} = 25\,m/s$ drives $s_{front} = 50\,m$ in front of the ego vehicle. Another object drives on the left lane $s_{left} = -35\,m$ behind the ego vehicle with a velocity of $v_{left} = 26.38\,m/s$.	Lane change left	✓

Fig. 19.10 Test cases for situation-based lane change testing

simulation. It provides tools for creating road networks and scenarios as well as a simulation backbone and rendering tools for the visualization. Moreover, it encompasses several simulation models for vehicle dynamics, driver behavior, and pedestrians. As a test bed, an ADTF filter graph is used to translate between VTD's simulation interfaces and the appropriate interfaces of the driving function.

Figure 19.11 illustrates scenario-based closed-loop testing showing the same scene of a scenario side by side in VTD's scenario editor and the ADTF driving function filter graph with an enabled visualization.

In Fig. 19.12, an exemplary highway scenario for lane changes is depicted. Initially, the automated vehicle drives with 35 m/s on the rightmost lane of a highway. It is approaching a slower vehicle in front of it. The dynamic disadvantage of following a slower vehicle motivates the automated vehicle to perform a lane change to the left (cf. subfigure (1) in Fig. 19.12). Since there is no slow vehicle

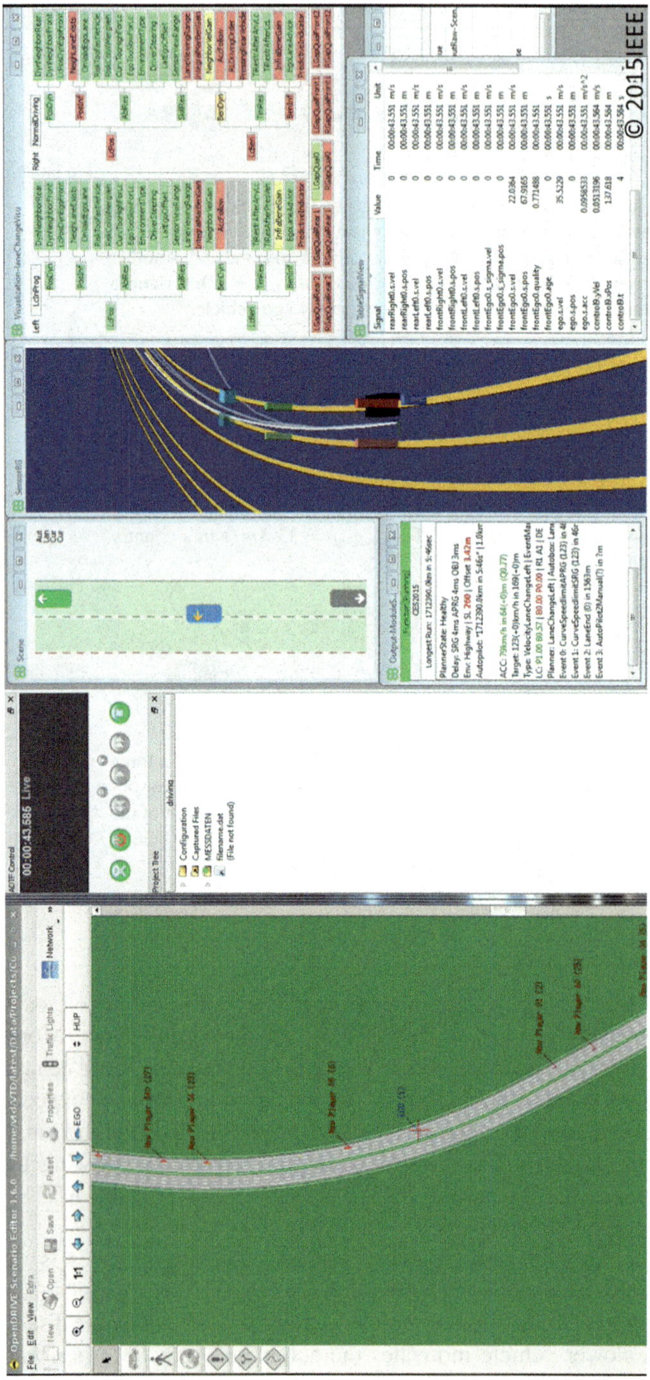

Fig. 19.11 Scenario-based closed-loop testing in Virtual Test Drive (VTD) and ADTF [17]

Fig. 19.12 Illustration of a highway simulation scenario to test a lane change left due to (*1*) dynamic benefits, (*2*) longitudinal gap adjustment, (*3*) a lane change right, and (*4*) exiting a highway to an exit ramp for an automated vehicle (*blue*)

in front to follow, the automated vehicle accelerates to reach the target velocity of $v_{\text{target}} = 35\,\text{m/s}$.

After passing several slower vehicles on the right lane by driving on the center lane of a three lane highway, the automated vehicle gets close to an exit, where it is commanded to leave the highway. Given the relative speed difference, the gaps on the right lane are too small for a direct lane change. Hence, the automated vehicle initiates a longitudinal gap adjustment to center itself to the best reachable and best sized gap. To achieve this, the automated vehicle decelerates to reduce the speed difference of itself and the traffic on the right target lane (gap adjustment). This is followed by a lane change to the right into the selected gap (cf. subfigure (2) in Fig. 19.12). Last of all, the automated vehicle exits the highway by changing once more to the right on a deceleration lane of the targeted highway exit (cf. subfigure (3) in Fig. 19.12). The automated vehicle adapts its speed to the upcoming curvature of the exit ramp (cf. subfigure (4) in Fig. 19.12).

Figure 19.13 illustrates selected state variables relevant for the lane change behavior planning in the before-mentioned scenario. The first three figures depict relevant state variables of the situation the automated vehicle is facing: namely, the (automated) ego vehicle's velocity longitudinally to the lane, the lateral offset to the center of the ego lane, and the distances to other objects directly in front of the automated vehicle in the ego lane and the right neighbor lane.

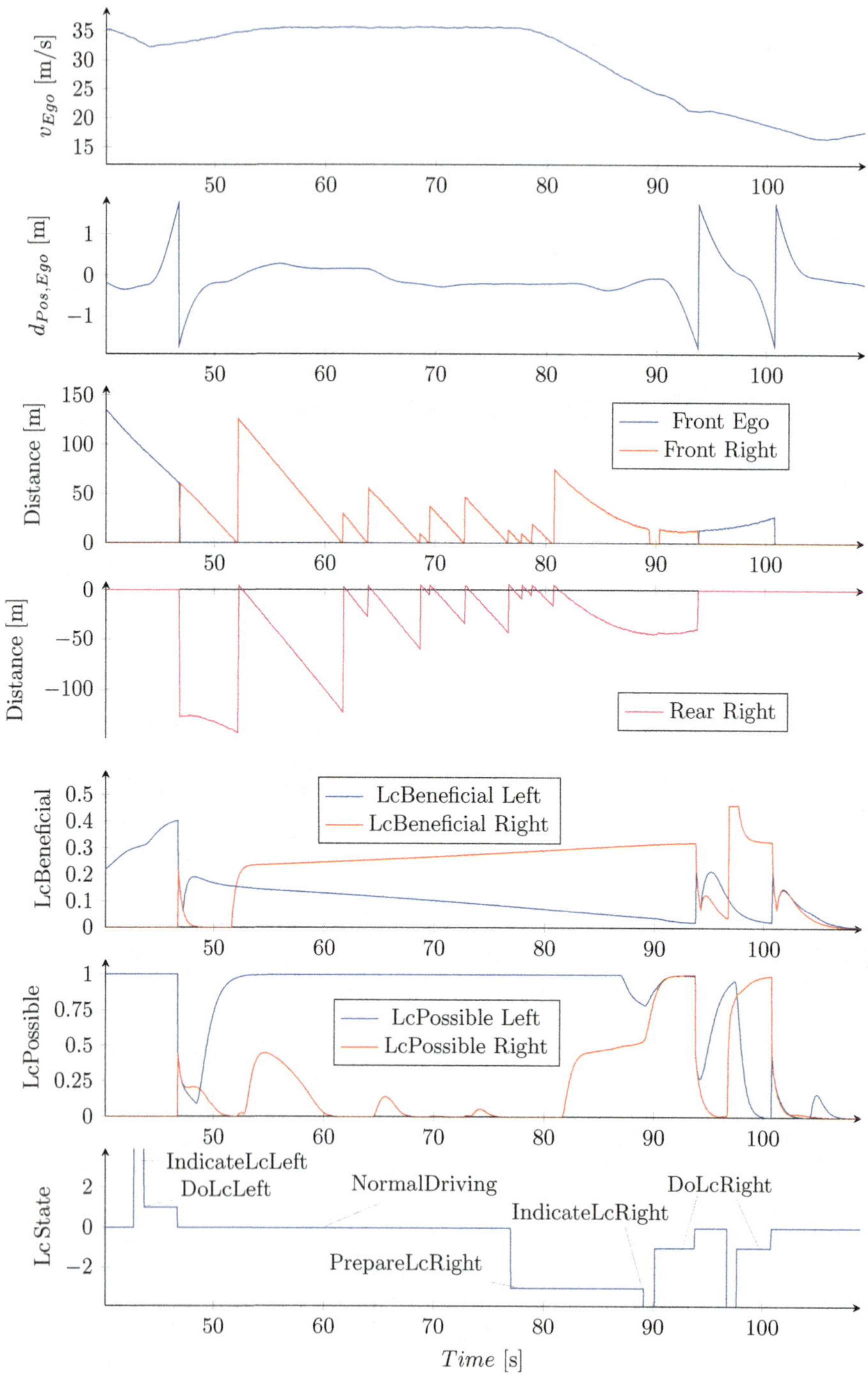

Fig. 19.13 Measured data for scenario-based closed-loop testing scenario on a highway

Moreover, Fig. 19.13 visualizes some hidden state variables resulting from a situation assessment (cf. Ulbrich and Maurer [17]). The two most relevant and here depicted ones are the lane change possible estimate and the lane change beneficial estimate. Last of all, Fig. 19.13 shows the lane change action resulting from the overall lane change planning process.

The earlier described scenario can be traced in the measured data. The longitudinal ego velocity plot visualizes the before-mentioned velocity profile. Initially, the automated vehicle drives with $v_{\text{ego}} = 35\,\text{m/s}$. After experiencing a marginal slow down by a vehicle in front of it, it activates the indicator to the left (cf. $\mathit{LcState} = \mathit{IndicateLcLeft}$ in last subfigure of Fig. 19.13), and executes a lane change by building up a lateral offset to the left (cf. $d_{\text{ego,pos}} = 1.8\,\text{m}$; subfigure (1) in Fig. 19.12). After crossing the left lane boundary, the lane detection changes the ego lane from the highway's rightmost lane to the center lane of the three lane highway. As the reference lane switches, it causes a jump in the lateral ego lane offset from $d_{\text{ego,pos}} = +1.8\,\text{m}$ to $d_{\text{ego,pos}} = -1.8\,\text{m}$. After this jump, the automated vehicle re-centers itself to the new lane. On this middle lane of the highway is no other vehicle closely in front of the automated vehicle. Hence it accelerates to reach the target velocity $v_{\text{Target}} = 35\,\text{m/s}$.

Several vehicles are passed or overtaken. This is illustrated by the distance of the immediate next vehicle in the ego lane and the right neighbor lane. Several slower vehicles are approached from behind (decreasing distance) until they are overtaken. In total, ten vehicles get overtaken by the automated vehicle.

After a certain time of driving the scenario requires the automated vehicle to exit the highway at an upcoming highway exit on the right. Therefore, the automated vehicle is required to change back to the rightmost lane of the highway. To achieve this, the automated vehicle activates a longitudinal gap adjustment at $t = 78\,\text{s}$ (cf. $\mathit{LcState} = \mathit{PrepareLcRight}$ in last subfigure of Fig. 19.13; subfigure (2) in Fig. 19.12). To simplify a lane change, the relative velocity difference between the automated vehicle ($v_{\text{Ego}} = 35\,\text{m/s}$) and the objects/gaps on the right lane ($v_{\text{Target}} = 22\,\text{m/s}$) has to be reduced. After a few seconds, a lane change becomes possible and the automated vehicle changes to the rightmost lane of the highway at $t = 90\,\text{s}$ by activating the indicator right and building up a lateral displacement to the right (cf. $\mathit{LcState} = \mathit{DoLcRight}$). At $t = 95\,\text{s}$, the lane change is finished and the automated vehicle is fully re-centered to the rightmost lane. After a few more seconds it reaches the beginning of the exit ramp (cf. subfigure (3) in Fig. 19.12). Even before the exit ramp is actually next to the automated vehicle, the right indicator is activated based on the information from an a-priori map that there is the exit ramp about to appear on the right.

This predictive indicator mode helps to prevent conflicts with other vehicles changing faster to the exit lane and possibly blocking the automated vehicle from being able to change to the exit ramp. After the exit ramp has reached its full width, the automated vehicle changes onto the exit ramp and slowly adapts its speed (cf. $v_{\text{Ego}} = 18\,\text{m/s}$; subfigure (4) in Fig. 19.12) to follow the right turn of the exit ramp.

19.6 Conclusions

In this article, the concept of situation-based open-loop testing and scenario-based closed-loop testing has been developed as a concept and illustrated in a case study. This has been founded on a terminology for relevant terms. The terms are used to present different levels of testing that proved to be useful for the test of driving function modules: unit tests, situation-based open-loop tests, scenario-based closed-loop tests, and real world driving tests as different steps for system testing and validation. The different advantages and limitations of the test methods are introduced. Moreover, a comparison between the introduced test steps and real world driving tests is given. The situation-based open-loop and scenario-based closed-loop test methods are demonstrated by a case study of a lane change behavior planning.

The article does not address a systematic approach to generate test-cases. So far, they were manually designed by a human expert. Moreover, the test-case evaluation is not automated for the scenario-based closed-loop simulation. At the moment a human expert needs to define and evaluate pass–fail criteria for each test-case. While this works well for testing during the development phase, it scales unfavorably for validation tests with thousands or millions of test-cases automatically executed on a simulation server farm.

The next step is to use the situation-based open-loop and scenario-based closed-loop test methods combined with a systematic test-case generation to automatically find system boundaries and scenarios, where the system under test shows a lower performance. This includes an automatic test-case evaluation based on eligible pass–fail criteria.

References

1. M. Arnold, M. Glinz, M. Erdmann, Survey on the scenario use in twelve selected industrial projects. Technical Report 98-7, Rheinisch-Westfälische Technische Hochschule Aachen, 1998
2. X. Bai, W.T. Tsai, K. Feng, L. Yu, R. Paul, Scenario-based modeling and its applications, in *Proceedings of the Seventh IEEE International Workshop on Object-Oriented Real-Time Dependable Systems (WORDS 2002)* (IEEE Computer Society, New York, 2002), pp. 253–260
3. P. Bergmiller, *Towards Functional Safety in Drive-by-Wire Vehicles* (Springer, Berlin, 2015)
4. Bibliographisches Institut: Situation. No. 21 in Meyers Enzyklopädisches Lexikon (English title: Meyer's Encyclopedia). Bibliographisches Institut AG, Mannheim (1977)
5. S. Geyer, M. Baltzer, B. Franz, S. Hakuli, M. Kauer, M. Kienle, S. Meier, T. Weissgerber, K. Bengler, R. Bruder, F. Flemisch, H. Winner, Concept and development of a unified ontology for generating test and use-case catalogues for assisted and automated vehicle guidance. IET Intell. Transp. Syst. **8**(3), 183–189 (2014)
6. International Organization for Standardization (ISO): ISO 26262:2011 Road vehicles - functional safety (2011)
7. S. Khastgir, S.A. Birrell, G. Dhadyalla, P.A. Jennings, Identifying a gap in existing validation methodologies for intelligent automotive systems: introducing the 3xD simulator. in *IEEE Intelligent Vehicle Symposium (IV)* (2015), pp. 648–653

8. R. Matthaei, M. Maurer, Autonomous driving - a top-down-approach. at - Automatisierungstechnik **63**(3), 155–167 (2015)
9. M. Maurer, EMS-vision: knowledge representation for flexible automation of land vehicles, in *IEEE Intelligent Vehicles Symposium (IV)* (2000), pp. 575–580
10. J. Ryser, M. Glinz, A scenario-based approach to validating and testing software systems using statecharts, in *12th International Conference on Software and Systems Engineering and their Applications (ICSSEA)* (1999). [Without pagination]
11. F. Schuldt, T. Menzel, M. Maurer, Eine Methode für die Zuordnung von Testfällen für automatisierte Fahrfunktionen auf X-in-the-Loop Verfahren im modularen virtuellen Testbaukasten (A method to assign test-cases for automated driving functions towards x-in-the-loop techniques in a modular virtual test framework). Workshop Fahrerassistenzsysteme 2015. Walting (2015), pp. 171–182
12. I. Sommerville, *Software Engineering*, 9th edn. (Addison-Wesley, Harlow, 2010)
13. H. Tadjine, K. Schulze, R. Roellig, H. Daniel, New methods and tools for the development and verification of safety functions during development of pedestrian detection systems, in *8th International Conference on Computing Technology and Information Management (ICCM)*, vol. 1 (2012), pp. 434–438
14. M. Touseef, Z.H. Qaisar, A use case driven approach for system level testing. Int. J. Comput. Sci. Issues (IJCSI) **9**(1) (2012), p. 78
15. W. Tsai, A. Saimi, L. Yu, R. Paul, Scenario-based object-oriented testing framework, in *Third International Conference on Quality Software* (2003), pp. 410–417
16. W.T. Tsai, X. Bai, R. Paul, W. Shao, V. Agarwal, End-to-end integration testing design, in *2001 IEEE Computer Society International Conference on Computers, Software & Applications (COMPSAC)* (IEEE, New York, 2001), pp. 166–171
17. S. Ulbrich, M. Maurer, Situation assessment in tactical lane change behavior planning for automated vehicles, in *18th International IEEE Annual Conference on Intelligent Transportation Systems (ITSC)* (2015), pp. 975–981
18. S. Ulbrich, M. Maurer, Towards tactical lane change behavior planning for automated vehicles. in *18th International IEEE Annual Conference on Intelligent Transportation Systems (ITSC)* (2015), pp. 989–995
19. S. Ulbrich, T. Menzel, A. Reschka, F. Schuldt, M. Maurer, Defining and substantiating the terms scene, situation and scenario for automated driving, in *18th International IEEE Annual Conference on Intelligent Transportation Systems (ITSC)* (2015), pp. 982–988
20. K.P. Wershofen, V. Graefe, Situationserkennung als Grundlage der Verhaltenssteuerung eines mobilen Roboters (Situation perception as basis for behavior control of a mobile robot), in *Fachgespräch AMS*, Munich (1996)
21. H. Winner, Device for providing signals in a motor vehicle WO Patent App. WO2002058975 A1 (2002)
22. H. Winner, Quo vadis, FAS? in *Handbuch Fahrerassistenzsysteme*, ed. by H. Winner, S. Hakuli, F. Lotz, C. Singer, ATZ/MTZ-Fachbuch (Vieweg+Teubner, Wiesbaden, 2015), pp. 1167–1186

Chapter 20
Safety Performance Assessment of Assisted and Automated Driving in Traffic: Simulation as Knowledge Synthesis

Thomas Helmer, Klaus Kompaß, Lei Wang, Thomas Kühbeck, and Ronald Kates

20.1 Introduction

20.1.1 Driver Assistance and Automation

Based on comprehensive, sensor-based detection of vehicle surroundings, the vision of self-driving vehicles has already become a reality in research vehicles (e.g., [1, 2]). Today, advanced driver assistance systems (ADAS) or partial automation, in which the vehicle assumes the role of a "copilot," can assist the driver in demanding situations, support him in complex or hazardous situations, and, if necessary, perform maneuvers automatically. Higher levels of automation can, at least temporarily, free the driver entirely from the driving task and enable him to fulfill other tasks. Thus, automation could cause a considerable improvement in the quality of individual mobility [2].

In addition to individual mobility, collective and socioeconomic impact of vehicle automation have a central role regarding further development and dissemination of those functions. In highly developed and industrialized countries, especially within metropolitan areas, satisfaction of the "mobility demands" often requires additional infrastructure and induces impacts on environment and quality of life. High percentages of automated vehicles could enable technological strategies that

This chapter is largely based on: Klaus Kompaß et al.: Fahrerassistenz und Aktive Sicherheit. Wirksamkeit—Beherrschbarkeit—Absicherung. expert verlag, Renningen 2015

T. Helmer (✉) • K. Kompaß • L. Wang • T. Kühbeck
BMW AG, Munich, Germany
e-mail: thomas.helmer@bmw.de

R. Kates
REK Consulting, Otterfing, Germany

D. Watzenig, M. Horn (eds.), *Automated Driving*,
DOI 10.1007/978-3-319-31895-0_20

increase the capacity of streets and the "mobility supply" and thus allow a more environmentally friendly and efficient use of resources.

Reliability and safety criteria are key factors for homologation and operation of advanced driver assistance systems and especially automated driving functions. Only when the overall safety level of a technology reaches (or exceeds) a generally accepted threshold should secondary criteria, such as comfort or efficiency, be considered.

Since the individual customer is hardly able to independently evaluate the safety features when purchasing a vehicle, he has to rely on trustworthy and objective evaluations by third parties. In addition to automotive manufacturers and suppliers, public decision-makers, regulatory agencies, insurance companies, and consumer protection organizations are key stakeholders in the evaluation of vehicle safety. In particular, demonstration of equivalent or superior safety in all operating scenarios is an important prerequisite for the introduction and homologation of novel technologies, such as automated driving functions (ADF).

Ideally, it should be possible to derive quantifiable predictions of safety impacts for new technologies in such a way that all stakeholders can agree on their relevance, objectivity, validity, and reproducibility—despite their sometimes divergent interests. A prerequisite for this is a careful definition of an appropriate safety metric with respect to the specific context.

Thus, for example, in the context of pedestrian protection, the reduction of pedestrian injury by mitigating or avoiding collisions is the primary objective. The quality of pedestrian protection system could be objectively quantified by a metric which relates to injury reduction (see below).

An objective metric of traffic safety also includes possible side effects [8, 13, 22, 33]: preventative pedestrian protection, for example, can result in occasional unwanted system actions due to the underlying physics. As a consequence, secondary risks can occur (e.g., rear-end collisions following an emergency braking maneuver). With the help of an objective metric, the ratio of desired safety improvements to unwanted side effects can be quantified. Based on this ratio, the function can be optimized with respect to traffic safety (Fig. 20.1).

Several important aspects of vehicle safety such as integrity of communication, data protection, and technical reliability are beyond the scope of this paper. The focus here is on the effects of ADAS and current as well as future fully operational automated systems on traffic safety.

Although ADAS are generally designed to minimize well-known accident risks or mitigate the consequences of unavoidable accidents, the optimal design of ADAS in terms of traffic safety is a complex task. ADAS usually rely on sensor-based decision algorithms, which often depend on certain parameter settings. However, any interaction with the driver or interference with the driver's task can in principle have a positive or negative impact on traffic safety, for example:

- *Lane departure warning*: A typical criterion for an alarm or intervention in case of impending lane departure would be "time to lane crossing" (TLC), calculated using sensor data. It could be assumed that earlier warnings (large

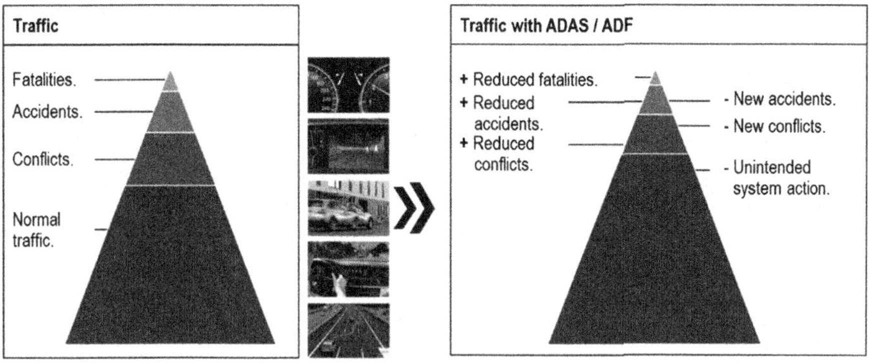

Fig. 20.1 Changes in traffic safety due to advances in driver assistance and automation (after [19])

TLC—thresholds) are "safer." However, too many unnecessary warnings could lead to negative feedback, such as a lack of acceptance.

- *Emergency braking systems*: It could be assumed that earlier system interventions avoid more rear-end collisions. However, some drivers might still execute an evasive maneuver, even if accident avoidance by braking was no longer possible (this is a well-known dilemma regarding braking and steering). Moreover, if the system reacts in such situations too early with a correspondingly strong intervention, the risk of accidents for the driver (loss of control) as well as for subsequent vehicles (rear-end collisions) can arise.

20.1.2 Assessment and Optimization of ADAS and ADF as Key Processes During the Development

The examples illustrate that design of advanced driver assistance systems or automated driving functions is a complex task, especially with regard to optimization of traffic safety.

One approach for structured analysis of impacts on traffic safety of existing ADAS is retrospective analysis of accident databases linked to vehicle equipment data (if feasible). However, large samples from accident databases and long observation times [8] are required for retrospective statistical evaluation of accident avoidance due to ADAS. Statistical challenges for evaluation of a particular ADAS arise due to multifactorial data characteristics: vehicles in the sample may differ in several aspects, not only regarding the equipment with the system in question. Vehicles from different manufacturers may have different functional characteristics, interaction concepts with the driver, or activation conditions. Differences of driver population or exposure to traffic scenarios can be correlated with the equipment. Such properties of retrospective studies act statistically as confounders and make an unambiguous interpretation of results and conclusions about causality challenging.

Comprehensive statements regarding changes in traffic safety ("effectiveness") are therefore often problematic [13, 22].

In addition, evaluation of the consequences of unwanted system actions is practically impossible based on retrospective studies alone, since the relevant accident types in general relate to a range of possible secondary effects.

Summarizing, retrospective studies can enable ex post evaluations, but can hardly be used for any optimization of ADAS or ADF, due to the long feedback loop of the development, retrospective evaluation, and redesign.

Interestingly, retrospective surveys are also considered as second best for assessment of new treatments or therapies in health care: the standard of quality is rather a prospective, randomized, controlled trial in a representative sample.

There is an urgent need for reliable and valid predictions of traffic safety for design and optimization of ADAS and especially for ADF. The evaluation and optimization of such systems using traffic safety predictions should be an integral part of the development process.

However, truly prospective, randomized, controlled, and representative studies of ADAS are not feasible on public roads due to ethical and practical considerations. This dilemma could be resolved using virtual, simulation-based traffic safety predictions of ADAS and ADF.

20.2 Overall Safety Assessment

20.2.1 Safety and Economy

Currently, passive, active, and integral safety functions contribute to the goal of improved vehicle safety [23, 25]. In the future, automation is also expected to exert a decisive influence on vehicle safety. It is increasingly necessary to assess and compare competing, technically feasible safety measures regarding their effectiveness in order to define appropriate priorities. Therefore, a comprehensive approach for safety assessment of new technical concepts is required.

Since automation affects the entire traffic flow, it can also have an impact on individual driving comfort and efficiency. Thus, in assessing ADF, comfort as well as social and economic aspects will play an important role in addition to safety.

20.2.2 Conflicting Objectives in Vehicle and Traffic Safety

In assessing effectiveness of ADAS and ADF, both positive influences as well as risks and side effects need to be considered and included [8, 13, 22, 33].

Considering active safety systems, some conflicting objectives are well known, for example, during the precrash phase: if a system classifies a risk situation

Table 20.1 Categorization of possible system actions (following [21])

		System action	
		Yes	**No**
Objective risk	**Yes**	True positive	False negative
	No	False positive	True negative

very late, "false negatives" or too late system actions can occur, reducing the effectiveness of the ADAS. Conversely, if a system responds to a potential hazardous situation very early, "false positives" can result (see Table 20.1): in these cases, the driver could possibly have avoided the accident even without system intervention. Excessive frequency of false positives can reduce acceptance and even result in negative feedback on traffic safety [8, 13, 20].

For future automated systems, even more complex trade-offs between conflicting objectives can be expected during the design phase.

The comparison of effectiveness and conflicting objectives is part of an integrated development and assessment process of advanced vehicle safety systems. Current methods based on single tests usually cannot properly take conflicting objectives into account [22, 33].

In order to quantify the effectiveness of ADAS in a target scenario by a virtual experiment (see below), an appropriate metric for characterization of the safety benefits is required. An ideal metric includes both avoided accidents and reduced injury severity as well as reduced fatalities in the remaining accidents. Probability models can be used to derive injury severity based on detailed accident characteristics in the target scenario. To calculate the metric, two steps are required: first, calculating the impact of the system on accident characteristics in the target scenario and, second, applying a conditional probability model of injury severity depending on these characteristics.

To model injury severity—quantified as MAIS (maximum abbreviated injury scale), for example—depending on accident characteristics, there are several complementary approaches. A commonly used method is the construction of statistical models (e.g., regression models) from existing accident databases (e.g., [11, 14–16]). Another approach is "co-simulation." Here, a representative sample of time series from accident simulations is generated and analyzed using a high-resolution crash simulation, which is capable of rapidly calculating injury indicators [9, 35].

As mentioned above, certain interventions of ADAS, such as automatic braking or evasive maneuvers, can induce serious side effects into traffic, for example, loss of controllability.

Milder side effects, for example, decrease in user acceptance, can also be caused by controllable interventions or warnings, if they are perceived as superfluous (false positive). Since ADAS can only fulfill their purpose once they are activated, a high alarm rate, lacking comprehensible justification for the driver, can lead to disproportionate non-utilization rates or frequent lack of response to warnings.

Such unintended side effects impact the overall quality of an ADAS and should therefore be considered as part of an assessment. For this purpose, it is necessary to classify false-positive actions and other unintended side effects regarding their frequency and severity.

A key characteristic, borrowed from medicine, is the NNT ("number needed to treat"). In terms of ADAS, NNT can be defined as the ratio between all system actions and true positives, i.e., desired system actions [5, 11, 31]. NNT can be calculated separately for each type of system action (warnings, interventions).

It is useful to distinguish false positives occurring due to technical limitations from those occurring due to situational uncertainties (such as a possible mitigation of a hazardous situation by other traffic participants). A camera-based sensor, for example, can cause false positives if a phantom object is detected due to light and shadow that apparently poses a hazard. Eliminating technically induced false positives may sometimes be accomplished without compromising effectiveness using, for example, test drives.

However, a portion of "unnecessary" system actions arises, as described above, due to essential conflicting objectives and cannot be eliminated without impact on effectiveness. Therefore, a systematic balancing in terms of "benefits" and "costs[1]" is useful during design of most ADAS: each system action is assigned a value for "costs," and each avoided adverse event (e.g., accident) is assigned a value for "benefit." The safety effect of an ADAS can then be optimized based on this balance (e.g., using NNT).

In practice, it is hardly feasible to carry out such an optimization based solely on empirical driving tests or other classical testing setups. Due to the rarity and variability of accident scenarios, it is virtually impossible to obtain a statistically reliable proportion of avoided accidents in the test time available. Therefore, it can be considered current practice to design advanced driver assistance functions to have a false-positive rate below a given threshold. The dependence of effectiveness on control parameters (such as threshold TTC value for triggering) may show trends in individual experiments; however, this approach does not allow optimization with respect to effectiveness and false positives. True ADAS optimization requires feedback between the assessment and development processes. Finally, the assessment of ADF will also require meaningful methods allowing an integration of optimization into the development process [8, 23, 33].

[1]"Costs" in this context is used as collective term for unintended side effects, not necessarily in the monetary sense.

20.3 Design and Optimization of ADAS Using Virtual Experiments

20.3.1 Paradigm of Design of Virtual Experiments

The concept of virtual experiment plays a central role in the comprehensive safety assessment of ADAS. A typical design of a virtual experimental uses the paradigm of a prospective, randomized, controlled, and representative trial. The difference to a "real" experiment is the substitution of the actual traffic flow with a simulated traffic flow and the representation of other "real" components by simulated components. Quality and validity are central requirements in this paradigm.

As in prospective, randomized trials, "treatments" (here one or more variants of an ADAS) are compared to a reference (control or "baseline")—for example a vehicle without the system. In the virtual trial design, target variables are statistically "captured" and used to calculate the relevant metric of traffic safety, analogous to empirical tests. Effectiveness of a system to the baseline can be quantified objectively using the metric. Just as in a randomized, controlled trial, the detection of small differences of effectiveness using statistical tests requires correspondingly high numbers of cases or long observation periods in virtual trials. Unlike in empirical studies or test driving, high virtual sample sizes can be produced with relatively modest resources.

20.3.2 Representation of Safety-Relevant Processes in Simulation

In principle, all relevant dynamic and human processes are represented as time series of states within the simulation. The simulated changes of those states can have both "deterministic" and "stochastic" characteristics. For example, the yaw rate of a vehicle on a dry surface is primarily deterministically dependent on steering wheel angle and speed, but slippage introduces an effectively stochastic component. The duration from a collision warning to braking by the driver can vary widely within a population of drivers and therefore must be regarded as stochastic. Simulation models must therefore be capable of representing not only deterministic but also stochastic properties.

Using safety assessment for the design and development of ADAS must therefore consider the stochastic properties of all relevant technical and human processes.

20.3.3 Knowledge Synthesis and Integration of Other Test Domains

All safety-relevant processes are modeled in sufficient detail in the simulation. The strategy is to capture the effects of ADAS on these processes. These processes are related to exposure variables, traffic flow, and dynamics of the driver-vehicle unit including human factors, technical systems, and so forth. There are many possible interactions within a driver-vehicle unit and between traffic participants and their surroundings. For ADF, the interaction between driver and vehicle may possibly be very limited (Fig. 20.2). Traffic safety is a result of all elements of the driver-vehicle control loop [26], together with other factors, such as infrastructure or regulations.

For technical systems, the entire process chain must be represented. The process chain generally consists of sensors, traffic environment modeling, algorithms (logic), and vehicle dynamics controllers and actuators (Fig. 20.3). Additionally, the system impacts and feedback loops on driver, vehicle, and traffic are modeled.

The models must in particular take the stochastic nature of physiological, psychological, and physical phenomena into account. Each warning or any direct intervention of an ADAS may trigger or affect a driver's individual response.

In this paradigm, simulation serves as a form of knowledge synthesis. The models are synthesized to obtain realizations of all process chains and finally the safety metric. The validity of any result from the virtual testing of ADAS will depend, of course, on the quality of the underlying knowledge base (i.e., the models). The development of an appropriate knowledge base therefore represents one of the central tasks of a holistic assessment approach.

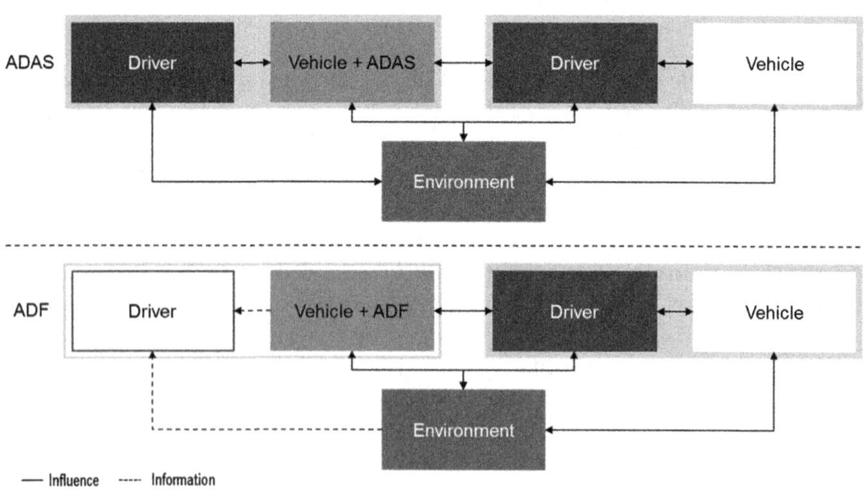

Fig. 20.2 Possible interactions of driver-vehicle-surroundings including ADAS or ADF; each including possible interactions with another human driver (after [3])

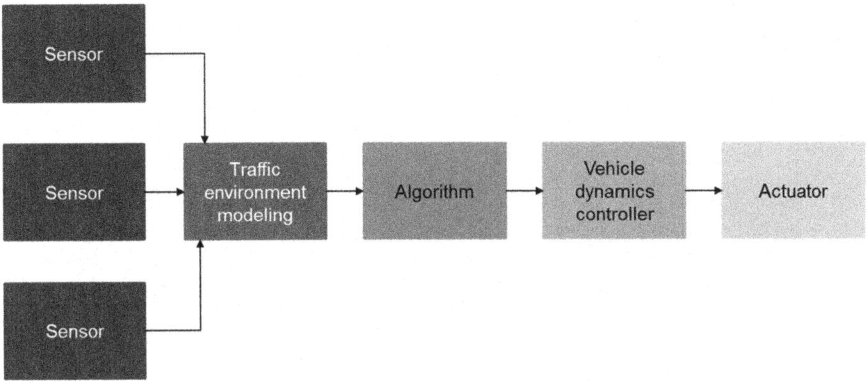

Fig. 20.3 Generic process chain of ADAS/ADF (without interaction with the driver) [24]

A wide range of existing data sources on vehicle, traffic, and human factors is used for modeling: accident databases and traffic surveys, NDS ("naturalistic driving studies"), FOT ("field operational test"), static traffic observations (e.g., highly instrumented research intersections), as well as classical testing methods such as tests on test tracks, laboratory experiments, and software, hardware, and "vehicle in the loop" (see, e.g., [7, 11, 12, 34, 36]).

In order to create relevant scenarios with a representative frequency in the simulation, "exposure models" are required. Exposure models, i.e., insights about the frequency of certain constellations of risk factors, are supported by traffic surveys, NDS, and FOT, for example.

Test methods are subject to certain ethical and practical limitations, for example, they need to be risk-free. Although classical testing methods alone hardly enable overall conclusions about effectiveness and the optimal design of ADAS, they provide valuable information to describe human or technical factors in a limited context. Functions can be tested on test tracks under varying conditions (road classes, weather) using a real vehicle with full experimental control. The obtained findings are used in virtual experiments to calibrate and to validate various simulation models (Fig. 20.4). Modeling can be performed empirically and/or on a theoretical basis.

For comprehensive assessments, representative scenarios are needed. From the set of relevant scenarios, individual situations are "sampled," i.e., created virtually. The values of all stochastic variables are drawn from appropriate distributions.

Sampling may be repeated or independent. In repeated sampling, each randomly generated scenario is simulated multiple times (e.g., with/without ADAS), whereas in independent sampling, new samples are drawn for each run.

Using simulation-based knowledge synthesis, it is possible to derive comprehensive and representative conclusions regarding effectiveness of ADAS.

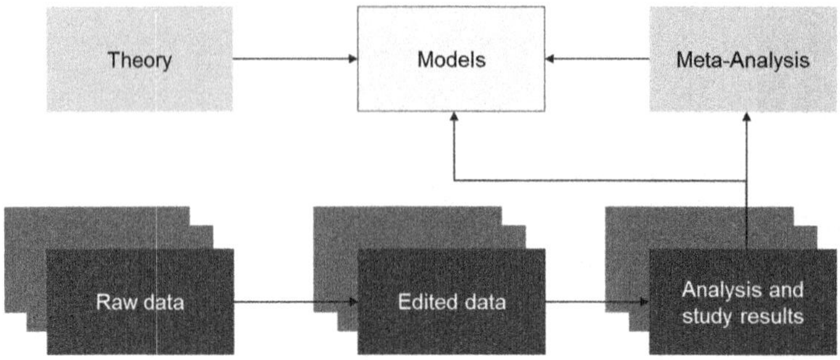

Fig. 20.4 Schematic process of modeling and validation [24]

20.3.4 Process Description for Assessing Pedestrian Protection

Some of the typical process models used for virtual testing can be illustrated in more detail using the example of a hypothetical camera-based ADAS for preventative pedestrian protection [18]. In an exemplary scenario, a pedestrian crosses a straight one-way street (one lane) between two intersections without occlusion. In this scenario, traffic flows with typical (daytime dependent) urban speeds, and there is no evasive maneuver possible for the vehicle.

If there is a conflict between the vehicle with ADAS and a pedestrian, the pedestrian protection system can issue a warning, reduce brake assist thresholds, or brake automatically.

In addition to models for the pedestrian protection system, process models are required for traffic participants, especially for pedestrians, for directly involved vehicles, and for traffic flow in general. The models describe exposure (i.e., how often certain constellations of variables occur), traffic environment, and dynamics of all road users.

As part of the exposure model for pedestrians, samples for the initial situation can be drawn from (possibly correlated) model distributions, including:

- Context variables: frequency of pedestrian crossings by age and sex depending on time of day and day of week
- Physiological attributes of pedestrians: height and weight as functions of age and sex; fatigue and blood alcohol depending on age, gender, day of week, and time of day
- Cognitive characteristics: attention, visual performance, and reaction skills (related to age, sex, alcohol, fatigue)

After a "virtual pedestrian" has been "drawn" and "generated" with his characteristics, the crossing processes are simulated using these features. Included are decision-making processes, such as gap acceptance, walking vs. running, selected destination and walking direction, initial velocity (which of course depends on the estimate crossing time), etc.

The modeling of a process such as gap acceptance often requires a considerable degree of complexity: high traffic density, for example, can increase the waiting time for a larger time gap; hence, some pedestrians become "impatient," with the result that they also accept smaller gaps after a certain waiting period. Various causes of cognitive estimation errors can lead to misjudgment of the time gap or the time required for crossing and thus provoke traffic conflicts.

Simulated safety metrics exhibit an influence of observable variables, such as age or impairment. The dependence on these variables can be compared to corresponding dependencies in published studies in order to validate detailed modeling aspects. For the assessment of ADAS, such detailed modeling aspects are at least indirectly relevant, because they determine the composition of the risk factors in conflicts and thus the potential effectiveness of ADAS.

Other stochastic process models describe perception of acute conflict, responses, and actions of the pedestrian (retreat, anticipation, etc.). The corresponding probability distributions of actions can, e.g., depend on the cognitive and physiological requirements and states at the time of reaction.

There are also corresponding process models for the driver. The cognitive state, for example, is essential for the driver's safety performance. The perception models consider, for example, geometric relations, environmental conditions, and dynamics of the situation. Following initial perception, reaction processes are modeled in accordance with established model paradigms [10, 17, 27, 28], depending on the element of surprise and cognitive status.

The actual simulation run includes dynamic models for pedestrian and vehicle movement, which in turn depend on the other process models. For both pedestrians and drivers, strong individual differences can be observed. Individual differences, such as reaction time of the driver or braking intensity, are especially important for the assessment of traffic safety. As a result, a random number of individual, but in sum representative, situations can be created on the basis of a traffic scenario (Fig. 20.5).

At the end, time series of all virtual state variables can be extracted from every run—including those variables that are difficult to detect in standard lab experiments. An example is the determination of which stimulus has led to the reaction and deceleration by the driver, i.e., his own perception of the critical situation or a system warning. The corresponding cognitive processes may occur in parallel. These findings can be incorporated directly into the design of a warning strategy.

20.3.5 Simulation of ADAS Effectiveness

In addition to the simulation of traffic flow, vehicle dynamics, other traffic participants, as well as human factors, such as perception and reaction, the process of virtual evaluation (Fig. 20.6) also requires an appropriate model of the driver assistance system.

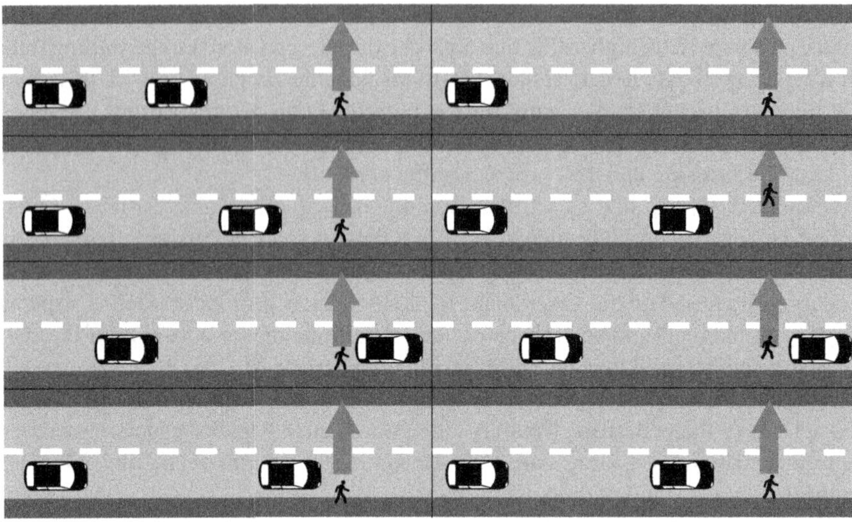

Fig. 20.5 Schematic representation of different individual situations as a result of stochastic processes in a traffic scenario [24]

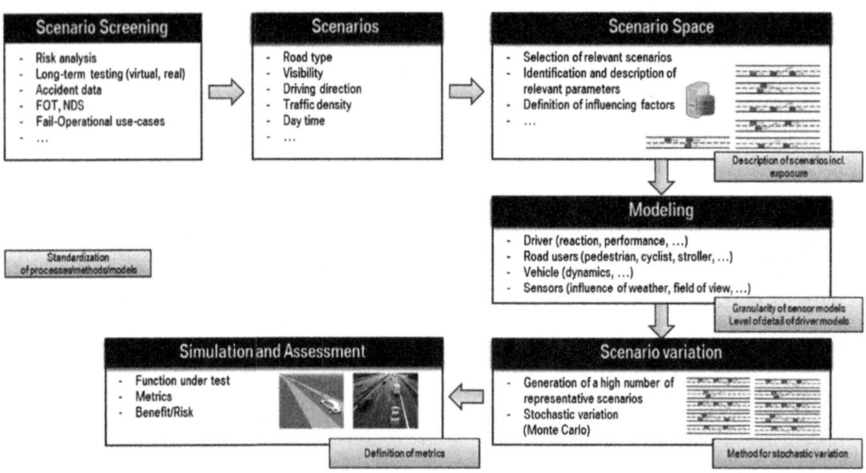

Fig. 20.6 Process of virtual assessment of ADAS: overview with example of preventive pedestrian protection (after [24])

The precise technical realization of an ADAS plays in general an indirect role for effectiveness assessment. For example, the detection of a pedestrian by means of a camera and the detection of the distance and the relative speed to a preceding vehicle are highly complex and device-specific technical processes. In order to assess the potential effectiveness of an ADAS, it is often useful to begin with an abstract model: system concepts can be analyzed at an early stage, without being limited by the exact technological realization.

Models of performance, detection characteristics (sensors, object classification, etc.), algorithm logic, and actuator models are in many cases important for simulation and assessment of ADAS.

The effectiveness of ADAS is, among other things, determined by the system boundaries. These can theoretically depend on the traffic context (such as vehicle speed or road class given by a digital map), environmental conditions (light conditions, weather), or on conditions for automatic system activation or deactivation. Furthermore, e.g., system activation by the driver or other human factors can influence effectiveness.

Detection systems in traffic are subject to system limits and to various uncertainties and latencies. As a consequence, in practice, detection systems often show stochastic characteristics. For example, the time for stable object recognition by means of a camera can depend on partial occlusion of the object or complexity of the traffic scene. The algorithms for situation detection and action usually rely on measurements from the detection systems; the derived characteristics (such as estimated "time to collision" for a detected object) are therefore also subject to corresponding latencies and uncertainties that require appropriate, mostly stochastic, modeling. Also, system actions often act indirectly by stimulating a driver reaction (e.g., warnings) or interact with (stochastic) driver actions (e.g., by lowering the threshold for brake assist). Overall, stochastic characteristics have a major impact on the overall safety assessment of ADAS.

With increasing complexity of the systems, an abstract representation of the functionality of an ADAS may require considerable effort. Also, verifying that an abstract system model actually behaves like the real technical system poses a challenge with increasing system complexity. To meet this challenge, alternative approaches are possible.

Instead of abstract models, real components of ADAS can be directly connected to the simulation using an appropriate test facility (e.g., hardware in the loop) [12]. In addition, findings from such test benches can be used for calibration and validation of relevant models in the simulation even without a direct connection. The actual code may be used in the simulation instead of an abstract model of the system logic. This results new challenges for the simulation and the models used due to the technical interfaces used.

20.3.6 Interpretation of ADAS Effectiveness

In the example scenario of "preventative pedestrian protection," millions of virtual crossings in the reference scenario (without pedestrian protection system) lead to several thousand collisions between pedestrians and vehicles. The attributes of these collisions are "known" from the simulation, i.e., speed, vehicle characteristics, collision constellation, age group, body height and weight of the pedestrian, etc. Using an appropriate model (see above), the distribution of the injury severity in collisions without ADAS can be determined. This distribution is used as reference for comparison.

A lower number of collisions occur in the virtual experiment after the same number of crossings with a pedestrian protection system. The properties of the collisions (such as reduced collision speeds) are changed due to the ADAS, and consequently the distributions of injury severity are modified. In this case the metric of injury severity represents the effectiveness of the preventative pedestrian protection system. The overall safety performance includes also NNT and an assessment of possible secondary risks, such as rear-end collisions in upstream traffic.

20.4 New Challenges for Virtual Assessment of Automated Driving Functions

20.4.1 Impact of Automated Driving Functions on Safety-Related Processes in Traffic

The monitoring function of systems of active safety is carried out in continuous operation; however, warnings and interventions are sporadic events. Apart from theoretically possible feedback, for example, through changes in user behavior, the perceived influence on the driving task usually remains very limited.

In contrast ADAS with regular control operations, such as active cruise control (ACC) or ADF, operate continuously. Safety assessment of these functions requires consideration of their effect on safety-related processes in traffic flow as a whole, not only in certain target scenarios. Overall traffic safety includes both positive effects and potential risks. This assessment requires a variety of additional exposure and process models.

20.4.2 Contributions to Safety Impact in Existing Risk Scenarios

A portion of possible positive contributions of ADF (similar to ADAS) comes from consideration of relevant and potentially hazardous scenarios where the advantages of ADF help avoid potential accidents or mitigate their consequences. "Scenario" in this context refers to all potentially hazardous traffic situations that can lead to a certain type of conflict.

Using virtual experimental design (as for ADAS), an appropriate reference sample of relevant scenarios can be defined and considered. The contribution to effectiveness due to ADF can be quantified using a sample of virtual experiments, once the scenarios and their frequencies are known.

Example, rear-end scenario on a freeway: Without ADF (or ADAS), an inattentive driver (failure "a") or a driver with unadjusted speed (failure "b") has an increased accident risk in a rear-end scenario. The rear-end scenario could be provoked by very slow or stationary downstream traffic behind a sharp curve or a hill. The driver experiences a surprising, sudden speed drop.

Lack of attention (a) is a typical consequence of insufficient driver activation, for example, after a long and strenuous journey in stop-and-go traffic. Inappropriate speed (b) can have various causes, for example, the latent danger of a freeway curve or a hill (due to the inherent visual restriction) might have been inadequately addressed by traffic signs, or the driver might have failed to recognize this hazard despite warning signs.

It seems likely that an ADF will not exhibit these two failure types and thus can avoid the impending rear-end collision—assuming adequate object detection by ADF. On the one hand, technology is always "attentive" (a). On the other hand, automatic adjustment of speed (b) is likely to be part of ADF and thus also reduces the accident risk in this scenario. In a virtual experiment, the safety effects of ADF in this scenario can be compared to the performance of human drivers in various reference situations, e.g., with and without emergency braking assistance.

20.4.3 Expanding the Spectrum of Safety-Relevant Scenarios

In general, automation has not only a potential for selective accident prevention from the perspective of the ADF vehicle, but could, given sufficient penetration, increase overall traffic safety due to collective effects such as harmonization of traffic flow (see, e.g., [32]). For example, traffic literature shows (e.g., [4]) that inappropriate speed is not only limited to individual drivers but may be a collective phenomenon of the traffic stream. A high penetration of traffic with ADF vehicles could thus also avoid accidents for non-equipped vehicles (due to the collective effects of early speed adjustment).

On the other hand, due to the continuous action of ADF in traffic, the spectrum of relevant scenarios is considerably larger than for most ADAS. As a consequence, fundamentally new issues and methodological challenges arise for the virtual safety assessment of ADF.

In general, traffic has a very high complexity—due to the numerous direct interactions between traffic participants and indirect interactions between individuals and the collective traffic flow. Nevertheless, traffic flow has several collective or "macroscopic" characteristics. Examples are "fundamental diagram" (empirical relationship between traffic flow and average speed on a freeway section) or "capacity" (characteristic traffic demand above which traffic flow tends to become unstable). Changes in macroscopic characteristics of traffic flow, in this context, due to ADAS or ADF, may in turn have effects on traffic safety.

Automated vehicles can affect direct interaction between vehicles, interaction between individual vehicles and traffic flow, as well as collective characteristics of traffic flow. Since all of these changes can affect conflict and accident probability, they must be included in traffic safety assessment.

To this end, many processes in normal, non-assisted, and nonautomated traffic will need to be reassessed. On the one hand, particularly relevant traffic processes include those in which potential conflicts are normally avoided by human anticipation, intuition, cooperative driver behavior, strategic defensive driving, etc., and thus do not necessary require emergency action. Moreover, human drivers frequently have a very high context sensitivity, which further increases traffic safety performance. These skills are a typical strength of human drivers (see also [6, 12, 18, 22, 25, 26]).

An open question and subject of the research and development is to what extent automated vehicles will have capabilities comparable to those of human drivers? Some safety-relevant characteristics, such as anticipation and defensive driving, could be even more pronounced with ADF than with human drivers.

ADF could have an impact on cooperative interactions between automated vehicles and individual human drivers. Cooperative behavior between human drivers can be illustrated on a freeway with a "keep-right" rule: consider a vehicle driving on the right lane behind a slow truck. On the adjacent lane, another driver recognizes this situation and strategically creates a gap to "let in" the other vehicle. The driver on the left lane intentionally slows down slightly; the other driver recognizes his intention, signals, and begins his lane change while monitoring the gap. Ideally, he can complete the lane change with minimal impact on steady traffic flow.

The ADF vehicle on the left lane must be able to complete two important tasks in order to fulfill the example of the lane change: first, it must recognize the situation of the vehicle on the right lane, and second, it must have an appropriate cooperative action strategy. Whereas cooperation between human drivers can be influenced by factors such as emotion or time pressure, these factors are hardly relevant for automation.

The rate of success of cooperative interactions between human drivers and their sensitivity to context is still subject of transport research. In addition, any influences due to automation are currently being explored. The modeling of cooperative processes in traffic flow and road safety is a current research topic that presents further challenges for simulation.

It is also important to model indirect interactions with impacts on cooperative driving. For example, in dense freeway traffic, automated vehicles might tend to keep larger time gaps than are currently common and thus might even be recognizable as automated vehicles. These larger time gaps could conceivably encourage individual human drivers to cut in more frequently. If automated vehicles react with corresponding braking decelerations to vehicles cutting in, traffic dynamics will change. The occupants of an automated vehicle may even get the subjective impression of an increase in travel time, which would theoretically affect acceptance and thus indirectly traffic safety. A high penetration of ADF might also lead to macroscopic consequences (fundamental diagram, capacity, platoon formation, frequency of shocks in traffic flow, etc.).

However, objective increases in travel time of automated vehicles are expected to be imperceptibly small and likely to be accepted. Moreover, the process of cutting in during dense "synchronized" traffic is not very rewarding anyways.

In general, the presumption seems natural that the overall effectiveness of automated functions regarding traffic safety will be positive. The objective is a quantification of this hypothesis by comprehensive assessment.

20.4.4 Philosophy and Procedural Approaches for Validation and Assessment of Automation

While design of ADAS usually involves balancing desired and unintended system actions with regard to traffic safety, more complex trade-offs between conflicting objectives can be expected in the development and design of ADF.

A proportion of positive contributions to the safety record of automated functions can be estimated using virtual experiments (analog to ADAS) in which existing, potentially hazardous traffic scenarios are investigated. However, due to the continuous operation of ADF, the overall safety performance could be strongly influenced by additional scenarios and constellations, taking into account low-probability events. These scenarios are not known a priori and may be "hidden" in the situation space. Furthermore, their relevance can also depend on context or function design (e.g., partial or high automation).

For example, the reaction of automation to lane keeping of other drivers might depend on local driving characteristics. If an automation function is designed for a traffic context with "very precise" lane keeping, as is common in parts of Central Europe, application to regions with less precise lane keeping could require an adaptation to local driving strategy. Otherwise, undesirable effects, for example, frequent braking in response to "vehicles cutting in," such as platoon formation, could increase.

The development of ADF presents considerable additional challenges: the situation space will be substantially larger than for current ADAS, and prediction of all relevant situations will be difficult. These issues complicate the assessment and optimization of key system characteristics, such as stability, robustness, and safety impact. A possible approach is an integrated and agile development and assessment process using a comprehensive tool chain.

Any methodology for continuous safety assessment during the development requires a comprehensive understanding of various existing methods and their specific role and contribution in such a complex process. Figure 20.7 provides an abstract overview of the roles of different testing instances within an integrated process.

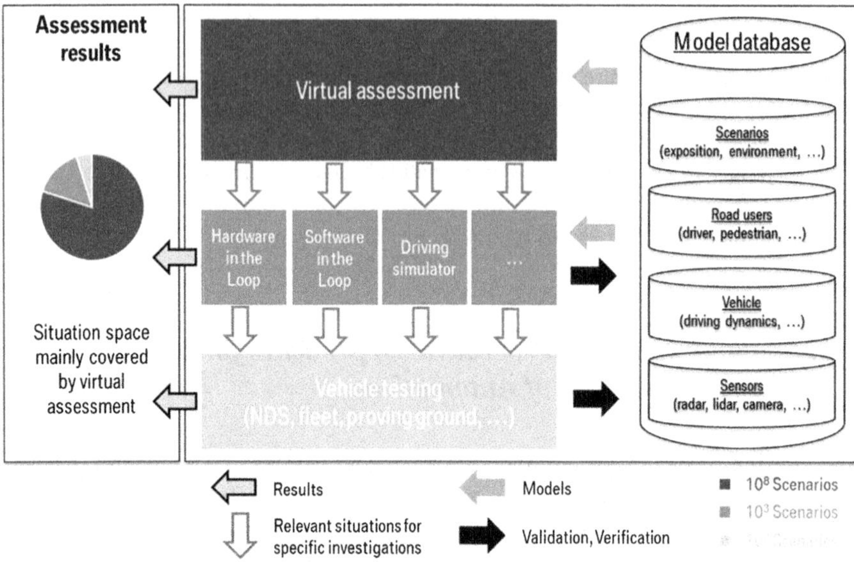

Fig. 20.7 Development and assessment process including exemplary testing instances

The core of the process is virtual simulation, i.e., software in the loop. The virtual test run can be executed for each major functional step in the development phase and tested against the generated set of scenarios. These virtual test runs are held on system level and identify the most critical, relevant, and failed scenarios which require a more detailed evaluation with a specialized testing instance (e.g., driving simulator, hardware in the loop, fleet testing, etc.).

Data and findings from empirical testing methods are used for modeling, as common for ADAS assessment. In particular, data from NDS, FOT, fleet tests, etc., could be used to form a database including an appropriate (possibly country-specific) model of exposure, behavior, and other key aspects.

Referring to Fig. 20.7, the scenario database as well as other models plays a key role for all development and test tools. The scenarios and their frequency (in terms of an exposure model) can be stored in a scenario database and reused for different test instances. In addition to traditional research methods (e.g., theoretical risk assessments, testing on the road, etc.), virtual continuous simulation offers new opportunities for "discovery" of relevant scenarios, especially, when it comes to combinations of factors which are rare and difficult to derive from the theory.

One approach to discovery of "unknown" scenarios is observing longer durations or distances by long-running continuous simulation: safety-critical scenarios are not explicitly generated (e.g., by certain given constraints), but arise spontaneously from a stochastic comprehensive model of traffic flow.

This approach implies challenging requirements for comprehensive modeling. These include reproduction of all relevant processes (also error processes) in traffic flow, with their respective frequencies. Modeling the frequencies of all relevant processes and scenarios essentially constitutes an advanced exposure model.

An initial set of scenarios can be specified using expert knowledge, field operational tests, and virtual test runs. Virtual testing generates many representations of stochastic processes in traffic based on models of traffic contexts, sensors, drivers, vehicles, and traffic dynamics. The objective is to provide a representative sample of the overall situation space taking into account the large number of potential scenarios including low-probability events [38].

In a virtual test operation, these scenarios could be checked automatically representing a kind of safety cycle. The frequency of scenarios could depend on different factors, for example, countries or environmental conditions. Virtual testing would fulfill the requirements of a safety assessment in this construct, as described above.

Virtual testing by simulation of a *single* scenario results in quantification of effectiveness of an automated system in *this* particular scenario. The safety performance of automation then results from the sum of effectiveness in *all* relevant scenarios weighted by their respective frequency (exposure).

A key issue concerns the validation of process models and, by extension, plausibility of simulation results. Validation of models involves utilization of appropriate testing procedures for each particular method in the development chain. Each method, for instance, test driving or a driving simulator, is used for validating the vehicle model or for MMI concepts. A validated model database increases the reliability of virtual testing; the quality of the models is of key importance for the development chain and the validity of the assessment result.

Verification of simulated system actions represents another important element of the validation process, by drawing samples from all simulations and testing these in recognized test institutes. Validation (and development) of process models could be accompanied by impartial scientific organizations.

Consequently, the objective would be an international consensus and implementation of scenarios, models, and the overall assessment approach by all relevant stakeholders in an international context.

20.5 Conclusion and Outlook

The task of safety assessment and optimization of automated functions raises new issues. In contrast to ADAS assessment, quality measures of traffic safety are principally related to all traffic scenarios in which a function is active. Since automation may change collective traffic characteristics, safety analysis must go beyond isolated human errors in currently occurring traffic processes and the impact of automation on these. Newly emerging, automation-related, scenarios have to be considered for a comprehensive safety assessment.

Validation and safety assessment of automated functions have to be understood as continuous and iterative tasks during the development, not as singular activities at the end of the development phase. Due to the variety of possible influences, the necessary assessment of automation approaches during the development would be extremely problematic based, for example, solely on fleet testing, since detection of rare effects requires correspondingly long observation periods. In addition, testing would have to be repeated, in principle, after every single change of the function.

The approach of simulation-based virtual experiments can be interpreted as knowledge synthesis. Still, some challenges for the assessment of automated systems arise. Relevant scenarios for automation are a priori unknown and can only partially be identified using existing methods. Due to the generally larger situation space involved in automation, modeling results in considerably more complexity as for the assessment of ADAS.

Quality requirements for traffic simulation are correspondingly higher, especially in terms of process models used. Traffic simulation will need to consider error processes and their resolution in normal traffic in more depth. The challenges include improved modeling of psychological processes, e.g., attention or activation (Yerkes-Dodson) [37]. An important aspect of the safety potential of ADF arises from avoidance of errors resulting from lack of driver activation and resulting attention lapses.

Despite sophisticated technology, systems will still be subject to system limits within the near future. Virtual experiments could make an important contribution to design and optimization of take-over requests to human drivers, in addition to safety assessment.

Critical traffic situations can require a decision among several unfavorable alternatives for action. Here again, virtual assessment can support the development of transparent decision algorithms. A general discussion of such alternatives has already begun in public [30]. Possibly all stakeholders can achieve consensus and agree on guidelines for the prioritization of alternative actions before market introduction.

Many (novel) projects, initiatives, organizations, and research activities are focusing on the effects of ADAS regarding traffic safety. So far, however, an international consensus on methodological issues in the context of an overall safety assessment of ADAS and ADF is still lacking.

Considering the importance and complexity of decisions and challenges, the initiative "Prospective Effectiveness Assessment for Road Safety" (PEARS) has the objective of developing a standardized and harmonized method for the overall effectiveness assessment of new systems, such as ADAS or ADF, which is accepted by all stakeholders [29]. Both benefits and potential risks should be quantified as part of the assessment. The objectives are, among others, a higher degree of legal predictability and adequate and objective consideration of individual and societal interests. This open platform provides important prerequisites for a global harmonization and standardization.

References

1. M. Aeberhard, S. Rauch, M. Bahram, G. Tanzmeister, J. Thomas, Y. Pilat, F. Homm, W. Huber, N. Kaempchen, Experience, results and lessons learned from automated driving on Germany's highways. IEEE Intell. Transport. Syst. Mag. **7**(1), 42–57 (2015)
2. M. Bahram, Z. Ghandeharioun, P. Zahn, M. Baur, W. Huber, F. Busch, Microscopic Traffic Simulation Based Evaluation of Highly Automated Driving on Highways, in *17th International IEEE Conference on Intelligent Transportation Systems - ITSC 2014*, ed. by IEEE (IEEE, 2014)
3. R. Bernotat, Anthropotechnik in der Fahrzeugführung. Ergonomics **13**(3), 353–377 (1970)
4. T. Connolly, L. Åberg, Some contagion models of speeding. Accident Anal. Prevent. **25**(1), 57–66 (1993)
5. R.J. Cook, D.L. Sackett, The number needed to treat: A clinically useful measure of treatment effect. Br. Med. J. **18**, 452–454 (1995)
6. C. Domsch, W. Huber, *Integrale Sicherheit - ein ganzheitlicher Ansatz für die Fahrzeugsicherheit. 17* (Aachener Kolloquium Fahrzeug- und Motorentechnik, Aachen, 2008)
7. A. Eskandarian (ed.), *Handbook of Intelligent Vehicles* (Springer, Heidelberg, 2012)
8. F. Fahrenkrog, A. Zlocki, L. Eckstein, Bewertung Aktiver Sicherheit: Vom Test zur Wirksamkeitsanalyse. ATZ **1**(116), 34–39 (2014)
9. I. Ferenczi, T. Helmer, P. Wimmer, R. Kates, Effectiveness Analysis and Virtual Design of Integrated Safety Systems Illustrated Using The Example Of Integrated Pedestrian And Occupant Protection, in *12th International Symposium and Exhibition on Sophisticated Car Occupant Safety Systems* (Fraunhofer-Institut für Chemische Technologie ICT, 2014)
10. M. Green, "How long does it take to stop?" Methodological analysis of driver perception-brake times. Transport. Hum. Factors **2**(3), 195–216 (2000)
11. T. Helmer, *Development of a Methodology for the Evaluation of Active Safety using the Example of Preventive Pedestrian Protection*, Thesis (Springer, Heidelberg, 2015), ISBN 978-3-319-12888-7
12. T. Helmer, T. Kühbeck, C. Gruber, R. Kates. Development of an Integrated Test Bed and Virtual Laboratory for Safety Performance Prediction in Active Safety Systems (F2012-F05-005), in *FISITA 2012 World Automotive Congress - Proceedings and Abstracts, 2012* (ISBN 978-7-5640-6987-2)
13. T. Helmer, M. Neubauer, S. Rauscher, C. Gruber, K. Kompass, R. Kates, Requirements and Methods to Ensure a Representative Analysis of Active Safety Systems, in *11th International Symposium and Exhibition on Sophisticated Car Occupant Safety Systems* (Fraunhofer-Institut für Chemische Technologie ICT, Pfinztal, 2012), pp. 6.1–6.18, ISSN 0722-4087
14. T. Helmer, R.R. Samaha, P. Scullion, A. Ebner, R. Kates, Injury risk to specific body regions of pedestrians in frontal car crashes modeled by empirical, in-depth accident data, in *Proceedings of the 54th Stapp Car Crash Conference, 2010*
15. T. Helmer, R.R. Samaha, P. Scullion, A. Ebner, R. Kates. Kinematical, Physiological, and Vehicle-Related Influences on Pedestrian Injury Severity in Frontal Car Crashes: Multivariate Analysis and Cross-validation, in *Proceedings of the International Research Council on Biomechanics of Injury (IRCOBI)* (IRCOBI, Hannover, 2010), pp. 181–198
16. T. Helmer, P. Scullion, R.R. Samaha, A. Ebner, R. Kates, Predicting the injury severity of pedestrians in frontal vehicle crashes based on empirical, in-depth accident data. Int. J. Intell. Transport. Syst. Res. **9**(3), 139–151 (2011)
17. W.E. Hick, On the rate of gain of information. Q. J. Exp. Psychol. **4**, 11–26 (1952)
18. W. Huber, J. Steinle, M. Marquardt, Der Fahrer steht im Mittelpunkt - Fahrerassistenz und Aktive Sicherheit bei der BMW Group, in *24. VDI/VW-Gemeinschaftstagung - Integrierte Sicherheit und Fahrerassistenzsysteme*, vol. 2048 *VDI-Berichte* (VDI Verlag, Düsseldorf, 2008), pp. 123–137

19. C. Hydén, *The Development of a Method for Traffic Safety Evaluation: The Swedish Traffic Conflicts Technique*, Bulletin 70 (Department of Traffic Planning and Engineering, Lund University, Sweden, 1987)
20. R. Kates, O. Jung, T. Helmer, A. Ebner, C. Gruber, K. Kompass, *Stochastic Simulation of Critical Traffic Situations for the Evaluation of Preventive Pedestrian Protection Systems* (VDI-Tagung Erprobung und Simulation in der Fahrzeugentwicklung, Baden-Baden, 2010)
21. D.G. Kleinbaum, M. Klein, *Logistic Regression. A Self Learning Text. Statistics for Biology and Health* (Springer, Heidelberg, 2010)
22. K. Kompass, C. Gruber, C. Domsch, Der Beitrag von Fahrerassistenzsystemen zur Aktiven und Passiven Sicherheit - die Integrale Sicherheit als Antwort auf die wachsenden Anforderungen an die Fahrzeugsicherheit, in *4. Tagung Sicherheit durch Fahrerassistenz, 2010.*
23. K. Kompass, T. Helmer, C. Blaschke, R. Kates, Ganzheitliche und integrale Fahrzeugsicherheit. ATZ - Automobiltechnische Zeitschrift **116**, 10–15 (2014)
24. K. Kompass, T. Helmer, L. Wang, R. Kates, Gesamthafte Bewertung der Sicherheitsveränderung durch FAS/HAF im Verkehrssystem: Der Beitrag von Simulation, in *Fahrerassistenz und Aktive Sicherheit. Wirksamkeit – Beherrschbarkeit – Absicherung* (expert Verlag, Renningen, 2015), pp. 45–66
25. K. Kompass, W. Huber, *Integrale Sicherheit – Effektive Wertsteigerung in der Fahrzeugsicherheit* (VDA, Technischer Kongress, Berlin, 2009)
26. K. Kompass, W. Huber, T. Helmer, Safety and Comfort Systems: Introduction and Overview, in *Handbook of Intelligent Vehicles*, ed. by A. Eskandarian (Springer, Heidelberg, 2012)
27. N. Lerner, Brake Perception–Reaction Times of Older and Younger Drivers, in *Proceedings of the Human Factors and Ergonomics Society Annual Meeting*, vol. 37 (1993), pp. 206–210
28. P.L. Olson, M. Sivak, Perception-response time to unexpected roadway hazards. Hum. Factors **28**, 91–96 (1986)
29. Y. Page, F. Fahrenkrog, A. Fiorentino, J. Gwehenberger, T. Helmer, M. Lindman, O. op den Camp, L. van Rooji, S. Puch, M. Fränzle, U. Sander, P. Wimmer, A Comprehensive and Harmonized Method for Assessing the Effectiveness of Advanced Driver Assistance Systems by Virtual Simulation: The P.E.A.R.S. Initiative, in *24th International Technical Conference on the Enhanced Safety of Vehicles (ESV 2015)*, number 15-0370, 2015
30. H. Prantl, Error. Süddeutsche Zeitung **37**, 13 (2015)
31. E. Schechtman, Odds ratio, relative risk, absolute risk reduction, and the number needed to treat – which of these should we use? Value Health **5**(5), 431–436 (2002)
32. C. Steinhoff, R. Kates, H. Keller, B. Färber, B. Färber, *Problematik präventiver Schaltungen von Streckenbeeinflussungsanlagen* (BMVBW Forschung Straßenbau 853, Bundesanstalt für Straßenwese, 2001)
33. A. Weitzel, H. Winner, C. Peng, S. Geyer, F. Lotz, L. Sefati, *Absicherungsstrategien für Fahrerassistenzsysteme mit Umfeldwahrnehmung* (BASt-Bericht F98, Bundesanstalt für Straßenwesen, 2014)
34. R. Wertheimer et al., *Ko-PER – Fahrerassistenz und präventive Sicherheit mittels kooperativer Perzeption: Partnerübergreifender Schlussbericht*, Schlussbericht (Bundesministerium für Wirtschaft und Technologie (BMWi), 2014)
35. P. Wimmer, A. Rieser, W. Sinz. A Simulation Method for Virtual Design and Evaluation of Integrated Safety Systems, in *NAFEMS World Congress*, 2013
36. H. Winner, S. Hakuli, G. Wolf (eds.), *Handbuch Fahrerassistenzsysteme* (Vieweg+Teubner Verlag/GWV Fachverlage GmbH, Wiesbaden, 2009)
37. R.M. Yerkes, J.D. Dodson, The relation of strength of stimulus to rapidity of habit-formation. J. Comp. Neurol. Psychol. **18**, 459–482 (1908)
38. A. Zlocki, L. Eckstein, F. Fahrenkrog, Evaluation and sign-off methodology for automated vehicle systems based on relevant driving situations, in *Transportation Research Board (TRB), 94th Annual Meeting*, Washington DC, USA, 11–15 Jan 2015

Chapter 21
From Controllability to Safety in Use: Safety Assessment of Driver Assistance Systems

Alexander Huesmann, Mehdi Farid, and Elke Muhrer

21.1 Introduction

Controllability of driver assistance systems is the topic addressed by the "Code of Practice for the Design and Evaluation of ADAS", and controllability in general is already mentioned in the Vienna Convention on Road Traffic of 1968 [1]. Nevertheless, in the automotive development context, the statement that drivers must at all times be able to control their vehicles can be interpreted very differently. Very often the thought of controlling the vehicle in a critical road scenario is present. However, controllability associated with assistance systems and automation has an entirely different facet. The aim here is, that the driver controls the vehicle in its nominal function, at the system limits and in case of system failures, meaning the driver is able to cope with traffic situations without harming himself or others.

In the context of automation, aspects of, the to-be-avoided, systems-over-confidence and mode awareness increase in importance and must be considered in the system composition.

The basis for the development of higher automation functions in the vehicle is the established methods and procedures for the controllability of driver assistance systems. Such approaches largely consider single scenarios isolated.

Due to the increasing complexity and networking of future systems, such singular considerations of single scenarios are not likely to be sufficient. Thus, a holistic consideration of the traffic situation is required.

This chapter is based on: Klaus Kompaß et al.: Fahrerassistenz und Aktive Sicherheit. Wirksamkeit—Beherrschbarkeit—Absicherung. expert verlag, Renningen 2015.

A. Huesmann (✉) • M. Farid • E. Muhrer
BMW AG, 80788 Munich, Germany
e-mail: Alexander.Huesmann@bmw.de; Mehdi.Farid@bmw.de; Elke.Muhrer@bmw.de

D. Watzenig, M. Horn (eds.), *Automated Driving*,
DOI 10.1007/978-3-319-31895-0_21

Initially, the controllability of driver assistance systems and the corresponding references will be addressed. Following this, the method of analysis of safety in use will be introduced and explained. This includes the definition of terms and their delimitations.

21.2 Controllability of Driver Assistance Systems

Nowadays, if controllability is mentioned in the context of driver assistance systems (DAS), it immediately is associated with two documents. The older one is the Vienna Convention on Road Traffic of 1968 [1]. Herein, the international rules of road traffic were agreed upon. The other is the code of practice (CoP) for the design and evaluation of advanced driver assistance systems (ADAS) [2]. The CoP emerged from the work of a series of European projects, in particular as a report in the RESPONSE 3 [2] project which was part of the integrated project of PReVENT [3].

The launch of the RESPONSE projects started out just before the new millennium, when the first driver assistance systems such as active cruise control (ACC) became available on the market and others in the automotive industry were under way. The central question was how to generate and verify the product safety of such assistance systems.

First, for formulating development standards, the field of application has to be determined. One of the questions in the project RESPONSE was what was meant by ADAS. For this purpose the three levels into which driving tasks can be divided were used: navigating, maneuvering, and stabilizing the vehicle [4].

It was agreed to observe rule-based systems under normal driving conditions, which are acting on the maneuver level, such as a lane change assistant. Furthermore, systems have been taken into account that are in transition between maneuvering and stabilizing, namely, assistance systems in emergency situations, such as an automatic emergency brake system.

A systematic assessment of emerging risks ultimately distinguished two categories: the technical safety and the safety of the human–machine interaction.

The technical safety of systems and how to develop appropriate safety concepts were excluded, because around the same time, the ISO 26262 standard for functional safety in the vehicle [5] was under development; hence, an appropriate set of procedural rules to ensure safety in E/E errors emerged.

The main focus within the research project RESPONSE 3 is the safety of the human–machine interaction, which is made possible by the controllability of the driver. Controllability is divided into three steps:

- The ability of the driver to perceive the criticality of a situation
- The ability of the driver to decide on an appropriate countermeasure
- The ability of the driver to also perform this countermeasure

Controllability is the focal point in the project RESPONSE 3.

Controllability has already been mentioned as a criterion in the Vienna Convention on Road Traffic of 1968 [1].

Thus, Article 8.5 states that:

- Every driver shall at all times be able to control his vehicle or to guide his animals.

And in Art. 13.1, it is written that:

- Every driver of a vehicle shall in all circumstances have his vehicle under control.

Herein it is implied that in the context of driver assistance, the driver also has to be able to manage traffic situations when using driver assistance systems. This includes handling the various traffic situations while driving with the support of the system, at the system boundaries, and in case of system errors.

A system error is considered as mentioned above in the standard for functional safety. Development guidelines for the nominal function and its system boundaries, in particular the human–machine interaction of assistance systems, are considered in the code of practice (CoP) for the design and development of ADAS which was published on the website of the ACEA and thus made publicly available in 2009 [2].

In the CoP the procedure for safety-oriented development and evaluation of safety aspects of driver assistance systems are described with a clear focus on the human–machine interaction. This is based on the identification of risks and the preparation of hazard and risk analysis as well as risk assessments. During the development process, measures aimed at ensuring the controllability of the assistance systems are derived from these actions.

When looking at the relationship of driver performance and automation [6], it is clear that with increasing automation, the controllability of situations, in which the driver has to take over the driving task or correct the trajectory, is not necessarily given. A common classification of automation levels in terms of requirements for the driver can be found in [7].

Here, the transition between partly automated driving and highly automated driving is essential. While partly automated driving, the driver needs to monitor the system permanently and has to be able to take over the driving task at any given moment. At the next higher level of automation, the highly automated driving, the driver no longer has to permanently monitor the system, and in case of a takeover request, the driver is given sufficient time to react. However, when the driver no longer needs to monitor the system, perceiving the criticality of a situation can no longer be fulfilled continuously, hence impairing already the first point of the abovementioned criteria of controllability. In addition, such functions would stand in contradiction to the Vienna Convention, where control of the vehicle is required. This triggered a proposal for amendment to the Vienna Convention [8] and was officially put to vote in 2014. This proposal also included an amendment to Art. 8, which states, that "a vehicle meets the requirements of Art. 8 paragraph 5 and Art. 13 paragraph 1, if it meets the requirements of the international certification regulations" (e.g., UN ECE rules). In addition, the requirements referred to in Article 8 and 13 are met if the driver is given the opportunity to turn off or override the assistance system. This amendment of the wording is intended to ensure the conformity of highly automated driving functions with the Vienna Convention.

It is becoming obvious that the criterion of controllability for a holistic safety assessment of functions with increasing levels of automation will not suffice any longer. Thus a holistic consideration of the traffic with all influencing components is required.

21.3 Safety in Use: A Holistic Consideration of the Driver, Vehicle, and Environment

In Sect. 21.2 we addressed the factor controllability. When evaluating the controllability, the overall situation of the driver, his vehicle, and its environment is of special importance (Fig. 21.1).

By that, controllability arises not only through the skill level and the performance of the driver but is heavily influenced by components of the traffic situation. On the one hand, every drivers' profile differs by the extent of long-term experiences and long-term knowledge, short-term intentions and states, and different perceptions of the environment; thus, every driver most likely reacts in a different way. The environmental conditions vary in weather, visibility, road conditions, and traffic density. Vehicles have different engines and different assistance systems on board. Depending on the chassis, vehicles also react differently in certain dynamic driving situations. This is not an exhaustive list of possible influences but is intended to show that all of the influencing components may take various forms. The driver steers the vehicle, reacts to the feedback of the vehicle, and moves it within its environment. A holistic approach of the traffic situations to be managed by the driver has to include these three interacting components. In other words: depending on the ratio of the components, the driver has to perform different maneuvers to safely navigate his vehicle through traffic.

For example, the driving task of paying attention and maintaining the speed level when driving on a route with low traffic density changes rapidly in a high-density environment (i.e., a traffic jam ahead of the driver), where stopping and starting

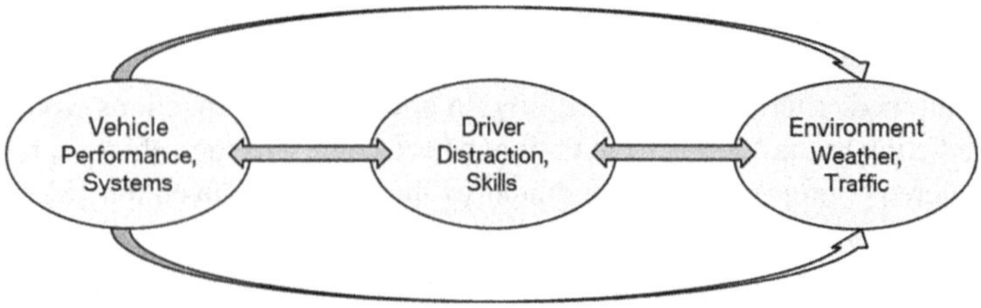

Fig. 21.1 A holistic traffic situation integrates the interaction of the driver, the vehicle and the environment

the vehicle is the main task. For these situations, different "levels" that have to be included in the risk assessment arise. Therefore, all factors in the assessment of risks in road traffic must be complied with. However, the complexity of traffic situations requires that the situations are systematically assessed. Without such, an exhaustive assessment of risks is not possible.

In this article, the methodology of safety in use is introduced. For this purpose the system behavior of an "error-free" function in regular use cases and at the system limits is systematically analyzed. Here, both the intended use and the likely misuse are mostly relevant. The term "error-free" case requires a more precise specification. This will take place in the following chapter. The aim of the analysis of safety in use is to know the total risk, which is evident from the use of a function in the different situations that may occur. Based on this assessment, the decision for specific actions in the system development can be made.

21.3.1 Systematic Analysis of Safety in Use

The initial focus of the safety assessment needs to be clarified first, in order to carry out a systematic analysis of whether a function is safe to use.

Safety must always be seen in the context of risks, where risks are described as the product of the likelihood of a negative event to occur and its damage severity.

The likelihood of occurrence of adverse events relating to road transport in the context of driver assistance is determined by the likelihood of occurrence of the respective traffic situation, the ability of an assistance system to suitably handle these situation, and, if necessary, the ability of the driver to master the situation. A holistic consideration does not limit the focus of technology and driver merely to the ego vehicle but may also take participating road users' skills into account.

The damage in traffic ranges from property damage to personal injury. The following analysis of safety in use is, according to the ISO 26262, oriented toward the avoidance of personal injury.

The well-known procedures of risk analysis and risk assessments form the basis for the analysis of safety in use. During this, the nominal function is considered as well as the anticipated misuse of it. What is not considered is the case of a technical error (E/E error according to ISO 26262). However, the limits of the sensors' performance are very well considered. In the drivers' perception, both cases show the same effect and are called "errors", for example, when the vehicle does not detect a risky situation in due time. The cause, however, can be different (e.g., the technical error resulted from an error in the control unit). By technical measures the likelihood of errors occurring can be downsized to the required extent. Any object not detected by the sensors because of its systematic blur is understood as a system boundary and thus part of the function in its nominal state, which includes the system limits.

A team of experts further combines their knowledge about the system development, the sensor development, the driver behavior, and the traffic situation, and as part of the risk analysis, the collective presents a list of likely events.

These events are evaluated in terms of their likelihood of occurrence in the traffic situation, their manageability by the technical solution, and the drivers' ability to control the system. If necessary, measures are derived from the evaluation to reduce the likelihood of occurrence, on the one hand, as a technical measure (e.g., as requirements for the sensory detection of the environment), and on the other hand, as a constitutive measure to increase the drivers' controllability of the system. Figure 21.2 shows an example of the structure of such an analysis of safety in use.

In some cases, a more detailed consideration of the events may be required. This is the case if there is ambiguity about the overall likelihood of occurrence and where the impact in terms of the damage is important for a holistic consideration of the safety in use. An example could be the event of personal injury when using the autopilot on the highway. In such cases it makes sense to disassemble the event into the parts that must occur collectively in order for the entire event to occur. Its methodological approach is the event tree.

The event tree is comparable to the error tree simply that the interesting situation was not an error but an event, as part of the regular use of the system.

In the disassembly the causes are identified and linked logically. For one thing, this approach can be used qualitatively to illustrate the cause mechanisms and to make them plausible. On the other hand, the individual events may also be quantified, thus creating greater transparency for the overall evaluation. A detailed sensitivity analysis is also possible on this basis.

Figure 21.3 shows an example of an event tree in the event of personal injury when using the autopilot system on the highway.

The event of the collision with a person on the highway is located at the top of the event tree. The level below describes the traffic situation (traffic jam) in which the event occurred. At the same level, the proportion of traffic jams within the overall operating time of the vehicle is taken into account. The next level describes how the event takes place. Here, the preceding vehicle performs a lane change maneuver very close to the obstacle to avoid a collision but does this too late for the driver to react in time and the on-board safety systems do not respond sufficiently.

The aspects that relate to the driver are displayed at the next level. At the bottom layer of the event tree, the likelihood to encounter a person on the highway can be specified for different causes for this particular traffic situation (i.e., traffic jam). By no means this example is complete but it is merely intended to illustrate the approach.

By this simplistic and incomplete example, it is obvious that a great deal of information, for the controllability of a specific traffic situation, is needed. This is information about the traffic situation, the technical performance of the system and its event chain, as well as the performance of the driver. Some approaches and examples, to obtain information in terms of the traffic situation and the drivers, are described in Sect. 21.4.

Evaluated Scenario : Functional characteristics in specific operating situations	Classification of the Situation	Accident Scenario	Assessment of the accident consequences	Averting of dangers	Assessment of the controllability		Justification for 'No' or proposal for action
Customer uses the function A in situation B...	*i.e. exceptional Situation, misuse of system, ...,*	Scenario may lead to accident XY ...		*consequences of the accident may be averted by XY*			*Proposed action n (1) Proposal A (2) Proposal B*
Customer drives with active highway traffic jam assistant in the use case. Vehicle leaves the lane and people are on the side track.	Vehicle leaves the lane and people are on the side track -> likelihood of occurrence very rare.	Vehicle leaves the lane and collides with people.	Collision with people: very high risk of injury.	The driver brakes or pedestrians leave the danger zone.	Situation for observant driver at low speed easily controllable.	yes	Based on a high automation trust prolonged response time can be expected: 1) avoid high automation trust 2) detection of pedestrians at highway traffic jam assistant.
Customer drives with active highway traffic jam assistant in the use case. Preceding vehicle changes lanes, since the current lane is blocked due to a broken down vehicle / accident vehicle	Preceding vehicle changes lanes because of an accident vehicle accident in its' immediate vicinity -> likelihood of occurrence is very rare.	Vehicle continues unrestrained, collides with person outside the broken down vehicle.	Collision with people: very high risk of injury.	Driver brakes or people leave danger zone	Situation for observant driver at low speed easily controllable.	yes	Based on a high automation trust prolonged response time can be expected: 1) avoid high automation trust 2) detection of pedestrians at highway traffic jam assistant.

Fig. 21.2 Exemplary presentation of an extract from a possible analysis of safety in use

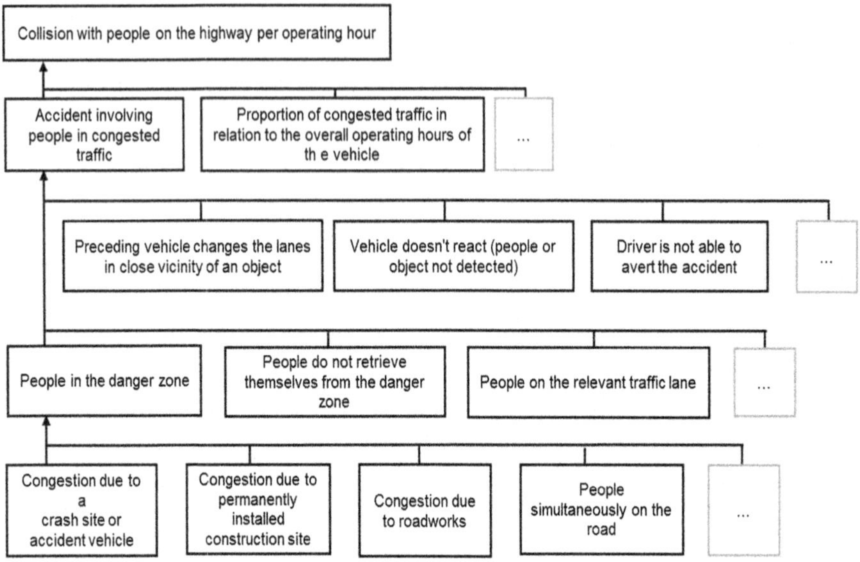

Fig. 21.3 Presentation of an incomplete event tree

21.3.2 Reference Values for Safety in Use

The previous section explained how to systematically arrive at a holistic risk assessment to determine the likelihood of personal injury when using driver assistance systems. In addition it was described how to employ the results as a basis for deriving safety-oriented measures.

So far, there is no reference for the regular use of an automobile. However, there are risk assessments in the area of mobility and consumer protection that could be used as a starting point. Two exemplary approaches that are considering the safety risk of accidental errors and of high-risk product characteristics shall be introduced here: first, the DIN EN 50126, which is used for the safety of the European railways and, secondly, the RAPEX procedure which develops guidelines for the rapid exchange of information on risks of general product safety in Europe.

The DIN EN 50126 was given the headline "The specification and demonstration of the Reliability, Availability, Maintainability and Safety (RAMS)" with the footnote "railroad application" [9].

Within the document, methods for managing the issues referred to in the title are given. Of particular interest here is the paragraph on risk and the assessment and acceptance of risks. Literally, it is written: "The acceptance of risks should be based upon commonly accepted principles", and in addition, as the criterion used in Germany, the minimum endogenous mortality (MEM) is specified.

In the Annex of the standard, the MEM is described in more detail by examples of risk acceptance principles. The MEM is based on the classification of death causes

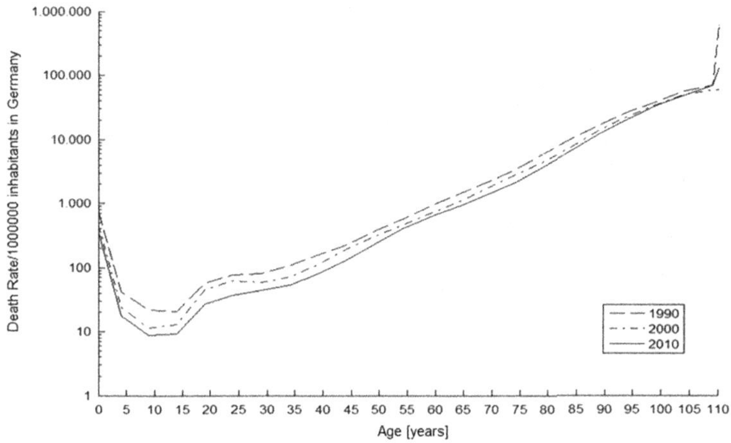

Fig. 21.4 Presentation of the age-related mortality [10]

including a group named "technological facts". This includes categories such as entertainment and sports, home improvement, engines and machines, and traffic. Not included therein are death events due to illness and congenital malformation. The term "endogenous mortality" refers to the proportion of a group to the whole extent of fatalities, which in turn can be broken down depending on age. In economically developed countries, the age-related minimum is the group of 5–15-year-olds (Fig. 21.4).

A normalized simultaneity of influencing systems is determined by the approach that new technical systems should not have a higher risk than already existing ones and that humans are always exposed to multiple systems simultaneously. This simultaneity factor is 20 and thus the reference value of the minimal endogenous mortality is again divided by 20 [10].

A second approach for a possible reference value provides the RAPEX procedure, which is described in the Directive 2001/95/EC on general product safety [11]—a community system for the rapid exchange of information among the European countries for product safety risks. This Official Journal of the European Commission also addresses risk assessment. The details of the RAPEX are laid out in the Commission's decision "on establishing guidelines for the management of the System for a Rapid Exchange of Information 'RAPEX' established under the Article 12 and the notification procedure under the Article 11 of the Directive 2001/95/EC on general product safety" [12].

Different degrees of risks are defined and can be determined from the sum of the degree of injury and the likelihood of occurrence. Cases which pose a serious health and safety risk to the consumer, i.e., the highest risks, are to be reported.

In the Annex of the document, injury scenarios, from bruises to death, are subdivided into four categories of injury. The likelihood of occurrence is expressed as a percentage value of the damage during the life span of a product and is further

Risk level from the combination of the severity of injury and probability

Probability of damage during the foreseeable lifetime of the product		Severity of injury			
		1	2	3	4
High	> 50 %	H	S	S	S
	> 1/10	M	S	S	S
	> 1/100	M	S	S	S
	> 1/1 000	L	H	S	S
	> 1/10 000	L	M	H	S
	> 1/100 000	L	L	M	H
	> 1/1 000 000	L	L	L	M
Low	< 1/1 000 000	L	L	L	L

S — Serious Risk
H — High risk
M — Medium risk
L — Low risk

Fig. 21.5 Classification of risks. Severity of injuries: (1 = Consequences completely reversible; 2 = Function affected for < 6 months; 3 = Function affected > 6 months or permanent loss of function; 4 = Death or disability > 10%) [12]

subdivided into eight levels between 50 % and 1×10^{-6}. Finally, for the evaluation and classification of the risk determined, a reference table is given, as shown in Fig. 21.5.

Rather than using existing procedures of dealing with safety risks occurring from errors or risky product features, the approach of a risk–benefit analysis seems also reasonable. Here, the security risk is assessed in relation to the gain of safety by using the systems. An effectiveness analysis could provide such a methodological approach [13]. The performance of the driver increases considerably in importance, and the issue of which performance would be recognized as a benchmark would be raised: here, the reference depends on whether the driver is averagely experienced or has more than average driving skills. In addition, results from the accident statistics could give orientation, when the question of "how often a driver error is responsible for an event" is considered. Also a legitimate approach in analogy to the abovementioned MEM, to not cause any adverse effect on the relevant risk group, would be the issue of the share of accidents related to the overall traffic situation.

It is obvious that there is still a need for the development of appropriate references and criteria.

21.4 Informative Sources for the Creation of an Analysis of Safety in Use

In Sect. 21.3.1, the systematic approach to the creation of an analysis of safety in use was described. On the one hand, all possible scenarios are collected; on the other hand, a quantification of the risk is tried to be achieved. Thereof, measures are derived in order to suitably contain the risk of damage.

Basically, there are three approaches to reduce risk:

The first category includes increasing the systems' controllability by the user and other road users. This, for example, can be achieved by enabling drivers to easily override the system or by other measures of the human–machine interaction, like takeover requests to improve the so-called mode awareness.

The expansion of technologies for situational awareness and control by the vehicle constitutes the second category. For example, the use of additional and redundant sensors for reliable and accurate object detection or the optimization of the detection algorithm in the classification of the relevant objects could be possible measures.

The third category includes functional adjustments, which limit or preclude the use of the system in potential risk scenarios. By that, the use of the systems in scenarios with a high risk of safety in use will be restricted. Examples of functional limitations are limiting of the speed reduction in automatic emergency brake systems to a level that is manageable for the following traffic or to limit the steering assistance of lane-keeping systems to scenarios where track lines are available and detected.

To ensure that the measures derived from an analysis of safety in use are necessary and can fulfill the desired outcome, the consideration must be very detailed and carried out with great care. This requires experts who have sufficient in-depth knowledge of the field.

At this point, it shall be briefly mentioned that studies from different scientific disciplines show that expert judgments that are not based on a sufficiently large and valid data base but are associated with significant inaccuracies concerning specific forecast predictions [14]. The inaccuracy of the forecasts increases with the increasing complexity of relationships within the observation unit. Driver assistance systems show an especially high complexity, since many interacting factors are involved (driver, vehicle, environment), each, which in turn show a high complexity of their own. Accordingly, for a reliable and sufficiently detailed analysis of safety in use, experts need data sources of which statistical data and facts can be derived as input variables for the event tree, and questions of the following type can be answered.

"Which transport scenarios with a potential safety hazard may occur during the system use?" or "How often do relevant traffic scenarios and relevant scenario parameters occur?" or "How well do different drivers and other road users handle the relevant traffic scenarios?" For a realistic assessment of the controllability, both

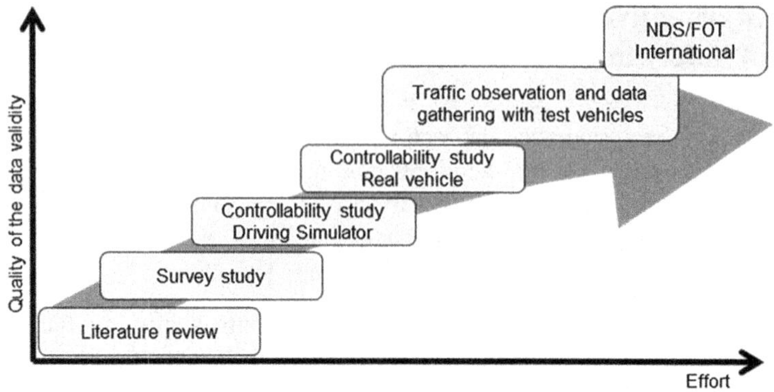

Fig. 21.6 Some methods for obtaining data in the field of controllability/driver performance likelihood of occurrence of traffic scenarios

the performance of the driver in the first encounter of the system and the medium- and long-term handling of the assistance system are of significance. A further question to be addressed is "How does the system perform in the scenarios under consideration?"

Thus, a key challenge in the conduct of an analysis of safety in use is to provide the data required in sufficient quality.

In principle two categories of data sources can be distinguished: the scenario-based and driver performance-related data sources. Some methods provide answers for both categories, while others only provide data for one. Figure 21.6 gives an overview of different methods for data generation. It is evident that the validity of the data increases depending on the complexity of each method.

Complexity here refers to the sum of the organizational, financial, and technical requirements for the preparation, implementation, and analysis of the data obtained from each method. For example, a literature review, based on a targeted analysis of "relevant" literature, is associated with much less effort in terms of the technical and organizational requirements than a driving simulator study with subjects.

On the other hand, a driving simulator study is, in turn, associated with far less effort than the so-called naturalistic driving study (NDS) or a field operational test (FOT). Both, the NDS and the FOT, involve several months to years of organizational and technical preparations before results can be obtained. However, the knowledge gained by a well-prepared and broadly based NDS/FOT study is tangible, reliable, and informative, because these methods are based on the observation of the driver behavior in dealing with one or more systems in a natural field environment and in a real vehicle.

Often, several of the methods described above must be used in the context of an analysis of safety in use in order to increase the reliability of the data to a sufficient level. In general, it is unlikely that a single method provides all the necessary data and facts. Different methods must therefore be used in parallel and complementary to close the knowledge gaps in the conduct of consumer safety analyzes. Below, the mentioned methods are briefly described and examples in the context of semiautomated driving functions are provided. It is important to note that a further method is the method of simulation, while these method is not addressed in this paper.

21.4.1 Data Collection: Literature Review

Literature reviews can be used for answering questions of relevant traffic scenarios and their frequencies, as well as questions related to the performance and controllability of systems by the driver. Particularly, in the field of driver performance and controllability, relevant literature sources are available. In the context of driver assistance systems, scientific contributions in the fields of general cognitive psychology, traffic psychology, ergonomics, human–computer interaction, and norms and standards of such may provide a first basis for evaluating the controllability and driver performance.

Starting with a limited number of relevant sources, the advantages of this method, if standard sources are used, lie in the limited effort and the general acceptance of the facts.

A disadvantage of the method is that the standard literature often provides only first generic advice and information for the use of safe design systems.

A disadvantage of the method is that the standard literature often provides only first generic advice and information of the safety in use of systems. Also, the reported facts of the assessed driver performance and controllability are often quite system specific. In addition, due to the increasing complexity of innovative driver assistance systems, they have not yet been studied in a high level of detail.

Consequently, the results of a literature review need to be interpreted carefully due to the lack of the informative value and reliability for a specific question for the system-specific safety in use. In regard to scenario-specific data, similar advantages and disadvantages, when conducting a literature review, apply. Factors for the evaluation of relevant traffic scenarios, such as the characteristics of the road and traffic, congestion time, average travel time and travel distance, weather conditions, velocity, and much more, have been studied and described in various sources. However, their codependency is often not considered (e.g., the likelihood of congestion depending on the weather conditions).

21.4.2 Data Collection: Questionnaires

Interviewing specially selected driver groups is another way to obtain basic data for an analysis of safety in use. Different groups can often give relatively good indication of the occurrence of relevant and potentially critical traffic situations. Especially where only few publications exist, a questionnaire study may provide first qualitative and quantitative indications. An advantage of surveys is they can be used to gain insight into specific experiences of a relatively large number of people. Additionally a survey can be carried out in different places and in different countries, thus facilitating a quick data collection of different regions' traffic situations. To the advantage that the necessary infrastructure for the preparation, implementation, and evaluation of such studies compared to other methods is relatively low. Due to possible different interpretations of questions, linguistic and cultural peculiarities, and the objectivity of the respondents, a survey study is often associated with uncertainties regarding the reliability of the results. Surveys in the field of assessing the controllability of critical situations are less advisable, as several studies have shown that respondents without experience of the situation in question often overestimate their own performance.

As an example of a questionnaire study, a study of the likelihood of certain traffic situations in China will be introduced (internal study of BMW). The background of the study was the need to acquire knowledge about the traffic situation in China for the safety in use of partially automated assistance functions. Due to the relatively short history of motorized road traffic and the ever-changing structures in the country, hardly any publication sources exist, which provide the latest and quantifiable picture of the traffic situation in China. The aim was to compare the frequency of safety-relevant traffic situations on Chinese motorways with those in Germany and to use them as an initial estimate for assessing the safety in use of assistance systems in China.

In preparation of the study, several interviews with German employees, working in China, and Chinese employees were conducted to identify the main differences of road traffic between China and Germany. It quickly became apparent that the opinion of respondents remained vague in interviews and did not allow for a quantification of the so-called exceptional situations. With exceptional situations such traffic situations were meant which pose a potential accident risk due to road users' unlawful conduct of their vehicle in traffic. Examples of exceptional situations are, among others, reverse driving on the motorway, wrong-way drivers, and cyclists and pedestrians on the motorway.

The development of the questionnaire and the implementation and evaluation of the study included several steps. First the definition of the questionnaires' main focus took place on the basis of interviews with Chinese experts. Subsequently, the questionnaire was developed, the rating scale selected, and the questionnaire was translated into English and Chinese language. A test phase in China was carried out, to optimize the questionnaire until it was finally distributed to selected participants in China and Germany. The analysis and interpretation of results marked the end of the study.

Three different samples of participants took part in the study: German employees working in Munich ($n = 25$), Chinese employees working in Beijing ($n = 34$), and German employees sent to work in Beijing ($n = 26$). In addition to the demographic items, the questionnaire included further questions about driver behavior and driving experience as well as the perception of the frequency of some exceptional situations.

The results of the survey revealed that Chinese respondents, on average, drive 260 hours per year on a highway in a traffic jam. Thus, Chinese participants experienced on average 23 times more often congestions on highways than the German participants. According to the results of the study, the German participants, on average, experience approximately 12 hours of traffic jam driving on German motorways, in 1 year.

The analysis of the questions regarding the exceptional situations has shown that all exceptional situations included in the survey were ten times more frequently experienced on Chinese motorways than on German highways, some even more often. Striking was that the results of the assessment of Chinese respondents in China partially differed from the assessment of the German respondents: German respondents who worked in China rated the likelihood of occurrence of the exceptional situations higher than the Chinese respondents.

Different perceptions of the criticality of traffic situations, due to cultural differences, driver training, or habituation effects of the local population, are just some of the possible explanations for this discrepancy. The different assessment of the frequency of traffic situations generally raises the question of how objective the results of survey studies are and highlights the need to validate the results. Nevertheless, such results provide an initial orientation or tendency of the differences in the two countries.

The example described shows how data about traffic situations and exceptional situations can be obtained with relatively little effort and in a short time through the use of questionnaires. However, the results also show that such questionnaire study must be validated or supplemented by more sophisticated methods.

21.4.3 Data Collection: Studies in the Driving Simulator or in Real Vehicles

Alongside with volunteer studies in the real vehicle, different types of driving simulator studies, assessing driver performance, controllability of driver assistance systems, and developing human–machine interfaces are increasingly being applied.

Both methods, with their advantages and disadvantages, can be applied in the course of the development and validation of the assistance system. Real studies hold the advantage of participants experiencing real driving dynamics in the examined scenarios, which is important for observing realistic driver reactions. However, the design of a real driving study is limited by safety issues and obvious limitations of the surrounding infrastructure.

Basically, to conduct a study with real vehicles, safety-oriented modifications have to be taken in order to reduce the risk of participants getting harmed. These include limiting the scenarios studied to low or medium vehicle dynamics, carrying out studies on safe and enclosed test routes and areas, excluding particularly critical traffic scenarios, and some other. Further, the influence of secondary and tertiary driving tasks on the driving performance can only be observed to a very limited extent, due to safety issues.

Another infrastructural limitation of real driving studies that, if at all, can be managed with great effort is the presentation of complex traffic scenarios that require several vehicles and other road users. Further, ensuring the reproducibility of the study traffic scenarios is often impeded because of external factors such as the changes in weather conditions. The conduct of real driving studies outside of the public traffic space may result in the examined traffic scenarios as being too artificial, leaving the issue as to whether the collected data of the driver' performance corresponds to the actual performance of the subjects.

An example of a study in the real vehicle, which provides basic data for an analysis of safety in use, can be found, for example, in [15]. The study examined two error categories to assess potential steering system disorders. Here, the controllability of the leap in the steering angle by switching off the active steering system and the controllability of the manipulated errors in different amplitudes was observed at various levels of velocity. Based on the analysis of the driver behavior, two different reaction phases can be distinguished: on the one hand, the initial response, i.e., the time interval until the end of the first driver engagement, and, on the other hand, the error compensation phase in which the occurring driving dynamics and lane tracking errors are compensated. Both the subjective assessment and the characteristics of the vehicle operation and the vehicle reactions show that a resulting leap from the system shutdown in the steering ratio is easy to control, even with demanding steering maneuvers. For the positioning error, there is apparent that up to a positioning error amplitude of 0.3° of the front wheel angle, no safety-relevant implications apply.

In [16], the drivers' performance was studied in approaching a stationary obstacle in the low speed range with an active ACC stop & go system. The study showed that all drivers were able to stop the ego vehicle in case of loss of the target object during the approach in time. In addition, strong learning effects were found in the study during the repetition of the test scenario.

In driving simulator studies, most safety-related and infrastructure-related constraints of a real vehicle study can be overcome, depending on the type of the simulator. The lack of driving dynamics in the static simulator or the limited dynamics in the dynamic simulators remains a significant weakness of the simulator methods. In recent years, a new method of volunteer studies has been deployed. It is called the vehicle-in-the-loop method. Here, the strengths of the simulator technology (e.g., the virtual driving environment) and the real vehicle (e.g., real dynamics) are combined. However, resulting from the combination, new challenges arise that need to be addressed [17, 18].

Simulator studies have been increasingly used in recent years for studies on driver performance and controllability of partial and highly automated assistance systems and provide valuable information that can be used in an analysis of safety in use.

In [19] a simulator study is described where the drivers' performance was studied in three different takeover request scenarios. The ability of the driver to take over is one main topic in automated systems. Depending on the scenario, subjects had to handle different driving maneuvers after the takeover request was given (e.g., stabilizing, maneuvering, and navigating). In the study, each scenario was run with takeover times of 4, 6, and 8 s. The driving speed during the highly automated driving was set to 100 km/h and test subjects had to simultaneously perform a tertiary driving task.

It has been found that a drop in the takeover period from 6 to 4 s leads to a particular high loss of comfort in subjective evaluations of the subjects. In contrast, no subjectively perceived comfort profit could be established when the takeover time was increased from 6 to 8 s.

In [20], a simulator study demonstrated the quality of the takeover by the influence of certain environmental parameters and different non-related driving tasks.

It was shown that visual–acoustic takeover request lead to significantly faster takeover times and improved lane-keeping performance in contrast to a strictly visual takeover request.

Depending on the request stimulus, the time the drivers needed to put the hands back onto the steering wheel, respectively, to take over the driving task while driving highly automated and given different complex takeover requests, was investigated in [21].

The examples described illustrate the advantages of simulator studies in the investigation of a variety of complex and reproducible traffic scenarios, non-related driving tasks, and risky takeover scenarios without endangering the driver or other road users.

21.4.4 Data Collection: Observation of Traffic

The observation of traffic can be performed either from a static or a dynamic observation point. In the case of static traffic observations, observation platforms are built at the relevant transport nodes, to capture the flow of traffic and the behavior of road users. In this case, road users refer to vehicles but also pedestrians and other, so-called, vulnerable road users. By analyzing the data of the observed traffic nodes, insight into the behavior of road users can be gained. A disadvantage of this method is that seldom other sensors than cameras are being used. Thus important parameters such as intervals between vehicles, velocity, and acceleration, among others cannot be determined sufficiently.

Alternatively to the static monitoring station, the detection of mundane road traffic situations can be realized from a first-person view by establishing one or more vehicles with sensors and environmental monitoring equipment. By this approach, the technical, financial, and organizational effort to carry out the measurements can vary greatly, depending on the amount of the used monitoring equipment, the number of vehicles, and the overall required length of the route. Often, this approach is used to selectively look for traffic events and their likelihood of occurrence. Since the effort of extracting the relevant events from the raw data can be extremely time

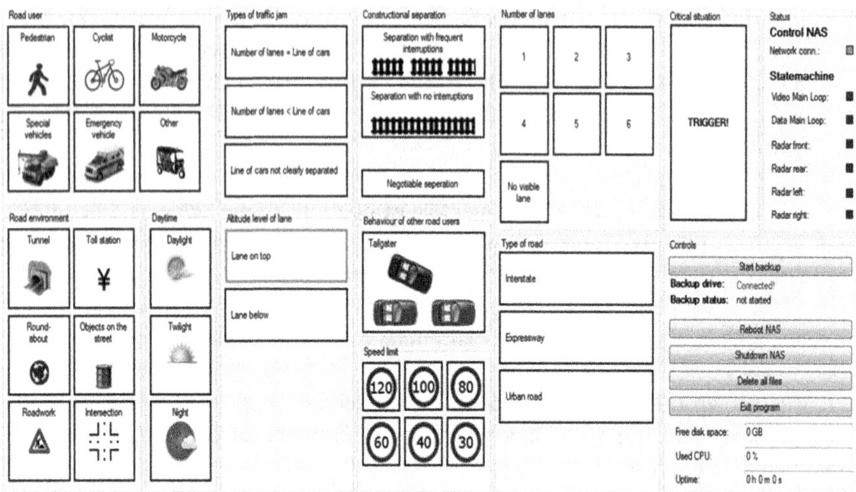

Fig. 21.7 The graphical user interface used in the above described study

consuming and costly, it is advisable to instruct drivers or passengers, to additionally code for trigger signals while driving. Subsequently, this facilitates the discovery of relevant situations.

As an example, a recent traffic observatory study was conducted in China and will be presented here. The study used test vehicles to gain insight into the traffic situation in China. As part of a pilot study, two test vehicles have been set up in order to acquire first data on the traffic situation in congested traffic in China. Each vehicle has traveled about 8 hours a day for 4 months, with much of the travel time consisting of congested traffic. Due to the built-in measurement technology, consisting of four cameras and four radars, the traffic situation in the front, rear, and side regions of the vehicle were recorded.

The vehicles were driven in two selected cities by two Chinese drivers. During the journeys, the codriver coded the relevant scenarios and objects that have been defined prior to the study.

For this purpose, the codriver was given a control panel with a graphical user interface (Fig. 21.7) to mark predefined road characteristics, such as the type of the road (urban, expressway, and interstate), speed limit, number of lanes, physical separation to opposing traffic, and observed objects, such as pedestrians, cyclists, and lorries. For a better understanding of the scenario, the codriver also recorded additional details, such as light conditions and of special sections of the road (tunnel, junction, roundabout, etc.).

During the analysis of the data, the search for the relevant situations was considerably facilitated by the trigger signals and captured characteristics of the situation set by the codriver. In addition, all site features tagged by the codriver were reviewed and, if felt necessary, corrected. 90 % of the codrivers' captured site features were identified and validated in the video analysis.

During the study an overall travel distance of 8400 km was recorded, with about two-thirds of the travels corresponding to congested traffic. A total of 4610 situations were marked by the codrivers. 1345 cases refer to pedestrian situations and 430 cases to cyclists' situations. If the number of captured pedestrian situations is set in relation to the distance and time traveled, a pedestrian situation occurs every 47 minutes on a Chinese motorway (interstate). The likelihood of occurrence of pedestrians on the expressways (city ring roads) is about ten times higher. Cyclists occur once, every 8.5 hours on interstate roads and on approximately every 15 minutes on the expressways.

By conducting a traffic observation with vehicles, the likelihood of occurrence of traffic situations, which have to be considered in an analysis of safety in use, can relatively well be objectified. The main disadvantage of this approach is that by the use of professional drivers in the test vehicles, in respect to the driver's driving style, a low variability arises. Also possible influences due to the drivers' behavior in emerging traffic situations cannot be excluded.

21.4.5 Field Operational Test

The following section is intended to present a specific method which is particularly useful for collecting in-depth knowledge of the driver, the vehicle, and the environment within their specific constellations. The field operational test (FOT) is a method where participants drive specially equipped vehicles under normal traffic conditions, in order to gain insight and to detect a drivers' natural driving behavior. When conducting a FOT, a large number of aspects must be considered. These aspects will shortly be described based on a study conducted by BMW. In contrast to the FOT, in the naturalistic driving study, subjects use their own vehicles. However, with both methods, the vehicles are equipped with the appropriate instrumentations to measure drivers' inherent behavior in the field.

Recently, a FOT was performed in Germany in order to collect information from daily traffic situations. This information can be used for the safety in use assessment and therefore for the safe development of assistance systems. Additionally, a lot can be learned about how drivers handle and interact with the assistance systems and how to incorporate this knowledge into a customer value in the development of future systems. The focus is on the questions of how often certain events (e.g., emergency lanes, congestions, etc.) occur in traffic, how critical they are (e.g., minimum distances to the front vehicle), and how drivers handle them. Here, the method is different to the aforementioned traffic monitoring: the ego driver is the essential factor of this measurement method.

In the current study, seven vehicles were equipped with appropriate measurement technology. They were used by selected drivers for a period of 3 months during the daily driving process. Before making such a time- and cost-intensive study, different factors must be considered and planned accurately. These factors are divided into the following working packages: the general experimental planning and design phase, vehicles and measurement technology, driving phase, documentation, and evaluation.

In the following paragraphs, these steps are explained in detail and the first results are displayed.

First, the aspect of the "general experimental planning and design" phase will be considered. This includes factors like question selection, sampling, and questionnaire development. A wide range of issues is the basis for all further decisions of the experimental design, e.g., the sample selection. Because vehicles are usually available only for a limited time period, you have to decide whether you want to investigate a lot of different drivers who drive only a short time or a few drivers who can collect long-time experience with the vehicle and the assistance systems. In this study, for example, it was of interest how drivers behave with the assistance systems for an extended period. Questions such as "Adapt drivers their behavior?" and "Can they deal with assistance system limits?" can be addressed.

For the selection of the drivers, certain criteria were applied. Thus, drivers should often take longer routes with a high traffic density, should not work in the development of driver assistance systems, and should not be a novice or a professional test driver. Seven drivers were selected (mean age 32 years; 2 women, 5 men). In addition to the sampling, the development of questionnaires is equally of importance. Therefore, understanding which questions cannot be detected by objective data and need to be covered with other methods are included in this phase. The occurrence of critical events, for example, is particularly important: "Which traffic situation led to the critical event?" and "How was the situation assessed by the driver?" are questions that can be covered with the method of questionnaires.

Next, we take a closer look to the aspect of "vehicles and measurement technology." This aspect includes factors such as "What kind of assistance systems the vehicles have on board?" and "Which measurement technology and/or additional sensors must be implemented?" Answering these questions and the decision for a specific measurement technology is mainly depending on the issues to be answered. For example, if additional sensors are required, the measurement technology has to ensure that this data will be integrated, and a synchronized recording of the vehicle data and the advanced sensor data is possible. For the questions of the current FOT, the traffic environment was particularly relevant. That is why in addition to the existing sensor technology of the vehicles, four side radars and four cameras were installed in the vehicles. The cameras were each directed to one side, forward, and to the driver.

If the technical modification completed, the sample selected, the test materials and questionnaires finalized, and the participants instructed, starting the driving phase of the field study is feasible. During the conduction the measurement technology should be checked regularly. In complex technical structures in which multiple components must be synchronized, it is advisable to consider not only the functionality of these components but also the quality of data to detect any failures and to correct errors. Another aspect is the support of the participants during they have the vehicle. In addition, the participants had to complete weekly and monthly

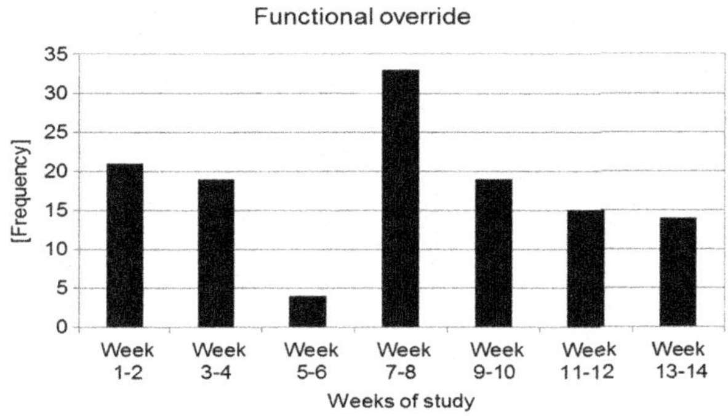

Fig. 21.8 Absolute number of measured functional overrides by the driver (From: Intern FOT conducted by the BMW AG, 2014)

given questionnaires, which were related to certain aspects of driving events or the evaluation of the experienced assistance systems.

In parallel to the execution of the driving, the analysis of the data available can be started. In the following some results will be shown.

A system often used by the drivers was the traffic jam assistant. This system helps the driver in longitudinal and lateral control and keeps the vehicle in traffic jams to 60 km/h within certain system limits in the lane. Especially of interest was in which cases the driver overrides this function? What are the causes for the drivers' wishes to drive differently than the system? To answer this question, a number of parameters can be used. On the one hand, the driving data shows us how many times the function has been overridden by the driver (Fig. 21.8).

This frequency of overrides seems high at the first glance. However, during the study, the traffic jam assistance function was 1812 times active, thus qualifies the absolute number of overriding. On the other hand, you can use the results from the questionnaires to assess how often the driver has experienced a function override. In Fig. 21.9, we can see how drivers evaluate these override moments throughout the study.

This result can be supported with the question of what was the reason for the override. Figure 21.10 shows that vehicles cutting in the drivers' lane are most often cited as responsible for overriding the function. With some distance, vehicles veering out and after those too little distances to the lane marks are stated reasons for an override.

These results can now be synchronized with the video data and the data of the car. So the information that is available is, for example, whether it was really necessary that the driver reacts or whether the situation could actually have been handled by the system. Using the example "function overdrive", one can see that it makes sense to refer to various parameters to answer questions. Over time, the drivers are able to detect system deficiencies and system behavior and to decide whether an override is necessary. Therefore, the observation of this parameter throughout the course of the experiment is particularly relevant.

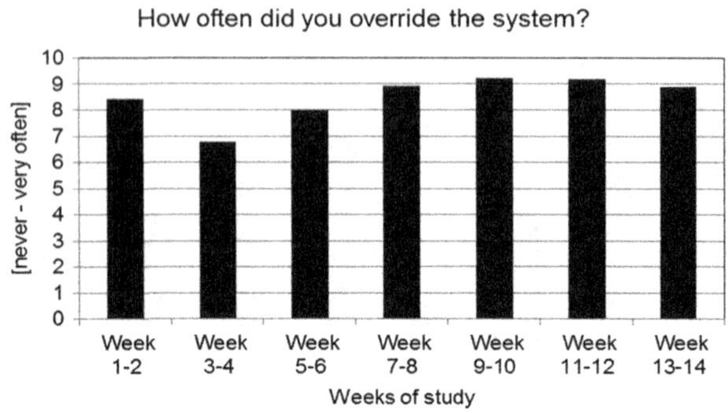

Fig. 21.9 Absolute number of functional overrides, estimated by the driver (From: Intern FOT conducted by the BMW AG, 2014)

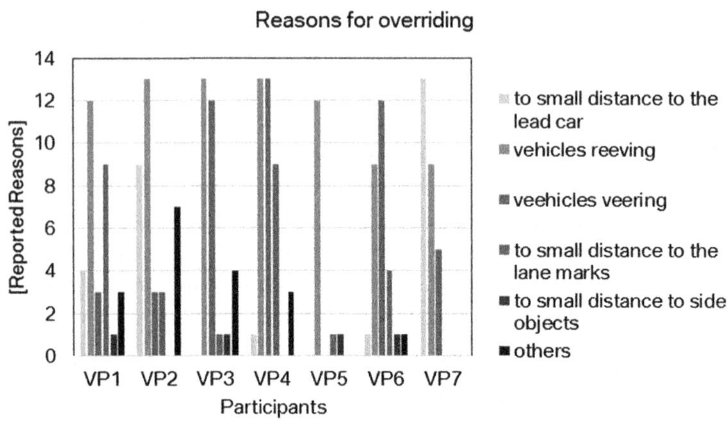

Fig. 21.10 Reasons for the functional override by the driver (From: Intern FOT conducted by the BMW AG, 2014)

In principle we can summarize that from such extensive studies, a variety of information can be obtained. These are—regarding the driver reactions and driver behavior—especially valid because they were obtained in the natural traffic environment.

21.5 Conclusion

With the method of analysis of safety in use, a systematic approach for a holistic assessment of possible risks has been presented, which can be applied with respect to automated systems in road traffic. An essential point is the difference of the concept of system errors in accordance to the ISO 26262 of functional safety. In the ISO 26262, the E/E failure of systems is considered, but not, for example, the limits of sensor performance. This often perceived by customers as a failure of the functions properties is in the system development per definition a known limit of the function or sensors and no error. Therefore, the system limit is also included in the analysis of safety in use.

For a detailed assessment of the likelihood of personal-hazardous events, extensive knowledge on driver behavior and driving situations is required. The more accurately the quantification of probabilities, the more basic knowledge must be known. For that reason, well-known methods of driver performance studies in the driving simulator to studies in regular traffic or field operational tests, respectively, naturalistic driving studies, are carried out. Aspects like country-specific factors and cultural differences increase the complexity of the analysis, hence, can increase the cost of data collection significantly.

The questions for accepted comparative values in the framework of the quantification of risks are still not answered in the context of automated driving today. In this paper, two approaches, the minimum endogenous mortality and the RAPEX procedure, were explained, which supply as a result of comparable values. Another approach could be to derive indications from a risk–benefit assessment. The development of such reference values and the scientific and social establishment is one of the most important tasks on the way to highly automated driving.

References

1. Convention on Road Traffic done at Vienna on 8 November 1968 (1968), http://www.unece.org/fileadmin/DAM/trans/conventn/crt1968e.pdf. Accessed 28 Sept 2015
2. Code of Practice for the Design and Evaluation of ADAS (2009), http://www.acea.be/uploads/publications/20090831_Code_of_Practice_ADAS.pdf. Accessed 28 Sept 2015
3. http://www.transport-research.info/web/projects/project_details.cfm?id=20297. Accessed 28 Sept 2015
4. E. Donges, Aspekte der Aktiven Sicherheit bei der Führung von Personenkraftwagen. Automobil - Industrie **27**, 183–190 (1982)
5. *ISO 26262, Road Vehicles – Functional Safety.*
6. Bengler K, Grundlegende Zusammenhänge von Automatisierung und Fahrerleistung, Beitrag in *Fahrerassistenz und Aktive Sicherheit*, 16./17.4.2015 (Haus der Technik, Essen, 2014)
7. *Rechtsfolgen zunehmender Fahrzeugautomatisierung*, Bericht der Bundesanstalt für Straßenwesen, Fahrzeugtechnik Heft F83, 2012
8. *Report of the sixty-eighth session of the Working Party on Road Traffic Safety, ECE/-TRANS/WP.1/145* (2014). http://www.unece.org/fileadmin/DAM/trans/doc/2014/wp1/ECE-TRANS-WP1-145e.pdf
9. *EN DIN 50126, Spezifikation und Nachweis der Zuverlässigkeit, Verfügbarkeit, Instandhaltbarkeit und Sicherheit (RAMS)*
10. *Human Mortality Database* (University of California, Berkeley and Max-Planck-Institut für Demographische Forschung, Rostock, 2015). www.mortality.org or www.humanmortality.de. Accessed 18 Feb 2015
11. Richtlinie *2001/95/EG des Europäischen Parlaments und des Rates vom 3.* Dezember 2001 über die allgemeine Produktsicherheit (2001). http://eur-lex.europa.eu/legal-content/DE/TXT/HTML/?uri=CELEX:32001L0095&from=DE

12. Entscheidung der Kommission, zur Festlegung von Leitlinien für die Verwaltung des gemeinschaftlichen Systems zum raschen Informationsaustausch "RAPEX" gemäß Artikel 12 und des Meldeverfahrens gemäß Artikel 11 der Richtlinie 2001/95/EG über die allgemeine Produktsicherheit (2009).
13. Kompaß K, Helmer T, Wang L, Kates R, Gesamthafte Bewertung der Sicherheitsveränderung durch FAS/HAF im Verkehrssystem: Der Beitrag von Simulation, Beitrag in *Fahrerassistenz und Aktive Sicherheit*, 16./17.4.2015 (Haus der Technik, Essen, 2014)
14. P. Tetlock, *Expert Political Judgment* (Princeton University Press, Princeton, NJ, 2005)
15. Neukum A, Krüger H-P, Fahrerreaktionen bei Lenksystemstörungen - Untersuchungsmethodik und Bewertungskriterien, in *VDI-Gesellschaft Fahrzeug- und Verkehrstechnik* (Hrsg.), Reifen-Fahrwerk-Fahrbahn (VDI-Berichte, Nr. 1791) (VDI-Verlag, Düsseldorf, 2003)
16. Neukum A, Lübbeke T, Krüger H-P, Mayser C, Steinle J, ACC-Stop&Go: Fahrerverhalten an funktionalen Systemgrenzen, in *5. Workshop Fahrerassistenzsysteme - FAS 2008*, ed. by M. Maurer, C. Stiller (FAS, Karlsruhe, 2008), S. 141–150
17. T. Bock, *Vehicle in the Loop – Test- und Simulationsumgebung für Fahrerassistenzsysteme. Audi Dissertationsreihe, Bd 10*, 1st edn. (Cuvillier Verlag, Göttingen, 2008)
18. G. Berg, *Das Vehicle in the Loop – Ein Werkzeug für die Entwicklung und Evaluation von sicherheitskritischen Fahrerassistenzsystemen*, Dissertation, Universität der Bundeswehr München, 2014
19. D. Damböck, M. Farid, L. Tönert, K. Bengler, *Übernahmezeiten beim hochautomatisierten Fahren*. Paper presented at the 5, Tagung Fahrerassistenz, München, 2012
20. Radlmayer J, Gold C, Lorenz L, Bengler K, *How Traffic Situations and Non-Driving-Related-Tasks Affect the Take Over Quality in Highly Automated Driving* (HFES, Chicago, 2014)
21. Naujoks F, Mai Ch, Neukum A. The effect of urgency of take-over requests during highly automated driving under distraction conditions. *Proceedings of the 5th International Conference on Applied Human Factors and Ergonomics AHFE 2014*, Kraków, Poland 19–23 July 2014

Chapter 22
Testing Autonomous and Highly Configurable Systems: Challenges and Feasible Solutions

Franz Wotawa

22.1 Introduction

Quality assurance has always been an important topic in the automotive industry leading to standards like ISO 26262 [1] where automotive safety integrity levels (ASILs) are defined based on hazard analysis and risk assessments. Often failure mode and effect analyses (FMEAs) are used to identify potential hazards occurring when particular system components fail. Alternatively, such systems might be analyzed accordingly to Leveson [2] where the author focuses mainly on a system theoretic perspective. In order to assure that the implementation does never lead to such dangerous situations, tests have to be carried out. Although the automotive industry has spent a lot of effort in quality assurance, there has been a lot of vehicle recalls during the past years, many of them caused by software bugs detected after deployment. As discussed by Altinger et al. [3], such recalls also have a huge economical impact. When considering that even today's quality assurance measures used in the automotive industry cannot prevent recalls, the question arises about consequences for safety of autonomous driving vehicles. For this purpose we discuss first the differences between ordinary cars and autonomous driving vehicles and second identify means for dealing with these differences.

Autonomous systems like autonomous mobile cars or robots have to provide a certain task, for example, driving from one place to another, while interacting with their surrounding environment. This interaction is based on observations such systems obtain from the environment using sensors and their internal current state from which actions are derived. These actions are executed using the attached actuators. In the case of autonomous driving vehicles, the information about the current position, the speed, other vehicles, and obstacles are obtained using attached

F. Wotawa (✉)
Institute for Software Technology, Technische Universität Graz, Graz, Austria
e-mail: wotawa@ist.tugraz.at

D. Watzenig, M. Horn (eds.), *Automated Driving*,
DOI 10.1007/978-3-319-31895-0_22

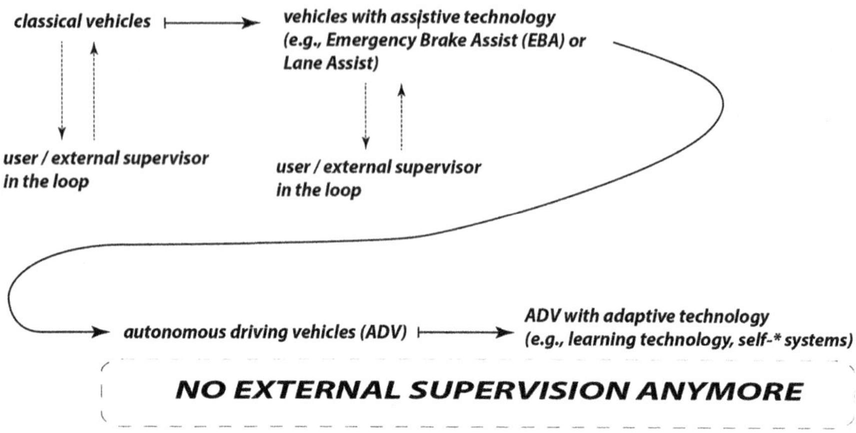

Fig. 22.1 From "classical" vehicles to autonomous driving vehicles and beyond

sensors like GPS, laser range finders, or vision systems. Given this information and the planned journey, such an autonomous vehicle decides on the next step like changing the lanes on a highway or initializing an emergency braking action. All such decisions may harm others, and thus such systems have to be thoroughly tested to assure that all safety critical requirements are fulfilled. This situation becomes worst in case of self-adaptive systems like systems that learn during operation or systems having a huge configuration space. It is interesting to note that even in the case assuming that all such systems behave deterministically from the point of view of the implemented algorithms and behaviors, there might be cases where small deviations of the measured input signals have a huge impact to decisions. Thus, from outside such a system might look like as behaving non-deterministically.

In Fig. 22.1 we depict the main difference between classical cars, vehicles with assistive technology, autonomous driving vehicles (ADVs), and ADVs with adaptive technology. Whereas ordinary vehicles and vehicles with assistive technology always have an external supervisor in the loop, i.e., the driver, there is no external supervision in case of ADVs and adaptive ADVs. Hence, any failure occurring during driving has to be detected by the ADV itself. Furthermore, appropriate countermeasures have to be taken into account immediately in order to prevent hazardous situations. For this purpose the ADV has to have access to necessary information from which a failure can be deduced. This includes not only the internal state of the vehicle itself but also its surrounding environment.

When allowing systems more and more autonomy, the question arises of how to ensure a proper behavior under all circumstances? Due to variations of sensory inputs, changes in the environment, e.g., change of lighting conditions, faults that might occur, or different configurations, the whole space of possibilities grows very quickly and it is very likely to not consider certain cases during development. For example, there are a huge variety of different traffic signs around the globe. Some of them are outdated but might be still found. Hence, the vision system of an autonomous vehicle responsible for detecting and classifying traffic signs has to be tested using many different versions of signs. Even worst the signs might be damaged, the color might not longer be the originally specified color, and so on. However, in any circumstances the autonomous vehicle has to react appropriately. From the stop sign example, we are also able to see that the variants depend on the region, which allows us to specify a certain context, e.g., the country where an autonomous vehicle is going to operate, in which certain rules, assumptions, or properties are valid.

In this chapter we assume that the subcomponents of the whole system already have been thoroughly tested and focus on the integration and system test. There we are in particular interested in assuring important requirements of autonomous vehicles under all circumstances. We will discuss the influencing parameters, which also entail consequences on how to test such systems. Depending on the degree of autonomy, we will see that testing during development might not be sufficient and thus require introducing monitoring at runtime for assuring meeting requirements like safety.

In the following we discuss related research of testing where we specially focus on autonomous adaptive systems. Afterward, we discuss the underlying problem and propose a testing method that takes the huge variety of configurations, parameters, and situations into account. Finally, we conclude the paper.

22.2 Related Research

Validation and verification (V&V) is an important part within the system development process. Under validation we subsume all activities necessary to check whether a developed system is the system that was indented to be developed, whereas verification is for checking that the system follows its specifications. Testing can in principle cover both parts of V&V. However, it is obvious that testing is only for detecting failures but not for ensuring that there are no faults in the system. Nevertheless, testing is currently the most important activity in V&V. In Fig. 22.2 on the left, the V process is depicted that is often used as a standard process in the automotive industry. In this process the different phases of the process like requirement analysis are paired with their corresponding testing activities.

For a general introduction into testing, we refer the interested reader to [4] and [5]. Testing as an activity generally speaking deals with search for interactions between the system under test (SUT) and a user in order to reveal an unexpected behavior. In general there might be an unfeasible number of potential interactions. Hence, testing as a discipline tries to identify the most relevant interactions with a SUT. Relevant in this context mean to cover sequences of interactions that if fulfilled assure that the SUT works as originally indented.

In order to reduce the testing effort, the automation of testing has been in the focus of research and development for several decades. Test automation comes with two favors: (1) automation of test execution and (2) automation of the generation

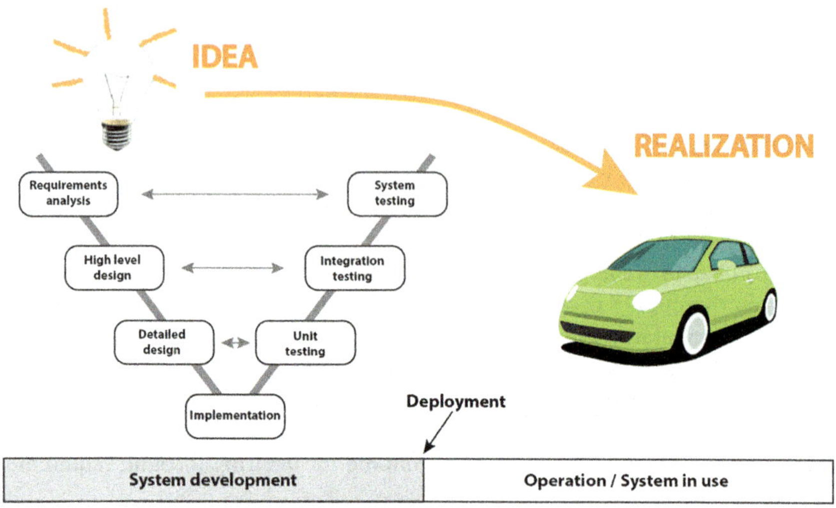

Fig. 22.2 From system development to its use

of interaction sequences. The former requires the implementation of frameworks that allows to program sequences of interactions with the SUT directly. The latter deals with the automated extraction of interaction sequences. This can be done from information about the input domains or models of the SUT. In case of model, we speak about model-based testing (MBT), which has gained a lot of attention. For an introduction of MBT, have a look at [6] or [7]. A model in the context of MBT is usually a state-space representation of the relevant parts of a SUT from which interaction sequences can be obtained via traversing the state space. In addition MBT also allows for checking whether the output generated by the SUT when stimulated with an extracted interaction sequence follows the expectations. Hence, there is no need to have an additional oracle, which allows for full test automation. Test automation including the use of models is very much in use within the automotive industry. For a detailed discussion on testing methods used, we refer to [3] where the authors present the results obtained from a questionnaire survey carried out in the context of automotive system development.

In the context of autonomous and self-adaptive system, quality assurance has become more and more important. Cámara et al. [8] compiled several articles dealing with methods for quality assurance of self-adaptive systems that are also of particular interests for ADVs. The main methods further outlined in [8] make use more or less of monitoring and runtime verification to identify critical situations to be handled directly whenever they occur after deployment of such self-adaptive systems. Nafz et al. [9] introduced the concept of allowed deviations from the expected behavior and propose a method that allows for guaranteeing that self-adaptive systems always do not exceed these boundaries. Steinbauer and Wotawa [10] described methods that make use of model-based reasoning to adapt in case of internal faults or changes in the surrounding environment. There the underlying idea is to reason from the models of the system and the environment to find a current configuration of the system that explains the current observations. In this way the given models describe more or less the behavioral boundaries of the system and the degree of adaption. In other chapters of [8], authors make use of runtime verification for identifying potential hazards. In runtime verification formal models of properties that are checked during operation are used to identify potential behavioral deviations. For more information about runtime verification and its use, consult [11] or [12].

In contrast to the mentioned previous related research, the methodology proposed in this chapter focuses on identifying potential hazards that might occur during operation at development time. We will see that such an endeavor requires dealing with multiple configurations and a huge parameter space in addition to dealing with the oracle problem in an appropriate manner. The oracle problem in this context can be solved when formalizing operational requirements and properties. For carrying out the tests, simulation environments have to be used for automation.

22.3 Problem Definition

Quality assurance of ADVs or adaptive ADVs has to consider two important issues. One is due to the fact that the ADV after delivery makes decision without human operators in the loop and thus requiring assurance that no unsafe decisions are taken at any circumstances. Therefore, testing has to consider all possible situations, which lead to a huge parameter space to be handled. Inputs for testing are formal descriptions of properties that have to be fulfilled always, e.g., safety requirements, the ADV configuration, external conditions, as well as internal faults that might occur during operation and that should be handled appropriately.

In Fig. 22.3, we depict some of the influencing parameters that fall into one of the three categories, i.e.:

- ADV configuration
- State of the ADV
- External conditions influencing the behavior of the ADV

All these categories have a lot of different parameters. An electrical ADV may have configuration parameters for the engine used, the availability of an air conditioner, the driving mode, and the used battery, all of them influencing the driving distance, the driving experience, and the functionality. The internal state of the ADV comprises information about internal faults, e.g., a broken wire, or even wrongly perceived external objects, all of them influencing the ADV's behavior. External environmental conditions include the current weather or light condition and also the context information, e.g., the country where an ADV is used. Here we might have to consider, for example, different stop signs during operation.

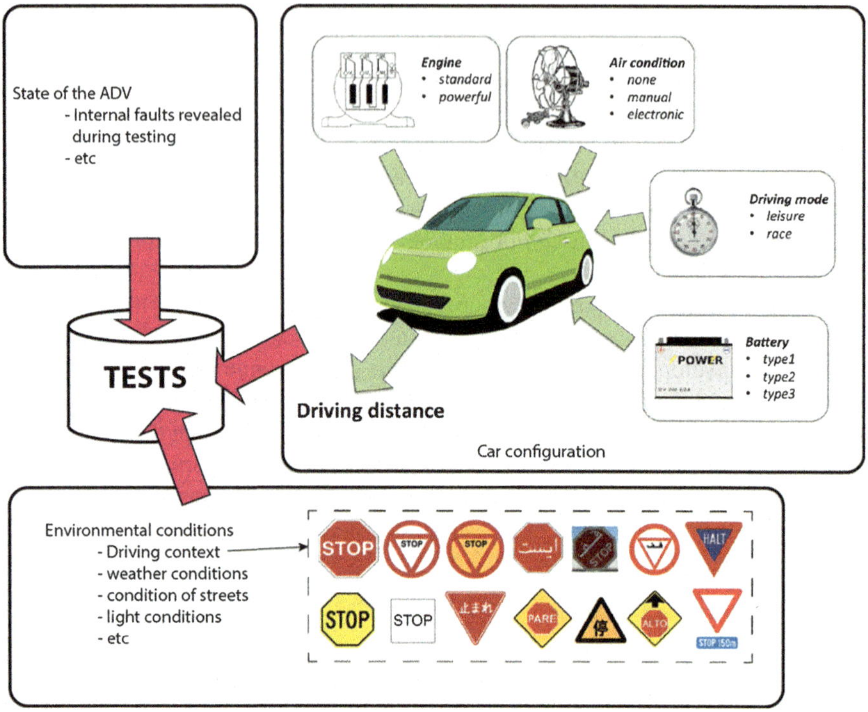

Fig. 22.3 The parameter space to be considered when testing ADVs

When considering n parameters each being able to take k values, we finally would have to consider k^n different combinations during testing. This is not feasible in practice and cannot be done within a reasonable amount of time. When assuming only two different values for all parameters and in total 100 different parameters falling in one of the three categories, we would require $2^{100} \approx 1.27 \times 10^{30}$ different tests. Assuming that each execution of one test needs one second, we need more than 10^{22} years to finalize testing. Hence, there is a strong need for a testing method requiring less tests but still assuring high quality.

Testing ADVs is somehow different from testing ordinary cars. Even in the case that there is also a high number of potential configurations and faults to be considered, we are able to test the different parts of the vehicle separately before performing an overall system test. In addition, because of knowing that there is a driver in the loop supervising the overall vehicle, testing can be more focused. This cannot be done in case of ADVs. Therefore, we have to consider both a high number of potential configurations and different situations that are influenced by the environment and its corresponding conditions. In the next part of this chapter, we introduce a solution to the testing challenge of ADVs that allows for keeping the testing overhead as low as possible. Besides complexity reduction the approach can also be fully automated.

22.4 Combinatorial Testing for Autonomous Adaptive Systems

In order to solve the testing problem for ADVs, we have to provide a solution that reduces the number of test cases and is also able to solve the oracle problem. The oracle problem in testing is the problem of classifying the output of a test as being correct or faulty. In case of MBT, the used models handle the oracle problem directly. In other test case generation approaches where the aim is to provide a set of input stimuli, solving the oracle problem is an important task for automating test execution. In the following we first describe how to solve the problem of generating test inputs. Afterward, we discuss how to solve the oracle problem, and finally, we bring all these parts together to provide a general and feasible solution.

22.4.1 Combinatorial Testing

Within the past decade researchers thought about the question of how to generate test inputs for highly configurable products or products having a large input space in a feasible way. For example, when providing software for mobile phones, someone has to take care of the different hardware devices as well as the different releases of the operating system among other important parameters like special settings of the phone network provider, the configuration of the software under test, or other software installed that might interact. Any combination of such parameters or factors might be relevant for revealing a faulty behavior. However, this cannot be done. Hence, someone thinks about the question whether combinations of two, three, or more parameters might be sufficient for fault detection.

From this deliberation researchers come up with *combinatorial testing* where only combinations of a fixed number of parameters are considered for test case generation. In Kuhn et al. [13], the authors give an introduction into combinatorial testing. For being self-contained, we briefly discuss the underlying ideas behind combinatorial testing. The basic idea is that a combination of values for t parameters is enough to cause the system to be executed in a faulty way that finally leads to an observable failure. Such a fault is called t-way interaction fault. Instead of considering all combination of parameters, we only consider all combinations for t parameters, which obviously reduce the number of generated test cases substantially. Another important finding is that the number of parameters t to be considered needs not to be too high to reveal all faults in software. Kuhn et al. [14] showed that for a larger variety of different programs, at most 6-way interactions have been enough to detect all faults. In many cases a lower number of interactions is sufficient to detect most of the faults. Hence, combinatorial testing is a good method for reducing the test suite size while keeping the failure detection rate high, which is a prerequisite for testing ADVs in practice.

In the following we formalize the concept of t-way combinatorial tests. We assume that we have a *model of the input space* comprising a set of *parameters* $P = \{P_1, \ldots, P_n\}$, where each parameter P_i has a *domain* D_i, i.e., a set of values.

A *test suite* TS is a set of tuples $(v_1, \ldots, v_n)$ assigning a value $v_i \in D_i$ to each parameter P_i. Such a test suite TS is a t-way *combinatorial test suite* if and only if the following hold: For any selection of t parameters, all combinations of values for these parameters are represented in TS. Note that there might be larger parameter sets in a t-way combinatorial test suite where all combinations are represented. Hence, a 2-way combinatorial test suite might capture 3-way interaction faults for some (but not for all) parameters.

Given a model of the input space and the parameter t, there are algorithms that compute all input tuples, i.e., the test cases, where the condition for t-way combinatorial tests hold. The ACTS combinatorial testing tool [15] developed jointly by the US National Institute Standards and Technology (NIST) and the University of Texas at Arlington can be used for this purpose. It is worth noting that ACTS also allows for specifying constraints among the input parameters. For example, when adding an air conditioner into a car, the car requires a larger battery. This information restricts the potential combinations. In a combinatorial test suite with constraints, all tuples not fulfilling constraints are removed, and some new might be added such that the condition for combinatorial test suites is always fulfilled.

The selection of the parameter t for combinatorial testing influences the number of interaction faults to be detected. In Kuhn et al. [14], the cumulative error detection rate reported ranged from almost 75 to more than 90 % for 2-way interaction faults and from 87 to 98 % when setting $t = 3$. Hence, in practice the value of t has to be chosen having the error detection rate into mind. It is worth noting that given practical requirements like a fixed budget for testing, a deadline for finalizing testing, or other restrictions, it is very unlikely to detect all faults during system development. Therefore, there is always a trade-off between those requirements and the number of tests to be carried out, which of course has consequences on the error detection rate. Whether a certain reduced error detection rate is acceptable or not relies on the application domain and within the domains on given standards and regulations.

Let us take the parameter space depicted in Fig. 22.3 and formalize the input parameter space such that we are able to extract a test suite. We take the car configuration as part of the example from [17]. There the electrical motor *emot* can take the two values *standard* or *powerful*. The air conditioner *ac* might be *none*, *manual*, or *electronic*. The driving mode *dm* can be *leisure* or *race*, and the battery *bat* has the domain {*type1*, *type2*, *type3*}. For this example, we ignore faults occurring. For the environmental conditions, we assume street conditions *strcond* to be *highway*, *city*, or *country* road and three different *stop* signs *stop1*, *stop2*, and *stop3*. The following table summarizes the parameters and their values:

Parameter	Values	Parameter	Values
emot	*standard*, *powerful*	*bat*	*type1*, *type2*, *type3*
ac	*none*, *manual*, *electronic*	*strcond*	*highway*, *city*, *country*
dm	*leisure*, *race*	*stop*	*stop1*, *stop2*, *stop3*

When using ACTS for computing a combinatorial test suite of strength 2, we obtain the following table:

emot	*ac*	*dm*	*bat*	*strcond*	*stop*
powerful	*none*	*race*	*type1*	*city*	*stop2*
standard	*none*	*leisure*	*type2*	*country*	*stop3*
powerful	*none*	*leisure*	*type3*	*highway*	*stop1*
standard	*manual*	*race*	*type1*	*country*	*stop1*
powerful	*manual*	*leisure*	*type2*	*highway*	*stop2*
standard	*manual*	*race*	*type3*	*city*	*stop3*
powerful	*electronic*	*leisure*	*type1*	*highway*	*stop3*
standard	*electronic*	*race*	*type2*	*city*	*stop1*
powerful	*electronic*	*race*	*type3*	*country*	*stop2*
standard	*manual*	*race*	*type2*	*highway*	*stop2*
powerful	*electronic*	*leisure*	*type3*	*city*	*stop3*

Instead of 324 test cases, which are all different possible combinations, the 2-way combinatorial test suite for the small example only comprises 11 test cases. A 3-way combinatorial test suite comprises 33 test cases and a 4-way one 83. The latter is still only one fourth of the total number of combinations, which saves a lot of testing effort while still being able to reveal potential faults.

For smaller input models used for obtaining a t-way test suite and a smaller number of t, the whole test suite generation time using available tools can be neglected. In Yu Lei et al. [16], the authors introduce algorithms that allow for computing 5-way tests for 20 parameters each having 4 values in less than a minute with their fastest algorithm leading to a test suite comprising 8606 elements. Note that the number of all possible combinations in this case is $4^{20} = 1.09951 \times 10^{12}$ and thus we obtain a much smaller test suite. However, depending on the test execution time, the number of test cases might be too high to meet given deadlines. In addition, for large values of t and a large domain of parameters, test case generation most probably takes too long when using current combinatorial testing algorithms. In such cases the combinatorial testing problem has to be partitioned into subproblems, to be handled independently. In each subproblem we take certain parameters and set them to a fixed value. The remaining parameters form the subproblem for which we are able to generate a combinatorial test suite. The selection of parameters can be done considering their likelihood for revealing an interaction fault. If two parameters are likely to not interact in an unwanted manner, their values can be fixed. Such information has to be obtained from domain experts.

When starting a test using the tuples specifying combinations of parameters, we are able to carry out specific parameterized tests. However, what is still missing is the question whether such test execution fails or passes. We discuss this issue in the following section.

22.4.2 Test Oracles

Carrying out a particular test case requires the execution of the SUT using the specified parameter values. The SUT execution in our context may require carrying out a simulation environment taking care of the specified parameters. The question now is how to distinguish a faulty behavior from a correct one in the simulation? One answer would be to have a user observing the simulation results and classifying a particular simulation run as passing or failing. However, this is not a practical approach because of the larger number of tests to be carried out. Hence, we need to automatize the test oracle.

One way of automatizing the test oracle is metamorphic testing (see [18] and [19]). The idea behind metamorphic testing is to utilize symmetries in functions or systems to be tested. For example, it is well known that $\sin(x)$ is equivalent to $\sin(x + 2\pi)$. Hence, we use the constraint $\sin(x) = \sin(x + 2\pi)$ as an oracle. Whenever this constraint is not fulfilled, we know that the test is a failing test. In general, we use all constraints that represent symmetries to form the test oracle. In the context of ADVs, we also might be able to define such symmetries and to use them as test oracles. However, metamorphic testing may require carrying out the test twice. For example, when checking whether $\sin(x)$ is equivalent to $\sin(x + 2\pi)$, we have to call the *sin* function twice. This would increase testing effort.

Another possibility, which is more appropriate in the context of ADVs, originates from the basic ideas behind runtime verification [11] where formal properties are checked when executing the SUT. This idea can be easily translated into our domain when formalizing safety properties and other requirements. For example, in none of the cases, we want an ADV crashing into an obstacle. Such a requirement can be represented formalized and also later on represented programmatically in the simulation environment. There the functionality of the simulation environment can be used, for example, to detect overlaps between the ADV and an obstacle and to determine that the crashing property is violated.

Hence, we assume that we are able to implement an oracle function that checks for safety property and requirement violations. Such an oracle function has to be closely integrated into the simulation environment carrying out the generated combinatorial test suite. It is worth noting that such an oracle is not always able to classify a test as passing or failing. In some cases the outcome might be inconclusive. For example, stating that a property must always hold cannot be proven within a limited simulation time. Only in cases where the property is violated we exactly know that the test is a failing test. Otherwise, there might be an interaction between the ADV and its environment not experienced so far that would be able to violate the property.

22.4.3 The Automated Testing Methodology

When combining the combinatorial testing approach with the automated testing oracle comprising information about properties, requirements, and metamorphic relations, we are able to come up with testing methodology for ADVs that can be fully automated. For this purpose we have to have an execution environment, e.g., a simulator, where each test can be carried out and where the testing oracle can be integrated. In the following we describe the proposed automated testing methodology in more detail. In Fig. 22.4, we give an overview of the testing methodology comprising the following three parts:

1. *Test case generation*: For the purpose of generating test cases, we assume that we know all parameters and their domains for the ADV configuration, e.g., the kind of battery or engine, the environmental conditions, as well as faults that might occur during operation. In addition we might also know some constraints between parameters that might limit potential combinations of parameter values. From the parameters, their domains, and the optional constraints, we use a tool (e.g., ACTS [15]) to generate a t-way combinatorial test suite where $t \geq 2$. The resulting test suite comprises tuples of values for each parameter that work as input for the test execution.

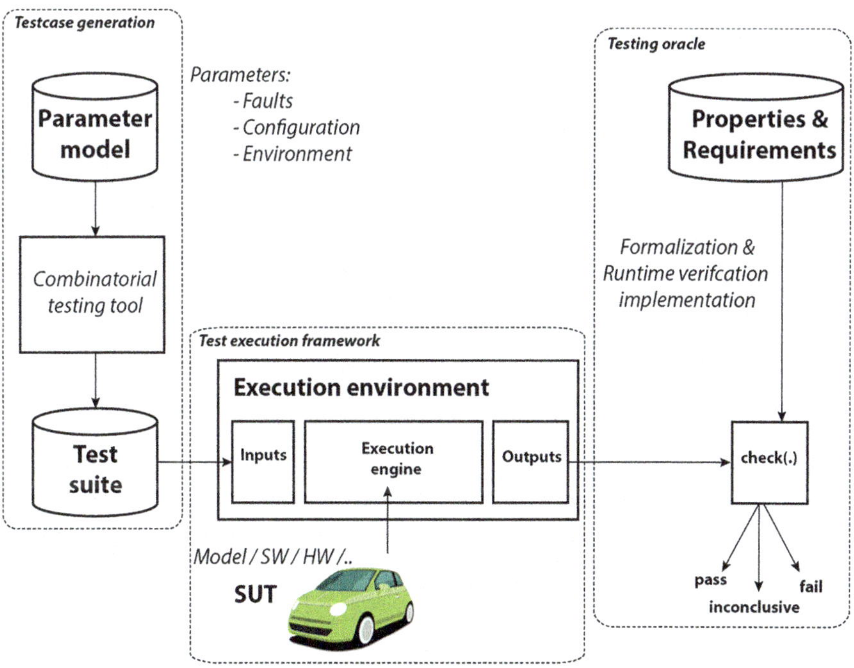

Fig. 22.4 Overview of the automated testing methodology for ADVs

2. *Testing oracle*: For generating the testing oracle, it is essential to have knowledge about safety property and other functional and nonfunctional requirements of ADVs. This knowledge has to be formalized. The formalized knowledge is used similar to runtime verification [11] to check the current state of the ADV during test execution. Here the idea is to implement a function *check*(.) that takes the values observed from execution and checks whether a failure occurs. For implementing *check*(.), we rely on previous work that has already been done in the context of runtime verification. After implementing *check*(.), we obtain a testing oracle that automates the task of judging the outcome of a test applied to the SUT. Note that *check*(.) returns not only *pass* or *fail* but also *inconclusive* in cases where properties cannot be finally verified because of restrictions due to the finite execution time.
3. *Test execution framework*: The third part of the framework for automating the system test for ADVs is the execution framework. Here we assume that the input from the test case generation component, i.e., a tuple of parameter values, configures a simulation run of the SUT, which might be itself represented as a model, software, or hardware. Note that there is a difference between a model representing the SUT and the input model used to generate the combinatorial test suite. A model representing the SUT has to capture the system's behavior, whereas the input model used for generating the test suite only comprises parameters, their domains, and an optional set of constraints. The execution framework executes this test and calls the oracle function *check*(.) each time new observations from the SUT are available. In case *check*(.) returns *fail*, the test is terminated and classified as failing test. In cases where the test is terminated due to given limitations on execution time, the *check*(.) function is called for finally classifying the test as *passing* or *inconclusive*.

The advantage of the proposed testing methodology is that the test generation and execution part can be completely automated. Moreover, combinatorial testing guarantees that important combinations of parameters with respect to their potential of detecting failures are considered during testing. This limits the number of combinations of test parameters substantially and makes the overall approach feasible even for a larger set of parameters and their corresponding domains. The disadvantage is that the input parameter space as well as the safety properties and other requirements has to be identified and formalized, which require effort. However, within the development processes usually used, the identification of such requirements is mandatory. Hence, the only additional effort is to formalize these requirements in order to allow utilizing runtime verification techniques directly. It is worth noting that the test execution environment has also to be extended in order to be able to cope with the testing oracle function. The required additional effort, however, should be insignificant.

22.5 Conclusion

In this chapter, we discussed the challenge of testing ADVs and adaptive ADVs. We identified the huge parameter space to be considered as one of the main reasons behind the complexity of testing ADVs. We also argued that ordinary vehicles even with attached assistive technology could be more easily tested due to the fact that we are able to define the scope of subsystems to be tested separately. Moreover, because of humans in the loop during execution, we have behavioral supervision available that further helps to restrict testing in this case. For testing ADVs we have to identify a method that (1) allows for reducing the testing parameter space while (2) keeping the potential failure detection rate high. In this chapter, we argue that combinatorial testing fulfills both criteria and make appropriate citations that support the used argumentation chain.

In addition to the generation of test cases, which are in our context the parameter values used as input to a simulation environment where tests can be carried out, we discussed the automation of the test oracle. A test oracle allows classifying tests as passing, failing, or inconclusive based on the given observations of variable values during test execution. In order to automate the test oracle, we suggested to use runtime verification that is based on safety properties and other requirements an ADV has to fulfill. Such information is always available in the context of vehicle development within the automotive industry. The properties and requirements have to be formalized in order to automate the test oracle.

The proposed testing methodology for ADVs combines the test case generation approach based on combinatorial testing with the automated test oracle based on runtime verification. For this purpose, we assume the availability of a test execution framework, i.e., a kind of simulation environment, which takes the tests as input and allows for integrating the test oracle. The proposed testing methodology is feasible because it requires only important combination of parameter values that capture faults that can only be revealed in cases of interactions between a fixed number of input parameter values occurring at the same time.

Considering only one testing methodologies for quality assurance is in general not a good idea, mainly due to differences in the objectives of the methodologies. Model-based testing aims at generating test suites in a rigorous and complete manner based on a model of the SUT. Combinatorial testing focuses on interaction faults that can only be revealed using the right combination of parameter values. Random testing tries to find interactions with the SUT that have not been foreseen and also interactions that are outside of the specifications. The purpose of manual testing is mainly on capturing possible interactions of humans with the SUT. The combined used of testing methodologies during development as well as other measures like coding guidelines is essential for quality assurance. An early use of the proposed testing approach within the ADV development cycles using models and software in simulation does not provide any guarantee to detect all faults. However, it would allow for detecting interaction faults earlier during development.

Acknowledgment The research work has been carried out as part of the 3CCar project co-funded by the Electronic Component Systems for European Leadership Joint Undertaking (ECSEL JU) grant agreement number 662192-3Ccar-ECSEL-2014-1 and the FFG grant agreement number 848715.

References

1. *ISO 26262:2011: Road vehicles functional safety* (International Organization for Standardization, Geneva, 2011)
2. N.G. Leveson, *Engineering a Safer World* (MIT Press, Cambridge, MA, 2011)
3. H. Altinger, F. Wotawa, M. Schurius, *Testing Methods Used in the Automotive Industry: Results from a Survey*. Proceedings JAMAICA'14, San Jose, CA, 21 July 2014
4. P. Ammann, J. Offutt, *Introduction to Software Testing* (Cambridge University Press, Cambridge, MA, 2008)
5. J. Glenford, *Myers, The Art of Software Testing* (Wiley, New York, 1979)
6. I. Schieferdecker, Model-based testing. IEEE Software **29**(1), 14–18 (2012)
7. M. Utting, B. Legeard, *Practical Model-Based Testing – A Tools Approach* (Morgan Kaufmann, Burlington, MA, 2007)
8. J. Cámara, R. de Lemos, C. Ghezzi, A. Lopes (eds.), *Assurances for Self-Adaptive Systems – Principles, Models, and Techniques*. LNCS 7740 (Springer, Berlin, 2013)
9. F. Nafz, J. Steghöfer, H. Seebach, W. Reif, Formal Modeling and Verification of Self-Systems Based on Observer/Controller-Architectures, in LNCS 7740 [8] (Springer, Berlin, 2013), pp. 80–111.
10. G. Steinbauer, F. Wotawa, Model-Based Reasoning for Self-Adaptive Systems – Theory and Practice, in LNCS 7740 [8] (Springer, Berlin, 2013), pp. 187–213
11. M. Leucker, C. Schallhart, A brief account of runtime verification. J. Logic Algebr. Program. **78**(5), 293–303 (2009)
12. C. Artho, H. Barringer, A. Goldberg, K. Havelund, S. Khurshid, M. Lowry, C. Pasareanu, G. Rosu, K. Sen, W. Visser, R. Washington, Combining test case generation and runtime verification. Theor. Comput. Sci. **336**(2–3), 209–234 (2005)
13. D. Kuhn, R. Kacker, Y. Lei, *Introduction to Combinatorial Testing*. Chapman & Hall/CRC Innovations in Software Engineering and Software Development Series (Taylor & Francis, London, 2013)
14. D.R. Kuhn, R.N. Kacker, Y. Lei, J. Hunter, Combinatorial software testing. Computer **August**, 94–96 (2009)
15. L. Yu, Y. Lei, R. Kacker, D. Kuhn, Acts: A combinatorial test generation tool, in *IEEE Sixth International Conference on Software Testing, Verification and Validation (ICST)*, 2013, pp. 370–375
16. L. Yu, K. Raghu, D. Richard Kuhn, V. Okun, J. Lawrence, IPOG/IPOG-D: Efficient test generation for multi-way combinatorial testing. Software Test. Verif. Reliab. **18**, 125–148 (2008)
17. F. Wotawa, I. Pill, Testing Configuration Knowledge-Bases, in *16th International Configuration Workshop*, 2014. http://ceur-ws.org/Vol-1220/06_confws2014_submission_13.pdf, pp. 39–46
18. T.Y. Chen, S.C. Cheung, S.M. Yiu, *Metamorphic Testing: A New Approach for Generating Next Test Cases*, Technical Report HKUST-CS98-01 (Department of Computer Science, Hong Kong University of Science and Technology, Hong Kong, 1998)
19. T.Y. Chen, J. Feng, T.H. Tse, Metamorphic Testing of Programs on Partial Differential Equations: A Case Study, in *Proceedings of the 26th Annual International Computer Software and Applications Conference (COMPSAC '02)*, Los Alamitos, CA (IEEE Computer Society, 2002), pp. 327–333

Part VII
A Sampling of Automated Driving Research Projects and Initiatives

Chapter 23
AdaptIVe: Automated Driving Applications and Technologies for Intelligent Vehicles

Aria Etemad

23.1 Project Overview

With 28 partners from across Europe and a budget of 25 million euros, AdaptIVe advances the technical performance of automated systems by developing and demonstrating new integrated applications for cars and trucks. AdaptIVe's results take automation to higher levels and support the goals of making driving safer and more comfortable and of reducing congestion and fuel consumption.

The project runs from January 2014 to June 2017 and is co-funded by the European Union with 14.3 million euros under Grant Agreement #610428. AdaptIVe tests and develops applications for typical traffic scenarios on motorways, in the urban environment and for close-distance manoeuvres, covering all levels of traffic complexity and speeds up to 130 km/h.

The functions will offer assistance and partially, conditionally and highly automated driving. A minimum risk manoeuvre will be implemented for all scenarios whereby the vehicle stops automatically in case of an emergency or if a malfunction occurs.

A. Etemad (✉)
Volkswagen AG, Wolfsburg, Germany
e-mail: aria.etemad@volkswagen.de; http://www.AdaptIVe-ip.eu

D. Watzenig, M. Horn (eds.), *Automated Driving*,
DOI 10.1007/978-3-319-31895-0_23

23.2.3 Close-Distance Scenarios

Improving the everyday driving experience starts within the lowest speed and distance range. A particular challenge is that close-distance manoeuvring requires sensors and algorithms that haven't been fully developed yet. Sensor sensitivity will be based on the traffic situation, allowing the vehicle to reliably detect other objects and free space over close distances and to navigate in this area by selectively giving priority to one direction over the other.

A robust perception platform is envisaged, taking into account the latest advances in embedded systems and communication and information technologies. This platform will support decision-making processes in complex situations. AdaptIVe also advances applications for automated parking at private homes and in outdoor environments, as well as in multilevel parking garages where a driver is always present. The Stop & Go function supports driving in close-distance scenarios.

Moving towards fully automated parking requires a learning car, whereby the car can train itself by becoming familiar with typical environments. The car shall then be able to drive and to manoeuvre within similar environments with a learned or provided map.

23.2.4 Urban Scenarios

Urban scenarios present special challenges due to the environment's complexity and dynamic behaviour. Traffic is dense, several types of road users or static obstacles are present, and the driving task includes negotiating traffic at roundabouts, intersections and merging manoeuvres.

AdaptIVe is developing embedded solutions to address the most demanding driving scenarios in a city: the supervised city control and city chauffeur functions. A key point for this development is the integration of existing and new functions into one system. Examples include automated braking, feedback on the gas pedal and steering wheel, automated cruise control and full supervised automated control. The level of support given to the driver ranges from correction and stabilisation of driver manoeuvres (in assisted mode) to automatic guidance (in automated mode). Communication with the infrastructure and other vehicles is being realised to anticipate the intentions of other road users and reduce the potential for conflicts.

23.2.5 Highway Scenarios

Highway scenarios demand a careful consideration of the different automation levels and the added value provided by cooperative approaches. Using the most up-to-date research, the project is pushing the limits of automated driving towards higher degrees of automation and incorporating cooperative driving functionalities.

The automated AdaptIVe vehicle will enter and exit highways, perform lane changes or filter-in manoeuvres and provide support in dangerous areas such as the end of a traffic jam. Other functions include the cooperative response to emergency vehicles on duty, also based on vehicle-to-vehicle (V2V) communication, and a speed and time-gap adaptation at motorway entrance ramps based on vehicle sensors. Additionally, predictive automated driving to reduce fuel consumption and CO_2 emissions will be implemented as well as basic driving functions such as following lane and vehicle, performing overtaking manoeuvres and handling stop-and-go traffic.

New cooperative technologies must be developed to enable a variety of automated cooperative driving functionalities. Drafts for a new radio transmission protocol for bidirectional V2V communication are being specified, implemented and tested, procedures that will enable negotiations between vehicles.

23.2.6 Evaluation

Existing evaluation methods for advanced driver assistance systems (ADAS) do not cover the requirements for the evaluation of automated driving functions. Therefore, new comprehensive approaches and test methods are required. AdaptIVe defines specific evaluation methodologies for automated driving functions in a comprehensive framework. The test and evaluation framework considers the technical, user-related and in-traffic behaviour evaluation as well as an impact analysis focused on safety and traffic efficiency. The framework thereby includes a specification of methodologies, test procedures, key indicators and experimental design with the applicable testing tools. The impact analysis is being conceived with a pan-European perspective. Ultimately, the framework and new methodologies will be applied to a set of selected representative functions in order to verify and validate the developed evaluation approaches.

23.3 Looking Ahead

AdaptIVe builds up eight demonstrator vehicles—seven passenger cars and one truck—to test and evaluate the AdaptIVe applications and functions. The project will showcase these systems during a final demonstration in 2017. At this point of time, automated driving will be far from being comprehensively researched. AdaptIVe will have laid the foundation for further research of the topic of automated driving.

Chapter 24
When Autonomous Vehicles Are Introduced on a Larger Scale in the Road Transport System: The Drive Me Project

Trent Victor, Marcus Rothoff, Erik Coelingh, Anders Ödblom, and Klaas Burgdorf

24.1 Introduction

During the past decades, road transport in urban areas has been dominated by cars, and this has resulted in increasing levels of traffic congestion and loss of time. As we approach 2020, there is substantial renewed interest in revitalizing urban road transport as one way of combating the challenges in metropolitan areas.

Autonomous driving opens up for possibilities for improved safety, improved fuel economy, reduced problems with congestions in urban areas, more efficient use of land for city planning, reduced emissions, and driving possibilities for the physically impaired. This new technology also opens up new possibilities to optimize the infrastructure in cooperation with the vehicle's ability to be more accurate controlled. It is generally accepted that, in about 90–99 % of all incidents and crashes, human behavior is partially or fully responsible [1, 2]. In the push toward reaching the goal of zero serious injuries and fatalities, there is a clear potential in automating the driving task, as it could take away the root cause of almost all accidents.

The debate is currently running high on the legality and desirability of having the car fully or partially assuming the control or if the driver should always be in control. For many of the involved parties, however, in assessing what is needed for reaching very low numbers of casualties, there is a shared view that more advanced assistance systems leading toward autonomous driving are desired or needed in reaching this goal. Analysis of a larger-scale deployment of autonomous vehicles is needed.

T. Victor (✉) • M. Rothoff • E. Coelingh • A. Ödblom • K. Burgdorf
Volvo Car Group, Gothenburg, Sweden
e-mail: trent.victor@volvocars.com

D. Watzenig, M. Horn (eds.), *Automated Driving*,
DOI 10.1007/978-3-319-31895-0_24

24.2 Problem Definition

The Drive Me project has a high-profile ambition to define and evaluate how autonomous vehicles will have a major importance for quality of life and achievement of a sustainable urban environment. To handle future challenges, society has to continue its positive development, and transport solutions have to be safe and environmentally sustainable as well as meet mobility requirements for citizens. Drive Me takes important steps toward vision zero in traffic safety (the vision that any loss of life is unacceptable) and the more ambitious future crash-free road transport system. In addition, a more efficient use of infrastructure and the space required for traffic are core areas in the project. A more sustainable mobility solution, based on increased automation of vehicles and optimization in the infrastructure, is measured on six main qualities: punctuality, capacity, robustness, usability, and traffic safety.

The Drive Me project is a research platform comprised of a number of contributing research projects and partner organizations. Volvo Car Group along with Swedish Transport Administration, Swedish Transportation Agency, the City of Gothenburg, Lindholmen Science Park, Chalmers University, and Autoliv are part of the Drive Me research platform. In addition, numerous additional projects and collaborating organizations will contribute to Drive Me.

Drive Me focuses on studying potential benefits when autonomous vehicles are introduced on larger scale in the road transportation system. The Drive Me platform is unique in international terms by focusing on the integration of autonomous vehicles with the infrastructure as well as the citizen. A holistic system approach is thereby possible, enabling optimization of transport at a whole new level. This is a possible paradigm shift within mobility and for creation of livable urban areas. This research initiative aims to combine knowledge from society, academy, and industry to create the sustainable mobility system of tomorrow. It aims to answer the following research questions:

1. How can *traffic safety* be improved?
2. How can *traffic flow* be improved? What is the impact on punctuality, capacity, robustness, and usability in the road transportation system?
3. How can *energy efficiency* be improved?

The ambition to put a fleet of 100 autonomously driving cars in the hands of real customers as advanced measuring probes on real road infrastructure is the largest project defined in the world today. These probes will be used to study the effect on safety, traffic flow, and energy efficiency. It is expected that rebuilt vehicles working as test probes with measuring equipment and test functionality for automated driving will provide sufficient information/data to evaluate the possible benefits and effects on the above research questions.

During the first stages of the project, estimated benefit potentials will be defined through computer simulations, desktop studies, and test-track studies. Later, test probes will be put into a real traffic environment (a section of about 55 km of motorway in Gothenburg, Sweden) to collect data for analyses, modeling, and quantification.

24.3 Measuring Probes: The Autonomous Vehicles

The objective with the autonomous vehicles, or test probes, is to allow ordinary Volvo customers to operate an autonomous vehicle on public roads such that the overall effects of these vehicles on the road transportation system can be studied. As opposed to modern production vehicles, these customers do not need to continuously supervise the vehicle operation during certain driving situations, and therefore they will be allowed to spend time on other activities. This of course puts strict requirements on the probe design, far beyond the requirements on ordinary production vehicles. All engineering details will have to be addressed in order to allow customers to use these probes in a safe way.

All test probes will be equipped with data logging tools such that the vehicle and driver behavior can be monitored continuously and its impact on the overall traffic can be analyzed.

24.4 Safety

The research on safety addresses the general question of how can traffic safety be improved? It focuses on (a) how the test probes will manage safety conflicts and pre-crash scenarios and (b) quantification of the impact of the implemented solutions on traffic safety. The safety-related functionality of the test probes will be developed in close interaction with an iterative evaluation of the functionality in virtual, test-track, and on-road environments. Safety-related impacts will not only result from the real-time avoidance behavior of the test probes in response to traffic conflicts (e.g., steering/braking to avoid a parked vehicle) but also result from compliance with rules and regulations (e.g., speed reduction in traffic flow), from functional safety (e.g., dependability and malfunction handling), and handling unforeseen consequences. A safety impact methodology will be used whereby safety benefits are expressed in terms of degree of crash avoidance and reduction of injury (e.g., [3]).

The following research questions will be addressed:

- How should the safety impact of the Drive Me test probes be quantified?
- Which safety conflict situations (load cases) should the test probes be tested on?
- How do the self-driving vehicle and the driver react in safety conflict situations?

- To what proportion and degree are crash types avoided or mitigated? If possible to determine, the main causes for the crash reduction and mitigation will be determined (e.g., rule compliance, types of human behavior).
- How can automation contribute to a crash-free road transport system?
- What are the main safety challenges with self-driving cars?

24.5 Traffic Flow

The traffic flow research will study how autonomous vehicles can contribute to a more efficient road transportation system, analyzing how autonomous vehicles can:

- Increase the capacity of the traffic system, i.e., allow more vehicles to pass through a given section of road infrastructure.
- Increase the robustness of traffic system, e.g., by minimizing incidents such from low-speed collisions or critical lane changes.
- Minimize travel time, e.g., by creating a smoother flow.
- Accurately predict time of arrival, thereby allowing travelers to optimize their trip.

This research will be conducted by taking two different perspectives. The first perspective is to analyze and modify the behavior of self-driving vehicles such that traffic efficiency is optimized. It is known that the behavior of the vehicle in terms of, e.g., acceleration profiles, distance keeping, and lane changing, affects the overall traffic flow. Vehicles driven by humans vary significantly such that the overall traffic system is never operated in an optimal way. When a significantly large set of vehicles is behaving in an optimal way, it will in the end affect the overall traffic system. The research questions are:

- How should the optimal behavior for an autonomous vehicle be defined?
- How does the amount of autonomous vehicles affect the overall traffic system?

The second research perspective is to modify the road infrastructure such that traffic efficiency is optimized. When a future road traffic system knows a significant amount of autonomous vehicles, one does not necessarily have to design the road infrastructure in the same way as today. We assume that, e.g., lanes can be made smaller as the road does not have to cater for all lateral oscillations that result from different behavior in the human driver population. This may mean that more lanes can be constructed or the remaining space can be used for nonmotorized traffic such as bicycle, hereby increasing efficiency. In a similar way, efficiencies may be obtained in the design of lightweight and low-cost overpasses and underpasses. The basic research question is:

- How should an efficient road infrastructure be designed, given a large amount of autonomous vehicles with optimal behavior for traffic efficiency?

24.6 Energy Efficiency

Sound data on the realistically achievable effect of autonomous vehicles on energy consumption and other environmental aspects is lacking [4]. While single potentials within autonomous vehicles are widely known and may give improvements up to 30 % in energy efficiency [5], there are currently no established methods to estimate the expected impact in real traffic situations. Aspects adding to the complexity of the subject are expected dependencies on surrounding traffic (density and patterns) as well as the necessity that the driver accepts changes in the vehicle behavior.

The main research question "How can energy efficiency be improved by autonomous vehicles?" has been detailed into the following research questions:

- Which energy efficiency potential can be realized with autonomous vehicles for individual mobility?
- Which functionality is required in the traffic control system to support efficient traffic control, and how can this effect be quantified by measurement?
- How large is the potential benefit for different degrees of automation and different shares of such vehicles in real traffic environment?
- How are environmental aspects as traffic noise and emissions affected by the introduction of autonomous vehicles?

A scientific method for the assessment of environmental aspects of automated driving will be developed and applied. This method is intended to help provide accurate, robust, cost-efficient, and practical information available to authorities, consumer organizations, and the automotive industry.

24.7 Conclusion

The Drive Me project has a high-profile ambition to define and evaluate how autonomous vehicles will have a major importance for quality of life and achievement of a sustainable urban environment. It focuses on studying potential benefits for safety, traffic flow, and energy efficiency when autonomous vehicles are introduced on larger scale in the road transportation system.

References

1. D.L. Hendricks, J.C. Fell, M. Freedman, *The Relative Frequency of Unsafe Driving Acts in Serious Traffic Crashes: Summary Technical Report* (National Highway Traffic Safety Administration, Washington, DC, 2001)
2. NHTSA, National Motor Vehicle Crash Causation Survey. Report to Congress. DOT HS 811 059 (2008)

3. M. Lindman et al., Benefit Estimation Model for Pedestrian Auto Brake Functionality, ESAR 2010 (2010)
4. International Energy Agency, *CO_2 Emissions from Fuel Combustion 2010* (OECD/IEA, Paris, 2010)
5. F. Faber, E. Jonkers, M. van Noort, M. Benmimoun, A. Pütz, B. Metz, G. Saint Pierre, D. Gustafson, L. Malta, Final Results: Impacts on Traffic Efficiency and Environment, EuroFOT 2012 (2012)

Chapter 25
Functional Safety and Evolvable Architectures for Autonomy

Rolf Johansson, Jonas Nilsson, Carl Bergenhem, Sagar Behere, Jörgen Tryggvesson, Stig Ursing, Andreas Söderberg, Martin Törngren, and Fredrik Warg

25.1 Introduction

Since October 2013, there is a national-funded research project running in Sweden called FUSE (FUnctional Safety and Evolvable architectures for autonomy). The motivating reason for the project is that tomorrow's vehicles should be capable to drive autonomously and that there is a number of questions to solve before this can happen in a safe and efficient manner. To a certain extent, this project contributes to the Drive Me project described separately in this book. However, FUSE has a longer time horizon and aims also to find solutions to vehicles with higher degrees of automation than Drive Me.

The FUSE project in particular focuses on system architectures and functional safety for autonomy. Current automotive systems and functional safety standards are evolving but have so far not considered autonomy. This implies that the limitations of current systems and the ISO 26262 standard [1] are currently unknown and

R. Johansson (✉) • A. Söderberg • F. Warg
SP Technical Research Institute of Sweden, Borås, Sweden
e-mail: rolf.johansson@sp.se

J. Nilsson
Volvo Car Corporation, Gothenburg, Sweden

C. Bergenhem
Qamcom Research and Technology, Gothenburg, Sweden

S. Behere • M. Törngren
KTH Royal Institute of Technology, Stockholm, Germany

J. Tryggvesson
Comentor, Gothenburg, Sweden

S. Ursing
Semcon, Gothenburg, Sweden

D. Watzenig, M. Horn (eds.), *Automated Driving*,
DOI 10.1007/978-3-319-31895-0_25

that investigations are needed regarding functional safety considerations and the scalability and cost-efficiency of architectures.

Functional safety refers to the ability to deliver services that can justifiably be trusted. In other words, failures that are more frequent and more severe than acceptable have to be avoided. While autonomy if realized correctly will improve safety, it also introduces autonomous control of the vehicle motion and thus new failure modes. The increasing complexity of the underlying embedded control systems requires new methods and architectures to be able to achieve cost-efficient and functionally safe autonomy.

Current technologies such as adaptive cruise control (ACC) always require the driver to supervise its operation. So in case something unexpected occurs, e.g., radar blockage or an animal crossing the road, then the driver has to take over control. Also in case an unlikely technical fault such as loss of braking pressure or steering assistance occurs, the driver is expected to take control. When a vehicle is driven autonomously, the driver cannot be expected to do this anymore, so alternative solutions have to be implemented and verified.

Figure 25.1 depicts three dimensions of autonomy and illustrates a challenge that is investigated in the FUSE project. The dotted line in the figure separates the left side, where there is a common understanding about functional safety, from the right side, where this understanding does not exist. In particular, there are three aspects needed to be addressed:

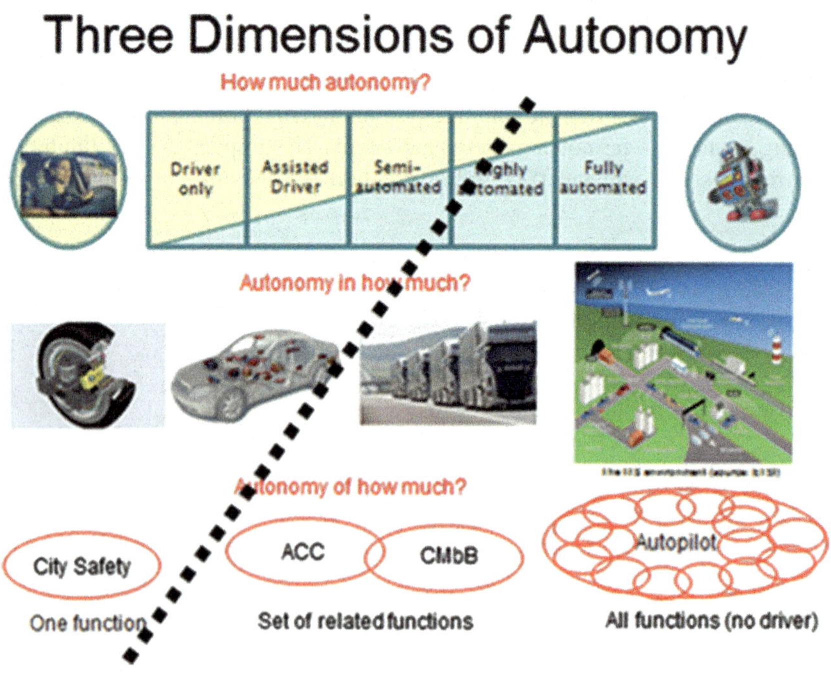

Fig. 25.1 Autonomy dimensions

- How to define functional safety for autonomy: Not addressed by ISO 26262.
- How to achieve functional safety: There is a demand for architectural patterns and division of responsibility among parties.
- How to prove functional safety: There is a demand for new compositional safety arguing.

This project is confined to autonomy within one vehicle and the environment in which it is operated. Cooperative driving functions such as platooning are outside scope. Still the three aspects are all relevant and need to be addressed in the project.

The development of dependable, safe, and autonomous vehicles requires the integration of research results from multiple domains, including embedded systems design, component- and model-based systems engineering, robotics, and artificial intelligence (AI). Many technologies that exist in advanced modern vehicles today borrow their concepts from the robotics and AI domains. However, a similar migration in systems architecture has not happened, despite the development of a wide variety of architectures within both the robotics and AI domains. There is a need to extract relevant principles from these domains and apply them to the automotive domain, keeping in mind the specific constraints arising from standards, regulations, and legislative mechanisms. The patterns and principles for autonomous automotive architectures can be consolidated in the form of function specific as well as overall vehicle reference architectures. Such reference architectures are missing in current states of the art.

In the following sections, some of the challenges addressed until mid 2015 in the FUSE project are summarized.

25.2 Why It Is Harder to Show Functional Safety for Autonomous Vehicles

Generally speaking, there are two kinds of consequences when changing from a vehicle where the driver is responsible to where the vehicle is supposed to take all necessary actions autonomously. Firstly, some things may become much more difficult and complex, and, secondly, they may become significantly different. When it comes to functional safety and road vehicles, we have identified both kinds of problems. The below problems have been presented and elaborated in [2–5].

25.2.1 Item Definition

The starting point of the ISO 26262 reference life cycle is the item definition, i.e., the confinement of the functional scope on which to apply functional safety. Traditionally, it is enough to consider one item at the time, and it is of no interest to look at the entire vehicle in an aggregated way. The underlying assumption is that whatever functionality is needed for the entire vehicle to behave safely is either performed by one of the items or by the driver. An implicit responsibility of the driver then becomes to complement the functionality of all items. As long as the items one by one are defined and assessed as safe, it is reasonable to leave to the driver to use a strategy for the driving where all unforeseen situations can be handled in a sufficiently safe way. The role for the ISO 26262 is then confined to the assessment of each individual item.

When reaching so high in the degree of automation that we no longer require the driver to be (quickly) responsive to the situations, this will imply that the responsibility to cover everything falling between the explicit items needs to be transferred from the manual driver to an autopilot. This means that all of a sudden, it has become an issue to make sure that the set of items is complete. This is because the autopilot also has to be covered by the set of items.

The item definition has both become much more complex and completely different in nature. The complexity increase is due to the much more complex task to solve. What makes the item definition completely different is the partly implicit definition. By implicit we mean that regardless of how we state the explicit item definition, it has to take care of a number of rare situations. We can no longer take the item definition as given but need to assure whether there is a need for more functionality in its scope, in order to make the behavior of the vehicle safe.

In the following section, we address the problem how to perform hazard analysis and risk assessment given that we cannot assume a complete and explicit item definition.

25.2.2 The Role of the Driver

When determining the ASIL attribute of a safety goal, we use as an input the attributes of the covered hazardous events (HEs): severity, exposure, and controllability. It is more or less obvious that we will see a change of the controllability factor when we change the role of the driver dramatically. The basic question is who is in control of the vehicle? If we only consider fully autonomous vehicles when there is no room for any person to interfere with the driving, we can simply conclude that all controllability factors should be C3, i.e., no possibility for a driver to compensate for failures. The problem is how to look at the situations when we on one hand are so high in automation that the driver is not requested to be (quickly) responsive and on the other hand we allow the driver to compensate for errors in the automation.

In traditional evaluation of the C-factor, the underlying assumption is that the activity of the driver can (more or less) compensate for problems caused by the electronic/electrical (E/E) implemented functionality. In higher degrees of automation, it might be hard for the driver to understand what a safe an efficient way to control the vehicle is. Even if the vehicle itself is doing perfectly well, the driver may think it is faulty and needs a compensating action. The consequence of the intervention of the manual driver might be that an accident is caused instead of avoided. Our conclusion is that once we introduce higher degrees of automation, it will no longer be valid to use anything but C3 and we should consequently not allow the driver to "compensate" for the automatic vehicle actions, unless judged as safe by the autopilot.

Another issue with the role of the driver is to always ensure a consistent view between the manual driver and the vehicle, regarding who is currently responsible to react (on unforeseen events). This problem is also known as mode confusion. There are two main possible categories of confusions: either both of them think they are responsible or none. If both the manual driver and the vehicle autopilot think they are the one responsible for taking full responsibility of the driving including handle unforeseen situations, we just have to select whom to give the last word. As we argue above, it is reasonable for lower degrees of automation to give the manual driver the possibility to override the vehicle-induced actions. Then the C-factor analysis is still relevant. For the higher levels of automation, we let the vehicle get the full control regardless what the driver tries to do.

The other case is trickier; when neither the manual driver nor the vehicle regards themselves as the finally responsible part. This kind of "underride" is foreseen to have a similar solution. The HMI should be designed in such a way that the manual driver stays responsible for the vehicle until there is a very explicit handover to the automatic system. Then the vehicle stays responsible until again there is a very explicit acknowledgment from the driver that she has accepted to get the responsibility back. The implication of this becomes that once a vehicle has got the responsibility for the driving, it needs to be capable to safely handle any situation without assistance of the driver for any decision or maneuver.

25.3 How to Perform Hazard Analysis and Risk Assessment

As concluded in the previous section, it will become more or less impossible to use a set of well-defined explicit items as the starting point for the ISO26262 activities when it comes to highly automated vehicles. From the FUSE project, we are proposing a slight modification of the early stages of the ISO 26262 reference life cycle. Instead of a firm item definition, we introduce a more loosely defined preliminary feature description (PFD) to act as input to the first step in an iterative process. This is the starting point of an iterative process as depicted in Fig. 25.2 below. In the following, we briefly introduce the steps.

25.3.1 Preliminary Feature Description

The difference between the preliminary feature description, PFD, and an item definition is that it does not need to contain the details required by an item definition

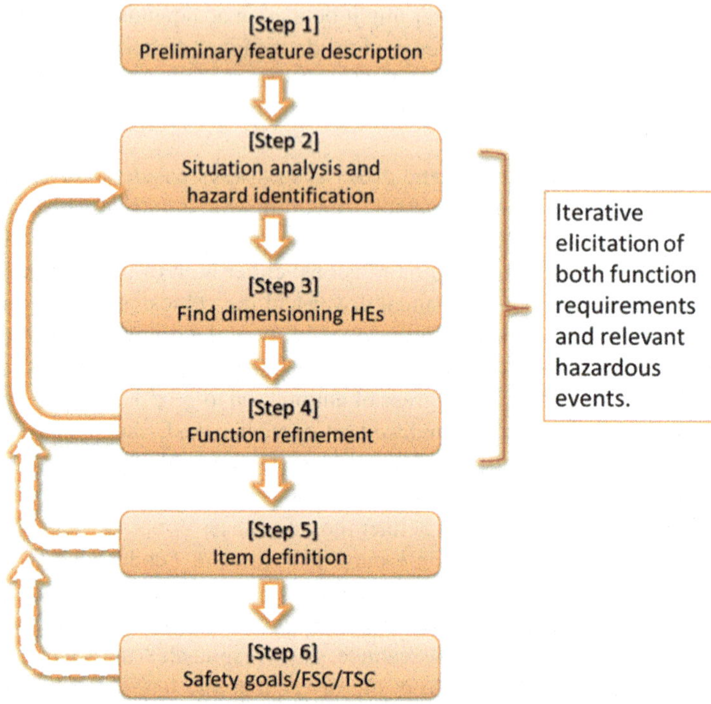

Fig. 25.2 Iterative process to conclude concept phase

according to ISO 26262. Rather, the aim is to describe the expected (customer) benefit of the proposed feature. From this starting point, the feature is iteratively refined until a point where it is possible to enunciate an item definition.

The key point of the PFD is to give a first workable description of what the proposed feature is supposed to do. The material could, for instance, be based on input from market research, previous projects, or whatever else might be available and deemed relevant to define the nominal function. It is an open question to which extent the quality and size of this material will affect the remaining steps in the process. On the one hand, the scope of the feature must become clearly defined during the iterative refinement process regardless of the initial input. On the other hand, it seems reasonable that more relevant input will make the rest of the process easier. A few possibilities for information in a PFD could be:

- A collection of a few use cases (high detail)
- A collection of a larger number of user stories (low detail)
- Very simple feature description

A use case describes a product use scenario with relatively high detail. It is described as a main scenario, possibly branching out in a number of alternate scenarios (variants of the main scenario). This content makes a use case, or a collection of use cases (depending on the complexity of the proposed feature), a promising candidate for information in a PFD. Since use cases have long been a popular tool, it might have the additional advantage of already being used in existing product research and development processes, which would make adoption of this format straightforward.

User stories are typically favored in agile development processes. The distinguishing features compared to a use case is that stories are much more lightweight and therefore quicker to develop. Instead, they need to be elaborated by the developers at some later development stage in order to become implementable.

We hypothesize that these features make user stories even better material for a PFD than the use case. The lightweight format makes it feasible to spend effort on creating a larger number of stories compared to use cases. In the context of autonomous vehicles, where the exact scope of a feature is not known at this point in development, this larger number of stories can aid in covering a larger part of the problem space compared to the use cases. This is illustrated in Fig. 25.3 where the user stories are smaller (i.e., less detailed) but has a better coverage of the space containing hazards relevant for autonomous driving.

Note that neither user stories nor use cases in any way guarantee coverage of the relevant hazards or provide any tools for this. Therefore, describing the function with user stories has the potential of providing a better starting point for hazard analysis that will help in the subsequent steps. But it will not in itself provide any proof of coverage.

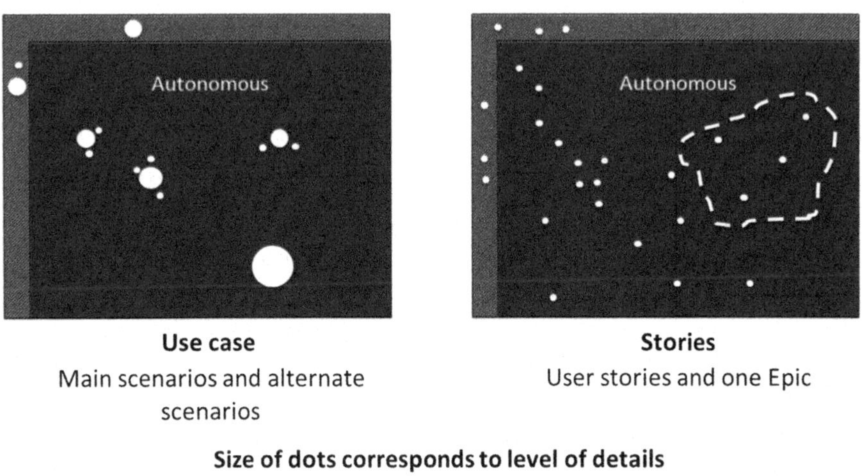

Fig. 25.3 Comparison of use case and user story format to describe a function

25.3.2 Situation Analysis and Hazard Identification

Situation analysis and hazard identification has a slightly different purpose when starting with a preliminary feature description compared with the traditional process where an item is already defined. It becomes part of an iterative process (steps 2–4) where the intended functionality and hazardous events are jointly elicited by stepwise refinement.

Compared to conventional analysis, there are two additional challenges: the function is not yet clearly defined, and for autonomous vehicles, the hazard analysis must assume the function is self-sufficient, i.e., does not rely on a driver as backup.

In order to start mapping the problem space, the idea is to use trees of generic operational situations and hazards as a starting point. For the first iteration, select a subset of situations from the table of generic operational situations and a subset of hazards from the table of generic hazards. The trees are organized after classes of factors, and in several levels where each sublevel has more detail than the previous one, making it possible to select the level of detail deemed necessary for each factor. This initial selection is based on what is reasonable to believe will affect the function based on the PFD. The operational situations and hazards are then combined to make up an initial set of hazardous events before moving on to step 3.

For each iteration, the situations and hazards can be made more detailed, where this is useful. That is, as the function becomes more clearly defined, it should also become more clear which situations need to be investigated in more detail.

Guiding factors for finding new or more detailed situations and hazards in order to advance the refinement process:

1. If it is hard to determine whether a previously defined HE falls within the scope of the function or not (see step 4), it needs elaboration to remove the uncertainty.
2. If the aspects of the function are not yet defined clearly enough to be able to write an item definition, these aspects need more analysis (see step 4).
3. Situations or hazards identified as potentially relevant in the previous iteration but left unused (i.e., did not become part of a HE) might indicate an omission. Otherwise, there should be a rationale why it was discarded.
4. Parts of the generic trees that have not yet been used should be revisited to see if there are any new situations or hazards that have become relevant.
5. Use rules for dominance and non-dominance of HEs (see step 3) to find potential new HE candidates.

For each iteration, new situations, hazards, and hazardous events are created in this manner. It should be stressed that the generic situation and hazard lists are a starting point but not an exhaustive list for all items and all contexts. As the function description matures and more known details about the context (vehicle) are taken into account, the generic lists have to be supplemented with context-aware specializations and additions, for instance, impact of performance characteristics of the target vehicle.

25.3.3 *Find Dimensioning Hazardous Events*

In order to keep the number of hazardous events manageable, HEs that will not contribute to the list of unique safety goals can be culled from the list. First, classification of hazardous events is performed according to ISO 26262, 3–7. During classification, each HE is assigned values for the exposure (E), controllability (C), and severity (S) factors, as well as a resulting integrity level (ASIL). Then, the list is processed using the rules for identification of dominance and non-dominance. The method is fully described in [5].

25.3.4 *Function Refinement*

The list of hazardous events becomes supporting material when trying to find requirements describing the nominal functionality. Normal requirements engineering is used to define the nominal functionality. The HEs from the previous step are considered. Any HE that may fall outside the scope of the function (perhaps because it is already handled by an existing item) is removed from the list of dimensioning HEs but kept as an aid to help define the boundaries of the function. A rationale for why the HE is outside the scope is noted.

After the first iteration, there should be a rough idea of what has to be included, and this can be assembled as a first step toward an item definition. Even if there are still many unknowns, this description can be used as input to the next iteration.

The process is illustrated in Fig. 25.4. After the first iteration, the function is delimited by a number of HEs, but the exact capabilities are still uncertain (illustrated by the fuzzy border separating the HEs within the function and those outside). For each successive iteration, it becomes more and more clear what the capabilities of the function must be.

When no more dimensioning HEs can be found using the guiding principles for finding HEs described in step 2, exit the iteration and continue to step 5. At this point, the function description should also have become detailed enough to write an item definition, and the assembled list of hazardous events will be transformed into safety goals.

25.3.5 *Item Definition*

The item definition is written according to ISO 26262. If this step uncovers new missing hazards, they shall be added to the hazard analysis and steps 2–4 repeated to include the new hazard.

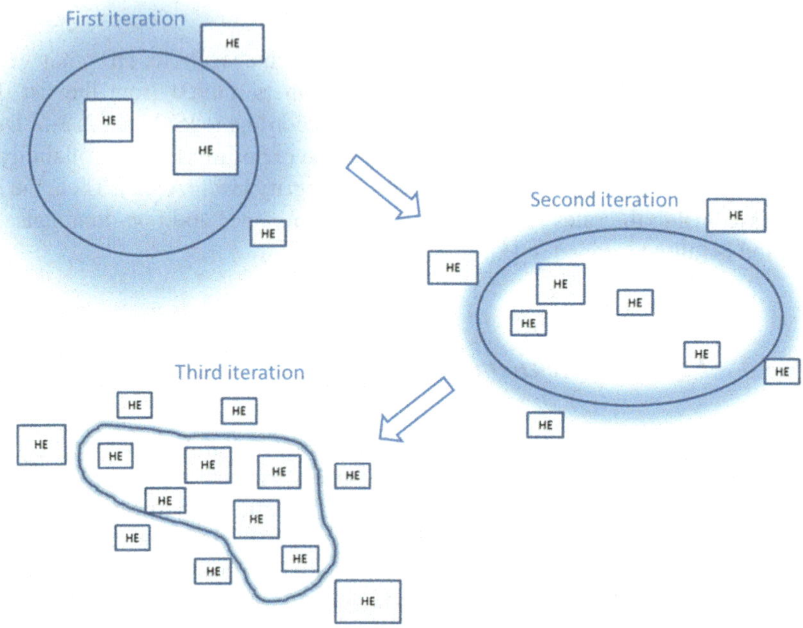

Fig. 25.4 Function refinement—view of hazardous events

25.3.6 *Safety Goals, FSC, and TSC*

From here on, the traditional V-model is followed. Safety goal, functional safety concept, and technical safety concept are constructed according to ISO 26262. In every step, if the refined design reveals any new hazards, the process is repeated from step 2 in order to see if any modifications are needed.

To further strengthen the confidence in coverage of the relevant hazardous events, the completeness can be challenged according to the idea in [4]. That is, to let a review team challenge the list of HEs looking for a candidate that is not covered in the list of safety goals, i.e., to find a new dimensioning HE.

25.3.7 *Consequences for Safety Case and Assessment*

The record and results of this iterative process can be used to strengthen the completeness argument for the HARA phase and should be especially useful in the context of autonomous functions. While not a formal proof of completeness, it is a systematic approach where it can be shown what issues have been taken into account in the analysis. The strength of the argument can be further improved by documenting a rationale for key decisions during the process, such as:

- The selection of operational situations and hazards and the level of detail used.
- Which classes of generic situations and hazards are determined not applicable for the function and why?
- If seemingly relevant (i.e., initially singled out), situations or hazards end up not being included in any HE.
- If it is decided that an initially selected hazardous event fall outside the scope of the final item.

The records and rationales from the iterative process, together with the results from the review team, should provide a solid base for a safety case.

25.4 How to Refine Safety Requirements

As mentioned in a previous section, there are things that become much more complicated when going up in degree of automation. In the previous section, we addressed the problem of identifying explicit item definitions and a complete and efficient set of safety goals. Even if these phases are mastered, we still have the problem of refining the safety requirements and allocate them onto the elements of the decided architectures. This problem is closely connected to the one of finding architectures on the different levels of abstraction, which is further addressed in a later section. The FUSE project proposes means for mastering the general problem of verifying that all steps of the safety requirement refinement are complete and correct.

In ISO 26262, there are general requirements that any lower level safety concept (safety requirements allocated onto elements of an architecture) needs to be verified and shown to be complete w.r.t. the higher level safety concept. Having a number of "partial" safety concepts spread among a number of companies, this verification task is certainly challenging. This is especially true for autonomous vehicles where the complexity of the functionalities is significantly higher than for manually driven vehicles. On the lower level, the safety requirements are assumed to be formulated similar to case when having manually driven vehicles. This means that there is a larger span of complexity to handle when the top-level complexity is higher, as for autonomous vehicles. Each step to be handled in the safety requirement breakdown chain is denoted the semantic gap. In Fig. 25.5 below, we depict the semantic gap between a safety goal and the underlying functional safety concept. Furthermore, the backward arrow that denotes refinement verification is depicted. This gap is present and calls for a refinement verification in phases of ISO 26262 refinement of safety requirements. During refinement, rationale, known as satisfaction arguments, shall be collected for the resulting composition, e.g., why is the refinement valid. The argument can contain, e.g., domain knowledge and design patterns. This is essential

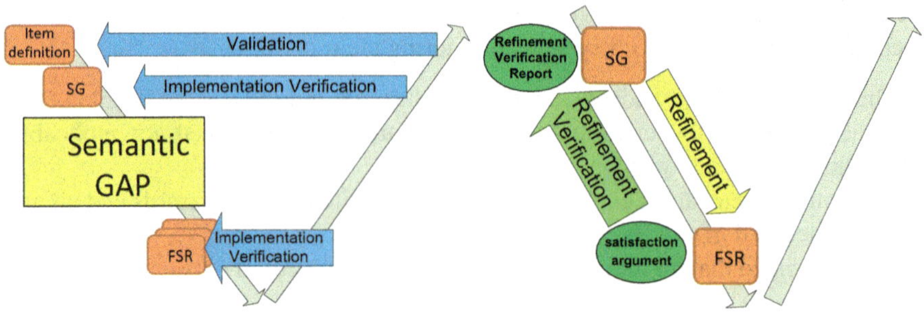

Fig. 25.5 The semantic gap

in almost every nontrivial refinement. The rationale justifies the "refinement path taken" through the semantic gap and improves to traceability.

In the FUSE project, we propose two main means for handling the complexity of the semantic gaps.

Firstly, we propose the possibility to introduce more steps in the safety requirement refinement chain. This is of course not forbidden today, but what we say is that ISO 26262 should explicitly introduce subphases for which the activities start and end on the same level of abstraction. This means that we can directly refer to work products that present the refinement (or the refinement verification) between a functional safety concept (FSC) and another FSC or between a TSC and another TSC. Introducing the formal ability of more steps would enable the lowering of the size of each semantic gap to overcome. In addition to this, we know that already today, for lower degrees of automation, a typical use case for a tier 1 supplier is to receive a TSC from the OEM customer and then break this down to an internal TSC. For both reasons, we need to complement the reference life cycle of ISO 26262 with subphases starting and ending on the same level of abstraction.

The second conclusion for how to safely master the refinement of safety requirements is that we need to be formal in the refinement verification. We argue the importance of having strong evidence, e.g., proof in a formal syntax for the correctness and completeness of every refinement verification. Furthermore, the assumptions and the domain knowledge acting as satisfaction arguments need to be explicit enough to serve as a part of this formalism. Large semantic gaps imply complex satisfaction arguments, which are hard to use in a convincing proof. These questions are further elaborated in [6].

25.5 What Functional Architectures Fit Autonomous Driving

When going up in degree of automation, we can regard this as the manual driver becoming more and more replaced by an autopilot. In the previous section, we addressed some of the responsibilities in terms of functional safety, which are transferred from the manual driver to the autopilot. If this was the question "what does the autopilot do?," in this section we address the question: "where does the autopilot hide?" As observed before, the functional safety issues are very much connected to the architectural ones, as the semantic of each safety requirement is partly dependent on the allocation on an architectural element. Beside this problem of enabling functional safety, there are a number of additional issues to cover when identifying what E/E architectures are efficient for autonomous vehicles.

Evaluation and refinement of a functional architecture for autonomous driving has occurred under the aegis of the FUSE project. The reference architecture includes (1) a categorization and description of the key functional components needed for autonomous driving, (2) rationale for the distribution of these functional components across the architecture, and (3) a three-layer architecture incorporating the described components. A comparison of the reference architecture with three representative architectures for autonomous driving has also been made, with a view of highlighting the similarities and differences.

The key functional components of the architecture are divided into three categories, (1) perception, (2) decision and control, and (3) vehicle platform manipulation. It is proposed that the perception and decision and control categories should be grouped into a "cognitive driving intelligence" layer of the reference architecture, whereas the vehicle platform manipulation category components should be included in a distinct "vehicle platform" layer. Such a split acknowledges the need for reusability of the architecture across different vehicle platforms (product portfolios), as well as the characteristics and development practices of the various research domains involved. The architecture also incorporates a layer for tele-operation or remote monitoring and management of autonomous vehicles. The identified functional architectural components have a substantial overlap with those found in other successful autonomous driving architectures. However, components for world modeling and semantic understanding are unique to the proposed architecture. The same holds true for an explicit component for abstracting the vehicle platform. The reference architecture has been applied to different vehicle categories and provides freedom to the component developers to test and deploy new algorithms while localizing the effect of any needed changes.

Details of the architecture are provided in chapter "Systems Engineering and Architecting for Intelligent Autonomous Systems." These questions are further elaborated in [7].

25.6 Conclusion

This paper presents the ongoing activities in the Swedish national-funded project FUSE. This project address a few of the questions needed to be addressed in the area of functional safety and E/E architectures, in order to enable autonomous vehicles. We conclude that the international standard for functional safety for road vehicles ISO 26262 needs to be updated in order to cover autonomous vehicles, and we give some suggestions to improvement. We also conclude that the issues of functional safety and E/E architectures are highly interconnected especially on the higher levels of abstraction, and we give recommendations for a functional architecture pattern suitable for highly automated vehicles. The project is ongoing, but some of the results discussed here have already been published. More details of the proposed solutions have been presented at a number of conferences as listed in the section of references [2–8] below.

Acknowledgment This work has been financed by the Swedish government agency for innovation systems VINNOVA in the FUSE project (ref 2013-02650). The participating companies in the project are Volvo Car Corporation, SP Technical Research Institute of Sweden, Semcon, Qamcom Research and Technology, KTH Royal Institute of Technology, and Comentor.

References

1. ISO, International Standard 26262 Road Vehicles—Functional Safety, ed, 2011
2. C. Bergenhem, R. Johansson, H. Sivencrona, Challenges in Applying Functional Safety for ADAS. Presented at SAE International Congress, 2014
3. R. Johansson, Automated Functions and Autonomous Vehicles. Presented at IQPC 5th International Conference ISO26262, 2015
4. R. Johansson, The Importance of Active Choices in Hazard Analysis and Risk Assessment. Presented at Third Workshop on Critical Automotive applications Robustness & Safety (CARS), 2015
5. R. Johansson, Efficient Identification of Safety Goals in the Automotive E/E Domain, Presented at ERTS^2, 2016
6. C. Bergenhem, R. Johansson, A. Söderberg, J. Nilsson, J. Tryggvesson, M. Törngren, S. Ursing, How to Reach Complete Safety Requirement Refinement for Autonomous Vehicles. Presented at Third Workshop on Critical Automotive applications Robustness & Safety (CARS), 2015
7. S. Behere, M. Törngren, A Functional Architecture for Autonomous Driving. Presented at the First International Workshop on Automotive Software Architecture (WASA '15), 2015
8. S. Behere, F. Asplund, A. Söderberg, M. Törngren, Architecture Challenges for Intelligent Autonomous Machines: An Industrial Perspective. Presented at The 13th International Conference on Intelligent Autonomous Systems (IAS-13), 2014

Chapter 26
Challenges for Automated Cooperative Driving: The AutoNet2030 Approach

Marcus Obst, Ali Marjovi, Milos Vasic, Iñaki Navarro, Alcherio Martinoli, Angelos Amditis, Panagiotis Pantazopoulos, Ignacio Llatser, Arnaud de La Fortelle, and Xiangjun Qian

26.1 Introduction

Automated maneuvering capability is expected to hold a key role in future mobility as it can probably provide safer driving conditions, improved comfort, and more efficient traffic management. The so-far relevant research undertaken by both industry and academic institutions mainly amounts to tackling different aspects of pure sensor-based automated driving such as sensing capabilities and V2X communications or vehicle control algorithms. Individual progress made along these threads has contributed to the deployment of integrated ADAS (advanced driver assistance systems) with increased levels of automated functionalities. However, highly automated vehicles are yet to come in mass-market deployment; the latter

M. Obst (✉)
Baselabs, Chemnitz, Germany
e-mail: marcus.obst@baselabs.de

A. Marjovi • M. Vasic • I. Navarro • A. Martinoli
École Polytechnique Fédérale de Lausanne, Lausanne, Switzerland
e-mail: ali.marjovi@epfl.ch; milos.vasic@epfl.ch; inaki.navarro@epfl.ch; alcherio.martinoli@epfl.ch

A. Amditis • P. Pantazopoulos
Institute of Communications and Computer Systems, Athens, Greece
e-mail: a.amditis@iccs.gr; ppantaz@iccs.gr

I. Llatser
Technische Universität Dresden, Dresden, Germany
e-mail: ignacio.llatser@tu-dresden.de

A. de La Fortelle • X. Qian
Centre for Robotics, MINES ParisTech—PSL Research University, Paris, France
e-mail: arnaud.de_la_fortelle@mines-paristech.fr; xiangjun.qian@mines-paristech.fr

D. Watzenig, M. Horn (eds.), *Automated Driving*,
DOI 10.1007/978-3-319-31895-0_26

calls for comprehensive investigation of the *complementarity* between onboard sensors, 5.9 GHz wireless communications, and distributed control algorithms.

Triggered by the so-far *limited convergence* between sensor-based automation and cooperative V2X communications, the AutoNet2030 project seeks to research and validate procedures and algorithms for 802.11p-based interaction control among cooperative (automated and manually driven) vehicles focusing on:

- Cooperative decentralized control system to realize fully automated vehicles and drive the advised maneuvering of manually driven vehicles
- V2X-message-based communications to (feed ETSI ITS standardization and) enable automated maneuver planning and traffic flow optimization
- Onboard sensor-based architecture to enable reliable positioning and lane-keeping automation

AutoNet2030 [1] intends to demonstrate how the combination of those three major automotive research threads will make lane-keeping/changing, maneuvering negotiations, and interaction between automated and manually driven vehicles more efficient and reliable. The prototyped cooperative automated driving system will be fully integrated into test vehicles and realistically demonstrated on a test track.

To realize those goals, AutoNet2030 has carefully selected and put together a cross-European consortium of high complementarity and distinct roles. The involved smaller partners such as SMEs (BASELABS, BroadBit Energy), research institutes (ARMINES, ICCS), and universities (EPFL, TUD) bring into the project their specialized high-tech competence in prototyping of automated driving technology, maneuvering control, and sensor data processing. On the other hand, large industrial partners (CRF, Scania, Hitachi Europe) will provide vehicle platforms and largely contribute to the AutoNet2030 system and vehicle integration.

26.2 Use Cases

In order to showcase scenarios with associated customer and societal value, the AutoNet2030 final demo presentations are designed to cover two diverse yet typical driving settings. One demonstration will be performed in highway-like conditions with two heavy-duty trucks and (at least) one manually driven passenger car focusing on (enhanced) variants of the convoy motion; the latter is one of the most promising methods for the introduction of automated driving on highways. The other demonstration is set in a more inner-city-like scenario where fully automated electric prototype vehicles will be used to (mainly) showcase automated functions for safe-fail distributed decision making and intersection coordination.

The set of the *high-speed scenarios* has been deliberately selected to demonstrate the efficiency of cooperative communications and perception capabilities as well as the realization of cooperative maneuvering in an inherently safe manner.

Scenario	Description
1. Joining 	One truck moves in high speed and a second one catches up in the same lane. When the second one approaches the first, they join the same convoy-like motion. After the convoy creation, automated driving functions assist in keeping speed/distance and lane position
2. Merging 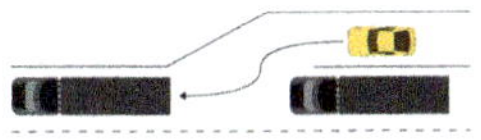	The two cooperative trucks of scenario 1 are approached by a manually driven car which seeks to merge into the same lane with the trucks. Automated driving functions recognize the merging and increase the distance between the trucks. Moreover, HMI in the car advices the driver to adjust speed and steering so that safe merging is possible
3. Leaving 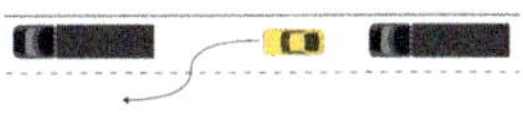	The three vehicles from scenario 2 are driving in cooperative mode in the same lane. The driver of the manually driven vehicle decides to leave by changing lane and pulling away from the trucks that remain under convoy motion
4. Lane changing 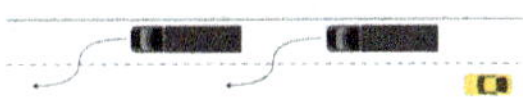	The two trucks drive in cooperative mode and a manually driven car approaches from behind in the left lane. HMI elements advice the car driver to maintain speed and position. Radar sensors on the trucks and a laser sensor on the car acknowledge that the left lane is free of conflicting objects and thus the trucks safely perform a lane change

The set of the low-speed scenarios has been deliberately selected to demonstrate the efficiency of cooperative communications, perception capabilities, and most notably the realization of cooperative decision making.

Scenario	Description
1. Car following 	Two vehicles drive on the same lane, maintaining a constant distance. First, the speed of the leading vehicle slightly varies, it performs an emergency brake, and then restarts. In both cases, we show how vehicles keep a perpetually safe relation
2. Merging 	Two vehicles that move in cooperative automated mode approach a merging point and are coordinated to merge into the main road. The vehicle having no priority decelerates to provide sufficient space. Using cooperative maneuvering, the merging can be performed in a safe and efficient way
3. Lane changing 	Two vehicles drive in different lanes at the same speed and one decides to change its lane. The vehicle on the targeted lane should decelerate to facilitate a smooth lane change of the other one. Using cooperative maneuvering, the lane change operation can be performed safely and efficiently
4. Intersection 	Two vehicles approach an intersection point with the same speed. A crossing order is decided through V2V messaging and cooperative decision making. Accordingly, the vehicles cross the intersection without collision. Information exchange and cooperative decision making are demonstrated

26.3 Human Machine Interface

Interacting with the vehicle driver is a challenge of major importance when implementing automated driving functions. The AutoNet2030 project has set its focus on the HMI design and development for partially automated vehicles, while it can occasionally (i.e., high-speed use cases) adopt its functionality to inform the users of automated vehicles. The project has relied on the multidisciplinary work of both engineers and cognitive HMI experts aiming to design a user-friendly interface that will facilitate advised maneuvering.

After having analyzed the considered AutoNet2030 use cases, a flow of road/vehicle events and actions (required by the user) has been identified. To cope with the specified requirements (of each use case), an innovative dual-display HMI system has been designed. It consists of a head-up display and a secondary display (i.e., Android device) as illustrated in Fig. 26.1. With this setup, the driver receives the most significant information and related maneuver advices over the HUD; the secondary display projects only informative messages (e.g., the reason why a certain advice is projected) and also provides the interactive capability (i.e., inputs by the driver) when safety conditions allow (e.g., a button to switch between manual and automated mode). In each case, a layout has been defined to cope with the different visual elements, the associated urgency level, and the way information is best presented to the driver. The displayed messages have been determined according to a predefined syntax structure and have been prioritized (for projection) with respect to a designed HMI logic that accounts for their significance.

Fig. 26.1 Indicative screenshots of the dual-screen AutoNet2030 HMI system: HUD (*left*) and Android (*right*)

26.4 Cooperative Control

26.4.1 Distributed Graph-Based Convoy Control

In terms of the distributed convoy case, the problem that we mainly seek to tackle amounts to following instance: We consider an unknown number of intelligent vehicles in a road which are able to communicate with each other locally and can localize themselves. The problem is how these vehicles can establish a dynamic multilane convoy which remains stable while allowing new cars to join and the current cars to leave.

Most of the works in group control of vehicles have been focused on single-lane convoy problems using reactive spacing control methods for consecutive vehicles. Point-follower and vehicle-follower, adaptive cruise control (ACC) [2], cooperative ACC (C-ACC) [3], and local controllers [4] are the main approaches for single-lane convoy control. In these strategies, the desired inter-vehicular spacing is maintained through basic control laws such that every controlled vehicle matches its distance and speed with the vehicle ahead.

AutoNet2030 approach for convoy control is based on the work by Gowal et al. [5], which proposed leaderless graph-based control for multilane convoy. An undirected graph $G = (V, E)$ is defined in which vertexes V correspond to controlled agents (vehicles in this case) and edges E correspond to inter-vehicle communication and relative positioning links. Built upon basic linear algebra, a stable solution to the formation control problem in two dimensions is given by

$$\mathbf{x} = (L \otimes I_2)\,(\mathbf{x} - \mathbf{b}) + v_{\mathrm{G}} \tag{26.1}$$

with $L = I \cdot W \cdot I^T$, where L (called Laplacian matrix) is obtained from the incidence matrix I that defines the edges of G and the weight matrix W which is a diagonal matrix used to tune the weights assigned to the edges. I_2 is simply a 2×2 identity matrix. The (x, y) absolute position vector for all vehicles is given by $\mathbf{x}$, and the desired offsets of the vehicles to the formation centroid are given by the bias matrix $\mathbf{b}$. The parameter v_{G} represents the desired velocity of the vehicles.

To make the graph-based formation control scalable and dynamic, we propose an approach in which graphs containing connections between vehicles are dynamically created, locally maintained, and automatically modified. The main ingredients of the dynamic graphs in our system are *local neighborhood* and *local identifiers*. We define a *local neighborhood* of a vehicle using topological distance, that is, distance measured in number of vehicles. *Local identifiers* allow each vehicle to enumerate the other vehicles in its vicinity using its local coordinate frame, by assigning ordered pairs (2 tuples) containing topological distances in x- and y-axes (in its local right-handed coordinate system). Finally, the Laplacian control (Eq. 26.1) is upgraded to a decentralized approach, assuming a connected (but not necessarily complete) graph, using only relative positioning information. Details are provided in [6].

26.4.2 Cooperative Intersection Management

Currently, traffic lights are equipped in traffic intersections to coordinate conflicting flows and ensure the road safety. However, the efficiency and safety of such system is doubted: 44 % of collisions in the USA are within the intersection area, and delays induced by traffic lights can be high. First proposed in [7], cooperative intersection management (CIM) allows autonomous vehicles to cooperatively cross the intersection without traffic light, fully utilizing the advanced sensing, communication, and maneuver capacities of vehicles. It is shown that CIM brings significant efficiency improvement (in terms of throughput, average delay, etc.) compared with traffic lights.

In AutoNET2030, CIM is one of the major research topics. We adopt a safety-oriented approach to tackle this topic. The goal is to design a mechanism for CIM that is not only efficient but also probably safe. We adopt the priority-based approach to separate the coordination problem into two subproblems: planning of vehicle priorities and brake-safe reactive control of vehicles. Vehicle priorities decide the relative orders of vehicles to cross the intersection. Brake-safe control of vehicles allows a vehicle to avoid collision with other vehicles having higher priorities (prior vehicles) even prior vehicles perform emergency brakes. Under mild assumptions, the overall safety and deadlock-free property of the proposed system can be mathematically proven. More details can be found in [8–10].

26.5 Cooperative Sensing and Perception Layer

26.5.1 Configurable Perception Layer

Building up reliable and accurate knowledge of the environment—often called an environmental model or perception layer—of autonomous vehicles is a crucial requirement in order to perform automated maneuvers in a safe and efficient manner. In general, the environmental model comprises static and dynamic entities which need to be observed and tracked over time. As both stages, decision making and control algorithms, directly depend on the robustness and the accuracy of the environmental model, this is considered a core part in automated driving.

Typically, several perception sensors such as radar, lidar, and camera are used in order to perceive the surrounding of the host vehicle. These asynchronous sensor information are continuously combined with data fusion algorithms in order to build a unified environmental model. The intuition behind this multisensor data fusion approach is usually twofold: First of all, as the field of view of a single sensor system is often limited due to physical constraints, combining multiple sensors that are mounted at different locations increases the surveillance area. Secondly, and this is usually more important, by using heterogeneous sensors, the overall robustness and performance can be increased as the combination of the particular sensor features (e.g., radar sensors can directly observe a radial velocity component, while camera systems usually give a proper estimate of the object's width) yields an improved environmental model.

Within the AutoNet2030 project, a full-scale 360° environmental model that integrates onboard perception sensors as well as communication information is required in order to safely perform and demonstrate the anticipated use cases for high- and low-speed scenarios as elaborated in Sect. 26.2. The perception layer has to be efficiently implemented to meet the real-time conditions and optimally exploit the computational resources of the embedded computers used inside of the test vehicles. Moreover, as AutoNe2030 deliberately uses several heterogeneous vehicle platforms, the 360° perception layer has to be easily adoptable to different vehicle configurations. For example, this includes the capability of easily replacing a radar sensor from one particular manufacture to another as well as the ad hoc configuration of sensor properties such as mounting position or sensor noise characteristics. In order to cope with the challenge of having a configurable 360° environmental model, a novel tool-based development approach [11] is used inside of AutoNet2030:

- **Prototyping**: In this stage, the environmental model is developed by performing a probabilistic data fusion among several sensors in the high-level programming languages C# and C++ with rich debugging capabilities. The data fusion is realized by using the probabilistic sensor data fusion framework BASELABS Create. This SDK leverages an incremental development process where each sensor can be added one by one until the final 360° configuration is reached. The resulting environmental model is tested and validated, both with recorded and online measurement data for each vehicle platform.
- **Vehicle integration**: In this stage, the already validated and tested environmental model is automatically transformed from high-level C# code to static embedded C-code that is appropriate for pre-series integration at ECU level. The automatic code transformation ensures that the C-code is functionally fully equivalent to the already tested C# version from the prototyping step.

26.5.2 V2X Communications for Automated Driving

Recent research activities [12] and successful field trials of V2X communication [13] are bringing the application of V2X communications to autonomous driving closer to reality. Cooperative autonomous driving is also the object of the European R&D projects i-GAME,[1] AdaptIVe,[2] and COMPANION.[3]

---

[1] http://www.gcdc.net/i-game

[2] http://www.adaptive-ip.eu

[3] http://www.companion-project.eu

V2X communications for autonomous driving represent a natural evolution of the communication system for cooperative vehicles. Initial V2X communication systems have been designed to provide driver assistance, which corresponds to level 1 in the definition of automation levels in SAE J 3016.[4] Higher levels of automation introduce new requirements that need the definition of new or enhanced messages, communication protocols, and their standardization for cooperative autonomous driving [14].

In order to practically implement V2X communications among heterogeneous vehicles, standardization is needed to guarantee their interoperability. For this reason, the ETSI Technical Committee in Intelligent Transport Systems (ITS) recently published the GeoNetworking standard, which defines forwarding algorithms for packet transport in VANETs. Furthermore, periodical Cooperative Awareness Messages (CAM) and Decentralized Environmental Notification Messages (DENM) allow vehicles to exchange information using a common language.

CAMs allow each road user participating in a V2X network to transmit periodically their station type, time, position, velocity, and many other parameters. This rather comprehensive description of an object state in combination with a communication range of up to 1000 m in perfect conditions makes CAM data appealing candidates for improved multisensor data fusion.

However, V2X communication in general and CAM transmissions in particular have some distinctive characteristics compared to classical onboard perception sensors. For example, this includes a global coordinate system, dynamic update rates according to the CAM trigger rules, as well as high latencies due to the communication channel. In AutoNet2030, the potential of V2X to complement the environmental model is investigated.

Furthermore, the AutoNet2030 project is researching new communication protocols and message types specifically designed to support the cooperative sensing and maneuvering among autonomous vehicles. In particular, new cooperative sensing messages, broadcast by every vehicle at a rate of 1 Hz to all its neighbors, are required to share the detected objects (such as vehicles, pedestrians, cyclists, etc.) among the cooperative vehicles. For the cooperative maneuvering functionality, several new types of messages are needed. For instance, the transmission of convoy messages coordinates the maneuvering of a multilane formation of vehicles, and the cooperative lane change service supports maneuver negotiations among vehicles aiming to perform a lane change, as well as relative space reservation. More details about these dedicated messages for cooperative autonomous driving are found in a recent paper [14].

[4] http://www.sae.org/misc/pdfs/automated_driving.pdf

26.5.3 Road Data Fusion Module

The road data fusion module aims at increasing the accuracy of the observations and detections of the road geometry as provided by individual vehicle perception components as well as the V2X sensor information. Road attributes as captured by lane markings, road boundaries, and map matched/refined GPS position are extracted from the available object tracking/image processing and positioning units and are fused using the map road geometry. The output of the module essentially increases the information availability and robustness of the system (e.g., lane information may be artificially reconstructed in absence of visible lane markings using only map data). It constitutes an accurate representation of the road geometry in the form of a road-segments list, each of one is described by clothoid model equations [15].

26.6 Conclusion and Outlook

We have presented the main parts of the AutoNet2030 body of work and notably the way the consortium has addressed the related automated driving challenges along the threads of the system architecture, the distributed control algorithms, and the cooperative perception capabilities. What is yet to be accomplished amounts to the standardized use of 5.9 GHz V2X communications for automated driving, the finalization of the AutoNet2030 software modules, and their integration in the available vehicle platforms. Test-track validation of the cooperative maneuvering control algorithms and overall system functionality will follow. With the successful completion of the above AutoNet2030 objectives, the project envisages to shape the path for cost-optimized and widely deployable automated driving technology.

Acknowledgement The research work has been funded by the European FP7 project AutoNet2030 (Grant Agreement NO. 610542).

References

1. A. de La Fortelle, X. Qian, S. Diemer, J. Grégoire, F. Moutarde, S. Bonnabel. A. Marjovi, A. Martinoli, I. Llatser, A. Festag, K. Sjöberg, Network of Automated Vehicles: The AutoNet2030 Vision, in *Proceedings of 21st World Congress on Intelligent Transport Systems*, 2014
2. L. Xiao, F. Gao, Practical string stability of platoon of adaptive cruise control vehicles. IEEE Trans. Intell. Transp. Syst. **12**(4), 1184–1194 (2011)
3. S.Y. Han, Y.H. Chen, L. Wang, A. Abraham, Decentralized Longitudinal Tracking Control for Cooperative Adaptive Cruise Control Systems in a Platoon, in *IEEE International Conference on Systems, Man, and Cybernetics*, pp. 2013–2018, 2013
4. K.Y. Liang, Coordination and Routing for Fuel-Efficient Heavy-Duty Vehicle Platoon Formation, Licentiate Thesis, 2014

5. S. Gowal, R. Falconi, A. Martinoli, Local Graph-Based Distributed Control for Safe Highway Platooning, in *IEEE/RSJ International Conference on Intelligent Robots and Systems*, pp. 6070–6076, 2010
6. A. Marjovi, M. Vasic, J. Lemaitre, A. Martinoli, Distributed Graph-Based Convoy Control for Networked Intelligent Vehicles, in *Proceedings of IEEE Intelligent Vehicles Symposium*, pp. 138–143, 2015
7. K.M. Dresner, P. Stone, A multiagent approach to autonomous intersection management. J. Artif. Intell. Res. (JAIR) **31**, 591–656 (2008)
8. J. Gregoire, S. Bonnabel, A. de La Fortelle, Priority-Based Coordination of Robots, CoRR, vol. abs/1306.0, 2013
9. X. Qian, J. Gregoire, F. Moutarde, A. de La Fortelle, Priority-Based Coordination of Autonomous and Legacy Vehicles at Intersection, in *Proceedings of the IEEE International. Conference on Intelligent Transportation Systems (ITSC)*, pp. 1166–1171, 2014
10. X. Qian, J. Gregoire, A. de La Fortelle, F. Moutarde, Decentralized Model Predictive Control for Smooth Coordination of Automated Vehicles at Intersection, in *European Control Conference* (ECC2015), 2015
11. N. Mattern, R. Schubert, A Hybrid Approach for ADAS Algorithm Development—From High-Level Prototypes to ECUs, in *Proceedings of the 10th ITS European Congress*, Helsinki, Finland
12. S. Kato, S. Tsugawa, K. Tokuda, T. Matsui, H. Fujii, Vehicle control algorithms for cooperative driving with automated vehicles and intervehicle communications. IEEE Trans. Intell. Transp. Syst. **3**(3), 155–161 (2002)
13. H. Stubing, M. Bechler, D. Heussner, T. May, I. Radusch, H. Rechner, P. Vogel, simTD: A car-to-X system architecture for field operational tests. IEEE Commun. Mag. **48**(5), 148–154 (2010)
14. L. Hobert, A. Festag, I. Llatser, L. Altomare, F. Visintainer, A. Kovacs, Enhancements of V2X Communication in Support of Cooperative Autonomous Driving, to appear in *IEEE Communications Magazine*, 2015
15. M. Tsogas, N. Floudas, P. Lytrivis, A. Amditis, A. Polychronopoulos, Combined lane and road attributes extraction by fusing data from digital map, laser scanner and camera. Inf. Fusion **12**(1), 28–36 (2011)

Chapter 27
Architecture and Safety for Autonomous Heavy Vehicles: ARCHER

Viktor Kaznov, Johan Svahn, Per Roos, Fredrik Asplund, Sagar Behere, and Martin Törngren

27.1 Summary of the Project

Machines are converging towards autonomy. The transition is driven by safety, efficiency, environmental and traditional 'robotics automation concerns' (dirty, dull and dangerous applications). Similar trends are seen in several domains including heavy vehicles, cars and aircraft. This transition is, however, facing multiple challenges including how to gradually evolve from current architectures to autonomous systems, limitations in legislation and safety standards, test and verification methodology and human–machine interaction.

One of the major challenges for developing a fully automated heavy vehicle is to design and develop a system with an acceptable level of system safety. New principles and methods for system architecture, safety analysis and system verification must be developed to reach the necessary safety level.

These challenges are particularly important in case of *fully* automated *heavy* vehicles, i.e. no human driver available in the vehicle, considering vehicle weight, size, life span, operational scenarios and the number of produced variants. The business case for *commercial* vehicles is also very different from passenger cars, whose prime purpose mostly is to transport the driver. *Commercial* vehicles are a transport tool, where the driver is not part of the cargo, and constitute roughly one-third of the transport cost.

The purpose of the project 'Architecture and Safety for Autonomous Heavy Vehicles—**ARCHER**' is to develop methods and principles for safety analysis of a fully automated commercial heavy vehicle, to create a reference architecture and

V. Kaznov (✉) • J. Svahn • P. Roos
Scania CV AB, Södertälje, Sweden
e-mail: viktor.kaznov@scania.com

F. Asplund • S. Behere • M. Törngren
KTH Royal Institute of Technology, Stockholm, Sweden

D. Watzenig, M. Horn (eds.), *Automated Driving*,
DOI 10.1007/978-3-319-31895-0_27

to develop methods and principles for test and verification of fully automated heavy vehicles. Also a proof-of-concept realization for overall verification and validation of the reference architecture and developed methods is included.

The project is building upon experiences and knowledge from finished and ongoing research projects partially funded by Swedish government, such as iQDrive (investigates semi-autonomous heavy vehicles where drivers are still in the loop), iQMatic (prototype autonomous transport system consisting of a fully equipped control room and at least one load carrying unit), FUSE (investigates self-driving cars and functional safety aspects) and ESPRESSO (develops and adapts model-based techniques that improve the quality and reduce the cost for development of embedded systems in trucks and especially safety critical systems) [34].

27.2 Background

The Swedish automotive industry makes up a large and significant part of Swedish exports. For Swedish vehicle manufacturers, the domestic market only represents a small proportion of global sales, whereas the majority of research and technical development work take place in Sweden.

A generational shift is now predicted, with vehicles moving from driver control to fully automated control. This shift has already begun—there are already active systems that intervene in extremely critical situations to avoid accidents and increase safety. Semi-automated functions such as 'Adaptive cruise control' are also already in place; the adaptive cruise control is an automated system whereby the driver hands over longitudinal control of the vehicle to the vehicle's autonomous systems. The adaptive cruise control offers considerable potential for development in terms of achieving improved energy and traffic efficiency. It is also well known that the underlying cause of many single-vehicle and workplace accidents involving vehicles is due to tiredness or distraction, in turn often arising from monotonous driving situations. By automating monotonous tasks, accidents of this type can be prevented, thus **improving traffic safety**.

The current emphasis and shift towards autonomy is enabled by cost-performance and maturity improvements in sensor, actuator, and semiconductor technology. As a result, the Swedish automotive industry has to adapt and change from a mechatronics engineering industry to becoming a full-blown Cyber-Physicals Systems engineering industry with a large and growing element of vehicle automation software operations as well as vehicle communication; see [1].

With regard to skills development in the field of automation, it is extremely important to launch Swedish projects where automated vehicle concepts are developed and tested in their natural environment. With the aid of tangible research projects applied to robust, autonomous research platforms, Sweden ought to be able to retain its position within vehicle automation at the same time as developing the skills required to cope with the generational shift initially facing work vehicles and subsequently commercial and private vehicles.

It is now clear that a number of activities need to be initiated in order to achieve greater breadth and depth regarding the autonomous vehicles skills area. This is partly a reaction to the ongoing development of society and transport systems and partly due to a clear desire on the part of visionary customers.

Areas such as 'sensor fusion', 'functional safety' and 'perception' are receiving strong attention by the automotive industry and in academic research, but there are limitations in efforts that explicitly address electrical system architecture, safety and verification—especially in the context of fully automated commercial heavy vehicles. The absence of a human driver creates specific challenges in case of heavy commercial vehicles considering vehicles weight, size, lifespan and complexity of managing a large number of variants. Furthermore, the development of fully automated heavy vehicles is strongly driven by a clear market demand, since, in most applications, there is no need of the human presence in a heavy vehicle, besides the safety aspects. The proposed project, 'Architecture and Safety for Autonomous Heavy Vehicles—**ARCHER**', complements existing Swedish and international projects on automation with its specific focus on (1) *architectures for* ***heavy*** *vehicles*; (2) ***full*** *automation, where a human is not available* and (3) ***verification and validation*** *of E/E systems with focus on safety.*

27.3 State of the Art

A brief account of state of the art (SOTA) is here given for safety, architecture and verification.

Safety The standard ISO 26262:2011 Road vehicles—Functional Safety [2] describes the prevailing best practices for achieving functional safety of road vehicles. Currently, it is only adopted for passenger cars, but it will likely be applicable for heavy vehicles within the time frame of the project. Adherence with the safety standard is important not only for the comprehensive coverage it provides throughout the product safety lifecycle (from concept to decommissioning) but also because demonstrating compliance with the prevailing best practices is an important factor for exclusion of liability in case of a system malfunction.

The ISO26262 process requires an early hazard analysis of the system, with the subsequent formulation of safety goals and the determination of Automotive Safety and Reliability Levels (ASILs) for each safety goal. However, it is debated whether the standard is immediately applicable to the development of intelligent, autonomous vehicle functions. One reason for this is that it relies on techniques for hazard and risk analysis in which human involvement, usually in the form of the vehicle's driver, is an important factor in the estimation of risk reduction levels and the subsequent calculation of ASILs for safety requirements [3]. For a truly autonomous vehicle, the involvement of a human being cannot be counted upon, as a mitigating factor in safety-critical operational situations. *Thus, either high ASIL levels will have to be accounted for (assuming no human controllability) or the existing standard needs to be upgraded to include new concepts of controllability, where machine intelligence replaces the human being.*

Functional safety considerations within ISO26262 begin with the analysis of risks and identification of hazards for relevant safety-related systems at the vehicle level (items). There exist a number of tools for low-level risk analysis, like Fault Tree Analysis (FTA) [4], Event Tree Analysis (ETA) [5], Failure Mode and Effect Analysis (FMEA) [6], probabilistic FMEA [7], Hazard and operability study (HAZOP) [8] etc. These tools and methods often provide sufficient support for low-level risk or hazard analysis. However, they typically require detailed design specifications and do not take the dynamic behaviour of systems into consideration. Known challenges for autonomous system design include the shifting of more workload to critical or work intensive time periods [9], tighter coupling making autonomous systems react faster [9], insufficient or inappropriate feedback on system modes [10], too much or too little feedback with regard to the specific user [11, 12], reduction of situation awareness [13], reinforcing of decision bias [13], enforcing limitations that are only appropriate in nominal situations also during extreme situations [12] and encouraging of reliance on automation to a degree at which manual skill decreases [13].

Several approaches have been suggested in the literature to address the lack of detailed design specifications at the early design phase. Of these, the Systems Theoretic Process Analysis (STPA) technique [14] has been shown to be promising for recognizing safety requirements and constraints of the system before detailed design [15, 16]. STPA also addresses the dynamic behaviour of a system, and several authors have reported positive outcomes from applying STPA on various systems [14, 17, 18], including automotive functions like forward collision avoidance [19]. This leads us to believe that STPA might be useful for safety analysis of autonomous systems. The application of hazard analysis techniques in the absence of human involvement remains an area that requires further exploration in the field of autonomous heavy vehicles.

On completion of a hazard analysis, the ISO26262 process continues with the generation of safety goals, ASIL classifications inherited from the identified hazards and the formulation of a Functional Safety Concept (FSC). The FSC is then refined into a Technical Safety Concept (TSC) which provides a basis for subsequent formulation of hardware and software safety requirements. At the various Safety Concept levels, aspects like fail-operational characteristics, redundancy, system monitoring and supervisory control enter the picture. A good overview of system monitoring and supervision concepts, especially for autonomous systems, is provided in [20]. The concepts of supervisory control and fail-operational modes are especially relevant for autonomous vehicles, since no generally safe states exist for an operational autonomous road vehicle. Therefore, it is necessary to investigate methods for supervisory control, the distribution of supervisors in the system architecture and their overall effect on safety argumentation for the system. This is where architecting and architecture comes into the picture as an important aspect of design for safety.

Architecture The traditional approach for electrical system architectures has been the usage of self-contained hardware platforms, referred to as Electronic Control Units (ECUs), connected in a communication network. These ECUs exchange information via coordinated signals, and this scheme, denoted 'Federated Architecture', allows for good separation of concerns and simplified verification of the integration. However, Federated Architectures face challenges of increasing functional complexity, non-optimal resource consumption, emergent behaviour and cost. To address these, both the avionics [21] and recently the automotive domains are transitioning towards the so-called 'Integrated Architectures' where multiple functions can be supported by one ECU, and one function can be distributed over multiple ECUs. Functions, subsystems or even library components can be developed by different organizations or vendors and may use different platform services while residing in a single ECU and sharing a limited set of communication channels [22]. This direction, however, will not eliminate the distributed nature of vehicle control, but may imply a new approach for stronger coordination of high-level control.

The transition to integrated architectures necessitates development of new methods, tools and techniques for the design and analysis of the architectures. This includes topics related to mixed criticality systems, multi-core and Network-on-Chip technologies, optimal mapping and allocation of system functionality among ECUs, assurance of overall system level technical properties like end-to-end timing and latency as well as extra-functional properties like system safety, reliability etc. This, in turn, necessitates development of architecture exploration techniques to evaluate and determine optimum architectures from the excessively large design space. Techniques like platform based design [23–25] provide early methodologies for systematic design of functions, architecture, their mapping, and implementation of the mapped architecture.

The introduction of autonomous functionality necessitates a critical look at system architectures and the architecting process. This is because autonomy as a system level property goes well beyond being 'just another requirement'. The robotics and artificial intelligence domains have traditionally led the development of autonomous system architectures. However, transitioning these architectures to the automotive domain is not a straightforward process, especially due to the automotive domain's requirements with respect to being safety-critical, incorporating legacy designs, differing development process and business cases and the sheer number of product variants involved. A thorough analysis of system architectures for autonomy in various domains is presented in [26], which includes a comparison of domain characteristics, commonalities and differences.

A promising way forward is the development of reference architectures for autonomous vehicle systems with the use of existing and new design patterns and principles that facilitate system integration (academically referred to, e.g. as composability and correctness by construction). The reference architecture approach has already been successful for individual autonomous functions like cooperative driving [27], and it should be possible to generalize it to the vehicle level, especially when the entire vehicle E/E architecture is controlled by a single organization. It is of particular interest to develop common reference architectures for both autonomous and non-autonomous variants of the same vehicle model. The use of novel autonomy patterns and principles are instrumental for reducing the complexity, thus reducing the difficult task of verification.

Verification Including Testing Traffic statistics from the United States of America show that the mean time between fatal crashes is about 2 million vehicle hours and the mean time between injury crashes is over 50,000 vehicle hours [28]. For an autonomous vehicle to be acceptable in society, it will need to be proven substantially safer than this baseline. This is a tough challenge when we consider that a fully automated commercial heavy vehicle would be a product that needs to be affordable to current customers, that needs to operate for the life of a heavy vehicle, which by far exceeds the life span of a passenger car, with a minimum of maintenance, and that is confronted with a highly stochastic operating environment, with hazards that need to be identified and mitigated in a fraction of a second. The extensive design redundancy and intensive preventive maintenance regimes that have made commercial aviation safe are not directly economically viable in the automotive sector. Furthermore, hazardous situations on public streets need to be dealt with more quickly than in the air, because of close proximity among other vehicles and pedestrians.

All these factors increase the testing and verification requirements on autonomous road vehicles. Simultaneously, the increasing complexity and exploding state space in autonomous vehicle architectures make it practically impossible to achieve total test coverage of both the vehicle architecture and its behaviour in all possible scenarios. Achieving sufficient test coverage is simply not possible within realistic product development timeframes, using the traditional approach of constructing a system or component prototype and running a sequence of tests on it. ISO26262 essentially says that the higher the risk, the more efforts and rigour should be spent for design and verification, with hints provided to all kinds of techniques, but without guidelines on how different techniques can be beneficially used and combined. *There is thus an imperative need to develop a new verification methodology*.

There exist a multitude of approaches to verification and testing, from manual reviews to automated testing. We focus here on model-based approaches which show great promise in dealing with verification and which complement the dominating industrial testing practices [29, 30]. We identify three main approaches for verification of complex systems: (1) *simulation*, (2) *formal verification* and (3) *verification problem formulation and formalization*.

The first two approaches involve the creation of models, i.e. formal representations of the system. In the simulation approach, the models are typically constructive whereas in the second case, the models are typically analytic; see, e.g. [29]. The strengths of simulation-based approaches include their use as constructive design models (where parts can be reused, e.g. for code generation and HIL testing), whereas the drawbacks relate to the impossibility to cover the large state space. Corresponding challenges thus include that of finding relevant test cases and understanding what constitutes relevant coverage. In the second approach, a formal representation is first created (or possibly generated from a constructive model) and is then analysed for the assurance of specific system properties or absence of undesirable properties. Analysis methods include model checking, theorem provers and other formal methods based on discrete mathematics, formal logic or hybrid formalisms. Strengths of formal verification are the possibility to cover complete parts of the state space and that the requirements and constraints are explicitly captured.

Challenges of formal verification include scalability to realistic size problems and their dependence on the formalized models and the tools required to process them. The approach requires specially trained users, even though there nowadays exists tools to facilitate model creation. The sheer effort in constructing models for a complex system has meant that this approach is not very common outside the development of safety-critical systems (practical usage in the automotive industry is so far limited). A hot research topic is to combine simulation and formal verification approaches; see, e.g. MBAT [31].

The third approach refers to how verification problems are formulated and formalized, involving issues such as the choice of the level and type of formalization of requirements capture, and how constraints and assumptions are captured. Safety incidents are known to frequently stem from incomplete understanding of assumptions and mismatches between models [14]. The formulation of requirements, constraints and assumptions is closely connected to verification by simulation, formal analysis and testing, since it will drive the overall approach to verification and validation, and be closely related to test cases. The use of contracts has emerged as a promising approach for the specification of the system and its components in such a way that the incremental assembly of the system is guaranteed to have the desired properties [32, 33]. Contracts enable explicit descriptions of the context in which components can be used and how they can be combined.

A common challenge for the three approaches is that of dealing with model management and integration, a separate research field in its own right, where ARCHER can draw upon results from the FFI ESPRESSO project and the iFEST ARTEMIS project, including dealing with challenges such as consistency management across multiple models [34, 35].

There is no silver bullet for verification of autonomous systems. The ARCHER hypothesis is that a combination of architecting (reducing complexity), combination of verification techniques (automating what is feasible and putting efforts on the critical and hard to formalize aspects), is the right way forward towards cost-effective verification.

27.4 Project Content

ARCHER is considered to be a first step as part of a longer term initiative. A continuation of the proposed project is planned to focus on verification and validation of results from ARCHER in demonstrational trucks and to progress towards higher TRL's and to work towards establishing an autonomy competence centre in close collaboration with the KTH ITRL lab.[1]

ARCHER contains three main focus areas in the context of fully automated heavy commercial vehicles: (1) methods and principles for safety analysis, (2) methods and principles for system architecture and (3) methods and principles for verification.

The first area is focusing on the safety perspective. As stated in the SOTA part of this chapter, the current safety analysis methods are largely based on a physically present human driver who can act as a last resort in the functionality degradation chain. For the automation levels without an operator, new degradation principles as well as new safety analysis methods and principles must be developed. To exemplify, a few relevant questions in this area are presented below.

- Where do available standards and methods for the safety analysis fall short when applied to fully automated heavy vehicles?
- How should available standards and methods for the safety analysis be improved to provide a reliable and effective methodology for development of fully automated heavy vehicles?
- How should the safety requirements be formulated to handle the trade-off between safety and availability?
- How will safety requirements change depending on the planned operational environment of a vehicle?
- How will principles for human independent diagnosis and diagnostic procedures be defined?

The second area is focusing on the system architecture perspective. A reference architecture (or a set of architectures that cover fundamentally different application scenarios and/or subsystems) will be developed, that fulfils the identified safety, test and operational requirements and addresses architectural bottlenecks, while considering realistic constraints on, e.g. infrastructure availability, legacy system integration and cost efficiency. To exemplify, a few relevant questions in this area are presented below.

- Which architectural principles and patterns should be used to provide an acceptable system safety and cost assuming no human driver presence at all?
- Which architectural principles and patterns should be used to be feasible for actual production of commercial vehicles?
- Which architectural principles and patterns should be used to make system feasible for testing and verification considering level of automation?

The third area is focusing on testing and verification. The verification aspect needs to be included already at design time or even at architectural design as a fully automated heavy vehicle cannot be verified by the current test methods. Principles for efficient testing to reach acceptable safety levels will be developed.

[1]The KTH and industry *Integrated Transport Research Lab*: https://www.itrl.kth.se/

A methodology which combines state of the art testing, simulation and formal verification methods will be developed. The methodology will consider and exploit characteristics in terms of faults/failure modes, behavioural and structure aspects (including architectural patterns), types of verification techniques and formalization including abilities for verification automation, as well as the criticality of faults/risks.

27.5 Project Targets

The main expected results from the ARCHER project are:

- methods and principles for safety analysis of fully automated heavy vehicles
- methods and principles and a reference system architecture enabling safe, secure and cost-efficient fully automated heavy vehicles
- methods and principles for test and verification and validation of fully automated heavy vehicles
- Three licentiate theses (half-way PhD thesis in the Swedish doctoral education, to be completed as PhDs in the follow-up project)
- Nine master theses
- increased collaboration between industry and academia

The target of the project is to develop a set of requirements, design principles, methodologies and a reference architecture for development of fully automated heavy vehicles. Given that the majority of industrial projects are at relatively low TRL levels, there is a need to consider architecture in order to deal with complexity, safety, availability and business considerations. Architecture, validation, verification and safety are strongly related, and there is thus a benefit in pursuing them simultaneously. With the proposed focus, ARCHER thereby paves the way for progressing towards higher TRLs.

References

1. CyPhERS deliverable D3.2. Market and Innovation Potential of CPS. Technical Report by the CyPhERS FP7 project, Aug 2014, http://www.cyphers.eu/sites/default/files/D3.2.pdf
2. ISO 26262:2011 Road vehicles—Functional safety (2011)
3. S. Behere et al., Architecture Challenges for Intelligent Autonomous Machines: An Industrial Perspective, in *Proceedings of the 13th International Conference on Intelligent Autonomous Systems (IAS-13)*, Padova, Italy, 2014
4. C.A. Ericson, Fault Tree Analysis–A History, in *Proceedings of 17th International System Safety Conference*, 1999
5. T. Tobioka, R.C. Bertucio, Use of event tree analysis in development of a LOCA test program. Trans. Am. Nucl. Soc. **39**, 590–591 (1981)
6. R. McDermott et al., *The Basics of FMEA*, 2nd edn. (Taylor & Francis, Boca Raton, FL, 1996)

7. H. Aljazzar et al., Safety Analysis of an Airbag System Using Probabilistic FMEA and Probabilistic Counterexamples, in *6th International Conference on the Quantitative Evaluation of Systems*, Hungary, 2009
8. J. McDermid et al., Experience with the Application of HAZOP to Computer-Based Systems, in *Proceedings of 10th Annual Conference on System Integrity, Software Safety and Process Security*, COMPASS, 1995
9. D.D. Woods, Decomposing automation: Apparent simplicity, real complexity, in *Automation and Human Performance: Theory and Applications*, ed. by R. Parasuraman, M. Mouloua (Erlbaum, Mahwah, NJ, 1996), pp. 3–17
10. B.N. Sarter, D.D. Woods, Pilot interaction with cockpit automation: Operational experiences with the flight management system. Int. J. Aviat. Psychol. **2**(4), 303–321 (1992)
11. R.D. Sorkin, Why are people turning off our alarms? J. Acoust. Soc. Am. **84**(3), 1107–1108 (1988). doi:10.1121/1.397232
12. R. Parasuraman, V. Riley, Humans and automation: Use, misuse, disuse, abuse. Hum. Factors: J. Hum. Factors Ergon. Soc. **39**(2), 230–253 (1997). doi:10.1518/001872097778543886. http://hfs.sagepub.com/content/39/2/230.abstract
13. D.A. Norman, The problem of automation: Inappropriate feedback and interaction, not over-automation, in *Human Factors in Hazardous Situations*, ed. by D.E. Broadbent, J. Reason, A. Baddeley (New York, Oxford University Press, 1990), pp. 585–593
14. N.G. Leveson, *Engineering a Safer World: Systems Thinking Applied to Safety* (MIT Press, Cambridge, MA, 2012)
15. T. Ishimatsu et al., Modeling and Hazard Analysis Using STPA, in *Proceedings of the 4th IAASS Conference Making Safety Matter*, p. 10, 2010
16. H. Nakao, M. Katahira, Y. Miyamoto, N. Leveson, Safety Guided Design of Crew Return Vehicle in Concept Design Phase Using STAMP/STPA, in *Proceedings of the 5th IAASS Conference,* pp. 497–501, 2011
17. S.J. Pereira, G. Lee, J. Howard, A System-Theoretic Hazard Analysis Methodology for a Non-advocate Safety Assessment of the Ballistic Missile Defense System, in *Proceedings of the AIAA Missile Sciences Conference*, Monterey, California, 2006
18. J. Thomas, N.G. Leveson, Performing Hazard Analysis on Complex, Software- and Human-Intensive Systems, in *Proceedings of the 29th ISSC Conference About System Safety*, 2011
19. S. Sulman et al., Hazard Analysis of Collision Avoidance System Using STPA, in *Proceedings of the 11th International ISCRAM Conference,* University Park, Pennsylvania, USA, May 2014
20. E. Baudin, J. Blanquart, J. Guiochet, D. Powell, Independent Safety Systems for Autonomy: State of the Art and Future Directions, Technical Report LAAS-CNRS No. 07710
21. C.B. Watkins, R. Walter, Transitioning from Federated Avionics Architectures to Integrated Modular Avionics, in *2007 IEEE/AIAA 26th Digital Avionics Systems Conference, IEEE*, Oct 2007
22. M. Di Natale, A. Sangiovanni-Vincentelli, Moving from federated to integrated architectures in automotive: The role of standards, methods and tools. IEEE Proc. **98**(4), 603–620 (2010)
23. A. Sangiovanni-Vincentelli, G. Martin, Platform-based design and software design methodology for embedded systems. IEEE Des. Test Comput. **18**(6), 23–33 (2001)
24. A. Sangiovanni-Vincentelli, A. Ferrari, System Design—Traditional Concepts and New Paradigms, in *Proceedings of ICCD*, 1999
25. A. Sangiovanni-Vincentelli et al., Alberto Benefits and Challenges for Platform-Based Design, in *Proceedings of the 41st Annual Conference on Design Automation—DAC '04,* pp. 409–414, 2004
26. S. Behere, Architecting Autonomous Automotive Systems: With an Emphasis on Cooperative Driving, Licentiate Thesis, KTH, Stockholm, 2005
27. S. Behere, M. Törngren, D. Chen, A reference architecture for cooperative driving. J. Syst. Archit. **59**(10), 1095–1112 (2013). doi:10.1016/j.sysarc.2013.05.014. Part C
28. S. Shladover, An Automated Highway System as the Platform for Defining Fault-Tolerant Automotive Architectures and Design Methods. NSF CPS Workshop Position Paper, 2011

29. M. Törngren et al., Model based development of automotive embedded systems, in *Automotive Embedded Systems Handbook*, ed. by N. Navet, F. Simonot-Lion. Industrial Information Technology Series (Taylor and Francis CRC Press, Boca Raton, FL, 2008). ISBN 9780849380266
30. A. Benveniste et al., *Embedded Systems Design, The ARTIST Roadmap for Research and Development*. Lecture Notes in Computer Science, vol. 3436 (Springer, Berlin, 2005). doi:10.1007/b106761. ISBN 978-3-540-31973-3
31. MBAT ARTEMIS project, http://www.mbat-artemis.eu/home/
32. P. Derler, E.A. Lee, M. Torngren, S. Tripakis, Cyber-Physical System Design Contracts, in *ICCPS '13: ACM/IEEE 4th International Conference on Cyber-Physical Systems*, 10 Apr 2013
33. J. Westman et al., Structuring Safety Requirements Using Contract Theory, in *SAFECOMP—32nd International Conference on Computer Safety, Reliability and Security*, France, 2013
34. ESPRESSO FFI project, http://www.vinnova.se/sv/Resultat/Projekt/Effekta/ESPRESSO/
35. iFEST ARTEMIS project, www.artemis-ifest.eu/

Chapter 28
Affordable Safe and Secure Mobility Evolution

Alexander Viehl, Udo Gleich, and Hendrik Post

28.1 Mobility Systems Evolution

Many of today's societal challenges such as global warming, tightening energy supplies, aging society, or security have an impact on transportation, be it automotive, rail, or aviation. ERTRAC, the European Road Transport Research Advisory Council, aims at a 50 % more efficient transport system by 2030 [1]. In terms of fatalities and severe injuries, the automotive industry targets a 60 % reduction until 2030.

In the premium market's segments, innovative functions are the most important factor to influence buying decisions. Future mobility solutions will increasingly rely on smart components that continuously monitor the environment and assume more and more responsibility for a convenient, safe, and reliable operation in parallel to an additional facilitation of energy consumption optimization and emission reduction. A major step for the evolution of autonomous systems is the transition

---

ASSUME Partners: For a full list, see https://itea3.org/project/assume.html

A. Viehl (✉)
FZI Forschungszentrum Informatik, 76131 Karlsruhe, Germany
e-mail: viehl@fzi.de

U. Gleich
Daimler AG, 89081 Ulm, Germany
e-mail: udo.gleich@daimler.com

H. Post
Robert Bosch GmbH, 70839 Gerlingen-Schillerhöhe, Germany
e-mail: hendrik.post@de.bosch.com

ASSUME Partners
http://www.assume-project.eu/

D. Watzenig, M. Horn (eds.), *Automated Driving*,
DOI 10.1007/978-3-319-31895-0_28

from assistance systems to automated hands-off systems. Public perception moves toward higher expectations with regard to the safety of highly autonomous systems. With hands-off systems, a failure rate clearly below the one of a human actor is expected. For automotive, the self-driving car is the next big revolution, and it is still unclear how functional and nonfunctional guarantees can be given probably zero defect for those new class complex-automated functions. Consequently, novel design and verification methods for such highly automated systems are needed to satisfy future safety-relevant systems requirements.

During the design process of a complex distributed system, system level requirements are broken down into more fine-grained technical hardware and software requirements and mapped to subsystem parts. During this requirement break down, various models are created to represent distinct aspects of the developed system. Hence, traceability between different abstraction levels of requirements and system parts must be established. The designs models and hence design decisions have to be verified with respect to their associated functional and safety requirements, and it has to be ensured that the implementation does not violate requirements.

One major limitation today is the unavailability of synthesis and verification tools for these specific models. While research prototypes show the general feasibility of formally proving the correctness of models or code with respect to given requirements, these tools have not been adopted by the industry widely.

Tools working on source code or implementation level that check for specific errors or design flaws are available and well applied today. However, higher-level requirements and the correctness of design decisions cannot be checked effortlessly and completely with high confidence.

In addition to that, further challenges arising with the shift to more autonomy are the increasing complexity and performance requirements of autonomous systems. On the way to realize this vision, the need for computing power will drastically increase far beyond what can be provided by conventional sequential single-core hardware. While the required efficiency and scalability compel future embedded microcontrollers (μC) to rely on multi- and many-core architectures, the change in hardware architecture also necessitates fundamental advances to software development methodology. Replacing today's essentially sequential technology by interconnected cores and omnipresent communication poses the colossal challenge in software development to identify and exploit means for concurrency still guaranteeing reliable and predictable behavior. One problem here is that analysis techniques and flows have to be extended to support parallelism and system complexity on several design and implementation levels.

Current analysis techniques are severely limited by the size and complexity of the embedded systems to be analyzed. Tools using abstract interpretation to prove the absence of runtime defects usually become difficult to use for code sizes above 200,000–300,000 lines of code depending on the used programming features and the code complexity. Model checking techniques are currently limited to much smaller sizes of the program state space since they enumerate overall possible program states without abstraction. Moreover, tools for the analysis of concurrent and multi-core software currently present a large number of false positives to the user. Consequently, an efficient assessment of analysis results for concurrent systems in an industrial environment is currently not feasible.

Combinations of different analysis methods and tools for concurrent systems are mostly in a premature state of research and not viable for industrial application yet. One reason for this situation is the absence of standardized interfaces between verification and modeling tools that support verification tasks. Formal verification tools typically only apply one technology and support one implementation language. A close collaboration of these tools is needed, allowing the exchange of analysis findings and given assumptions across modeling languages and tooling borders.

28.2 Objectives

The main goal of the ASSUME project is the affordable, standard-compliant development and verification of highly automated, safety-relevant, and performance-critical mobility systems. A strong focus is on development methods for concurrent systems and static verification techniques. The ASSUME algorithm portfolio will be the key technology to bring innovative solutions from sandboxes into consumers' daily lives. ASSUME provides a seamless engineering methodology to overcome this roadblock. The problem is addressed on the constructive and on the analytic side. For efficient construction and synthesis of embedded systems, the project provides new tools, standards, and methodologies to cover most of the challenges by design. In addition, ASSUME provides a well-integrated sound static analysis solution that allows proving the absence of problems even in a multi-core environment. New algorithms will be integrated in exploitable tools. New interoperability standards and requirement formalization standards will facilitate cooperation between different market players.

The major field of innovation for ASSUME's industrial partners (end users) resides in the model-based parallel software engineering for multi- and many-core processors. The project enables the effective use of formal verification and synthesis technology along the design flow. Methods and tools are developed that support the safe migration of sequential programs into parallel paradigms. Safety in parallel execution environments will further be ensured by the extension of formal methods to support concurrency for multi- and many-core CPUs.

This engineering methodology, which will be supported by code generators and static analyzers, enables system/software developers to move from the current engineering for sequential functions to a concurrent realization, in order to benefit from the increased performance and hardware integration that result from the use of modern parallel processors.

Sound static analysis holds the promise to perform fully automated and exhaustive detection of important classes of errors (such as runtime errors) with full control and data coverage. Despite important past progress, related tools still suffer in many contexts from limitations in precision and in domain of application. ASSUME delivers analysis methods and tools with improved precision and efficiency. The currently available technologies for the verification of nonfunctional properties (absence of runtime errors) are extended to larger and more complex systems. Moreover, through improvement and combination of analysis techniques, the verification of functional properties (such as requirements) through static analysis will be implemented. This is possible through a complete traceability of formalized requirements to their implementation. It is made possible to find example scenarios that might violate a requirement. These examples are very valuable information to identify the cause of defects and errors in the models and implementation. Static analysis techniques are extended and improved for the industrial use of concurrent software on multi- and many-core platforms. The ASSUME project proposes to extend existing static analyzer tools and prototypes for concurrent software along several lines:

- The sound support for true concurrency offered by scheduling on multi-core hardware
- The precise handling of the associated models of memory consistency
- The efficient migration of sequential software to parallel applications
- The support for modern embedded operating system

Technological innovations are driven by the industrial use cases provided by industrial partners to achieve high precision through specialization.

28.3 Expected Outcomes

The expected outcome of ASSUME is the design of a static analysis platform (SAP), the first of its kind, able to check for both classic and concurrency-related runtime errors on large embedded industrial multi-core software, with a high efficiency and low rate of false alarms.

ASSUME advocates the idea that static analysis of software can be extended to provide affordable means to prove the absence of defects. The newly developed static analysis platform (SAP) will extend the state of the art in the following aspects:

- SAP scalability is improved to obtain a high enough precision-runtime tradeoff for future software complexity.
- SAP supports fully parallel software running on multi- and many-core μC.
- SAP integrates a new framework for the formalization of safety and security requirements.
- SAP allows for seamless traceability and impact analysis of functional and extra-functional properties in model-driven and conventional development.

The key advancement of the ASSUMEs analysis platform is the improved integration of verification tools across language and tool boundaries. The ASSUME platform will facilitate various available models along the embedded development chain to improve traceability and associate requirements at different levels, solutions, and validation and verification activities. The main benefits of this integration are higher analysis precision and the ability to verify larger and concurrent systems, to check functional properties, and at the same time decrease the verification effort. The versatility of the platform will be achieved by the provisions of exchange formats but is not limited to this. The SAP will allow the implementation of meta-algorithms for an intelligent (re-) combination of algorithms to get overall improved analysis results. Another valuable feature is the improved exploitation of available hardware resources by the development of methods and tools that enable efficient parallelization due to the more precise estimation of the worst-case execution time (WCET) and other properties in parallel systems.

Last but not least, the widespread use of multiprocessor systems-on-chip architectures (including multi-/many-cores) imposes significant changes to the models and methods used to perform scheduling and code generation for embedded platforms. The adoption of multi-/many-core architectures is driven by scalable performance arguments (concerning speed, power, etc.), but this scalability comes at the price of increased complexity of both the software and the software mapping process (including scheduling and code generation). Part of this complexity can be attributed to the steady increase in the size of software that is run by a single system. But there are also significant qualitative changes concerning both the software and the hardware. In software, more and more applications include parallel versions of signal processing, simulation, and control algorithms, which are best modeled using dataflow models (as opposed to independent tasks). Providing functional and real-time correctness guarantees for parallel code requires an accurate control of the interferences due to concurrent use of communication resources. Depending on the hardware and software architecture, this can be very difficult. There are two main reasons to this. The first one concerns communications: as the tasks are more tightly coupled and the number of resources in the system increases, the on-chip networks and shared memory banks become critical resources, which need to be explicitly considered and managed during scheduling. The second reason concerns automation: the complexity of large many-cores and of the (parallel) applications mapped on them is such that code generation, allocation, and scheduling must be largely automated. Formal verification techniques for these automated tools are in great need.

ASSUME proposes a set of methods and tools that enable the efficient use of synthesis techniques in the design flow of parallel software. The common denominator of these techniques is formalization and full automation. The technical focus is placed on formal compiler verification and on correct real-time implementation for parallel applications. In both cases, hardware modeling is recognized as a major issue and dedicated specific attention.

Figure 28.1 shows the proposed fields for technology innovations of the ASSUME project along the development chain of embedded systems.

The circular arrow represents the system development or more precisely the design flow. The proposed technological contributions of ASSUME are arranged around and in the center of this circle. The development chain differs from classical

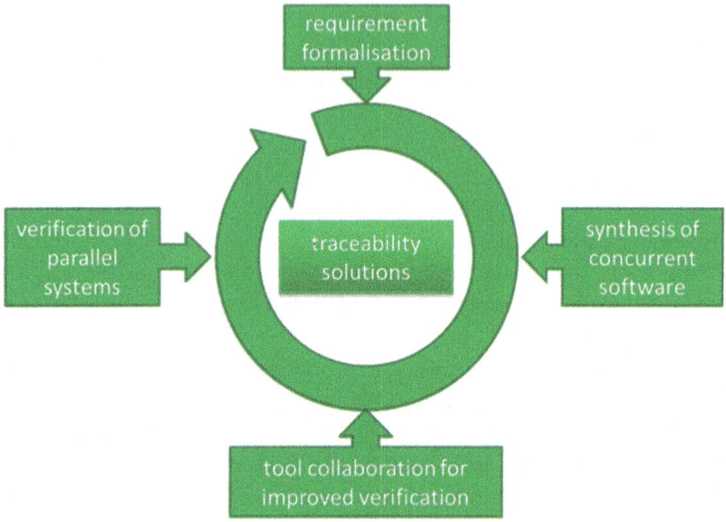

Fig. 28.1 ASSUME fields of technology innovations

software development in terms of distribution of development roles as contained in the market value chain but also technically due to the close integration into physical processes. Because of the severe consequences of failures in these systems, development processes are strictly regulated in the specific standards for the several mobility domains.

Acknowledgments This work was partially supported by BMBF under grant number 01IS15031.

Reference

1. ERTRAC Strategic Research Agenda 2010, Towards a 50% More Efficient Road Transport System by 2030, May 2010, www.ertrac.org

Chapter 29
UFO: Ultraflat Overrunable Robot for Experimental ADAS Testing

Hermann Steffan, Christian Ellersdorfer, Andreas Moser, and Julian Simader

29.1 Introduction

Our method, which was developed over the last years with increasing importance, is the application of autonomously driving platforms, called "UFOs" (cp. Fig. 29.1) [1], which are GPS controlled. They form the high-end motion platforms for different targets, as they allow full 2D motion, independent of the infrastructure. No adaptation of the infrastructure is necessary. The top speed is currently limited to 80 km/h.

These UFOs are designed in a way that the height of the platform is lower than the height of the ground plate of the vehicle under test (VUT). Even in case of a collision, no damage is caused to the VUT or the platform itself. In addition they are surrounded by ramps, which allow a smooth overrun of the platform by the VUT, even at high speeds.

29.2 Structure of the UFO Platform

The internal structure of an UFO is depicted in Fig. 29.2. It consists of several subsystems which are described in the upcoming section.

The **propulsion system** of the platform typically consists out of one or two electric motors. Their power depends on the required performance of the platform. Typical power of the whole drivetrain is 5–15 kW. Mostly the electric motor is used

H. Steffan (✉) • C. Ellersdorfer
Institute of Vehicle Safety, Graz University of Technology, Inffeldgasse 23/1, 8010 Graz, Austria
e-mail: h.steffan@tugraz.at

A. Moser • J. Simader
Dr. Steffan Datentechnik, Salzburgerstrasse 34, 4020 Linz, Austria

D. Watzenig, M. Horn (eds.), *Automated Driving*,
DOI 10.1007/978-3-319-31895-0_29

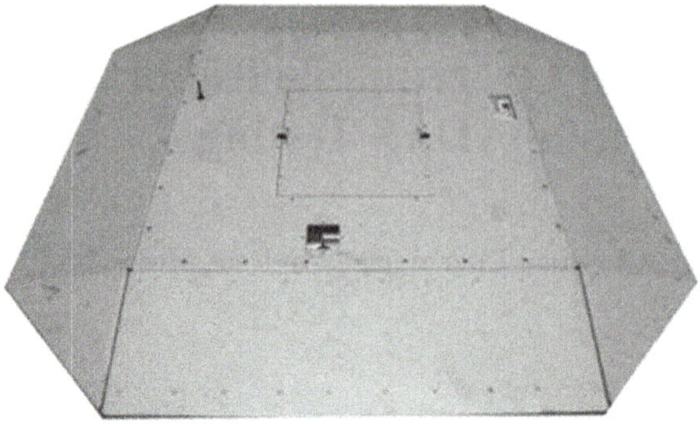

Fig. 29.1 The UFO platform (without dummy)

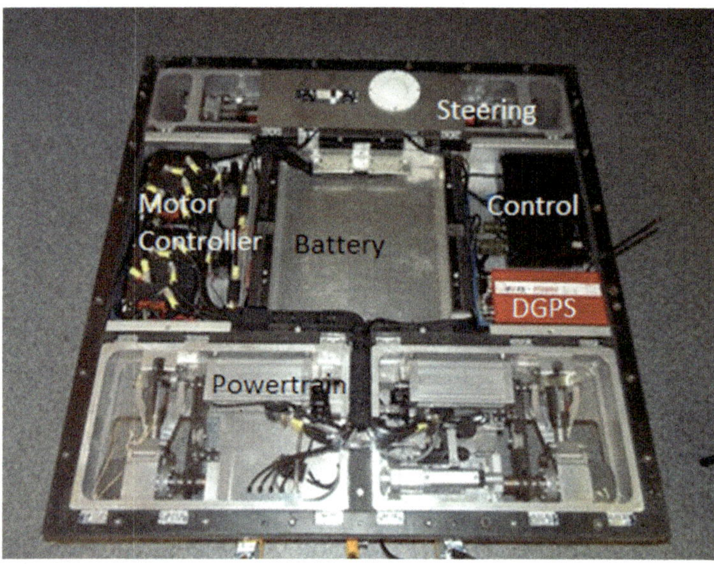

Fig. 29.2 Interior of an UFO platform

for braking. Nevertheless, as decelerations up to 0.7 g are requested, the platform also includes a hydraulic braking system.

The **energy supply** is provided by Li-ion batteries, which are provided as exchangeable packs.

As **navigation system**, high-performance differential GPS (DGPS) is used. It combines DGPS-corrected satellite signals with inertia systems, using a Kalman filter. To control the platform, only systems providing this real-time synchronization of inertia system and DGPS can be used to archive the precision necessary for complex test scenarios. Position and heading, the two main controlling parameters, can be derived with a typical accuracy of ±5 cm.

As **DGPS systems** cannot be used indoor, it currently limits the system to outdoor use. At the moment, there are alternative navigation systems in evaluation, which can also be used indoor.

Using **WLAN communication** between car, platform, and base station, the timing and the target information is transmitted to the platform. In this way, the platform can also be attached to the VUT using a master-slave relation. The steering algorithm of the control system will calculate the necessary parameters to follow the target path.

Based on these systems, the **control system** inside the platform derives the signals for steering and engine control/brake by comparing target and real position. When the navigation signal is temporarily missing, the inertia platform can extrapolate the missing positions. The inertial system also reduces the noise of the DGPS signal with respect to the heading information.

There are two ways to steer the platform: either there are one or two wheels which are rotated by a steering servo to drive the platform along the target path.

To simulate, e.g., pedestrian movement, allowing also on spot rotation, steering control can also be performed by prescribing different velocities to the left and right wheels/chains. In this way, the platform can also rotate on the spot.

29.3 Communication Infrastructure

Figure 29.3 shows a typical communication setup to allow all kinds of necessary information transfer:

As basis for the communication, wireless modules (Wi-Fi) are used, running on 2.4 and 5 GHz. For the DGPS correction signal, 434/868 MHz frequencies are used.

As Wi-Fi is not intended to run real-time applications, every message must imply universal time codes, generated by time synchronous clocks included in every unit. The synchronization is guaranteed by using GPS time in every unit.

The Central Control Unit (CCU) is dedicated to control all devices, involved in the test setup. As also multiple vehicles or platforms may be involved in one scenario, this CCU has to provide this functionality. Within this CCU, the individual paths of the controlled vehicles and platforms are generated. The may be generated either within GPS coordinates or also relative to the VUT. In this way, the VUT can also act as a master controlling the platforms as his slaves. As an alternative, all involved partners can also run according to fixed paths with a fixed timing.

When the VUT acts as master, there are different methods to control the platforms:

The simple method is using a light gate triggering the start of the platforms. In this way, the platforms are following predefined paths, but their start is triggered by the light gate. Typically timing and velocity of the VUT are measured simultaneously for an optimum performance to allow more accurate testing, when the VUT is driven manually.

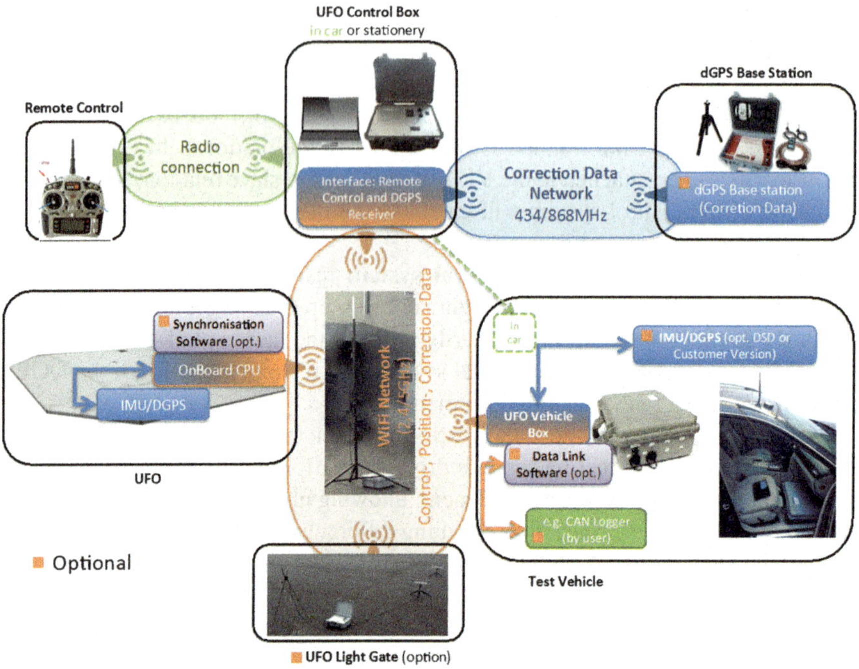

Fig. 29.3 Communication backbone of the UFO test equipment

As an alternative, the VUT can also be equipped by a DGPS, which reports the position, velocity, and direction of the VUT to the base station. Based on this information, the velocity and path of the platforms may be adapted in real time. The advantage of this method is a very high degree of accuracy and flexibility. The disadvantage is that the VUT must be equipped with a DGPS and communication module. The integration of this equipment into the VUT requires typically 20 min setup time.

29.4 Definition of Test Scenarios

Typically these platforms can be used in infinite number of scenarios. These situations are either driving scenarios like priority violation, turning left, overtaking, or lane changes or parking scenarios. Many additional scenarios are derived from the evaluation of accident databases, where scenarios with repeated accident potentials and accident occurrence are mainly used for evaluation. In addition, critical driving scenarios from normal driving are repeated [3].

Out of these accidents, specific scenarios are derived and typically varied to ensure the functionality of the system not only in one exact scenario but in a set

Fig. 29.4 Reconstruction of a pedestrian accident scenario

of scenarios with similar characteristics. Various different databases are currently in use to develop these scenarios. One of the most commonly used accident databases is the GIDAS [2] database which is handled by the Medical University of Hannover and the Technical University of Dresden. They collect more than 2000 accident cases per year, where a special accident evaluation team goes onside the accident scene in selected areas. The selection criterion for the accidents, added to the database, is the occurrence of at least one injured person.

Based on the evaluation of this data bases, critical scenarios and there possible evaluations are derived. These scenarios are also used to define the boundary conditions and criteria for the sensor systems and algorithms.

After reconstruction of the accident, a set of reference scenarios is derived and can be exported to the platforms and steering robots. It can be rerun on a selected test area. The environment can either be neglected or represented by selected stationary obstacles (Fig. 29.4).

29.5 Summary

This paper presents an overview of the UFO test platform. Currently it is primarily used to reconstruct accidents. Additionally it can be utilized to evaluate ADAS functions. In the future, it could also be used for testing autonomous driving functions.

References

1. German accident database, www.vufo.de/forschung-und-entwicklung/gidas/. Accessed 14 Jan 2016
2. J.M. Wille, M. Zatloukal, Volkswagen Group Research rateEFFECT—Effectiveness Evaluation of Active Safety Systems
3. Schmidl, ADAS Testing with the UFO Testing System, in *Autonomous Vehicle Test & Development Symposium*, Stuttgart, 2015

Chapter 30
Intelligent Transport Systems: The Trials Making Smart Mobility a Reality

Patrick Pype, Gerardo Daalderop, Eva Schulz-Kamm, Eckhard Walters, Gert Blom, and Sasha Westermann

As urban populations increase, issues such as traffic jams, pollution and road fatalities will grow in tandem. More than half of the world's population now lives in urban areas (54 %), and growth is expected to accelerate in years to come.[1] The cost of congestion to the global economy is also growing and is expected to cost Europe and the USA $4.4 trillion between 2013 and 2030.[2] Better connected vehicles and infrastructure (Intelligent Traffic Systems—ITS) presents a viable solution to these issues, creating an intelligent road network that is safer, greener and more efficient.

V2X —vehicle-to-vehicle (V2V) and vehicle-to-infrastructure (V2I)—is one of the key technologies that will underpin ITS. V2X works by enabling ad hoc data exchange between the vehicle and environment via wireless Internet networks—in other words, allowing vehicles to interact with each other and the surrounding infrastructure like traffic lights and road signs within a 2000-m range. Vehicles can then alert drivers to potential traffic issues, even beyond the line of sight, so they can adjust their driving accordingly to avoid accidents or congested areas. Alerts could include blind-intersection collision, road condition hazards, road works, presence

[1]According to the World Health Organization http://www.who.int/gho/urban_health/situation_trends/urban_population_growth_text/en/

[2]According to the Centre for Economics and Business Research http://www.cebr.com/reports/the-future-economic-and-environmental-costs-of-gridlock/

P. Pype (✉) • G. Daalderop • E. Schulz-Kamm • E. Walters
NXP Semiconductors, Leuven, Germany
e-mail: patrick.pype@nxp.com

G. Blom
City of Helmond, Helmond, Netherlands

S. Westermann
Hamburg Port Authorities, Hamburg, Germany

D. Watzenig, M. Horn (eds.), *Automated Driving*,
DOI 10.1007/978-3-319-31895-0_30

of emergency vehicles, stationary or slow moving vehicles, traffic jam, accident warnings, as well as traffic signals or signage indicators.

This might sound like a highly futuristic scenario, but V2X chipsets became available for mass production last year (from NXP Semiconductors) and are already being deployed by automakers, usually alongside complimentary technology such as advanced driver assistance systems (ADAS), i.e. radar. V2X-equipped vehicles could therefore be commercially available as soon as 2016. The potential of this technology is already being taken seriously by governments across the world, with numerous high profile V2X trials being implemented.

30.1 ITS Corridor: Austria, Germany and The Netherlands

In 2013, governments in Austria, Germany and the Netherlands signed a memorandum of understanding to create Europe's first ITS Corridor. The project, which is due to be delivered in 2016, will see 1300 km of motorways between Rotterdam, Frankfurt and Vienna fitted with intelligent transport systems.

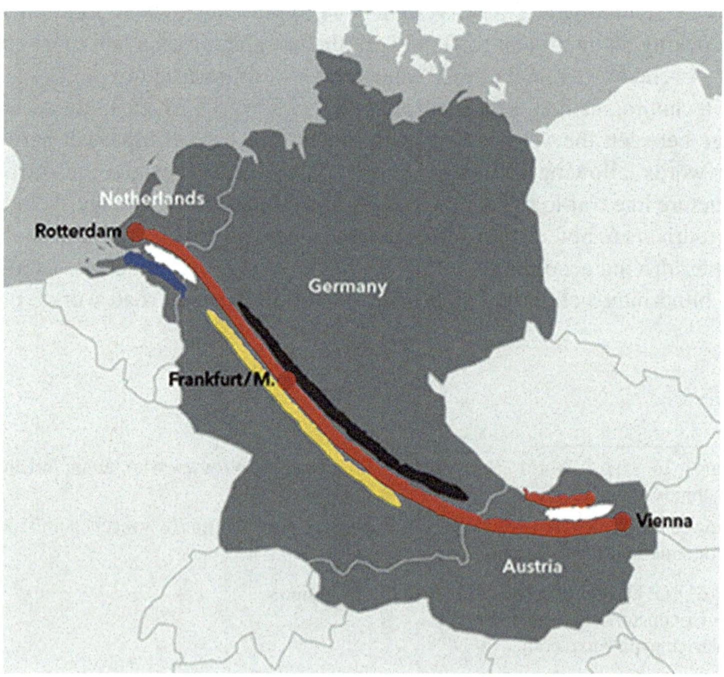

The cross-border project is focused on the use of technologies that warn drivers, via onboard units, that they are approaching road works. Drivers can then take an alternative route or reduce speed, improving both road safety and traffic flow. Cars in the corridor that are fitted with new, in-car equipment can also pass on real-time road traffic information to traffic control centres. With an exact location from GPS systems, centres will have an accurate and up-to-date picture of the traffic situation in specific areas and can manage processes more effectively.

In 2014, technology partners in the project—including Siemens, NXP Semiconductors, Cohda Wireless and Honda—joined forces with politicians and highways agencies across the three countries to run a 'Communicating Cars' trial along the corridor, showcasing the benefits of the project. Demonstrations in test fields at Munich, Vienna and Helmond showed how the new technology could alert drivers to warnings such as a slippery road surface, pedestrian crossings, slow vehicles, etc., as well as upcoming road works, oncoming emergency vehicles, pending speed limits and braking of vehicles ahead, all allowing drivers to take the necessary precautions and avoid unnecessary accidents.

On the Dutch part of the ITS-corridor, the Dutch government is working with various partners such as NXP Semiconductors to prove the value of ITS technology in alleviating a core problem experienced on many highways with dense traffic: 'phantom' traffic jams caused by driver behaviour and braking rather than accidents or incidents.

'Spookfiles', as they are called in Dutch, account for 20 % of all of traffic jams in the Netherlands. Prevention or at least mitigation of this kind of congestion could very much support Dutch mobility policy goals to better use the existing infrastructure instead of building new roads.

30.2 Helmond

Also in urban areas, optimising the use of existing infrastructure is one of the main goals for using ITS. The city of Helmond is involved in Compass4D, a 3-year European project designed to show concrete benefits of Intelligent Traffic Systems for citizens, city administrations and companies. The trial involves a pilot fleet of more than 600 vehicles—including buses, taxis, emergency services vehicles and private cars—across Helmond and six other European cities, all equipped with interoperable onboard units that can 'communicate' in real time to road side units. Drivers of these vehicles receive alerts from the units to improve energy efficiency and increase road safety. The services fall into three main categories:

- **Red light violation**—if another vehicle has or is about to violate a red light (including emergency services vehicles) or if they are at risk of violating a red light. Drivers are also warned about other nearby vehicles and vulnerable road users (pedestrians and bicycles) that are also acting on a green light.
- **Road hazard warning**—either static hazards such as road works or 'dynamic' which could include a car suddenly braking up ahead.
- **Energy-efficient intersection**—provides information on traffic light sequences such as 'time to green' or 'time to red'.

Armed with this information, drivers can be much more aware of their surrounding environment and have more time to react to potentially hazardous situations to avoid collisions. Improving reaction times at traffic lights and knowing to cut the engine in a case of a long wait can also reduce congestion and pollution. The pilot will end in December 2015, but Helmond as well as most of the other Compass4D-project partners already decided to continue the services after the project phase.

Helmond also has been involved in a European trial to improve fuel efficiency and reduce CO_2 emissions of trucks by 25 %. Freilot uses V2X to allow 14 traffic lights along Helmond's major through road (Europaweg Kasteel-Traverse and Deurneseweg) to communicate with onboard devices in trucks and fire brigade vehicles. The vehicles are given priority passing at traffic lights and issued with speed advice based on surrounding traffic to ensure optimum efficiency. The system also allows truck drivers to book loading spaces in cities with heavy traffic to save valuable time. Pilots began in 2010 and have been so successful that the city decided to keep it going. Lyon, Bilbao and Krakow are also involved in the project.

30.3 Hamburg

V2X systems are a core enabler of safe and efficient platooning, where freight trucks travel very close together, optimising airflow and creating a slipstream for the vehicles to move in, saving energy and fuel consumption. The technology enables vehicles to communicate with each other, so if the first vehicle brakes, the others automatically do the same, without driver intervention. The vehicles can then travel at optimum distance to drive efficiency (40–50 ft), even at high speeds on highways, without sacrificing safety.

Of course, vehicles need to stay together for the entire journey, which can be particularly difficult through traffic lights. Also, for platooning, one vehicle needs to take a lead. A lot of traffic flow enhancements can already be performed with convoys. A trial at Hamburg Port demonstrated how this could be achieved. A convoy of five freight trucks was fitted with onboard V2X units from NXP that could communicate with traffic lights around the Hamburg Port, as well as each other. This meant that traffic lights 'knew' when a group of vehicles were together and could ensure the entire group passed before the lights turned back to red. Truck drivers were permanently informed about the presence of other trucks in their vicinity by means of a small display.

The trial also demonstrates how intelligent transport system could protect vulnerable road users like school children. By integrating RFID tags into the school uniforms, the road side units were able to detect when school children were crossing the road, change lights accordingly and send alerts to the vehicles with onboard units.

While this trial only took place over the course of 1 day, Hamburg Port is planning to implement permanent systems in the near future. This will start with roadside units at one crossing and 30 trucks with onboard devices, extending to more than 10 crossings and up to 200 trucks in the coming years. Furthermore, NXP is in talks with a school uniform manufacturer in Hamburg to integrate RFID tags into the uniforms. This would be a citywide project which could significantly improve road safety for school children. The authors would like to thank the Hamburg Port Authority (HPA) and the Hamburg Ministry for Economy Transport and Innovation for their continuous support in these trials.

Internet convergence and the birth of intelligent traffic systems is a pivotal moment for the automotive industry and society. Smart mobility will, without a doubt, improve road safety by reducing human error (which causes 90 % of traffic accidents today), reduce congestion and improve energy efficiency which costs the global economy billions every year.

Integrating automotive systems with the Internet does of course have its risks. If systems are hacked and fed the wrong information, or manipulated to perform certain tasks, there are potentially fatal consequences. The industry therefore needs to work together to ensure the quality and integrity of data. Privacy also needs to be protected so individual driver behaviour cannot be tracked. These are key issues that will determine the trust of consumers and wide-scale adoption of connected, self-driving cars needed for safer, greener and more efficient road networks. With its security and identification technology, NXP is perfectly suited to circumvent attacks and to make the traffic safer while protecting users.

As the trials already underway prove, ITS presents too many benefits to ignore. The industry now needs to work towards standardisation and securing trust of manufacturers and consumers so that full benefits can be realised. Groups like the Car2Car consortium, ETSI and the High-level ITS Advisory Group to the European Commission will play a driving role in this process, as well as the successes experienced from ITS trials around the world.

Chapter 31
A Sampling of Automated Driving Research Projects and Initiatives (EC Funded, National)

ARTEMIS Industry Association

31.1 Introduction

ARTEMIS Industry Association is THE association for actors in Embedded Intelligence within Europe. As private partner, the association represents its members—industry, SMEs, universities, and research institutes—in the ECSEL Joint Undertaking and continuously promotes the R&I interests of its members to the European Commission and the Public Authorities of the participating states.

The main tasks of ARTEMIS-IA are to:

1. Create cooperative research and innovation projects throughout Europe for the benefit of its members and the European economy.
2. Represent industry in the best way possible and explain to policymakers the need of research and funding in Embedded Intelligence.
3. Contribute to improving coordination between the European, national, regional public, and private R&D activities in Embedded Intelligence and thus pave the way to greater "Europeanization" of the R&D scene in key areas.
4. Implement the ambitious European Industrial Strategy created by the ARTEMIS-IA/ETP working group SRA (ARTEMIS SRA 2016 [3]).
5. Build a self-sustaining Innovation Environment for European leadership in Embedded Intelligence.
6. Full development of the innovation potential of SMEs in Embedded Intelligence.

ARTEMIS Industry Association—The association for actors in "Embedded Intelligence" within Europe

ARTEMIS Industry Association (✉)
Eindhoven, the Netherlands
e-mail: info@artemis-ia.eu

D. Watzenig, M. Horn (eds.), *Automated Driving*,
DOI 10.1007/978-3-319-31895-0_31

7. Address European strategic priorities in addition to the individual interests of companies and countries, upstream and downstream.
8. Strengthen European industry and address societal challenges, as, for example, by issuing the "High-Level Vision 2030 [4]" authored by ARTEMIS-IA & ITEA.
9. Aligning research agendas for Embedded Intelligence.

31.2 Mission of ARTEMIS-IA

ARTEMIS-IA believes that targeted investment in innovation in Embedded Intelligence is crucial to support Europe's ambitions and ensure our #1 position in producing advanced complex and safety relevant products.

ARTEMIS Industry Association nurtures the ambition to strengthen Europe's position in Embedded Intelligence and to attain world-class leadership in this domain to support the European Industry. ARTEMIS innovation strategy is to challenge the application contexts, based upon exploitation of European strengths and opportunities by:

1. **Building on the leading positions where Europe is strong**, in specific technologies and in various application domains, particularly for the safety critical high reliability real-time applications in the field of automotive, aeronautics, space, and health sectors.
2. **Innovate to create new opportunities** for Europe to be positioned at the forefront of new or emerging markets with high potential growth rates to become among the world leaders in these domains and particularly target process industries, smart cities and energy efficient buildings, environment, food, and agriculture.

31.3 Structure of ARTEMIS-IA

ARTEMIS Industry Association is a membership organization with 180 members and associates from all over Europe. The multidisciplinary nature of the membership provides an excellent network for the exchange of technology ideas, cross-domain fertilization, as well as for large innovation initiatives. Members can vote in the general Assembly and elect representatives to join the Steering Board. ARTEMIS-IA is one of the three private members inside the ECSEL Joint Undertaking (Electronic Components and Systems for European Leadership), which is implementing the common strategy via regular call for R&I project proposals (Fig. 31.1).

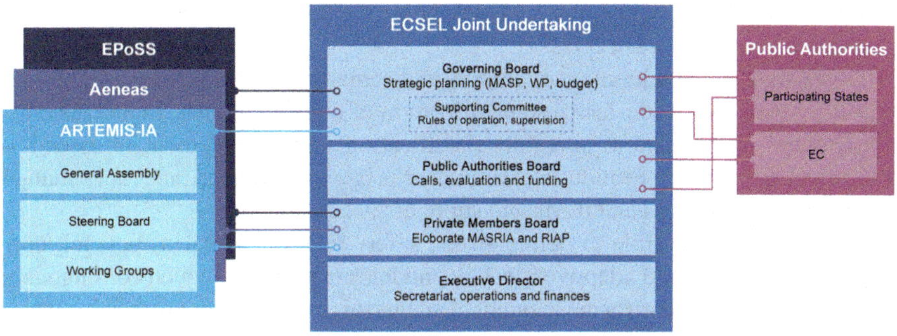

Fig. 31.1 ARTEMIS-IA as private member in the ECSEL Joint Undertaking

31.4 Automated Driving and ARTEMIS-IA

Automated Driving has quickly invaded many research agendas. The contribution of ARTEMIS-IA to the MASRIA2016 dedicates a separate section to this topic.

Significant breakthroughs have recently been made in advanced driver assistance systems by European vehicle manufacturers and suppliers. In order to swiftly proceed towards highly automated driving and flying, where the system relieves the driver from steering, accelerating, and monitoring of the vehicle environment, three steps can be foreseen in the automotive domain. First, by 2020, conditional automated driving (SAE Level 3) is expected to be available in low speed and less complex driving environments, e.g., in parking lots and in traffic jam situations on one-way motorways. Second, by 2025, conditional automated driving is expected to be available at higher speeds in environments with limited complexity, e.g., on highways. And third, by 2030, (conditional) automated driving is expected to be available in most complex traffic situations, i.e., in cities.

In closed and secured environments (e.g., factory floor, precision farming, business and leisure parks, etc.), it is likely to be proposed first. While the embedded systems therein will probably also be closed and carefully tailored, support for open environments will follow and impose much more critical demands: embedded hardware and software will have to be updated on a regular base to follow, e.g., legal requirements, respect the latest standards, introduce new security aspects, services, and features, and to finally stay compatible with the latest vehicle technology.

Eventually, vehicles with different levels of automation will be built on advanced driver assistance systems and cooperating components as well as on detailed driver status monitoring and environmental perception. Such systems will have to be validated under virtual, semi-virtual, and real world conditions. This requires dependable solutions for advanced sensors and actuators, data and ontology fusion, efficient computation and connectivity, security, precise location, time and velocity detection, detailed scalable low cost and dynamically updated maps, precise lateral vehicle control, novel man–machine interfaces and human interaction technologies, cyber physical systems integration, black box recorder for near incident data, and (real-time) simulation concepts.

To separate the development of sensors and actuators from control strategies and trajectory planning, a (de-facto) standardization of object handling, object descriptions, scene interpretation, situation classification and management is essential. Therefore, the creation of industrial frameworks is recommended, and an exchange of test procedures between OEMs and suppliers is encouraged.

As it seems impossible to define all the safety relevant scenarios upfront, new "learning" concepts and adaptive lifecycle models are required, which continuously analyze real-world data for near incident scenarios, evaluate the potential impact, modify the control software or strategies, validate the improved systems, and update all related vehicle components (maps, control software, information on road conditions, etc.) in a highly dependable way, i.e., safe, secure, and in real-time.

Traffic and fleet management systems are crucial for highly and fully automated systems. Dependable communication networks with a wide coverage and high availability and data links among vehicles as well as between humans, vehicles, and the infrastructure will be fundamental for traffic management systems. This will allow cooperative decision making in vehicle guidance and benefit from high performance computing systems.

In-vehicle and inter-vehicle standardization has to go hand in hand with technology development. Similarly, development of advanced traffic infrastructure is mandatory to provide a frameset for automated transportation systems.

Automated driving is enabled by innovations in embedded systems, as 70–90 % of all innovations in vehicle development currently are based on embedded systems. More and more societal economic, ecologic, and individual added value will be based on embedded systems in cars and trucks that are interfacing with data, software, and embedded systems in their local and global environment.

Automation, reliability, security, and affordability of such enhanced driving services depend upon highly innovative hardware and software solutions, their rapid market volume penetration and future-prone modularity, open architectures, and standard interfaces.

To achieve the targets in the automotive industry, nearly all new concepts and functions rely on superior functionality implemented in embedded systems and applied in nearly all major components of future vehicles (Fig 31.2).

ARTEMIS members participate already in European projects that focus on creating Embedded Intelligence in automotive applications, as depicted in the ARTEMIS SRA Matrix (Fig. 31.3). Partners in industry closely work together and realize state-of-the-art intelligent solutions supporting (future) market demand. Two core contributing projects in the Automated Driving Roadmap are DESERVE and CRYSTAL.

With the market of Advanced Driving Assistance Systems expected to grow rapidly over the next decade, project DESERVE aims to build an innovation ecosystem in embedded systems for assisted and automated driving with partnering automotive R&D actors and possible applications in other industrial domains.

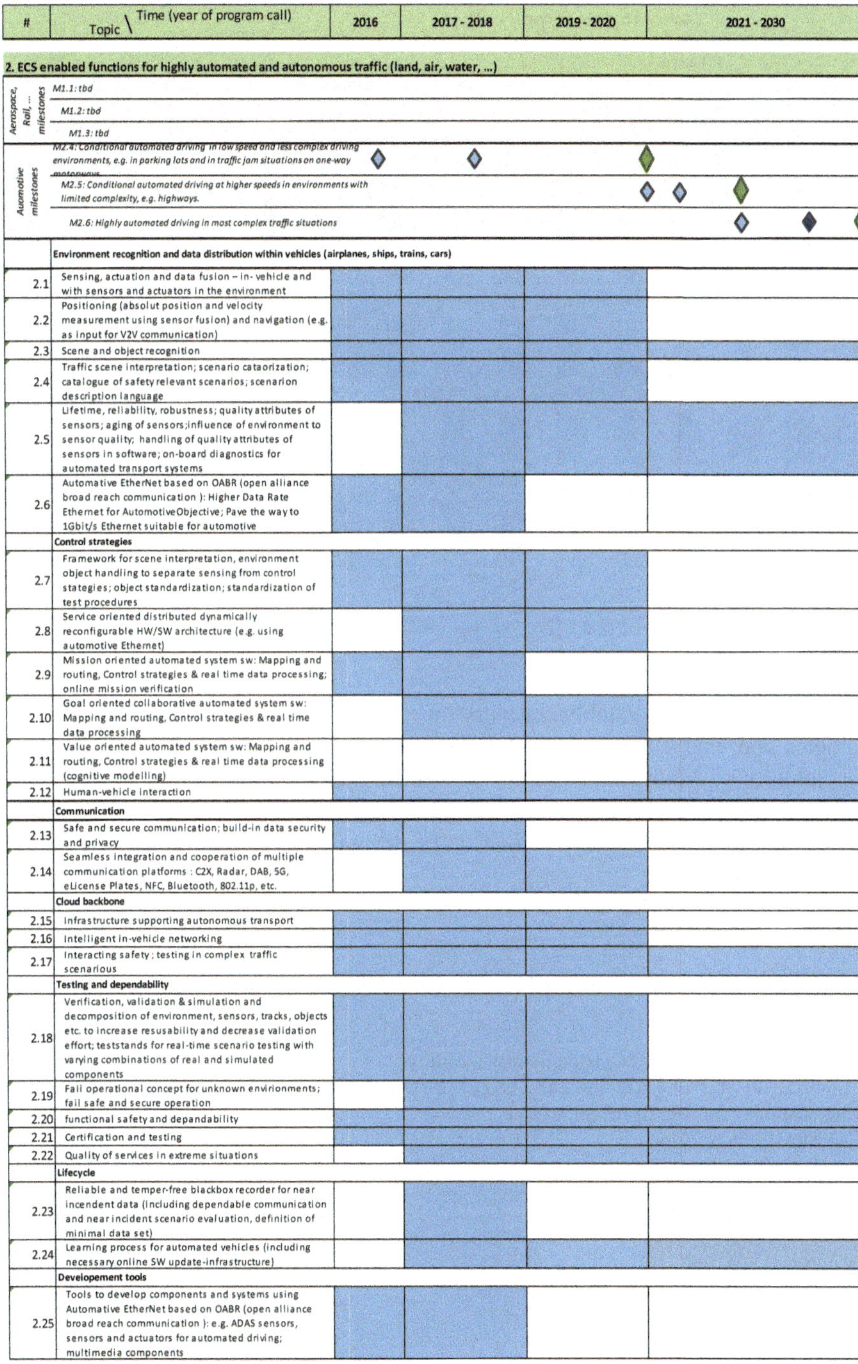

Fig. 31.2 Research topics in Automated Driving [1]

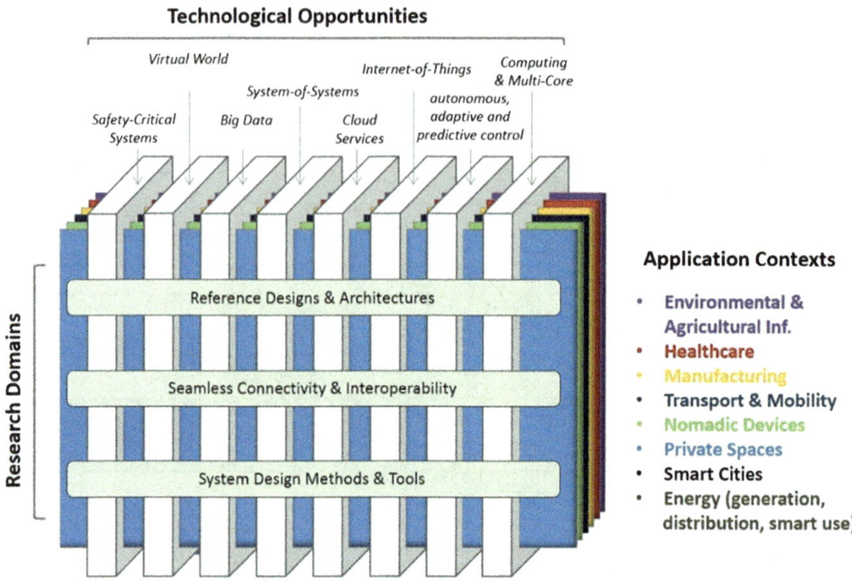

Fig. 31.3 ARTEMIS SRA Matrix [2]

Project CRYSTAL establishes workflows based on current and emerging technologies and enable their use in the engineering environment of, among others, the automotive industrial domains to reduce system design costs through the improvement and smart integration of system analysis, safety analysis, and system exploration tools as well as a reduction of design cycles by developing reusable technological bricks.

References

1. ESCEL-JU MASRIA2016-chapter-smart-mobility
2. Embedded/Cyber-Physical Systems ARTEMIS Major Challenges: 2014-2020. 2013 DRAFT addendum to the ARTEMIS SRA 2011
3. ARTEMIS SRA 2016
4. High Level Vision 2030

Chapter 32
ERTRAC: The European Road Transport Research Advisory Council

Josef Affenzeller

32.1 Introduction

ERTRAC[1] acts as European technology platform for road transport. The members of ERTRAC are representatives from all the stakeholders of the road transport sector, including private and public organizations involved in research, and gathering also administrations from both European and national levels. The main tasks of ERTRAC are to:

- Provide a strategic vision for road transport research and innovation in Europe.
- Define strategies and roadmaps to achieve this vision through the definition and update of a strategic research agenda (SRA) and implementation of research roadmaps.
- Stimulate effective public and private investment in road transport research and innovation.
- Contribute to improving coordination between the European, national, regional, public, and private R&D activities on road transport.
- Enhance the networking and clustering of Europe's research and innovation capacities.
- Promote European commitment to research and technological development, ensuring that Europe remains an attractive region for researchers and enhancing the global competitiveness of the transport industries.

[1]ERTRAC, 66 Avenue de Cortenbergh, 1000 Brussels, Belgium, www.ertrac.org

J. Affenzeller (✉)
AVL List GmbH, Graz, Austria
e-mail: josef.affenzeller@avl.com

D. Watzenig, M. Horn (eds.), *Automated Driving*,
DOI 10.1007/978-3-319-31895-0_32

- Support the implementation of Horizon 2020, the European framework program for research and innovation

ERTRAC has established several working groups which consist of recognized experts coming from the ERTRAC members. The ERTRAC documents prepared within these working groups include scenarios, the SRA, and the research roadmaps. The working groups organize workshops on a regular basis, with the possibility to invite external experts. Currently, there are five working groups:

- Urban mobility
- Long distance freight transport
- Energy and environment
- Road transport safety and security
- Global competitiveness
- Connectivity and Automated Driving

The Working group on Connectivity and Automated Driving was established 2016 based on the importance of this topic for the automotive industry.

ERTRAC is supported by FOSTER-Road, a coordinated support action (CSA) financed by the European Commission as well as by an industry-led association (supporting institutions group) called ERTRAC SIG. By bringing together relevant stakeholders, ERTRAC supports the establishment of consensus-based plans and roadmaps addressing the key societal, environmental, economic, and technological challenges in the areas covered by the ERTRAC working groups. CSA activities also include project monitoring, strategic research agendas, business case models, innovation plans, and coordination of research on European and national levels. They include multimodal issues as well as comprehensive dissemination activities. ERTRAC SIG finances the office, dissemination activities, events, as well as publication of promotional materials.

32.2 Mission of ERTRAC

Road transport plays a vital role in the European economy and society. The road transport sector involves a wide range of industries and services from vehicle manufacturers and suppliers to infrastructure providers, mobility management, communication technologies, energy companies, and many others. Road transport, integrated with the other modes of transport, enables mobility for people and goods across Europe.

- Overall, road transport related industries provide employment to more than 14 million people in Europe and directly contribute 11 % to the European gross national product.
- Road transport has a major impact on our daily lives, as it is one of the primary means of access to employment, services, and social activities.

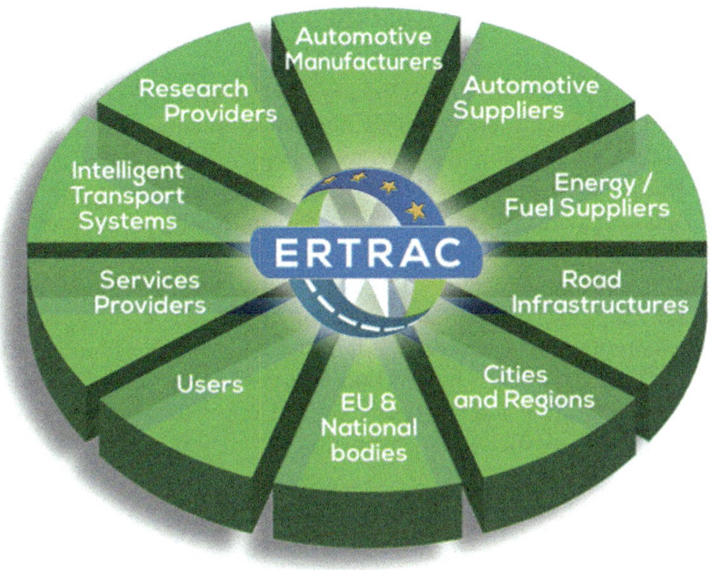

Fig. 32.1 ERTRAC members represent all the actors of the road transport system

- Road transport creates links, and these links are a key factor in developing social, regional, and economic cohesion within the European Union.
- Road transport contributes to the quality of life of every European citizen.

Because of the importance of the role of road transport in Europe, an accelerated development of sustainable, integrated transport solutions is necessary. The mission of ERTRAC is to provide a framework to focus the coordinated efforts of public and private resources on the necessary research activities.

32.3 Structure of ERTRAC

ERTRAC members represent all the actors of the road transport system as illustrated in Fig. 32.1. ERTRAC consists of five bodies: the ERTRAC plenary, the executive group, the working groups, the ERTRAC office, and supporting institutions group (SIG). ERTRAC SIG is a nonprofit association, legally established in Belgium.

32.4 Automated Driving Roadmap of ERTRAC

Automated driving is seen as one of the key technologies and major technological advancements influencing and shaping our future mobility and quality of life. ERTRAC is fully aware of this emerging technology, and its task force "connectivity and automated driving" has recently released a technology roadmap providing the current status of automated driving with regard to the implementation in Europe [1]. The ERTRAC roadmap is based on available documents for automated driving. The overall objective is to identify challenges for implementation of higher levels of automated driving functions. A lot of work has been done on this topic by various stakeholders and multi-stakeholder platforms (e.g., iMobility Forum,[2] EUCAR,[3] CLEPA,[4] ERTICO,[5] EPoSS[6]) and in European research projects. Therefore, it is essential to avoid any duplication of activities and concentrate on the missing items, concerns, and topics for future implementation.

Reference

1. Automated Driving Roadmap, European Road Transport Research Advisory Council (ERTRAC) (2015)

[2]iMobility Forum, http://www.imobilitysupport.eu, accessed: August 31, 2015.

[3]European council for automotive R&D (EUCAR), http://www.eucar.be, accessed: August 31, 2015.

[4]The European association of automotive suppliers, http://www.clepa.eu, accessed: August 31, 2015.

[5]Europe's intelligent transportation system organization, http://ertico.com, accessed: August 31, 2015.

[6]European technology platform on smart systems integration (EPoSS), http://www.smart-systems-integration.org, accessed: August 31, 2015.

Chapter 33
SafeTRANS: Safety in Transportation Systems

Jürgen Niehaus

33.1 Introduction

SafeTRANS[1] is a German not-for-profit association,[2] comprising stakeholders from various sectors of the transportation domain: Avionics, Automotive, Rail, and Maritime. SafeTRANS' members are OEMs, suppliers, and system operators in the transportation domain—such as Daimler, Airbus, Deutsche Bahn Netz, Siemens, AVL, Bosch, Hella, and Safran Engineering—, tool vendors and software houses, which support the development process for electronic components and systems in the transportation domain—such as AbsInt, BTC Embedded Systems, Esterel, Symtavision, and TTTech—, and research organizations and universities with a track record of technology transfer in this area—such as DLR, Fraunhofer Institutes, and OFFIS Institute for Compute Science.

SafeTRANS provides a communication and knowledge exchange platform for pre-competitive research and development activities of its members. Main activities initiated and conducted by SafeTRANS are:

- Initiation of theme-oriented round tables and working groups, identifying cross domain needs and objectives in pre-competitive R&D areas, leading to common R&D strategies and roadmaps harmonized between organizations and across domains.

[1]SafeTRANS e.V., Escherweg 2, 26121 Oldenburg, Germany, www.safetrans-de.org

[2]Organized as a "gemeinnütziger eingetragener Verein" under German law.

J. Niehaus (✉)
SafeTRANS, Oldenburg, Germany
e-mail: juergen.niehaus@safetrans-de.org

D. Watzenig, M. Horn (eds.), *Automated Driving*,
DOI 10.1007/978-3-319-31895-0_33

- Dissemination of harmonized R&D strategies and roadmaps, thus providing a harmonized view of R&D priorities to public authorities, national and European funding programs, and the community.
- Project incubation: Support the implementation of harmonized R&D strategies by initiating suites of R&D projects conducted by member organizations.
- Support sustainability of project results by paving the way for follow-up projects and by pushing standardization activities.
- Leverage these activities to a European level, harmonizing roadmaps with European partner cluster organizations, disseminating R&D strategies on European level, and initiating large-scale European R&D projects.
- Provide knowledge sharing facilities like workshops and conferences, newsletters, etc., allowing members from each transportation domain to learn from each other.

SafeTRANS is funded to a major part by its members. In addition, SafeTRANS participates in public funded projects, mostly on the level of Support Actions, e.g., on roadmap development and similar.

33.2 R&D Strategies and Roadmaps

In 2009, SafeTRANS coordinated the creation of the "National Roadmap Embedded Systems" [1], a strategy document detailing on how Embedded Systems technology can and will contribute to solving the big societal challenges Europe faces. This document was, and still is, one of the major reference documents of the funding program *Embedded and Cyber-Physical Systems* of the Federal Ministry of Education and Research BMBF [2]. The roadmap conceived in there has found its way into European Funding Programs like the Joint Undertakings ARTEMIS [3] and ECSEL.

Based on this roadmap, SafeTRANS significantly contributed to the Integrated Research Agenda Cyber-Physical Systems [4], a strategy document by acatech, the German National Academy of Science and Engineering, which in turn has been one major input to the "Industrie 4.0" Initiative of the BMBF [5].

In 2015, SafeTRANS, together with the *Gesellschaft für Informatik* GI and the *Verband der Automobilindustrie* VDA, published the Automotive Roadmap Embedded Systems [6]. In a scenario driven approach, where the scenarios are derived from analyzing future challenges and opportunities in the automotive industry as well as market trends and societal changes, gaps between the current capabilities of Embedded Systems technology and the capabilities needed to overcome these challenges are identified and the corresponding R&D needs characterized. This analysis is complemented by a corresponding survey focusing on the design process of Embedded Systems in automotive applications.

Roadmapping and the corresponding project incubation also are a major part of SafeTRANS' European activities. As a founding member of EICOSE, the European

Institute for Complex Safety Critical Systems Engineering, SafeTRANS cooperates with major clusters all over Europe, especially the French Pôle de Compétitivités Systematic-Paris-Region and Aerospace Valley and the Austrian ARTEMIS Austria Embedded Cluster. Together, these clusters have provided major input to the Strategic Research Agenda of the Joint Undertakings ARTEMIS and ECSEL as well as to the Horizon 2020 ICT programme and the Eureka programme ITEA3.

33.3 Working Group on Highly Autonomous Systems: Safety, Testing, and Development Process

SafeTRANS has installed a Working Group on *Highly Autonomous Systems: Safety, Testing, and Development Processes* (AK HAS) at the beginning of this year. In this working group, experts from member and nonmember organizations—airplane, car and maritime OEMs, System Integrators, Suppliers, research organizations—exchange knowledge and best practices on how to develop, analyze, test, and certify highly automated systems, with a special focus on safety aspects, testing activities, and development processes and its tool support. For selected aspects, the working group will develop concepts for generic solutions and initiate appropriate R&D projects. These aspects include situation awareness and situation interpretation by automated systems; user/operator modeling for highly automated systems; (generic) system architectures and execution platforms/middleware; and methods, processes, and tools for designing such systems, spanning the whole range from the requirement phase to testing and deployment. Special consideration is given to the following challenges

- Which artifacts of the environment have to be detected with which level of confidentiality? How can systems project the future evolution of a traffic situation with a high enough level of confidentiality? How can we ensure the integrity of such "internal world models" and which modeling techniques are appropriate?
- How can we cope with uncertainty caused by sensor limitations and different trust levels assigned to information stemming from other traffic participants or from the cloud? How can we ensure safety even though internal world models contain uncertainties?
- Which verification, validation and test methods can be used to cope with the enormous complexity and evolution of the system's context/environment, ensuring functional safety of highly automated systems?

As a cross-cutting challenge common for all domains and applications, *modelling of the system environment* has been identified as a high priority research need. Such models have to cope with four types of cooperations and interactions, namely (a) system to environment, (b) system to system, (c) system to human, and (d) system to information networks (cloud), and consider evolutions of the environment over time as well as uncertain knowledge.

As any SafeTRANS Working Group, AK HAS is open to members and non-member organizations.

33.4 Sustainability and Standardization

SafeTRANS is dedicated to ensuring sustainability for results of R&D projects that have been identified as having a key impact on safety critical systems engineering. One such result is the establishment of a so-called Interoperability Specification (IOS), which is a standard for ensuring interoperability of tools used in the development process.

A Reference Technology Platform (RTP), as it is understood here, is a "tool-box" for the development, analysis, and test of embedded and cyber-physical systems in various application domains. It contains processes, methods, meta-models, and interoperable (IOS-based) software tools, which describe and support the complete development process for critical embedded systems. These components can be combined to form domain- and application-specific development processes supported by software tools according to specific methods and adhering to specific standards, as needed for the particular domain and the particular application. The RTP for Critical Systems Engineering has been created in a long standing strategic public–private partnership on a national and European level, combining more than 25 global companies in the domains Aerospace, Automotive and Railway Systems, as well as various Tool Venders and Research Organizations, with an effort of more than 100 Mio Euro.[3]

To further drive the formal standardization of the IOS, SafeTRANS together with major IOS stakeholders has initiated and coordinates a project called CP-SETIS (towards Cyber-Physical-Systems Engineering Tool Interoperability Standardization) under Horizon 2020, which will support the existing and future IOS/RTP-projects in establishing the IOS as a formal standard. The main objectives of CP-SETIS are

- The alignment of all IOS-related forces within Europe to support a common IOS Standardization Strategy, aiming at a formal standardization process of the IOS.
- The definition and implementation of sustainable IOS Standardization Activities supporting both, formal standardization of "stable" IOS versions as well as extensions of IOS, if possible within existing structures that survive the lifespan of single projects.

Thus, CP-SETIS will drive the formal standardization of the IOS, and—even more important—align all IOS stakeholders, derive processes for handling IOS

[3]Amongst others in Projects CESAR (JU ARTEMIS, 60 Mio Euro Effort), IPs SPEEDS (FP 6), Sprint, DANSE (FP 7), MBAT (Artemis), SPES-XT (BMBF), CRYSTAL (ARTEMIS), most of which have been incubated by SafeTRANS members.

standardization and extensions in a project-independent way, and implementing these processes within existing structures (i.e., nonprofit organizations, like the ARTEMIS Working Groups, EICOSE, or SafeTRANS).

33.5 Conclusion

SafeTRANS is the major information and knowledge exchange platform for Safety Critical Systems Engineering in Transportation in Germany. Its main activities are the initiation of round tables and theme-oriented working groups, roadmap development, and project incubation, as well as supporting sustainability measures for project results and providing the link to similar clusters in Europe, to European funding programs and to national public authorities.

References

1. W. Damm, R. Achatz, K. Beetz, M. Broy, H. Daembkes, K. Grimm, P. Liggesmeyer, Nationale roadmap embedded systems, in *Cyber-Physical Systems*, ed. by M. Broy (Springer, Berlin, 2010), pp. 67–136
2. http://www.softwaresysteme.pt-dlr.de/de/embedded-cyberphysical-systems.php. Accessed 28 Sept 2015
3. ARTEMIS-IA (eds.), Embedded/Cyber-Physical Systems ARTEMIS Major Challenges: 2014-2020. 2013 Draft Addendum to the ARTEMIS-SRA 2011, ARTEMIS-IA, 2013
4. Eva Geisberger and Manfred Broy (Hrsg.), *agendaCPS: Integrierte Forschungsagenda Cyber-Physical Systems* (Springer, Berlin, 2012)
5. https://www.bmbf.de/de/zukunftsprojekt-industrie-4-0-848.html. Accessed 28 Sept 2015
6. Gesellschaft für Informatik, SafeTRANS, Verband der Automobilindustrie (Hrsg.), Eingebettete Systeme in der Automobilindustrie: Roadmap 2015-2030, 2015

Chapter 34
A3PS: Austrian Association for Advanced Propulsion Systems

Wolfgang Kriegler and Stefan Winter

34.1 Objectives and Tasks of A3PS

The Austrian automotive industry—or more precisely—automotive supply industry represents a significant value for Austria. Austria exports higher values in automotive parts and components than imports new, complete vehicles. Furthermore, the automotive sector has the highest share of researchers—about 14 %. Austria's universities and research institutions in this field enjoy high international reputation.

In order to maintain this favorable position and to secure Austria's competitiveness in this field, the industry, research institutions, and the responsible Austrian authorities need to collaborate very closely. The common goal is to support the successful market launch of innovative, advanced vehicle technologies including new energy carriers. Therefore, the bmvit founded the A3PS in 2006 to support an active technology policy of the ministry and to strengthen Austria's research and development activities.

Since its foundation, A3PS has developed into a well-established strategic public–private partnership (PPP), serving as a reliable partner for the ministry as well as for the partner companies and scientific institutions.

The key priorities in the area of road transport are to support clean, sustainable, affordable, and safe mobility. A3PS helps the officials understand the current technology trends and the R&D requirements of the Austrian stakeholders in this field of expertise. This flow of information provides valuable input for the Austrian technology and funding policy. On the other hand, for the A3PS community, the understanding of Austrian policies are an essential basis for their long-term research planning which provides them with planning security even in technologically risky areas.

W. Kriegler • S. Winter (✉)
Austrian Association for Advanced Propulsion Systems, Donau-City-Straße 1, 1220 Vienna, Austria
e-mail: stefan.winter@a3ps.at

D. Watzenig, M. Horn (eds.), *Automated Driving*,
DOI 10.1007/978-3-319-31895-0_34

A3PS addresses all advanced drive train and vehicle technologies (e.g., advanced ICE technologies; hybrid-, battery electric-, and fuel cell vehicles; as well as advanced fuel technologies including bio fuels, active safety measures like ADAS) and supports the whole innovation cycle (research, development, deployment).

Objectives and Tasks

- **Cooperation:** Regular joint activities to enable cooperation and common projects for member institutions.
- **Networking:** Stimulating R&D cooperations embedding the Austrian industry and research institutions into new national and international value chains in leading positions.
- **Information:** Strengthening the competence of Austrian enterprises and research institutions by collecting, compiling, and disseminating information on advanced propulsion systems and new energy carriers. Information for the public on the potentials and the state development of advanced propulsion systems.
- **Competence Presentation:** Presenting the Austrian technology competence to national and international conferences and initiatives.
- **Representation of Interests:** Supporting the representation of Austrian interests in international committees and initiatives of the EU and the IEA.
- **Orientation:** Establishing a common view between industry, research institutions, and technology policy by developing a common strategy, roadmaps, and position papers for reinforcing technology development.
- **Advisory function:** Providing fact-based consultancy and recommendations for policy makers to support the optimization of their policy instruments (funding programs, regulations, standards, public procurement, etc.) and to inform the public of the opportunities and perspectives of these new technologies.

34.2 ADAS in the Technology Roadmap of A3PS

Requirements on future vehicles will become more demanding than ever before. On one hand, they will need to comply with stringent future emission regulations (e.g., EU6c) under more challenging conditions (WLTP, RDE). On the other hand, it seems to be certain that the European legislature will head for CO_2 emission targets between 68 and 75 g/km in 2025. Additionally, social aspects that are difficult to predict such as changing consumer behavior or new mobility concepts must be taken into account. From the present A3PS members' point of view, the development and production of future vehicles will be driven by aspects as summarized below:

- Environmental impact
- Efficiency
- Safety (zero fatality)
- Demographic change
- Limited fossil fuels and raw material shortage

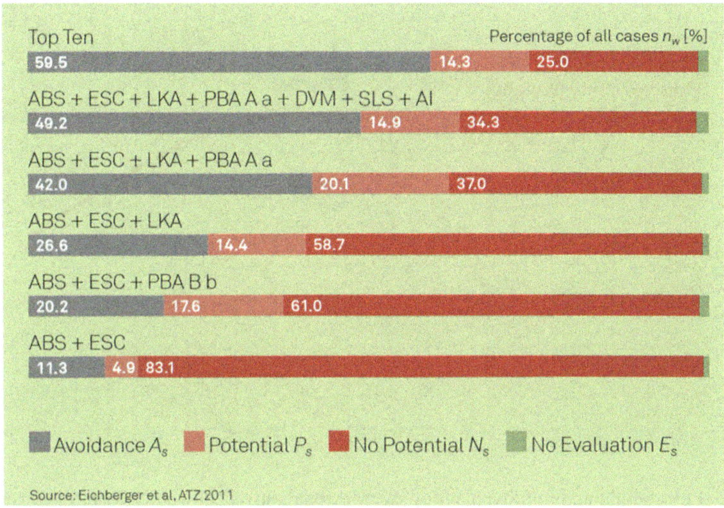

Fig. 34.1 Potential effectiveness of combined systems

Those drivers, in the short and medium term, will cause the development of a variety of alternative vehicle technologies and fuels, which ultimately correspond to the respective application purpose and vehicle class.

Besides energy efficiency and emissions, zero fatality must be a goal of good governance. From today's perspective, only a combination of passive safety measures and advanced vehicle control systems can achieve this scenario.

Advanced vehicle control systems mainly aim to increase energy efficiency and safety as well as to improve comfort and enable the communication between vehicle and infrastructure. Since human factors causes the majority of all accidents, advanced vehicle control systems have the potential to avoid those accidents and, therefore, save human life. The chart below shows that an accident avoidance of over 50 % is possible for a combination of ABS, ESC, lane keeping assist (LKA), predictive brake assist (PBA), automated emergency braking (AEB), driver vigilance monitoring (DVM), speed limiting systems (SLS) and alcohol interlock (AI) (Fig. 34.1).

Experts in automated driving around the globe expect a dramatic reduction of vehicle collisions, accidents, and fatalities in the range of minus 90 % once these functionalities are deployed into, e.g., 90 % of the vehicles on the road. Assuming that a worst case crash happens at a maximum speed of 10 km/h (around 3 m/s) compared to today's regulation of Euro NCAP5 [equal to 50 km/h (15 m/s)], the safety concept of all vehicles will have to be redrafted, enabling the application of lightweight structures, reducing the crash buffer, and finally resulting in less energy consumption and better propulsion performance (Fig. 34.2).

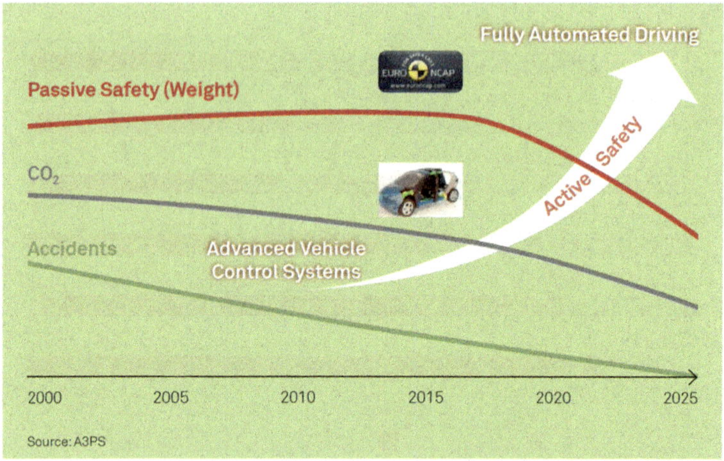

Fig. 34.2 Less weight and improvement of safety through advanced vehicle control systems

Still, the demand for individual road vehicles is growing on a global scale, whereas road infrastructure capacity can neither balance this demand today nor can it be extended in line with the number of vehicles. Therefore, automated vehicles are a key element for an efficient future road transport system.

The A3PS roadmap drafts the path for the radical change from conventional vehicle concepts (SAE automation level 0) to fully automated driving vehicles (SAE automation level 5) in the long term. Actually, huge effort is being expended in academic and industrial R&D, launching numerous research projects and prototype developments. A key issue will be system reliability. When system reliability is granted, these technologies can lead to an "electronic revolution" inside the vehicle.

The A3PS members keep track by monitoring the development in the field of advanced vehicle control systems. Not only because of its relevance to energy efficiency and emission behavior but to push innovation in overall vehicle technologies and to increase the chances for the Austrian industry. This also applies to many companies and institutions in the area of vehicle electronics and software development.[1]

[1] A3PS Technology Roadmap "Eco-Mobility 2025 plus".

CPSIA information can be obtained
at www.ICGtesting.com
Printed in the USA
LVOW06*0115231017
553353LV00001BB/108/P

9 783319 318936